Francis Day

Report on the Sea Fish and Fisheries of India and Burma

Francis Day

Report on the Sea Fish and Fisheries of India and Burma

ISBN/EAN: 9783337228118

Printed in Europe, USA, Canada, Australia, Japan

Cover: Foto ©Andreas Hilbeck / pixelio.de

More available books at **www.hansebooks.com**

REPORT

ON THE

SEA FISH AND FISHERIES

OF

INDIA AND BURMA.

BY

SURGEON-MAJOR FRANCIS DAY, F.L.S. & F. Z. S.,
INSPECTOR GENERAL OF FISHERIES IN INDIA.

CALCUTTA:
OFFICE OF THE SUPERINTENDENT OF GOVERNMENT PRINTING.
1873.

INTRODUCTION.

1. THE following report is the result of investigations made since 1867 into whether a wasteful destruction in the method of working the fisheries of India exist, and the present condition of the fishermen. It is, in fact, a continuation of my Fresh-water fishery report,* but limited to the sea and estuary fisheries.

2. The questions circulated were as follows:—*For Collectors.* (1).—Are breeding and immature fish destroyed in the tidal estuaries of your district to any extent? (2).—Could the salting of sea-fish be increased in your district, and if so, how? (3).—Would the proposition in paragraph 7 be advisable or practicable in your district? [The paragraph adverted to was as follows:—" It is unlikely that the local sale of fish could be much increased, but the captures might be salted and sent elsewhere. * * It may not be amiss to suggest, that if large enclosures were made near favourable localities (away from large towns), and where fish could be salted, would the following be impracticable? That salt might be sold inside them, at rates just remunerative, for the *bond fide* salting of fish. This question seems worthy of consideration, and on which the opinion of the local civil officers is highly desirable."]

3. *Those for Tehsildars or Native Officials on the coast were,*—(1).—What is the selling-price of salt per maund in your district? (2).—Is Government salt used for curing fish? (3).—Is salt-earth or sea-water employed for this purpose? (4).—Has the practice of selling fish increased or decreased of late years? (5).—In old times, had the fishermen caste any peculiar privileges they do not now possess? (6).—Are there any headmen of the fishing castes; is such hereditary, or how obtained, and what are his duties and emoluments?

---

* Dated Madras, December 5th, 1872.

(7).—Does any one claim any rights in respect to the sea-fisheries? (8).—How are the fishermen supplied with boats or nets? (9).—Have the sea-fishermen increased, decreased, or remained stationary?

4. Unfortunately the answers and enquiries into the fresh-water fisheries showed that a destructive and exhaustive plan of working them had by degrees crept in throughout Hindustan, destructive both to the fisheries and fishermen, and demonstrating that British rule has been nearly the ruin of these industries. It is to be regretted that the same result appears in the sea-fisheries, although from a totally different cause.

5. General fishing has been permitted inland by every poaching practice, and as the opportunity has been afforded to the people to poach the waters, they have not been slow to avail themselves of such. But in sea-fisheries, although every one may (as a rule) fish as he pleases, salt has been subjected to so heavy a duty that it is virtually unobtainable by the fish-curers at a price which would permit salt-fish being sold to the general public.

6. Ruin to the fish-curers' trade has reacted on the fishermen, due to curtailing their market, and so cutting off the stimulus for labour.

7. It appears evident that to render the sea-fisheries useful for providing wholesome salt-fish inland, nothing is necessary but cheap salt to the fish-curers. If the fish-curers' trade improves, an augmented demand for the fresh article must spring up, and then thousands of fishermen will be raised from poverty to comfortable circumstances, and that, not only without entailing on Government any pecuniary loss, but what appears of far more importance, affording a wholesome animal food for the people inland, and striking at one of the roots of those diseases which affect the maritime districts.

CALCUTTA;
November 15*th*, 1873.

TABLE OF CONTENTS.

PAGE.

REPORT ON INDIAN SEA-FISHERIES 1
Origin of investigation into sea-fisheries of India—Respecting those who have formerly written on this subject—The sea-fisheries of India to where they extend. The phenomenon of mud banks in the ocean—No cause to apprehend that any decrease in the amount of sea-fish exists—Migratory ones may be temporarily absent even for several consecutive seasons—If no great demand for fish exists, the fishermen take those most easily captured, as the small ones, or the fry These previously were the bait, the larger sorts followed; due to their capture, the latter stay further out to sea—Localities worked by the sea-fishermen—The deep-sea fishermen. Netting which in the deep necessitates a considerable outlay, and how they have to borrow money at exorbitant rates of interest for the same—Deep-sea nets:—For sharks, mackerel, &c., used near large towns. Hooks and lines for large fish as sharks, or for smaller ones as sea perches, &c. Generally much capital not required for hook-and-line fishing. Fish are sometimes salted out at sea, and the reason—Fishing in the shallows. Nets, stake-nets, cast-nets, hooks-and-lines—Fishing in estuaries, creeks, and mouths of rivers, weirs, &c.—Deep-sea-fishing most expensive; vessels, however, employed in the coasting trade. If only line-fishing is carried on, such may not require great outlay.

THE SEA-FISHERMAN 7
Sea-fishermen, probably, in old times, were divided into those who fished the deep sea, and those who only fished the shallows, but now owing to depression of their trade both have taken to working the shallows—Patriarchal customs amongst sea-fishermen—In the Madras Presidency fishermen have three classes of headmen; the superior or priestly, who is hereditary, exercises influence over large tracts of country; the second class also hereditary, over a few towns or villages; and the elective headman, who is only over one street or village—Madras fishermen induced to settle in Bengal—Privileges and importance of the fishing tribes in olden times—Present organization of fishing classes—The present condition of the sea-fishermen in Sind—The present condition of the sea-fishermen in the Bombay Presidency—The present condition of the sea-fishermen in the Madras Presidency—Present condition of the sea-fishermen in Bengal—Present condition of the sea-fishermen in Burma—Present state of the sea-fishermen generally, and the emoluments they receive for fishing.

THE FISHES OF THE INDIAN SEAS 23
Natural division of the Marine fishes—Non-predaceous marine fish. The gregarious and the non-gregarious—Migratory non-gregarious fish—Predaceous sea-fishes, the gregarious, and those not so—How small fish are the baits which lure larger species to certain localities—Non-migratory sea-fishes—The sea-fishes of the sub-class Teleostei—Fishes of the Acanthopterygian or spiny-rayed order. Families *Percidæ*, *Pristipomatidæ* and *Squamipinnes*—Families *Nandidæ*, *Mullidæ*, *Sparidæ*, *Cirrhitidæ*, and *Scorpænidæ*—Families *Teuthididæ*, *Berycidæ* and *Polynemidæ*—Families *Sciænidæ*, *Xiphiidæ*, and *Trichiuridæ*—Families *Acanthuridæ* and *Carangidæ*—

THE FISHES OF THE INDIAN SEAS.—(*Continued*) . . . 23
Families *Stromateidæ, Coryphænidæ*, and *Scombridæ*—Family
Trachinidæ—Families *Batrachidæ* and *Cottidæ*—Family *Gobiidæ*
—Family *Blenniidæ*—Families *Sphyrænidæ, Atherinidæ, Mugili-
dæ* and *Ophiocephalidæ*—Families *Pomacentridæ* and *Labridæ*
—Order of *Anacanthini* or spineless fishes. *Gadüdæ, Ophidüdæ*,
and *Pleuronectidæ*—Order *Physostomi*. Families *Siluridæ* and
Scopelidæ—Family *Scombresocidæ*—Family *Pseudoclupeidæ*—
Family *Clupeidæ*—Family *Clupeidæ* continued—Family *Muræ-
nidæ*—Order *Plectognathi*: family *Tetraonidæ*—Sub-class *Chon-
dropterygii* or cartilaginous fishes: family *Carcharüdæ*—The
Batoidei or rays and skates—Peculiar localities inhabited by fish.

THE MIGRATIONS OF SEA FISH 37
Migrations of sea-fish and cause of their being often found in fresh
water which is not connected with the sea—Breeding of sea-fishes
and where they deposit their eggs—Breeding of sea-fishes, continued—Immediate loss of condition after spawning not a necessary
result in sea-fishes—Diverse modes in which the young are produced—How some male cat-fishes carry the eggs about in their
mouths until hatched—Breeding of sea 'cat-fishes,' continued—
Breeding of cat-fishes in fresh waters different from those in the sea
—Cartilaginous fishes and how their young are produced.

FISH IN AN ECONOMIC POINT OF VIEW 42
Sea-fisheries ought not only to be serviceable to those living in their
vicinity, but also to the inland residents—A very large proportion
of the people inland would be consumers of salt fish could they
obtain it at a reasonable price—The local demand of fish close to the
sea is generally well met during the calm months of the year, but
where salt is dear the fish is often sold and eaten putrid—Modes in
which fish are cured or prepared for future consumption—The
distance inland uncured fish can be conveyed to market in the
plains of India—How sea-fish are cured along the coasts—How is
salt-fish prepared? with the comparative amount used in curing
with monopoly salt or salt-earth—Mode in which excised salt is
employed in curing sea-fish—How fish are cured with salt-earth.
The cost and what will be the result of subjecting the salt-earth to
an excise—Result of interfering with the collection of salt-earth to
both the public, the fish-curers, and the fishermen's trade—Comparative, economic, and sanitary qualities of fish cured with good
salt or salt-earth—Fish cured with excised or monopoly salt—
Quality of fish prepared with salt-earth—The incidence of the salt-
tax directly on the fish curers' occupation—Salt tax continued, and
how the price of salt re-acts on the Sind fisheries—Effects of the
salt-tax in the Bombay Presidency on the fisherman's trade—Returns from the Madras Presidency most complete—Imports and
exports of salt and dried fish by sea on the Western Coast continued—Fish-curers' trade flourishing in western, languishing in
eastern, coast of the Madras Presidency—Exports by sea of salt-fish
from western India, and the sales of salt, the latter having but little
connection with the former—Why augmented salt-fish curing has
not increased the sale of Government salt in western India—Cause
of present depressed state of some of the sea-fisheries—How the
salt-tax affects the health of the people, and ruins both the fish-
curers' trade directly and the fishermen indirectly—Salt-fish is imported duty free into India, and this depreciates the trade of those
who employ Government excised salt—Why under present circumstance it does not pay to enter largely on the fish-curers' trade in
the Bombay Presidency, &c.—Minor obstructions to the fisherman's
trade—Local taxes continued; the 'Rajah's cat,' 'curry fish,' &c.

	PAGE.
SEA-FISHERIES OF GREAT BRITAIN	66

The sea-fisheries of Great Britain—Enquiry into the fisheries of Great Britain—Reasons why every body should be permitted to do just as they please.

OBJECTIONS TO ANY REMEDIAL MEASURES 68

Objections to any remedial measures in Bombay—Objections to any remedial measures in Madras—Objections to any remedial measures in Bengal—Objections examined in detail.

REMEDIAL MEASURES PROPOSED 74

Remedial measures proposed—Remedial measures proposed in Bombay—Remedial measures proposed in Madras—Remedial measures proposed in Bengal—Analysis of remedial measures proposed.

RESULT OF THIS INQUIRY. 82

Result of this inquiry.

REMEDIAL MEASURES 83

Remedial measures—Incidence of the salt tax—Curing fish with salt-earth most important to the poor—Selling salt at or about prime cost of the fish-curers—The collection of salt-earth for fish-curers used inside enclosures to be permitted—Salt disposed of inside enclosures—The collection of salt-earth from this purpose no loss to the Government revenue—Result of cheapening salt to the trade of salt-fish.

APPENDICES.

BELUCHISTAN i

Sea-fishery at Gwadar: how it has sprung up of late years. Import duty on salt-fish into India abolished; salt-tax in India increased—Gwadar, its fisheries: mode of preparing the captures.

SIND iii

Sea-fisheries of Sind—Favorable prospects of the fishermen—Salting of fish increasing—Government salt exclusively employed : price Re. 1 a maund—Fishermen's operations unrestricted—Headman of caste—Ruinous mode in which the fishermen provide themselves with nets and boats—The fishermen represented by the Tehsildar as being in a very impoverished condition—Reasons for doubting the correctness of the Tehsildar's opinion—Taxes which formerly affected the sea-fishermen—Annual licensing of boats followed by a return to the auction system—Present plan of levying a license of Rs. 5 a ton annually—Sea-fish very abundant—Mode of disposing of the captured fish—How the fish are prepared—Fish-oils—Salted fish—Varieties of fish—Opinion of the Commissioner of Sind—Opinion of a Tehsildar—Amount of fish carried inland by rail.

PAGE.

BOMBAY xi
Sea-fisheries of the Bombay Presidency—The supply of fish only equal to the demand along the sea-coast districts—The local salt-fish trade has decreased or entirely ceased, the article being merely sun-dried—Four-fifths of the native officials consider that the local trade has decreased—The Government and the retail price of salt in the various districts—Government salt not employed by the fish-curers—Sea-water and salt-earth used, also drying fish in the sun—Four descriptions of Indian-cured fish—(1) with good salt, (2) salt-earth, (3) sea-water, (4) sun-dried—Fish cured with good salt—Fish cured in the British territory with salt-earth—Fish prepared with sea-water—Sun-dried fish—The present rate of duty on salt, perhaps, also combined with the abolition of the import duty on foreign-cured fish, has apparently ruined the trade in most parts of this Presidency—Five-sevenths of the reporters consider the fishermen have decreased of late years—Conclusions as to reports from the Bombay Presidency—Respecting remedial remedies which have been proposed—Imposing an import duty of 10 per cent. *ad valorem*. The proposition examined—Whether salt might not be sold cheap inside enclosures erected in suitable places for the *bond fide* curing of fish—Opinion that if anything is done to stimulate fishing in the sea, such would be tantamount to holding out inducements for the destruction of the fisheries—Opinion of the Collectors of salt revenue and their assistants—Opinions of the European officials of Broach—Opinions of European officials in Kaira—Opinions of European officials in Ahmedabad—Opinion of the Collector of Tanna—Opinion of the Collector of Colaba—Opinion of the Collector, &c., of Ratnagiri—Opinion of the Collector of Kanara—Opinion of natives respecting the fish and fisheries—Opinions of the native officials of Broach—Opinion of native official in the Kaira Collectorate—Opinion of native official at Ratnagiri—Opinion of the native officials of Kanara—Salt-fish carried inland by the Bombay railway.

MADRAS xxix
Sea-fisheries of the Madras Presidency: those on the Eastern or Coromandel Coast differ from those on the West or Malabar side—The supply of fish, as a rule, is locally sufficient during the fine weather—It is difficult, if not impossible, to decide whether the supply of marine fish is the same as in former years, because certain species migrate and often remain away for several consecutive seasons—If there is a sufficient market, the fishermen capture the deep sea-fish; if the market is bad, they fish along the shore for small fish, and scare away the larger sorts that are coming for their prey—Salt-tax in the Madras Presidency: when it may not directly affect the curing of fish—Salt-earth permitted to be collected free for fish-curing in some districts—How fish are cured in districts where salt-earth is prohibited—How fish are cured in those districts where the use of salt-earth is permitted, and effect of the salt-tax on this trade—How the salt-tax acts?—Has it improved the prospect of fishermen where salt may be collected free of duty, and ruined it for those who have to buy monopoly salt?—Sea-fisheries Government property—The fishermen: (1) those who ply their trade in the sea, but owing to decreased trade have to go as sailors, &c., elsewhere; and (2) the estuary fishermen, who very probably in old times were the sellers and curers of fish, but whose occupation has been ruined by the salt-tax—Patriarchal customs among fishermen—Breeding fish and fry destroyed—Fixed engines for capturing fish, some legitimate, others most unfair; and destruction sometimes wasteful—Moveable engines for capturing sea-fish—Conclusions—Remedies which have been alluded to—Answers to

                                                                    PAGE.
MADRAS.—*Continued*. xxix
a proposition as to whether salt might not be sold cheap to
fish-curers under certain restrictions—Observations of the Revenue
Board on the fishermen, and a proposal to extend the excise on salt
to Malabar and South Canara—Collector of Malabar on the salt excise
—Application for returns of exports and imports of salt-fish, &c.—
Injury to coast-fisheries occasioned by irrigation weirs—Opinion of
the Collector of South Canara—Opinion, &c., of the Collector of
Malabar—Salt-fish carried by the railway inland—Return from
Travancore—Replies from the Dewan of Cochin—Opinion, &c., of
the Collector of Madura—Replies from the Collector of Tinne-
velly—Opinion of European officials of Tanjore.—Opinion of the
Collector of Trichinopoly—Replies from European officials of South
Arcot—Opinions of European officials of Madras—Replies of Nellore officials—Opinion of the Collector of the Kistna district—Opinion of the Collector of the Godavery district—Opinions of European officials in the Vizagapatam district—Reply of the Collector of
Ganjam—Observation of the Collector of Bellary—Replies of the
Collector of Kurnool—Opinions of native officials in South Canara—
Opinions of native officials in the Malabar Collectorate—Opinions
of native officials in the Madura Collectorate—Opinion of native
officials in the Tinnevelly Collectorate—Opinion of native officials
in the Tanjore Collectorate—Opinions of native officials in the South
Arcot Collectorate—Opinion of native officials in the Madras Collectorate—Opinions of native officials in the Nellore Collectorate—
Opinions of native officials in the Kistna district—Opinions of
native officials in the Godavery Collectorate—Opinions of native
officials in the Vizagapatam Collectorate—Opinions of native officials
in the Ganjam Collectorate.

CEYLON cviii
Price of salt in the island of Ceylon: and table of imports of salt-
fish and salt—Investigations into the state of the sea-fisheries of
Ceylon—Modes of sea-fishing in Ceylon—Native opinions respecting the fisheries—Bennett's remarks on the sea-fisheries.

BENGAL cxv
Bengal sea-board not adapted for fisheries, due to some local causes,
its sea and estuary fisheries—The fishermen—The fisheries, how
they are unworked—The high price of salt and its relationship to
fish-curing—Other causes adduced for diminished supply of fish—
Remedies that have been proposed—Respecting the sale of salt
at a reduced rate within enclosures for the purpose of salting fish—
Opinions of European officials in Orissa—Opinion of the Collector
of Cuttack—Opinions of officials in Pooree—Mr. Geddes' opinions—
Opinion of the Collector of Balasore—Opinion of European officials
in the Presidency Division—Bhaugulpoor Commissionership. Opinion of Collector of Monghyr—Opinion of the late Dr. Cantor—
Chittagong Division. Opinions of European officials—Presidency
Division. Opinion of native officials.

BURMA cxxvi
Sea-coast and fisheries of Burma, how they are now but little worked—
Fishing increasing—Price of salt—Opinion of the Chief Commis-
sioner—Modes of preserving fish—Nga-pee, two kinds—Exports of
salt-fish and salt from British into Upper Burma—Burmese not
good sea fishermen—Arracan. Opinions of the European officials—
Personal observations—Tenasserim. Opinions of native officials,
Rangoon. Opinion of native officials.

ANDAMAN ISLANDS	cxxxviii

Situation, &c., of Andaman Island—Opinion of Chief Commissioner—Fisheries of the Island.

ISINGLASS	cxlv

Isinglass, what it is, and where procured—Discovery of the export of this substance from India. Dr. Royle's pamphlet on the subject—Its uses, and the forms that are brought to market—Indian fishes from which this substance is obtained—The Polynemi fish which is useful in this manufacture—Other estuary or marine fishes from which isinglass is prepared—Isinglass from siluroid fishes—Indian isinglass in the home market—How it is manufactured in Russia—Characteristics of East Indian isinglass—Exports.

FISH-OIL	cxlvi

Fish-oil exported from India—Different varieties of Indian fish-oils—Medicinal fish-liver oil—Its manufacture at Calicut—The cost of its manufacture—The reason why the cost has augmented—The amount prepared—From what the oil is prepared—How the fish are captured—Process of manufacture—Common fish-oil, how prepared—Common fish-oil continued—Burmese fish-oil.

SEA-FISHES OF INDIA AND BURMA	cliii

The sea-fishes of India, Burma, and Ceylon. A compilation of those recorded.

Sub-class Teleostei . . . cliii

Order Acanthopterygii . . cliii

Family	Percidæ	cliii
,,	Pristipomatidæ	clxvii
,,	Squamipinnes	clxxvii
,,	Nandidæ	clxxxiii
,,	Mullidæ	clxxxiii
,,	Sparidæ	clxxxvi
,,	Cirrhitidæ	cxc
,,	Scorpænidæ	cxci
,,	Teuthididæ	cxcv
,,	Berycidæ	cxcvi
,,	Kurtidæ	cxcviii
,,	Polynemidæ	cxcix
,,	Sciænidæ	cci
,,	Xiphiidæ	ccv
,,	Trichiuridæ	ccvi
,,	Acanthuridæ	ccvii
,,	Carangidæ	ccix
,,	Stromateidæ	ccxxi
,,	Coryphænidæ	ccxxii
,,	Nomeidæ	ccxxiii
,,	Scombridæ	ccxxiv

		Page.
Family	Trachinidæ	ccxxvi
,,	Batrachidæ	ccxxix
,,	Pediculati	ccxxx
,,	Cottidæ	ccxxxi
,,	Cataphracti	ccxxxiv
,,	Gobiidæ	ccxxxv
,,	Callionymidæ	ccxlvii
,,	Blenniidæ	ccxlviii
,,	Sphyrænidæ	cclii
,,	Atherinidæ	ccliii
,,	Mugilidæ	ccliv
,,	Aulostomatidæ	cclvii
,,	Centriscidæ	cclviii
,,	Trachypteridæ	cclviii
,,	Pomacentridæ	cclix
,,	Labridæ	cclxiv
Order Anacanthini		cclxxv
Family	Ophidiidæ	cclxxvi
,,	Pleuronectidæ	cclxxvii
Order Physostomi		cclxxxii
Family	Siluridæ	cclxxxiii
,,	Scopelidæ	cclxxxviii
,,	Scombresocidæ	ccxc
,,	Pseudoclupeidæ	ccxcvii
,,	Clupeidæ	ccxcviii
,,	Symbranchidæ	cccvii
,,	Murænidæ	cccviii
Order Lophobranchii		cccxii
Family	Syngnathidæ	cccxiii
Order Plectognathi		cccxiv
Family	Sclerodermi	cccxv
,,	Gymnodontes	cccxix
Sub-class Chondropterygii		cccxxii
Order Plagiostomata		cccxxii
Family	Carchariidæ	cccxxii
,,	Scylliidæ	cccxxvi
,,	Pristidæ	cccxxvii
,,	Rhinobatidæ	cccxxvii
,,	Torpedinidæ	cccxxviii
,,	Rajidæ	cccxxix
,,	Trygonidæ	cccxxix
,,	Myliobatidæ	cccxxxii

REPORT

ON THE

SEA-FISHERIES OF THE INDIAN EMPIRE.

I. THE origin of the present investigation into the

ERRATA.

Page clxxiii, for 'DATNOIDES' read 'DATNIOIDES.'
„ cxcviii, note 2 belongs to *Rhynchichthys ornatus*.
„ ccxxxi, erase *Genus* MINOUS and Species *Monodactylus*.
„ ccxxxiii, 19 lines from the top, for '4/13' read '2/13.'
„ ccxlv, 10 lines from the bottom, for 'scales cycloid' read 'scales ctenoid.'
„ ccci, for '726. Clupeai' read '726. Clupea.'
„ cccvi, 8 lines from bottom, for 'abdomen not serrated' read 'abdomen is serrated.

or specimens sent to Europe. *Dr. Jerdon*, in 1851, in the 'Madras Journal of Literature and Science,' wrote a paper entitled 'Ichthyological Gleanings in Madras. *Dr. Cantor* also published some remarks on the sea-fisheries of the Bay of Bengal. *Dr. McClelland* observed upon a few of the more useful fish, and *Dr. Helfer* reported on the fisheries of Mergui.

III. The sea-fisheries of India and Burma are those in
<small>The sea-fisheries of India to where they extend. The phenomenon of mud banks in the ocean.</small> the open sea, along the coast, extend up large rivers so long as they are within tidal influence, or exist in the back-waters and estuaries, more especially on the western coast. In Malabar, and also in certain spots along the Coromandel coast, vast mud banks are present in the sea wherein many kinds of fish find abundance of food, immunity from much disturbance in the surrounding element, and an excellent locality in which to breed. The exact cause of the existence of these large tracts of sea

* Dated December 5th, 1872.

REPORT

ON THE

SEA-FISHERIES OF THE INDIAN EMPIRE.

I. THE origin of the present investigation into the condition of the fisheries of India is fully entered upon at the commencement of my Fresh-water Fishery report,[*] of which this, solely relating to the sea-fisheries, may be considered the second portion. Irrespective of the various reports formerly adverted to, this contains in addition the result of a tour through some of the districts of Madras, and a condensation of the answers received to the questions circulated in 1872.

Origin of investigation into the sea-fisheries of India.

II. There have not been many ichthyologists who have directed their attention to the sea-fisheries of India. *Russell*, who as Naturalist to the Madras Government spent some years at Vizagapatam, published figures of 200 varieties of fish captured in that locality, and several authors have described species as existing in the seas of India, either from drawings or specimens sent to Europe. *Dr. Jerdon*, in 1851, in the 'Madras Journal of Literature and Science,' wrote a paper entitled 'Ichthyological Gleanings in Madras. *Dr. Cantor* also published some remarks on the Sea-fisheries of the Bay of Bengal. *Dr. McClelland* observed upon a few of the more useful fish, and *Dr. Helfer* reported on the fisheries of Mergui.

Respecting those who have formerly written on this subject.

III. The Sea-fisheries of India and Burma are those in the open sea, along the coast, extend up large rivers so long as they are within tidal influence, or exist in the back-waters and estuaries, more especially on the western coast. In Malabar, and also in certain spots along the Coromandel coast, vast mud banks are present in the sea wherein many kinds of fish find abundance of food, immunity from much disturbance in the surrounding element, and an excellent locality in which to breed. The exact cause of the existence of these large tracts of sea

The sea-fisheries of India to where they extend. The phenomenon of mud banks in the ocean.

---

[*] Dated December 5th, 1872.

wherein mud remains in solution is still a mystery, but anyhow the ocean is so smooth that, even during the height of the south-west monsoon, vessels can run for shelter into their midst, and once there are as safe as when inside a breakwater. If the surface is so still, of course so is the water below, and such spots seem to be well suited to the siluroid fishes, which appear to be especially constituted to reside in muddy waters, wherein they find their food as much (or more) apparently by the use of their feelers as by eyesight. Excluding these local phenomena, we have the following localities which are frequented by fish: (1) the deep sea; (2) the shallows; and (3) the estuaries.

IV. Although in different years there may be a considerable difference in the amount of fish present at certain seasons, due to the irregularity with which the mackerel, oil-sardine, and other varieties arrive or refrain from coming, still it does not appear that there is any paucity of fish in the sea, or that there is reason to fear that the present race of fishermen will ever do any injury to the sea-fisheries by over-fishing. There is rather cause for supposing that, due to some influence or other, the harvest of the sea is not being reaped, that man is often in a state of semi-starvation whilst in his vicinity exists a supply of excellent animal food which is practically inexhaustible, and only requires effort to capture. But before we condemn the apathy of the fishermen, we must enquire whether a market exists were he to capture the fish, or is his trade so trammelled directly or indirectly by laws and regulations, that it is impossible he can carry it on in a profitable manner.

No cause to apprehend that any decrease in the amount of sea-fish exists. Migratory ones may be temporarily absent even for several consecutive seasons.

V. If he has no market, the fisherman naturally only carries on his trade where he can do so with the least amount of expense and toil, and this will always be by capturing the smaller sorts, which can be taken and dried with ease. It is patent to all that the smaller kinds of fish prefer the vicinity of the coast, and seek their food close in-shore. In the same way many of the fry of larger species come up the estuaries, backwaters, and mouths of rivers. It is much easier to capture these by weirs, fixed engines and traps, than to take the larger, more predaceous and strictly deep-sea fishes. But by disturbing the shore

If no great demand for fish exists, the fishermen take those most easily captured, as the small ones, or the fry. These previously were the bait, the larger sorts followed inshore: due to their capture, the latter stay further out to sea.

(3)

and destroying the small fish, prawns, &c., the fisherman is capturing the food which previously decoyed the larger and more predaceous ones in, and thus he scares away what would be the natural supply, and subsequently asserts that the fish have decreased. This, however, does not signify to him so long as the market is supplied and sufficient obtained for his family requirements. Neither does it occasion much injury if there is only a small local demand, and salt is not available for preserving the surplus.

VI. The following localities are those which are most *Localities worked by the sea-fishermen.* frequented by the salt-water fishermen—(1) the deep sea; (2) the shallows; and (3) the estuaries: whilst the effect of the seasons on the fishing has likewise to be taken into consideration.

VII. The deep-sea fishermen, or rather those who ply *The deep-sea fishermen. Netting which in the deep necessitates a considerable outlay, and how they have to borrow money at exorbitant rates of interest for the same.* their occupation beyond the shallow water, do so in one of the following ways:—either by nets or hook-and-line. Deep-sea netting in most parts of the coasts of India is not carried on to any extent, partly due to the insufficiency of the demand to render such remunerative, and partly to the expense incurred in the manufacture of the necessary nets and the cost of building seaworthy boats; for most of this class of people have to borrow money at exorbitant rates of interest wherewith to supply themselves with the requisites for their work. In Sind, the Tehsildar of Kurrachee observes that the fishermen borrow money from merchants and others to buy boats and nets. A net (description not recorded, but probably for deep-sea fishing) costs about Rs. 400 or 500. The old net being worn out every year, a new one is generally made. A boat costs about Rs. 1,000, and generally lasts for some years. The fishermen sell all the fish which they get to the persons from whom they borrow money for the purchase of boats and nets at half the ruling rates in payment of the sum borrowed, there being a clause to that effect in the bond: for instance, if a fish is worth one rupee, the fishermen give it them for eight annas. But it is worthy of note that a good market exists for the sale of any amount captured, perhaps, as I shall afterwards show, due to cheap salt being procurable wherewith to cure the excess which is over and above that required for local consumption. In fact here the fishermen can even afford to pay a tax per ton on their fishing boats.

VIII. Off Sind the large shark net come under the head of deep-sea nets, whilst along the Malabar coast during the mackerel and sardine seasons float-nets, having a mesh suited to the size of the species it is desired to capture, are employed for taking these two descriptions of fish, also for the seir-fish and horse-mackerel, but not expressly for any other sorts. Likewise in the vicinity of large towns, or where there exists a good demand for fish, deep-sea fishing with nets is sometimes engaged in. The usual mode, however, of carrying on this fishing is by means of hooks-and-lines; these again may be divided into two descriptions, *first*, the larger ones, which are fastened by a chain to a strong cord, and employed for sharks and other predaceous fish; *secondly*, the smaller kinds of hooks used in catching sea-perches, sciænas, polynemi, and other edible or valuable forms, and these are usually most esteemed as food. In this description of fishing a large capital is not usually necessary, at least in those districts where catamarans or rafts are employed. If, however, line-fishing is carried on off coral reefs, as the Andamans, large numbers of hooks are lost, because the hooked fish often dashes into or below the coral, and the line becomes divided and lost. For line-fishing in some places, as off Kurrachee, moderately-sized boats are used, prawns being found to be the most killing baits. A suitable bank out at sea having been selected, the boat is anchored, and each fisherman uses one line. The fishes captured vary with the season of the year, but may be considered to average about 1lb. or 2lbs. each : the nearer inland, the smaller the size of the fish, whilst the best descriptions appear to be furthest from the land. I may here mention that in some boats fish are opened, cleaned, and salted whilst at sea, in others the whole of this process is carried out on shore. This is especially the case along the western coast of India, because the fishermen can purchase salt at Goa or other foreign settlements at three or four annas a maund, take it out to sea, capture and salt their fish there, and then run in and dispose of them in a British settlement.

Deep-sea nets:—For sharks, mackerel, sardines, &c., used near large towns. Hooks and lines for large fish as sharks, or for smaller ones as sea perches, &c. Generally much capital not required for hook-and-line fishing. Fish are sometimes salted out at sea, and the reason.

IX. Fishing in the shallows, or rather close to or within a moderate distance from the shore, is carried on in many ways and with numerous descriptions of nets and

Fishing in the shallows. Nets, stake-nets, cast-nets, hooks-and-lines.

fixed engines. Occasionally a number of fishermen join together the pieces of net which belong to each, and thus make a very long drag-net, one end of which is kept on shore, and the other taken round a considerable circumference of water, and as the tide makes it is slowly hauled. Nets are also attached to stakes, but this is more frequent inside harbours or estuaries. Others again use cast-nets either from the shore or small boats, and this is very successful when the shoals of sardines arrive. Sometimes the fishermen go singly, at other times several join together. Hook-and-line fishing is also carried on in the shallows, not only at the mouths of rivers, but likewise in the surf along the coast: in this way the Polynemi and cat-fishes are often captured.

X. Fishing in the estuaries, creeks and mouths of rivers is carried on either by means of fixed or moveable nets, weirs and fixed engines of various forms and shapes (see appendix), and also by hooks-and-lines. The following description of one mode of obtaining fish along the sea-coast is by Mr. H. S. Thomas, Collector of South Canara:—"There are marshes by the river side that are flooded at every high tide; the fry of the sea-fish frequenting the estuaries are in the habit of coasting along the very edge of the rivers and running into all shallow places. When the tide rises over these marshes, the fry go in with it, probably finding more insect food among the swamp grass and on the freshly inundated land. But when they think to return with the ebbing tide, they are met by long lines of close wattle and fine basket-work that allows the water to pass but not the fry. At every tide in the day time the fry are thus waylaid and then left high and dry, thickly strewn in long lines, whence they are carried away in basket loads. The mullet suffer much in this way. They are a desirable sea-fish, and the wholesale destruction of their fry in this way should be prevented." (Appendix, p. xliv.) From the Malabar Collectorate we have much the same accounts. "At high tide many young fish and a small number of breeding ones at the estuaries of Darmapatam and Mahé rivers enter into the marshy grounds along their sides, where they are either netted or caught without the use of any apparatus, when the water recedes at low tide, and the fish are left on the surface of the ground. As these rivers seldom, during the hot season (January to May), overflow their banks at high tides, the fishermen, in order to get the fish into the low marshy grounds enclosed by them for this purpose, often cut open the banks and thus enable the water to flow in, carrying fish with it into these enclosed spaces. As soon as a good supply of fish is collected, that part of the bank left open is closed by a valve made of split bamboos, generally so contrived as to let the water flow back through its interstices, and at the same time bar the egress of the fish. When the water has so receded, the fishermen simply have the trouble of picking up the fish. A good deal is, however, wasted, inasmuch as only such fish as are capable of being used are taken, while the rest, generally very small ones, are left to perish" (p. lxxvii). The same wasteful mode of fishing is adverted to as existing in Ceylon (note, p. cx). Only nets with small meshes are employed, whilst the complaint is that natives of India introduced such into that country.

XI. In sea-fisheries, where fixed engines are not employed, the occupation necessitating most capital is for the deep-sea, both on account of the superior description of boats required and the great outlay necessary for the purchase or manufacture of the nets. Many of these vessels, however, are employed on the coasting trade, such being found to be more profitable. In Bombay, the Assistant Commissioner in charge of the salt works and ports of Guzerat, observes of the fisheries between Damaun and Surat :—" The sea-fisheries are not extensive, as far as I can learn, only about 200 vessels compose their fleet, but the number depends on the coasting trade, such having the preference as being the most remunerative. The carrying capacity of each vessel ranges from 20 to 40 candies. They remain out for about a week, but the actual period of their stay is regulated by the success they meet with. They take small quantities of salt to cure the fish they catch. The fish thus obtained is taken to port and sold to dealers, who dispose of it for local and inland consumption. * * The fishing season lasts for ten or twelve weeks between January and March" (p. xx). Line-fishing may also in certain localities, due to local causes, entail some considerable outlay, especially where large boats are used for this purpose, as off Sind. Or it may require very little capital, as on the Coromandel coast, where it is chiefly carried on from catamarans, or slight rafts ; also in rivers on the western coast or in Burma.

Deep-sea fishing most expensive : vessels, however, employed in the coasting trade. If only line-fishing is carried on, such may not require great outlay.

THE SEA-FISHERMEN.

XII. In commencing an examination respecting this class of people, I must begin with observations which, though matters of conjecture, are doubtless susceptible of proof or disproof. It appears probable (some fishermen have asserted that it was so) that in olden times the fishermen who plied their calling in the sea or within tidal reach were divided into two distinct classes, (1) those who captured fish *in the deep sea* or beyond their own depth, and (2) those who fished *from the shore* or in backwaters. The remains of this division may still be found in the Ganjam District and elsewhere: but due to the present depressed state of the fisheries, owing to the want of demand for fish, the deep-sea fishermen have taken to the cheaper calling of plying their occupation close in-shore. This first class of people appear on beaching their boats seldom to carry their fish, leaving this to the women and children, by whom also curing fish may have been carried on, as the drying or partial curing they now obtain is. It appears, if one may accept the statements of the fishermen, that the class which fished the shallows and backwaters are those who more immediately carried on fish-vending and fish-curing.* These conditions, however, are, so far as I know, based upon the statements of these people, and though I believe them to bear the appearance of truth, I only advance them as being probably so.

Sea-fishermen, probably, in old times, were divided into those who fished the deep sea, and those who only fished the shallows, but now owing to depression of their trade both have taken to working the shallows.

XIII. In several parts of India, but more especially in the Madras Presidency, the fishermen have customs of a patriarchal nature, which are more strictly observed on the Coromandel than on the western coast, but they are reported to be falling into disuse. In *Sind* there are four divisions of the fishermen castes, each of which has its own head, which headmanship is hereditary. The duty of the headman is to settle caste disputes and other matters of a trifling nature, and to conduct the religious ceremonies connected with marriages and deaths. On the occasion of marriages a cloth or 'lungi' is presented to the headman, its value varying with the circumstances of the parties undergoing such ceremonies. It was, so late as a year ago, the practice (now stated to be abrogated) to give

Patriarchal customs amongst sea-fishermen.

---
* The *Madras Revenue Board* observe that "the estuaries, however, are fished by a distinct class, who have most probably no other support."

a fish to the headman of their own division when returning from fishing (page ix). In the *Bombay Presidency*, in the *Junjura District*, there are hereditary headmen to the fishing castes; they possess Yakoob Khan's sunnud, authorising them, their heirs and successors, to exercise all the authority and duties of the Sir Patell and Chogala of kolies: all Sircar orders are to pass through their hands, and the settlement of disputes amongst the kolies are decided by them (page xxvi). At *Broach*, *Jambusar*, and *Hansot*, the fishermen have also headmen, but the office is elective and not hereditary (page xxvii). The same is reported from *Kaira*, but the Thakore allows him a few acres of land, and his duties are to settle disputes amongst the fishermen (page xxvii). In *Ratnagiri*, in the fishing castes, there are persons distinguished by the names of 'hodekur' and 'patell;' their duties appear to be those of mediating between parties engaged in any small disputes, of which there are a good number. There do not seem to be any emoluments attached to the office, but the 'hokedur' used to be exempted from the poll tax (page xxvii). In *Kanara*, likewise, there are headmen to the fishing races; their duties are confined to settling caste disputes, for which they have no regular emoluments (page xxviii).

XIV. In the *Madras Presidency*, in the *South Canara Collectorate*, at *Mangalore*, the fishermen have headmen called Guricars; among Mogers (one of the fishing castes) the office is hereditary. His duties are to make enquiries regarding the observance of caste rules amongst the members of his sect.

In the Madras Presidency fishermen have three classes of headmen; the superior or priestly, who is hereditary, exercises influence over large tracts of country; the second class also hereditary, over a few towns or villages; and the elective headman, who is only over one street or village.

He is entitled to get the usual honors and betel-nut, &c., on the occasions of marriage and such ceremonies, but derives no other emolument (p. lxxii). At *Kasargod* and some other places the fishermen have no headmen (p. lxxii). At *Udipy* the Mogers and Karves are the only two castes who follow fishing as an occupation. The Karves are limited in numbers; they have a common place of residence, which is styled a kéri (row of houses). For each such kéri there is a headman called Guricar, who investigates caste matters among all the people who reside there. The office is hereditary, and no emoluments accrue to it. In a similar way, the Mogers live in groups of houses which are termed 'patna' (a town). For every such patna there is a headman who is termed Guricar; he decides caste questions, but does not appear to derive any emolument, except that he is entitled on auspicious or inauspicious occasions to precedence in receiving betel-nut. His office is not hereditary, and his election or removal depends on the will of the people. Over the Mogers of all the Collectorate (except Kundapur) there is a spiritual preceptor termed Mangal Pujary; he resides at a place called Benne Kudru, near Barkur. His duties are to frame rules in regard to caste matters, to see if the people conform to them or not, and to impose penalties on those who infringe them, &c. The people of the caste raise money for him; his office appears to be hereditary (page lxxiii). In *Malabar*, and in fact other localties where Native Christians are the fishermen, they have no headmen, but in some other places the priests appear to settle their disputes. It is curious that in *Ceylon* the Roman Catholic Church has appropriated the right to the sea-fisheries, which Government very properly resigned to the fishermen, and it is observed of the mode in which they collect the rent, that "if the share is not paid, the rites of the Church are refused" (page cxiv). At *Tellicherry*, in Malabar, the fishermen do not appear to have headmen, but should any dispute respecting fishing arise, the matter is laid before certain wealthy men of their own caste, whose decision is final. In reality there are no recognised headmen among the fishermen here, but the owners of boats and nets have certain respect paid them among this class of people. Some of the wealthy among them hear all complaints arising between themselves, and settle all caste disputes, &c. In some cases these arbitrators are remunerated for their trouble, but no fixed compensation is given them: such depends on the importance of the case. Their meetings for purposes of arbitration are held in a house built by subscription for this purpose, their assemblies being generally held at night. It is a noticeable fact that so thorough is their belief in the integrity of their arbitrators that their decisions are always implicitly submitted to (page lxxviii). In the *Tanjore* Collectorate, the Tehsildar of *Negapatam* replies that there is "one man styled 'Nambian' as the head of the Pattanavans of the fishermen castes, inhabiting the villages on the coast between Cuddalore and Vedaranien. His place of residence is also called 'Nambian Cooppam;' his office is hereditary, and on his death all the fishermen unite together, and appoint (? acknowledge) his heir as their headman. His main occupation is to settle disputes arising amongst the Pattanavans. Sometimes he uses a net of his own, and employs coolies who catch and sell fish for him. He goes in a palanquin to the villages inhabited by the said Pattanavans to enquire into matters of custom; the villagers come in advance to meet him, and present their respects to him, and conduct him to the village. During his sojourn in a village his expenses are borne by the fishermen of that village: he enquires into the offences committed, punishes the offenders, and collects the fines, &c. If a large net is nearly ready to be used a present of 7 pons, an ancient coin, and a cloth is made to the headman, who gives a written receipt for the same, and it is only thereupon that the net is used. If a marriage takes place in a house, it is not to be performed without a present of 200 betel-nuts and as many leaves, and of two fanams (five annas) being first made to the headman. In the event of a marriage being celebrated in the village where the 'Nambian' resides, rice and vegetables are to be presented to him, besides the aforementioned presents; the fishermen who live in the same village as the 'Nambian' are in the habit of giving him fish for his diet as 'Valaikari.' These are his emoluments, in addition to the income derived from using his own net. A document to the above effect executed by all the people of the fishermen caste to the 'Nambian' of the old days is still in his possession (page lxxxv). At *Muthipettah* it is stated "the fishermen of this place are Sanagars, who are divided into three factions, *viz.*, Periakatchi, Sinnakatchi, and Nadukatchi, each of which party has a headman of its own styled 'Marakayar.' The headman of each faction settles disputes as to relationship, &c., arising among the Sanagars, and takes precedence by the chief men on the occasions of marriage and such-like ceremonies: he, however, receives no kind of income from fishermen as emolument for his post" (page lxxxvi). The Tehsildar of *Tanjore* states that "there are headmen of the fishermen caste. There are certain degrees of headmen, the highest is styled 'Nambiar' or 'Puttun Kattigal,' who has authority over a number of fishing villages along the coast and whose word is supreme: his office is hereditary, and carries with it emoluments in the shape of a percentage upon the fish captured. The duties of the office, like most hereditary ones, are light, chiefly of a patriarchal nature, consisting of the settlement of disputes amongst themselves, attending the celebration of marriages, &c. The 'Nambiar' or 'Puttun Kattigal' has the privilege of receiving the first betel-nut amongst this class of men. The next degree of headman is called Nattamaikkaran, who is the recognised head of the village, and whose duties and emoluments are similar to those of Nambiar, though on a small scale, whilst the office is not hereditary" (p. lxxxvii). At *Tritrapundi* "there are headmen of the fishing castes; their post is hereditary, and when all the heirs are extinct it is bestowed on a competent man, selected for the purpose by the residents of each hamlet or street. Each headman determines the labour, &c., to be performed by men subject to his jurisdiction, fixes the rate of wages, &c., and gets such work done by them. He obtains for his services as much as each of those working under his control receives as his wages, and another extra share as a special remuneration for his headmanship. He also settles ordinary disputes regarding caste and custom that may arise amongst those classes. His emoluments cannot be ascertained otherwise than in the aforesaid manner" (p. lxxxvii). At *Myaveram* "there are headmen called 'Nettameigars;' the post is hereditary. On the occasion of marriages and funeral ceremonies the fishermen caste people act up to his orders. On marriages he has eight annas to one rupee according to their ability. This is all his income: he gets nothing for funeral ceremonies" (p. lxxxviii). At *Thealli* "there is one headman for each 'Cuppam' or small village on the sea-coast; the rank is hereditary. His duties are to settle the disputes amongst them regarding their caste, and to be the chief for carrying out marriage as well as funeral ceremonies. His emoluments are, that he receives from each family a fee at the rate of eight annas at each marriage" (p. lxxxix). At *Cuddalore* the fishermen caste " have headmen; the post is hereditary. They attend marriages and other ceremonies occurring in the caste, and distribute betel-nut to the people on the occasion, for which they receive from four annas to two rupees, according to the circumstances of the parties" (p. xcii). In the *Nellore* Collectorate the " fishermen have one religious headman to whom they give at every marriage four annas, with 2½ seers of rice and other grains. There is another named priest in their caste to whom also they give at marriages rupees two, with 4½ seers of rice and other grains. The latter man (priest) has frequently to visit the coast and other places where fish are taken, but does not go with fishermen. Their headman in religion has to decide their disputes, and if he is unable to settle them, the priest has to pass the final orders. The head in religion, and also the priest, are hereditary offices" (p. xciv). The Superintendents of Sea Customs in this Collectorate report that "the fishing castes in this district are four—(1) Palle, (2) Tuli, (3) Patapu, and (4) Chambadi. Each caste, and in fact each village or hamlet, where a number of them club together and reside (forming what is termed a 'Palliem'), has its own headman, called in some cases 'Pedda capu,' and in others 'Pedda Arkattu.' This office is hereditary, and on the failure of heirs, the community join together and select one from among their number to be their future headman. The duties are mostly honorary. The headman presides at all marriage and religious ceremonies, for which he receives certain 'russooms' or fees. He settles all petty quarrels and disputes in his 'Palliem:' he is looked up to, and his word obeyed with greater respect than any one else in that 'Palliem.' He has the privilege of being exempted from work. If a vessel strands or comes off his hamlet in distress, he gathers together all the able-bodied men, and gives help: so likewise at any Government call he furnishes help and collects labour, but is exempt from personal work. The duties of the office are not defined, but recognised merely by custom and long usage. The emoluments likewise are not regular, but consist of contributions or fees paid by the people from long acknowledged habits" (p. xcvi). The Tehsildar of *Strikarikota* states that "amongst the fishing castes there is one Adimulam Setti at Madras, who is the principal headman among the fishermen of the Pattapuvandlu castes: there are two others, Dalavaya Venkatraya Setti and Mantrichina Venkatraya Setti as headmen at Puliyenjeri Kuppam in this division; this headmanship is hereditary. The one at Madras is regarded with priestly reverence, while the other two are looked to for the settlement of religious disputes. These headmen fish like others for their maintenance: at marriages a fee of $6\frac{1}{4}$ annas is paid to the headmen, as well as fines for breach of religious rites. Half of these collections go to the headman at Madras, and the remaining portion is enjoyed by the said two headmen. But this practice is said to be gradually falling off, because some pay on the occasion whilst others get their wants attended to without payment" (p. xcvii). In the *Kistna District* the Sea Customs Superintendent at *Bandar* observes that "the fishermen possess the same privileges they had formerly: those who are thus referred to being privileged to catch fish are headmen. Though they are not privileged by any competent court, yet they are enjoying that privilege as hereditary through the favour of influential members for the time being. Their privileges are to catch fish and to obtain emoluments at festivals and other happy occasions. The headmen of the villages claim a right to fish in the sea; they do not allow others to catch fish" (p. xcviii). At *Vizagapatam* "there are headmen, and the post is hereditary; in default of any who have ceased to be headmen, others are appointed from among the fishermen by the authorities. Their duties are to perform acts connected with their religious duties and Government work, such as exporting and importing goods, &c. At feasts, &c., they receive at the rate of eight or four annas. They are allowed to set up stake-nets in rivers, and a share is allowed to them out of the fish caught by the other fishermen. Their emoluments are small, but the exact amount is not known" (p. xcviii). " There is a headman at China Ganjam and Peda Ganjam. He is supreme in matters of religion and festivals, but derives no emoluments: whilst the moturpha tax existed he was exempted from it. Among the Pattapu caste, people who catch fish by employing boats, there is a headman who settles their family disputes and religious customs: he is annually allowed eight annas for every boat employed in fishing" (p. xcix). At *Bapatla* "among the fishing castes each village has one or two headmen: the fishermen do not understand how the headmen were formerly appointed; they do not possess any certificate or patta to show by whom they were appointed. The office is hereditary; the headmen decide family disputes, and direct the fishermen to furnish supplies to Government when required. At marriages they receive betel-nut; on festive occasions they are asked prior to relatives. During festivals of the village goddess the headmen perform the ceremony, the cost being paid by the fishermen. First the headmen's sheep are sacrificed, subsequently those of other people" (p. c). In *Repalli* "each village has a headman who is termed Pedda Capoo: he receives four annas at a marriage. When the moturpha tax existed he was exempt: he has to obtain boats when required for Government service. When the fishermen captured fish he used to have a share" (p. ci). At *Coconada*, it is observed that "there are two kinds of headmen of the fishing castes, *viz.*, Kulapeddah (head of the caste), and Jattupeddah (head of an assembly). The first sort of headman is hereditary, whilst the latter is conferred on some one by all the inhabitants of the village.

The Kulapeddah will be headman of the caste for two or three districts, and such headmen employ themselves in settling religious disputes, in conducting such public affairs as may have to be performed on behalf of fishermen, &c., and in disposing of cases, such as adultery, &c., if committed in these castes. They have neither land nor other emoluments. Presents are given them at times of marriages" (p. ciii). The same customs respecting headmen are reported as existing in *Coringa* (p. civ), also in the *Vizagapatam* Collectorate at *Bimlapatam*, where the headman is termed 'Pillaho' (p. cv); likewise at *Vizagapatam*, where "they have hereditary headmen, whose duty it is to settle caste disputes. His emoluments are—(1) if he goes and asks the fishermen when they catch fish they give him two or four pies worth; (2) in marriages he receives three annas for putting a turband on the head of the bridegroom" (p. cv).

XV. In *Bengal* very little information has been recorded respecting the fishermen. At *Poree* there is a settlement of Madrassees who were induced to go there. These Telinga Lulliyas "cultivate no land, but live on the sands of the sea-shore, and are boat-men as well as fishermen: none but these men are capable of managing surf boats." The inducement to settle held out "was the promise of certain employment in boating, salt for four months, and free leave to fish in the sea and collect shells for burning into lime for the remaining eight" (p. cxvii), promises most deliberately made, but concerning which faith with the fishermen does not appear to have been considered worthy of keeping. In the *Chittagong* district, at *Noakhally*, "there are headmen of the fishing caste termed 'Sirdars,' who possess an hereditary right to the title. A vacancy occurring on the death of an heirless headman is generally the cause of much dispute between those who consider themselves entitled to the right of succession, and is filled up on the decision and nomination of their zemindars. The duties of the headmen are to preside over marriages, religious ceremonies, and feasts, and to decide all social disputes, for which they receive from one to four rupees, and at times both money and a cloth according to their rank" (p. cxxiv).

XVI. The fishing castes in olden times appear to have had a much more important standing than they at present possess. They seem to have had their chiefs, and to even been ready to join in military expeditions. The Samorin in 1813 sent a deputation to Portugal, where the ambassador was induced to become a Christian, and was knighted by John III under the name of "John of the Cross." However, on his return to Calicut, he was banished from the Samorin's Court in disgrace, as a renegade from his father's faith. In 1532 he joined the fishermen or Parravers, and appears to have been installed as their chief, as he headed a deputation of 85 of them to Cochin, imploring the assistance of the Portuguese against the Mahomedans. The whole of these fishermen, 85 in number, are said to have been converted. A Portuguese fleet was then sent to their relief, and 20,000 are reputed to have immediately consented to be baptised. Ten years subsequently Xavier organised a church for them.

XVII. It appears probable that the present organisation of the fishing classes is the remains of some ancient system, for on no other supposition can the existence of persons holding such an extensive sway be accounted for. The village or patriarchal system of an *elective headman* to a caste inhabiting each street or hamlet, is only what is seen elsewhere amongst other labourers: so likewise is the *hereditary headman* over several villages. But amongst the fishermen there exists priestly chiefs, two of whom are to be found on the Eastern Coast, one being at Madras, and the other at Cuddalore, the territory of the former existing up the Coromandel Coast, the other being more south of Madras. The third* is in South Canara, where he appears to have spiritual control over a large district. These persons, whose posts are hereditary, claim or receive fees and fines from those of their caste living in large tracts along the sea coast, and are the final referees in all cases of caste or family disputes or squabbles. The next grade is also hereditary: they, however, only hold sway each over a few villages: the duties are the same, and some of the emoluments appear to have to be transmitted to the priestly superior. On the death of this last description of headman without heirs, a fresh one is usually elected by the people of his caste. The *elective headman* is chosen by the residents of a single hamlet, and his duties are to decide disputes, be present at marriage and religious ceremonies, often fix the work, and assist in certain Government duties. His emoluments seem to be trifling.

Present organisation of fishing classes.

---

* It is by no means improbable that others exist, but no notice of them has been recorded.

As the duties and perquisites of each of these three grades of headmen have been fully detailed in paras. XIII to XV, I have not considered it necessary to recapitulate them in this place.

XVIII. The next question must be, *what is the present condition of the fishermen, and how do they work?* In *Sind*, after having paid a license of Rs. 5 a ton on their fishing boats, they may work as they please. The fishermen borrow money from merchants for purchasing boats and nets, and dispose of their captures to the mortgagee at half the ruling market rates (p. x), but whether this goes or not towards liquidating the original debt is not stated. The *Commissioner* observes that "the fishermen are well off" (p. ix).

<small>The present condition of the sea-fishermen in Sind.</small>

XIX. In *Bombay*, the Assistant Commissioner in charge of the sea-shore salt works and ports of *Guzerat*, observes that "fishermen are, as a rule, poor. The precariousness of their occupation and the uncertain profits derived from it often compel them to accept service as sailors in coasting vessels, laborers, and in fact anything that will ensure them a steady and certain means of living. These remarks apply to the whole body of fisherman engaged in the various kinds of fisheries spoken off" (p. xx). In the *Junjura district*, "the fishermen supply themselves with boats and nets: six or ten club together, build a boat, make a net, and divide the produce into shares. The sea-fishermen have decreased" (p. xxvi). In *Broach*, "the fishermen make nets themselves, and sometimes purchase them from their caste-men. They have generally boats of their own, and those persons who have no boats get them on hire from others. The number of fishermen appears to have decreased" (p. xxvii). In *Kaira*, "boats are little used; the nets are made by the fishermen themselves. The fishermen have decreased with decreasing trade" (p. xxvii). In *Rutnagiri*, "the practice of salting fish has decreased within the last fifteen years, in consequence of the increase in the price of salt. * * The fishermen supply themselves with boats and nets. The sea-fishermen have decidedly increased in numbers" (p.p. xxvii and xxviii). If the practice of curing fish has decreased, and the fishermen decidedly increased, such must either be due to a much augmented demand for fresh fish, or else the fishermen from increased numbers will be considerably poorer. An error has probably occurred in the answer

<small>The present condition of the sea-fishermen in the Bombay Presidency.</small>

From *Kanara* the same reply has likewise been received from the tehsildars (p. xxviii). *Mr. Commissioner Pratt* observes that " at present no larger number of men are engaged on fisheries than are required to provide an amount of fish sufficient for local consumption. * * The practice of curing fish has to a great extent diminished, owing, partly to the falling off in the amount of fish usually captured, and also the duty charged on salt in British territory" (p. xviii). In *Kaira*, it is stated " that the fishing has greatly fallen off of late years. The supply of fish is now too scanty to render the adoption of Dr. Day's suggestion necessary" (p. xxvii).

XX. In the *Madras Presidency*, in the *Tinnevelly* Collectorate, " as a rule, the fishermen of the coast are a very miserable lot of people, and excessively poor: the way in which they now work is by a system of advances from their 'chummaties' or headmen,* a few of whom reside in each village, and supply nets, lines, boats, &c., for the use of which a certain share (one-third) of all the fish caught is taken by the chummaty. Sea-fishing is the daily employment of a large number of the inhabitants living on the sea-coast: these men have certain contracts to supply fish with headmen of the 'Paraver' (fisher) caste, distinct from the chummaties" (p. iv). In the *Nellore* Collectorate, " no one claims any rights as regards sea-fisheries, but different villages are extremely tenacious of particular local limits, within which they claim exclusive rights of fishery. The fishermen purchase their own boats and nets, which are often pledged to .the contractor or soucar who advanced the purchase money. The number of fishermen has remained stationary" (p. lxii). The following remarks have been received from the *Native officials*. In the *South Canara* Collectorate, the tehsildar of *Kasargod* observes that the fishermen buy boats and sometimes nets, but usually manufacture these last themselves. The sea-fishermen have increased of late years (p. lxxii) : the same answer comes from *Udipy*, where the augmentation of this class of people between the last and the present census is set down at 15 per cent. (p. lxxiii). At *Kundapur* the fishermen " who are well-to-do have their own boats or hire them : nets they make themselves from hemp they grow or purchase ; these people appear to have increased of late years" (p. lxxiii). In the *Malabar* Collectorate, the Superintendent of the Sea Customs, *Cochin*, observes " the Native

_{The present condition of the sea-fishermen in the Madras Presidency.}

---

* These are not the elective or hereditary headmen previously alluded to, but traders.

C

Christians who engage in sea-fishing here are not of the fisherman caste in the proper acceptation of the term. * * About 30 of the fishermen here possess boats and nets : the owner of a boat has generally sufficient members in his family to man the boat. Where they fall short of the required number of hands, neighbours make up the deficiency; the latter get an equal share each of the fish captured" (p. lxxiv). At *Ponany*, " the majority of the fishermen have their own boats and nets, which others of more limited means obtain on hire. There is an annual increase in the number of fishermen" (p. lxxiv). At *Cannanore*, " the fishermen, boat and net-owners are ' Mukuwars', a low caste of Hindus, the ' Collakars' or Native Christians, and the Moplahs. The fishermen are supplied with boats and nets and other requisites for fishery by the owner of the boats and nets, who also advances them a certain sum of money (charging them no interest for the same) to ensure their services. The money thus advanced is not deducted from their daily labor. It is generally refunded by them, should they be unwilling to work for the party advancing the money, and in some instances in case of death, inability to work from extreme old age or infirmity, or in cases of desertion, the money advanced is a loss to the boat-owner. The owners remain on shore while the fishermen go out, and on their return to the shore the owners of the boats and nets sell their captures. Such as remain unsold are taken by the boat-owners (for salting purposes*) at the average rate at which the portion sold realized, and the sale being thus completed, they divide the proceeds equally between themselves, *viz.*, one-half to the owners of the boats and nets, and the other half to the fishermen : but should the latter prove unsuccessful and capture only sufficient to realize their expenses for the day, the boat and net-owners surrender their share in favour of the fishermen" (p. lxxvi). At *Tellicherry* " the rich fishermen are the boat and net-owners ; they do not go to sea themselves, but supply the poor among them with fishing apparatus, and pay them besides for working them. Such contracts are often reduced to writing. The remuneration is half the supply of fish captured, the other half going to the boat-owner, unless the take is very small, when all goes to the fishermen" (p. lxxix). The fishermen " have, however, by consent, made certain rules which are strictly observed. The most noticeable amongst these rules is the right of the first discoverer

---

* The use of salt-earth untaxed is permitted.

among a lot fishing together off a shoal of fish. In this case the man who first saw the fish is allowed to capture them without hindrance from the others, even though at the time when the fish were discovered he was not prepared to launch his net" (p. lxxviii). In the *Madura* Collectorate, "fishermen are supplied with nets by the better-to-do of them, called 'Sammanothy.' The fish caught is divided equally between the owner of the boat and the fishermen, but the amount is regulated by circumstances. The sea-fishermen have increased on the whole, as the lower castes on the coast have taken to it. The aboriginal fishermen castes, 'Paravars' and 'Karayans,' have decreased, many of their families having emigrated to other parts of the coast : a great part of the Paravar population have given up fishing, and betaken themselves to sea-faring. The Pallavarayan and Kadayar castes have remained stationary" (p. lxxxiii). In the *Tinnevelly* Collectorate, the tehsildar of *Ottapiduram* estimates "their average daily earnings at two annas taking all the year round, excluding costs" (p. lxxxiv). In *Mungery* "the contractors generally allow the workers one-fourth of the captures, the average daily earnings being from 1 to 6 annas" (p. lxxxiv). In *Tenkarei*, "the remuneration paid by contractors to fishermen is one-third share if large, one-half if they are small, and their average daily earnings vary from 2 to 8 annas" (p. lxxxv). In the *Tanjore* Collectorate at *Muthipettah*, "the fishermen procure small boats and nets at their own cost : but those who have neither the one nor the other, join those who have them, and go along with them for fishing, the income derived from the fish captured is divided into as many shares as there are men engaged in the job, with 1½ extra share (one for the boat and half for the net), thus the share of each man being equivalent to that allotted to the boat. The number of sea-fishermen have decreased" (p. lxxxvi). In *Tritrapundi*, "the fishermen procure nets and boats at their own expense, and those who cannot afford to do so get a loan thereof, while some join those who are possessed of nets or boats in catching fish ; in the latter case, fish caught are divided into three parts, of which two form the share of the owners of the nets and boats, the third part going to those who actually catch them. The fishermen as to numbers have remained stationary in Topputurai and thereabouts, but decreased in Mutupettai" (p. lxxxviii). In *Myaveram*, "all the fishermen have not got nets : some eight or ten persons engage themselves as coolies under a net-holder." The amount of share the coolies receive is not stated, except that it is a proportion of the fish, and "every coolie carries at once his share to other places, and by selling them for grain or cash earns a livelihood" (p. lxxxviii). In the *South Arcot* Collectorate, " men of the Carriar, Patnaver, and Pullie castes fish in the sea, whilst it is solely Pullies who do so in the back-waters ; they are known by the name of Shemdavers. * * The seafishermen on the coast earn between ten and twelve annas a day. The fish on the coast are said to have diminished, and the cause is attributed to the bad seasons we have had of late. The fish appear to have receded into the deep sea" (p xcii). In *Cuddalore*, "fishermen supply themselves with their boats and nets : these men have been on the decrease of late years" (p. xcii). In the *Madras* Collectorate, " the fishing population are decreasing " in *Chingleput* (p. xciii) : the other tehsildars report an increase. In the *Nellore* Collectorate " some retail traders do not fish, but merely purchase to resell. * * The rule is that the fish are hawked about and are generally exchanged for grain." In four talooks the fishermen are said to have decreased, in three to have increased, and in two to be stationary (p. xciv). It is also observed "that Pullie people generally fish in salt rivers and Tuli people in the sea, but people of both castes give their fish to traders on contracts, or sell them personally, and use what remains for themselves. * * Most of the Tuli people are very poor, their daily earnings only just covering their expenses. Whenever they may be in need of boats, or nets, or at least of repair for their old ones, they borrow money from traders, and give them fish every now and then in satisfaction of their debts. Some fishermen also obtain money by exporting grain in their own vessels, or giving their vessels for freight. Fishermen in Gundur talook come with their families to the coast of Ongole about the month of January, and quit the coast by the end of June, during which period they export quantities of fish to their own district, and besides take home with them the remaining lot" (p. xcv). Respecting the fishermen, " some are of opinion that the means of livelihood have generally decreased, as the demand for salted fish has generally diminished. Others again think, where the trade in salt-fish has increased, the income of the fishermen has likewise improved" (p. xcvi). "The fishermen of each hamlet or pollien are very tenacious as to their peculiar rights to fish within certain limits, whether in the sea or in any tidal creek or estuary. These limits have never been defined or recognised by any authority, but have been admitted by long established usage among the fishermen themselves from time immemorial; and if the fishermen of one hamlet are found transgressing their limits and plying their trade within the limits of a neighbour, the result is a never-ending source of dispute and quarrel among the men of both hamlets. But the quarrel seldom goes further than themselves, and receives no countenance at the hands of any authority" (p. xcvi). In *Striharikota*, " during the last four years, the quantity of fish taken has been less than in previous years, consequently the sea-fishermen are in a poor condition" (p. xcvii). In *Handukur*, " the practice of salting fish has decreased, as fewer fish are captured" (p. xcviii). In the *Kistna District*, the tehsildar of *Bapatla* reports that "the fishermen living along the coast from Peraly up to Peda Ganjam fish for four miles out to sea, obtaining perches. Other species are caught along the shore in great quantities, whilst those taken in the sea are few" (p. c). In the *Bunder* talook " the fishermen have decreased since the cyclone, in which many were washed away : about 20 boats go to sea for fishing purposes, which, as well as the nets, have decreased" (p. c). In *Repalli*, " fishermen report that they are decreasing in numbers, as are also their boats and nets. * * The fishermen go 1½ a miles out to sea for fish, which they salt, but there is only a demand for the small sorts. * * The fishermen being unable to purchase salt for salting fish, take advances of money for their livelihood from fish merchants coming from Bunder, &c., to whom they deliver their captures, and the merchants have them salted by coolies employed by them. * * For the last two years fish have been scarce" (p. ci). In the *Godavery* Collectorate, the tehsildar of *Ramachendrapur* reports "the daily earnings of those who fish in the sea will be one rupee, while the other fishermen who fish in canals, &c., earn four or two annas a day, which is not more than sufficient for subsistence. The supply is not equal to the demand. The fishing population has decreased in consequence of a few having resorted to Moulmein and other coasts for carrying on their trade" (p. cii). At *Poddapurum*, " the fishing population has increased of late years" (p. cii). In *Coconada* they have " decreased of late years" (p. ciii). At *Pittapur* they " have not increased, because a few have embarked for Moulmein on account of famine, while some have died of cholera" (p. ciii). At *Coringa*, " the average daily earnings of sea-fishermen are about one rupee, but of those who fish in the rivers perhaps four annas" (p. civ). The Deputy Collector at this place observes that "there are many who claim rights respecting sea-fisheries opposite to their huts or places of residence. Such disputes give rise to civil actions" (p. civ). In the *Vizagapatam* Collectorate, at *Bimlapatam*, "the villagers residing on the sea-shore consider they have a claim to cast their nets before outsiders" (p. cv). In the *Ganjam* Collectorate, in the *Berhampore* talook, "fishermen working as coolies are not paid in money, but receive half the fish captured in fresh waters, and one-third of those taken in the sea goes to the owners of the nets and two-thirds to the coolies, who earn about two annas a day" (p. cvi).

XXI. In *Bengal*, in the *Pooree* Collectorate, "the fisher class come from southern districts,

Present condition of the sea-fishermen in Bengal.

and are rather strangers here, so that their customs are not very easy to get at in this district" (p. cxx). The people appear to have been induced to come from Madras districts to settle, and promises made seem to have been broken (p. cxix). This is a reason why one must feel averse to recommend such a plan for adoption again. So long as the civil officer who took an interest remains, all goes well: a new one arrives, and some personal views respecting the law of supply and demand, which have no bearing whatever on the subject, induce him to curtail the promised privileges.

XXII. In *Burma* the sea-fishermen appear to be well off: whilst the sea, which is swarming with fish, is not properly worked.

Present condition of the sea-fishermen in Burma.

This is due to two causes—*first*, the mode in which they prepare their fish (*Nga-pee*) is not suited to the Indian market; *secondly*, that they prefer fishing along the shore to venturing further out to sea. Respecting the proposition to introduce Madras fishermen (p. p. cxxvii & cxxviii) to capture the fish and export them to India, I cannot help thinking (see my remarks p. cxxviii, note) that the plan as proposed will fail. Likewise, now that is self-evident that the seas of India could provide sufficient fish were the indigenous salt within the reach of the fish-curers, why should a foreign trade be stimulated prior to opening up the home one.

XXIII. If we examine the foregoing replies respecting the present state of the fishermen and their mode of working fisheries, we find, as I shall have to observe

Present state of the sea-fishermen generally, and the emoluments they receive for fishing.

that, well off in Sind, they are, unless in the vicinity of large towns, miserably off in the Bombay Presidency. Well off down the western coast of Madras,* but once round Cape Comorin, they become, as observes the Collector of Tinnevelly,† a very miserable lot of people, and such is the same account, except near large towns, all the way up the Coromandel Coast. In most of those places where the fishermen are said to be poor they are also reported to be decreasing in numbers, due to cholera, emigration, or taking service as lascars in coasting vessels. In localities where they are poorest they appear also to be most litigious, as up the Coromandel Coast, where they claim the fishing before their huts, and irrespective of which they generally state the fish are decreasing, although in some places it is asserted they have migrated to the deep sea. Their mode of working fisheries is either employing their own boat (manned by relatives or hired coolies), or borrowing the money for the purchase of the same. The mode of remuneration is generally by a division of the spoil; and wherever the fishermen are most prosperous, it is not the rule that they receive the highest remuneration. Along South Canara and Malabar the boat and net-owner receive half the captures, probably due to the demand for fish being unlimited, the coolies being always readily able to dispose of their captures. As Cape Comorin is rounded the boat-owner receives from one-fourth (Mungery) to one-third if large, half if small (Tenkarei), or even half (Madura), one-third (Tripundei, and also in the Godavery Districts), or the whole of the captures are divided into shares as follows: if six men go in one boat the fish are sub-divided into $7\frac{1}{2}$ shares, one being for the boat and half a one for the use of the net. The emoluments received by the fishermen being thus returned: Tinnevelley 2 annas a day, Mungery $1\frac{1}{2}$ annas a day, Tenkarei 2 annas 8 pies a day, Godavery one rupee. In many districts they accept advances for their season's work, or contract to supply fish at a certain rate and in a certain quantity (if procurable) to merchants and others who re-sell it fresh or prepare it salted, either by themselves, the aid of hired coolies, or persons of the fishermen caste.

THE FISHES OF THE INDIAN SEAS.

XXIV. The marine fishes may be divided into three great natural (not zoological) classes. *First*, those which are more preyed

Natural division of the Marine fishes.

---

* Salt-earth is allowed to be gathered by the fish-curer untaxed.
† The collection of salt-earth becomes a penal offence.

upon than predaceous, and which may be sub-divided into the gregarious and those which are only partially so. Those which arrive in vast shoals, or the gregarious forms, evidently approach the coast at certain seasons for breeding purposes; they are generally destitute of any considerable means of defence, but become preyed upon, not only by other fishes and predaceous animals, but form in a large degree food for man. Those which are not gregarious or only partially so evidently likewise come for breeding purpose, and also to follow some particular form of animal food, which is present off the shores probably for breeding. The seasons when these fishes arrive mostly correspond to the cold months of the year, when the sea is not so agitated by the monsoons. *Secondly*, we have the predaceous fishes, which may also be in shoals, following those on which they are able to prey: or they may be non-gregarious or only so to a slight extent. *Thirdly*, there are the non-migratory forms, some of which are also predaceous: these live along the shores and backwaters, and some in the deep sea.

XXV. Amongst those classes of *non-predaceous** *fish*
Non-predaceous marine fish. which arrive at certain seasons, there
The gregarious and the non- are, as observed, the gregarious and
gregarious. non-gregarious forms. The gregarious, or those which appear in vast shoals at certain seasons off the coast for breeding purposes, may again be sub-divided into those which breed in the fresh-waters, and the remainder which do so in the sea. In my fresh-water fishery report I have adverted to the enormous ascents of hilsa *(Clupea palasah)* up all the large rivers for breeding purposes, mostly during the south-west monsoon (June and subsequently), and it is a most important circumstance that they are almost invariably as plentiful in one season as they were in the preceeding year, provided no impediments in the rivers exist entirely, barring their ascent. If one examines the varieties of fish taken along the sea-coast throughout the year, these hilsa will be found extending their range to whereever food is plentiful; they will only be missed during the breeding season, and even then young ones will be present. This would appear to show that they never migrate any great distance from the shore. In fact this fish is not so capricious in its arrival as are the more marine forms. Thus a periodical supply of food is afforded to people far inland un-

---

* This term, of course, is only employed in a comparative manner, to those fish which are exceedingly predaceous, as nearly all forms are more or less so.

less man in his greed impedes or entirely arrests their ascent by means of fixed engines and weirs, and so annihilates the supply. The second sub-division of the gregarious forms are most important in an economic point of view, provided man turns this harvest of the sea to a proper account. These strictly marine shoals are much more capricious as to the years of their advent than are the hilsa. The forms most generally known in the Indian waters are perhaps the mackerel, *Scomber kanagurta*, and the oil sardine *Clupea Neohowii*, which are generally so abundant along the Western coast, Ceylon, and the Andaman Islands. Besides these there are many species of anchovies which arrive in vast shoals. Surely it is taking a very narrow view of what the uses these droves of fish should be put to, if we consider them exclusively arriving for the benefit of the dwellers in maritime districts. Nature here provides every requisite for rendering them useful to distant localities, vegetable substances from which the nets may be constructed, wood for boats, and salt for the preservation of the captures, and we shall have to enquire whether full advantage is taken of all these provisions of nature, and if not, what is the reason for this apathy?

XXVI. Secondly, amongst these non-predaceous fishes we find certain forms which are certainly not so gregarious as those already alluded to, as they are only abundant at certain seasons of the year. There appears to be a constant migration of some kinds, probably not solely for breeding purposes, but perhaps they may be pursuing some especial article of food. For it is not only the fish, but many invertebrata that are constantly appearing merely for a short period of time. The little *Chætodon pretextatus* has only been recorded in the Malay Archipelago and Cochin, two widely separated localities; and in the latter place I have merely taken it during the first fortnight of the south-west monsoon. The curious form and beautiful colours of this little fish are so remarkable that it is highly improbable, if they were present during any other period of the year, I should not have procured them. Another instance occurred whilst at the Andamans at the end of 1869 and commencement of 1870, when I remarked that even during my brief sojourn, the beautiful *Acanthurus lineatus*, which was numerous at the period of my arrival, could not be obtained at the time I left. The migrations of other non-predaceous and non-gregarious fishes will be alluded to under their respective genera.

XXVII. Amongst the *predaceous* fishes the gregarious are hardly so numerous as those which are not so. They are mostly found in the order Acanthopterygii as well as among the Chondropterygii. Amongst these fish which appear to be usually found in shoals perhaps as well known as any is the Bonito (*Thynnus*). I was fishing one day off the new breakwater at Kurrachee, where we obtained a very large number of oil-sardines *(Clupea Neohowii)*, a vast quantity of which had arrived in the harbour. All of a sudden a great commotion commenced, and a shoal of large fish dashed in amongst them, evidently causing the greatest consternation. I told the fishermen to make a cast over one, as I required a specimen. They begged to be excused, as they feared their nets would be broken. These were evidently Bonito; they came quite close to the break-water where I was standing, and that evening some were captured in the large seine nets. The fishermen assert that these predaceous fishes, if they cannot tear through the nets by main force, frequently bite the meshes, and they consider the *Chrysophrys* as very much addicted to this practice. Having made an arrangement with the fishermen, we pulled over to a rocky island where these fish were reputed to be found. A cast was made, the net enclosed a *Chrysophrys*, it made one dash, was through the meshes like a shot, spinning right into the air, and certainly the net appeared as if it had been cut, not torn. I mention this instance here to show how much more difficult it may be to net the predaceous fishes with their cutting teeth, than the non-predaceous ones that do not possess any teeth that could be employed for biting nets. The *non-gregarious* predaceous fishes are exceedingly numerous and of very varied characters. Amongst the most voracious of these must be placed the sharks, rays, and allied forms. They follow the shoals of the migratory non-predaceous fishes, and as a consequence the presence of the food on which they live is a cause of their appearance in the same waters.

XXVIII. It is in fact evident that the oil-sardines, small though they be, lure the larger fishes into the vicinity of the shores which they themselves frequent; consequently any reason which may induce these fish to go elsewhere must cause a paucity of supply in the seas which they desert. This, however, is not only seen on this large scale, but is likewise apparent in another form. Thus the larger forms of sea-fishes, as perches, &c., come towards the shore to prey on immature or small fish, crustacea, &c. If this their natural food is destroyed to any very great extent they have not that inducement to come near the land which they would otherwise possess. The result, therefore, of denuding the coasts of small fishes is, that the superior sorts do not come so close in-shore as they otherwise would, and the fishermen are reduced to capture merely the non-predaceous or inferior kinds, or the fry.

XXIX. Amongst the non-migratory forms there are likewise very great differences, and some are much more predaceous than others. Many of these, although termed non-migratory, certainly change their places of residence at certain periods of the year, either to follow their food, or to escape from rough into quiet waters.

Non-migratory sea-fishes.

XXX. In the sub-class *Teleostei*, the spiny-rayed or *Acanthopterygian* order furnishes about 616 of those forms which frequent the seas and tidal waters of India and Burma. Amongst these are some which attain to a large size, and are very predaceous in their habits. Some reside in the deep sea or off marine banks, and when captured are generally of excellent sorts for the table, but bait-fishing appears to be best suited for their capture. These do not appear to be furnished with accessary breathing organs, as was observed (F. W. F. report, p. 15.), enabling those frequenting inland waters to migrate from place to place, such an addition would appear to be unnecessary.* As a full description of the *families and genera* has been given in the Appendix (from p. cliii,) a detailed account here of where each are to be found appears uncalled for.

The sea-fishes of the sub-class Teleostei.

XXXI. Amongst the *Acanthopterygians* and in the family of perches (*Percidæ*), we find a great diversity of size to which they attain. Thus Cantor mentions one (*Serranus horridus*, or the adult of the *S. lanceolatus*) the weight of which exceeded 130 lbs.; and Hamilton Buchanan observes that the usual size of another (*Serranus coiodes*) in the Ganges is between four or five feet in length. The *Mesoprions* belonging to this family are,

Fishes of the Acanthopterygian or spiny-rayed order. Families Percidæ, Pristipomatidæ and Squamipinnes.

---
* That it may be required in sea-fishes, see Genera *Clupea*, *Chatoëssus*, and *Chanos*.

however, of usually greater immediate value in the production of food, because they appear to come further inland to breed, and are consequently much more frequently captured: some also attain to a very large size. Then we have the little species of *Ambassis* and *Apogon*, which frequent places more inland, and so probably assist in tempting the larger forms to come nearer in-shore.* In the family *Pristipomatidæ* are many large and excellent fish, but they are as a rule inferior to the Percidæ, some in fact, as the *Therapons*, being usually rejected by Europeans as food. The air-vessels of both the families mentioned are considered as affording a good quality of isinglass, but an inferior quantity. In the family *Squamipinnes* those of the genera *Chætodon*, *Chelmo*, *Ephippus*, and *Holacanthus* are very inferior, but when of sufficient size members of the *Heniochus* and *Drepane* are eaten by Europeans, whilst the *Scatophagus* is likewise by the very lowest classes.

XXXII. The family *Nandidæ* do not afford sufficient marine species to render any allusion to them necessary. Merely one example of *Plesiops* appears to have been captured in the Indian seas, and that only at the Andamans. The family *Mullidæ* are largely represented, but it is remarkable that they are held in no estimation by the Europeans, although the far-famed 'Red mullet,' or 'Woodcock-of-the-seas,' as it is termed, is considered so excellent in Europe. In the family *Sparidæ* there are scarcely sufficient of the genera *Crenidens*, *Sargus*, *Lethrinus* *Sphærodon*, or *Pimelepterus* to render them of much economic moment. The genus *Pagrus*, however, affords the *P. spinifer*, which is abundant in places and excellent for salting. Amongst the members of the genus *Chrysophrys* are many that are held in great esteem whether eaten fresh or salted. The family *Cirrhitidæ* likewise only afford small fishes, which are not held in any esteem, whilst those in the family *Scorpænidæ* appear to be almost universally rejected.

Families *Nandidæ*, *Mullidæ*, *Sparidæ*, *Cirrhitidæ*, and *Scorpænidæ*.

---

* I have already, para. XXVIII, mentioned how the oil-sardines coming inside a harbour may be said to bait the locality for larger and more valuable forms, as the 'Bonito.' I will here mention a very interesting circumstance which occurred in South Canara when I was with the Collector, Mr. H. S. Thomas. We were going out live-bait fishing early the next morning, when I suggested, where are the baits? We will take them *en route* I was informed. At 5 A. M., the next morning, we started, still no baits. We came to a minute rill, when every one commenced hunting for the required live-baits. First I saw numerous little *Haplochili* (see P. W. F. Report, p. cclxxvi) that had ascended to capture the almost microscopic insects. Soon, however, first one walking fish, *Ophiocephalus gachua*, then another were perceived hunting the *Haplochili*; being air-breathers, they could ascend after their prey in water which did not cover their backs. This appeared to me a good illustration of the vast comprehensiveness of nature's plans.

XXXIII. The family *Teuthididæ* furnishes a vast number of fish, and although they are not considered as of the most superior quality, still are freely eaten fresh, and also salted and dried. The family *Berycidæ* do not afford fish which are esteemed as food. The *Kurtidæ* likewise are not of any great value. Amongst the family *Polynemidæ* are some of the most valuable of the sea fish, all as food, and two species on account of the large amount of isinglass they afford (see Appendix, p. cxlii). Some of these fishes attain a very large size, especially in estuaries, where they appear to thrive best, as Ham. Buchanan observes that *P. teria*, "in the Calcutta market, is often found six feet long. I have been assured by a credible native that he saw one which was a load for six men, and which certainly, therefore, exceeded in weight three hundred and twenty pound, avoirdupois." This genus also affords the celebrated mangoe-fish, which at the breeding season swarms into the estuaries and rivers of Bengal and Burma.

Families *Teuthidida, Berycidæ,* and *Polynemidæ.*

XXXIV. The family *Sciænidæ* is valuable for two reason : first the isinglass, which is obtained from most of its members, which, although said to be of a rather inferior quality to that furnished by the *Percidæ* and *Pristipomatidæ*, is of a larger amount. Secondly, the middle and lower classes employ them very extensively as food, besides salting them for distant markets. The *Xiphiidæ*, or sword-fish family, are at times numerous off the Coromandel coast during the cold season of the year, and are employed as food. The scabbard fishes, *Trichiuridæ*, are held in very diverse estimation in different parts ; off the Meckran coast, where salt was plentiful and cheap, the fishermen did not care even to bring them to land asserting that they were useless as food. Along the coasts of India, on the contrary, they are prized, as owing to their thin ribbon-shape, the natives can dry them in the sun with the use of a very small amount of salt, or even none at all.

Families *Sciænidæ, Xiphiidæ,* and *Trichiuridæ.*

XXXV. The ' Lancet fishes', *Acanthuridæ*, are held in about the same estimation as the *Teuthidic'æ*, already adverted to. The ' Horse-mackerels,' *Carangidæ*, are amongst the most important families present in the Indian seas, not only numerous in species, but also in numbers. They are much more common along the Coromandel than the western or Malabar coast, whilst all are excellent, both fresh or salted. Even the *Chorimemus* is esteemed as a fish well adapted for salting, but some

Families *Acanthuridæ* and *Carangidæ.*

object to it cooked fresh as rather too dry; the same may also be said of the *Trachynotus*. The little *Equulas* might be thought to be of but slight economic value, owing to their generally small size and almost absence of any amount of muscle. But it is owing to this very cause they are so sought after; they can be soaked in the sea and dried in the sun, and thus as dried-fish are considerably sought after, especially as they are cheap. The *Lactarius* is esteemed either fresh or salted, and for the latter purpose it appears to be well adapted.

XXXVI. The 'Pomfrets,' *Stromateidæ*, are very justly celebrated both amongst Europeans and natives; extensively distributed, they are common almost everywhere from Sind to the most southern point of Burma. Besides being excellent fresh, they also salt well. The 'Dolphins,' *Coryphænidæ*, are common off the Coromandel coast in the cold weather, and though rather dry are still considered good as food. The little *Mene* is useful in the same manner as observed upon regarding the *Equulas*. The 'true mackerels,' or *Scombridæ*, are very valuable, not only on account of the amount of food they furnish in their fresh state, but because they are extensively salted : the common Indian mackerel, *Scomber kanagurta*, can be preserved to a certain extent with salt-earth, and this is employed for the purpose in many places, but when it is proposed to send the article to distant markets, as from the western coast either inland to the coffee estates or elsewhere as to Ceylon, a certain amount of good salt is usually employed. This family also possesses other fish which are excellent either fresh or salted, as the 'Bonito', *Thynnus*, the 'Seir-fishes', *Cybium*, whilst some natives likewise eat the *Elacate*, but not, I believe, the 'Sucking-fishes', *Echeneis*, which are usually taken from off captured sharks, to which they are adhering with great tenacity.

XXXVII. The family *Trachinidæ* may be thus subdivided, first into those genera which are rejected or but little esteemed as food, *viz.*, *Uranoscopus*, *Anema*, and *Percis*; and *secondly*, the highly esteemed genus of whiting, *Sillago*. These last are very extensively distributed, and are equally consumed in a fresh or salted state.

XXXVIII. The 'Frog fishes', *Batrachidæ*, are so repulsive in their appearance, that they seem to be everywhere rejected as food : the family *Cottidæ* have no better reputation, in fact some species are asserted to be so very poisonous that they are even rejected as manure. However, the 'crocodile fishes,' *Platycephali*, are occasionally eaten, but they are so dreaded, due to the spines they are armed with, that the first thing a fisherman who has captured them does, is to knock them on the head.

XXXIX. The Gobies, *Gobüdæ*, have very few represen-
Family *Gobüdæ*. tatives that attain any size, but still they are numerous and consequently of considerable use. Being mostly littoral species, they are of great service in decoying larger fishes in-shore, and thus as baits are of considerable consequence. They may be divided into two divisions, *first*, those as the *Gobius*, *Euctenogobius*, *Apocryptes*, *Gobiodon*, and *Eleotris*, &c., which die soon after they are removed from the water; *secondly*, those which live for a long time if kept moist after they have been taken from their native element, as the *Periophthalmus* and *Boleophthalmus*, which may be seen migrating about over moist pieces of tidal-covered mud, or even up rocks and pieces of wood in pursuit of insects and such like food, but instantaneously diving out of sight on the fancied approach of any source of danger. Large numbers of the *B. Boddaerti* are brought alive to the Bombay markets in baskets covered with a wet cloth.

XL. The Blennies, *Blenniidæ*, are only found on rocky
Family *Blenniidæ*. coasts as a rule, where other and superior edible forms exist; they may well be considered as food for larger fishes. Clambering over moist localities, often temporarily made captive in small basins in the rocks which retain water, they become imprisoned until the succeeding tide relieves them; they collect small animal substances as their food, and are in turn themselves sustenance for the larger kinds.

XLI. The voracious and dangerous *Sphyrænas*, termed
Families *Sphyrænidæ, Atherinidæ, Mugilidæ,* and *Ophiocephalidæ.* 'sharks' in some places, due to their destructive powers, are all used as food, and many are taken along the coasts, often of a considerable size. The little *Atherinidæ* come in large shoals along the shores, and often go up the estuaries; they are extensively dried in the sun. The 'Mullets,' *Mugilidæ*, are abundantly represented, and most useful as food; but when at Mangalore this year, I was informed that the higher classes of Hindus resident there have an aversion to eating them, because they allege their heads resemble those of serpents, whilst the same people do not refuse

the 'snake-headed fish' or *Ophiocephalidæ*, which are esteemed as highly nourishing. The lower classes likewise had no such prejudices, whilst they were extensively salted. During the breeding months of the year large quantities of these fish are taken, and their roes dried or salted ; in fact, Cochin fish-roes, which were prepared from these fishes, used to bear a very good reputation in the markets.

XLII. The families *Pomacentridæ* and *Labridæ* do not
<small>Families *Pomacentridæ* and *Labridæ*.</small> exist in sufficient numbers off the low coasts of India and Burma to render them of much economic importance, besides which they are generally held in but small estimation as food. They are more numerous off the Andamans than any other locality I have personally investigated.

XLIII. The spineless fishes or *Anacanthini* possess
<small>Order of *Anacanthini* or spineless fishes. *Gadūdæ, Ophidūdæ*, and *Pleuronectidæ*.</small> some descriptions which are exceedingly valuable. They are divided into two sub-divisions, those more or less allied to the cod family, but which have only insignificant representatives on the seas of India, *viz.*, the little fishes of the genus *Bregmaceros*, and some of the family *Ophidiidæ*. The second sub-division, or those allied to the flat fishes or 'soles,' *Pleuronectidæ*, are numerous and attain a considerable size. Coloured only on the upper side, and swimmers close to the bed of the sea, it must be evident that clear water is not so suitable to their existence as where it is discoloured. They are found most numerous and attaining the largest size where great rivers as the Indus, the Ganges, the Irrawaddy and the Salween debouch into the sea; in the Madras Presidency they are as a rule finer and more common on its Malabar than on its Coromandel coast.

XLIV. Amongst the Physostomi are the scaleless cat-
<small>Order *Physostomi*. Families *Siluridæ* and *Scopelidæ*.</small> fishes. *Siluridæ* are mostly found in muddy waters and frequenting the mouths of large rivers and estuaries. Thus in the clear water at the Andamans they are but rarely taken, although up some of the creeks they are occasionally captured. They are common from Bombay down the western coast, and from the mouth of the Kistna along the muddy coasts of Bengal and Burma, whilst in the Mergui Archipelago they are most abundant. The largest sized ones are captured at the mouths of rivers, especially when such localities as the Sunderbunds exist there. It is evident that the number of highly sensitive and muscular barbels with which they are provided must be for the pur-

pose of feeling their way about in waters that are almost too dense for vision. The mode in which these marine forms guard their egges will be alluded to further on (para. lx, et. seq.). These fish are mostly armed with a strong bony and toothed pectoral spine, a most formidable weapon both for offence and defence. Some years since at Cochin a sea snake was brought to me alive as a curiosity, it being armed with a serrated spine on one side of its neck. I opened the reptile and discovered that it had swallowed a cat-fish, *Arius*, and though its capacious jaws could permit its entrance, the pectoral spine of its prey had perforated its neck, and thus protruded externally, giving the appearance as if the snake were armed. The estimation in which the marine forms of these fishes are held as food varies; they are as a rule rejected by Europeans, and natives commonly consider them as inferior. But last December when at Cochin I met a fisherman returning from his day's work, his little boy was dragging two large skates by their tails; having purchased them, I enquired what other fish he had. He displayed a few mackerel which I did not require. After some considerable trouble I obtained a sight of his remaining spoil, which consisted of 'cat-fishes' *(Arius)*; being at that time investigating this family, I at once decided he must sell them, but it was with the greatest difficulty I obtained his consent; he asserted he was going to carry them home for family consumption, and strongly urged that the mackerel should be taken instead. In the report from the Superintendent of Sea Customs at Tellicherry (p. lxxvii), it is observed that "cat-fishes are cured to a very great extent and exported to Colombo, as well as to some parts of the Tinnevelly district." Under the head of isinglass (p. cxl) I have described the appearance of the fish sounds procured from fishes of this family (p. cxliii), and how they differ in shape from those obtained from the more percoid-form of fishes. In the 'scaled-siluroids,' *Scopelidæ*, there exist a few fishes which are excellent eating, but rarely found in very large numbers, if we except the 'Bomaloe' or 'Bombay Duck,' *Harpodon nehereus*. This fish is numerous in Bombay, becomes rare all down the Malabar coast, a few are taken at Madras, but as we proceed towards Vizagapatam they increase largely and are common all along the coast of Bengal and Burma, ascending large rivers within tidal influence. Owing to their almost gelatinous character and absence of any large amount of muscle they are easily dried.

XLV. Amongst the *Scombresocidæ* are many marine fishes which are most valuable in an economic point of view. The Gar-fishes, *Belone*, are exceedingly abundant at Bombay, but not so numerous elsewhere along the coasts of India. These fishes are frequently seen in the Bombay markets three feet or even more in length, and are there generally termed the 'Lady-fish,' why, I have not been able to ascertain. They are exceedingly voracious, and it is curious to see them pursuing other small fishes. The terrified little anchovies will dash out of the water along the surface, but the gar-fish follows, and appears to skim along the waves after its prey. The curious *Hemiramphi*, with, as a native expressed, only a lower jaw, come in vast droves, and are everywhere considered excellent for food, although with the exception of *H. far* none attain to any large size. Their roes are much prized on the western coast during the cold months of the year. Flying fishes, *Exoceti*, although so very abundant down the Red Sea, and often seen in quantities off the coasts of India, are not captured to any extent as food, although they are excellent eating. They are more often obtained when the weather is a little windy, on board vessels at sea at night-time, apparently partially attracted by the lights.

Family *Scombresocidæ*.

XLVI. The Pseudoclupeoids, *Pseudoclupeidæ*, contain some fishes of considerable economic importance. The salmon-like 'Milk fish' has been introduced into tanks in South Canara, and attains a large size. The *Elops saurus* likewise, besides being eaten fresh, is salted to rather a large extent.

Family *Pseudoclupeidæ*.

XLVII. It is, however, amongst the true 'herrings,' *Clupeidæ*, that perhaps the most important of Indian fishes are to be found, not only as themselves affording a great amount of food for man, but as being an inducement to other larger and better species to migrate to the seas which they frequent. In the genus *Chirocentrus* the single species is salted to a great extent: however, like most fish which have a large external row of teeth, it is exceedingly vicious, and the fishermen assert that when captured it lays hold of the nearest object, even if such is merely a piece of wood. The delicate little sardine-like *Dussumieria* are usually present in large numbers in the cold season throughout the coasts, but it is amongst those termed *Clupea* in the Appendix that the largest shoals are seen. The *Clupea fimbriata* is very common, but

Family *Clupeidæ*.

not nearly so much so as the true oil-sardine, *C. Neohowii*, whose apparently capricious advent or disappearance has been adverted in at p. cl. Doubtless were the objects on which they prey examined into, it would be found that the absence of the latter was the cause of the non-appearance of the former. In the same way a complaint was made that at Calicut the amount of sharks' livers required for the manufacture of the medicinal fish oil could not be procured, and on investigating the cause it was evidently due to the oil-sardines which form their principal food being absent. Doubtless they had followed the shoals.

XLVIII. The little *Clupea melanura* I have seen in vast quantities at the Andamans.

Family *Clupeidæ*—continued.

The hilsa has been already referred to as a breeder in the fresh waters. The very similar *C. toli* does not appear ever to ascend rivers, but curiously enough disappears from the coast at the time the hilsa ascends into fresh water, but to where it goes has not been ascertained. Irrespective of the foregoing, there are vast numbers of *Pellona*, *Opisthopterus*, and *Raconda* captured along the sea-coast, and either consumed fresh, dried, or salted. The genus *Engraulis* is most prolific in fish, and these are very extensively consumed either fresh or dried. Another form, the *Chatoëssus*, is also found along the coasts, more especially of western India, coming into the backwaters, but in South Kanara I observed that the existence of large parasites upon them the rule, a healthy state the exception. Whether this was due to their residing in these localities, or whether they had come there to rid themselves of their tormentors, may well be questioned.

XLIX. The 'Eels' of the seas of India are neither in such number nor so esteemed as to be of any material consequence in the food supply, except those of the genus *Murænesox*, of which I have seen specimens ten feet long in the Bombay markets, where they are cut into transverse slices and thus disposed of. In fact of the others it is more usual to see them cast up by the tide during the monsoon months than for sale in the local bazars.

Family *Murænidæ*.

L. Amongst the order *Plectognathi* none are eaten in India, so far as I am aware, except by the poorest of the poor. Some of the *Tetrodons* are however relished by the aborigines of the Andaman Islands.

Order *Plectognathi*: family *Tetaonidæ*.

LI. The sub-class of cartilaginous fishes furnish some which are eminently useful to man as food, although none are what are considered of the most superior description. The true sharks if large are mostly captured for their fins, which are removed, dried and exported to China, or else for their livers to manufacture into fish-oil (see p. cxlix). Some of the medium size are cut up and salted, whilst the little ones are eaten either fresh or salted.

Sub-class Chondropterygii or cartilaginous fishes; family Carcharüdæ.

LII. Salted shark is considered very strengthening by the people of Malabar, but the Bengalis, except the very lowest castes, will not touch it. In Bombay the Africans may be seen every morning purchasing small sharks for their own consumption. Their skins are used for covering sword belts in some places, also for smoothing down wood. The hammerheaded sharks are equally employed with the true ones.

Sharks as food: or their skins how used.

LIII. Amongst the sub-division *Batoidei* are many sorts equally useful with the sharks. The 'saw-fishes'* livers are employed in the same manner for their oil, whilst the ' skates' are not less useful. The electric ray, however, which absolutely swarms along the Meckran coast, appears nowhere to be of any value. the rays are eaten in many places by the poorer classes, but they do not appear to be anywhere held in estimation.

The Batoidei, or rays and skates.

LIV. The localities in which some of the marine fishes take their abodes are sometimes very peculiar. Wherever there is a wreck, some species are certain to be inside the vessel, evidently in search of food, and we can easily understand how others may be found inside corpses, as there they are finding sustenance. But their reason for inhabiting some places can hardly be thus accounted for. When at the Andamans some of the aborigines showed me how the beautiful little *Glyphidodon anabaloides* could be captured in any numbers. Diving from off the side of the canoe, they brought up large pieces of coral, amongst the branches of which these fishes were very numerous and packed in exceedingly closely. It appeared to me that as sharks, rays, skates and other predaceous fishes were numerous, these little fellows sought shelter from their foes by rushing in amongst the coral branches, from whence their pursuers could not extract them. But more curious still is it to find fish inside other living animal substances,

Peculiar localities inhabited by fish.

* I have procured the *Pristis Perritotteti* in a river far above tidal influence.

and which they employ as their home. At Gopaulpore on the Coromandel coast I was surprised at observing inside a large Medusa brought to shore in a fishing net a living *Therapon servus*, Bl., and on enquiring into the circumstance was informed that this was by no means an uncommon place where they were to be found, and several more were shown me in the same curious habitations. At the Andamans Captain Hamilton observed to me that a few days previously he had remarked some small fish living inside a polype at North Bay. One day he dug out one, dragged it to the shore, and captured three little fish from its interior : replacing them in the sea, they appeared to be doubtful what to do, swimming round and round as if searching for something. The living polype was now returned to the water, and they at once swam towards it, following it as it was dragged back to its original locality. The species of fish was *Amphiprion bifasciatum*, Bl. A few days subsequently when fishing on the 'Jolly Boys' island, one of the aborigines brought a specimen of the pretty yellow and white banded *Amphiprion percula*, Lacép., and on being told it was 'good,' we were taken to see where others could be easily captured. They detached *Actiniæ* from the coral by inserting the hand behind them, and on shaking it two or more of these little fishes fell out. They told us the rule was for a pair to reside inside each of these large sea nettles. We saw this process carried out thirteen times, and the results were 27 healthy little fishes.

THE MIGRATIONS OF SEA FISH.

LV. The irregular way in which the true marine breeding and gregarious fishes come and go has been already referred to, but there still remains one point which requires a few remarks. It is very common to find true marine fish in pieces of fresh or brackish water not far removed from tidal influence, and this is due to their having entered these places whilst the monsoons were at their height, the rivers and swamps full, and owing to which a communication existed between them and the sea. As the waters subsided this channel becoming interrupted, the fish have to remain there until the next year's monsoon re-opens the communication. A few months since, at Kasargod near Mangalore, on the western coast, I observed numerous *Medusæ* present in a piece of water between which and the sea all entrance or exit had been barred for many weeks. These gelatinous

Migrations of sea-fish, and cause of their being often found in fresh water which is not connected with the sea.

creatures may be perceived swarming up the rivers and backwaters of the western coast during the south-west monsoon, and doubtless predatory fishes pursue them and other kinds of food into localities where they run the risk of finding their exit cut off as the body of the water falls. It does not follow, however, that this is necessarily a cause of their death, as it is very common to capture such fish in these places even after the water has ceased to be saline. Thus the Milk-fish, *Chanos salmoneus*, is found in a tank at Cundapur in South Canara, and report says that Hyder Ali introduced it for his own use. This is certainly a marine species, although now it breeds and thrives in the fresh or rather slightly brackish water, attaining to 20 or 30 lbs. in weight. It appears probable from local circumstances that when this tank was dug it communicated with the sea; that the fish were thus introduced, and gradually the communication must have been closed. Along with the Milk-fish is a sea perch, *Mesoprion rangus*, also a *Chrysophrys* and a Mullet. Centuries since the Romans knew that some sea-fish could be kept in fresh water if the change were gradual.*
Whilst many sea-fishes, as has been already observed, ascend long distances up fresh-water rivers in pursuit of their prey or for breeding purposes.

BREEDING OF FISHES IN SALINE OR BRACKISH WATER.

LVI. The question of where the sea-fishes deposit their ova has been a fruitful source of contention for a long period. Whether it floats about and is then vivified and subsequently hatched, or left for this purpose on mud banks in the ocean or along the shore or in estuaries, are still open to investigation. One would suppose that fish-eggs would hardly be left to float about at the mercy of the waves for several reasons,—*first*, all the eggs of fresh-water fish sink in water, so do those of all

<small>*Breeding of sea-fishes, and where they deposit their eggs.*</small>

<small>* " M. Felix Plateau has undertaken a number of experiments to determine the question whether the cause of the death of fresh-water animals when removed to sea-water, and of marine animals when removed to fresh-water, is the difference in the density or in the chemical constitution of the water. His observations were made mostly on various species of Articulata; he found that those fresh-water species which possess an aerial respiration can survive the change to salt water, while those which possess only a branchial and cutaneous respiration die quickly. By experimenting on water made denser by the solution of sugar, M. Plateau came to the conclusion that the density of the water is not the destructive agent, but a portion of the salts held in solution. The chlorides of sodium, potassium, and magnesium, he found to be very quickly fatal to fresh-water species, but the sulphates of magnesium and calcium had no prejudicial effect. In the same manner the death of marine animals in fresh water appeared due to the giving off of sea-salt from their bodies to the surrounding fluid. All these facts he believes explicable from the laws of endosmose and diffusion."- · 'Nature', August 1871.</small>

the marine forms which I have examined, except perhaps some of the cartilaginous fishes, but then there exist filaments to the horny case which envelopes the ova, apparently for the purpose of attaching it to seaweed or other substances. Irrespective of this, if fish ova floated about at the mercy of the waves they would doubtless be extensively destroyed by sea birds and fishes, whilst the young would be hatched wherever currents took the eggs, and probably to localities entirely unsuited to their condition. All descriptions of substances, especially if gelatinous, which are floating about or brought up by the nets from the bed of the ocean, have been considered by fishermen as the eggs of sea-fishes.

LVII. Wherever eggs are deposited still we find many modes in which the sea-fish propagate their kind in India. It is perfectly certain that the vast shoals which at certain seasons arrive off the coast do so for breeding purposes. Full of roe when they arrive, some being males (with soft roes) other females (with hard roes), deficient in roe prior to their disappearance, and many young ones being found which previously were not present, no other conclusion can be arrived at than that they are bred off the coasts of India. Thus the oil-sardine, *Clupea Neohowii*, appears off the western coast of India, Ceylon, &c., and although full of almost mature ova they are deficient in fat, and it is not until their breeding has been completed that they commence to fatten; consequently by October, sometimes before, and for a couple or three months subsequently, they are well adapted for extracting oil from (see Appendix, p. cl).

Breeding of sea-fishes—continued.

LVIII. I have already pointed out in my Fresh-water Fishery Report that the hilsa, *Clupea palasah*, as soon as it has completed breeding, becomes lean and lanky, and does not fatten until it has regained the sea, but this rule does not obtain in the oil-sardine, or rather as it always breeds in the sea, it has the circumstances around it which permit its at once fattening on the extrusion of its ova. It appears to be the rule that those fishes which breed in these are not subject to such a loss of condition (unless due to disease) after depositing their eggs, as are those marine forms which emigrate from the sea into the fresh waters for this purpose.

Immediate loss of condition after spawning not a necessary result in sea-fishes.

LIX. Amongst the non-migratory sea and estuaries fishes there appear to be many diverse modes in which the young are

Diverse modes in which the young are produced.

produced: I have already adverted to the common one, but whether we possess any monogamous *Teleosteous* sea-fishes we have as yet no decisive evidence, perhaps such may be found amongst the *Gobüdæ*.

LX. I observed a peculiar way in which the young of some of the marine *siluroids* or 'cat-fishes' are hatched, *viz.*, by the eggs being deposited in the mouth of the male, who carries them about in safety until the little ones obtain their exit. At the end of April this year (1873), owing to the great assistance I received from Mr. H. S. Thomas, the Collector of South Canara, I was enabled to make an interesting investigation of some of the coast fisheries in his district. Upwards of one hundred specimens of the 'scaleless cat-fishes, *Arius* and *Osteogeniosus*, were obtained, and amongst them were some eggs loose in the baskets in which they were carried; these averaged half an inch in diameter, and were similar to those I have at other times observed in some marine siluroids. In the mouths of some of the males of each of these species there were from 15 to 20 eggs, some of thése they dropped on being disturbed. These escaped eggs were evidently some of those which were loose in the basket. On examining these eggs some were in an early stage of development, other more advanced, whilst a few had almost perfected embryos inside, in fact, in one the young was hatched, but with the bag still adherent. These eggs filled the cavity of the mouth, and extended backwards to the branchiæ, while on dissecting the fish all were males, and food was entirely absent from the intestines of those that had eggs in their mouths. The fishermen were positive that they were thus carried about and hatched, and certainly appearances led to that conclusion. The proportion of sexes appeared to be five males to one female.

How some male cat-fishes carry the eggs about in their mouths until hatched.

LXI. Next the females came under examination: on tracing up the ovi-sacs it appeared that very large numbers of eggs were present, but not all of one size. Furthest removed from the external orifice the eggs were of full size, about fifty in number, whilst other batches of much smaller ones existed, evidently to take the place of the large ones when they should have been excluded. The full-sized eggs were each attached by a pedicle of varying extent to the inside of the ovi-sac, giving them a considerable resemblance to the

Breeding of sea 'cat-fishes'— continued.

eggs of the crocodile, *C. palustris,* or of common fowls. The ventral fins in the females were considerably enlarged, reaching to well over the commencement of the anal, the rays having a deposit of fat on them, whilst the innermost ray had a large pad of fatty matter behind it. These fins can be expanded into a cup-shaped surface, and it does not appear improbable that this is a provision of nature for the reception of the eggs as extruded. Perhaps the eggs received into this receptacle are here vivified by the male, who then removes them in his mouth. Although the males at this interesting period are reputed to fast, evidently such must be the case if they carry eggs in their mouths without swallowing them; the females do not do so; their intestines were found to be full of food. In fact food is a necessity to the females for the production of the young. These fishes are said to continue breeding for several months, and they apparently lay their eggs in batches.

LXII. Why the eggs of marine siluroids should be carried about by the males until they are hatched it is difficult to conjecture. I have examined into the question of the size of the eggs of fishes in the closely allied genus *Macrones,* who breed in fresh water, and find they are rather minute. Although unable to give a reason why the eggs of the marine forms are large, it is clear that if small they could not be conveyed about by the males without the chance of their being swallowed. Living amongst constantly shifting mud banks, swarming with predaceous fishes, probably the eggs would never be permitted to come to maturity, but protected by their parent whose osseous serrated dorsal and pectoral rays render it a most dangerous opponent, this plan for the preservation of the species may be the most efficacious one.

Breeding of cat-fishes in fresh waters different from those in the sea.

LXIII. Amongst the Cartilaginous fishes, which are highest in the scale, the young are produced in a different manner from those in the genera already adverted to. Many of the sharks and their allied extrude living young, which are formed in an enlarged portion of the oviduct, where they either lie free and are surrounded by an albuminous fluid, or else possess a distinct placenta. Some, however, deposit ova, which are enclosed in quadrangular horny cases furnished with filamentous prolongations at either extremity for the purpose of attachment to suitable spots, and thus prevent the

Cartilaginous fishes, and how their young are produced.

f

sea current carrying them away to localities where they would probably be destroyed. The hard horny covering in due time allows the exit of the embryo at its anterior extremity. These cases, often seen thrown up by the waves, are termed 'sea purses.'

FISH IN AN ECONOMIC POINT OF VIEW.

LXIV. The sea-fisheries of India are or should be exceedingly valuable as affording an inexhaustible supply of animal food both to persons living in their vicinity as well as inland, when means exist to transport the fish either in a fresh or cured state. It surely is a very narrow point of view to look at this question, as some civil officers in maritime districts have considered it, *viz.*, "my sea-coast districts are supplied, what more can be desired;" surely it is desirable that these fisheries should be useful to persons situated far from the coast. Irrespective of mere food, they may be serviceable directly or indirectly in trade, as producing isinglass, fish-oils, or manure, as well as requiring materials for the building of vessels, the manufacture of nets or hooks-and-lines, whilst the fact they are the nursery of the native sailors should not be lost sight of.

Sea-fisheries ought not only to be serviceable to those living in their vicinity, but also to the inland residents.

LXV. The first question for investigation is, *what proportion of people would be consumers of fish could they obtain it?* and this divides itself into the necessary enquiries who consume it fresh and who prefer it salted? In Sind the use of fish as an article of diet is almost universal, whether fresh or salted: but as the freshwater fish are cured to a considerable extent, there is not consequently that demand for sea-fish for home consumption that exists elsewhere. In the Bombay Presidency (see Fresh-water Fishery Report, p. xxxiii), the majority of the inhabitants of the inland districts are consumers of fish when they can procure it. There appears to be an unlimited demand inland for sea-fish either salted or dried, and the supply does not equal the demand. In Madras, the great majority of the population are not precluded by their religious prejudices from eating fish, but it is evident that in many places it is only the lower castes and the Christians who consume it when salted; this of course may also be due to the inferior quality of the article as at present prepared. In Bengal we have a slight difference, there can be no doubt, that the fresh fish is more

A very large proportion of the people inland would be consumers of salt-fish could they obtain it at a reasonable price.

largely in demand than the salted. This, however, I think, I shall have reason for showing may be partly accounted for by the repressive duty on salt, which renders it practically unobtainable for the fish-curing classes, consequently they simply dry the fish. Thus the article is inferior, whilst the Bengalee is bigoted. We may sub-divide this Presidency by first enquiring what the Ooriahs prefer. The Collector of Balasore (p. cxxi) observes that "even in the town of Balasore, only six miles from the sea, the fish sold in the markets is so stale that no European would touch it, and much of it is putrid. * * The people in this district do not salt their fish: they dry it in the sun, and eat it when it is quite putrid. They like it in this way, and there is no reason why they should be interfered with." The foregoing fact that it is thus sold in Balasore I have personally witnessed, but the Collector's reasoning is, I think, open to discussion. A zemindar of the same locality, observing on the sanitary condition of the fishermen, remarked, "Cholera seems made for these people" (p. cxv), and that putrid fish frequently engenders disease* is too well known to require dilating upon. But we arrive at the conclusion that salt is rarely employed in the preservation of fish eaten by the Ooriah. The dried fish is also exported inland to the hill people. Dr. W. Hunter observes of this often famine-stricken population that "all castes in Orissa, below the Karans or writers, would gladly use salted fish: and at this moment they consume great quantities of fish imperfectly cured in the sun and more or less rotten. With the Chasas, or peasant population, who form the great body of the people, this is a favourite article of food: indeed, almost the sole relish which they can afford to their monotonous rice diet. The husbandman stores up his supply of dry fish in reed baskets,

---

* It may be well to decide whether it is humane or even prudent in a sanitary point of view to make the price of salt so excessively high that it cannot be used to preserve fish with, and thus compel the people to go without or consume it putrid or rotten. We read that "in Bergen there are two large hospitals devoted exclusively to the treatment of patients suffering from a peculiar form of disease, brought on by eating badly-cured fish; the disease is a mixture of leprosy and elephantiasis" (both common in Orissa). In Ireland in 1645 we are told that the leprosy was driven out of Munster by the English: the disease being due to the people eating foul salmon or those out of season. This was prohibited and the prohibition enforced, "whereby hindering these barbarians against their will to feed on that poisonous meat; they were the cause of that woeful sickness which used so mightily to reign among them hath in time been almost abolished." The Collector of *Ratnagiri* (p. xxv) states that "the high duty on salt is undoubtedly a source of epidemics and other serious illnesses induced by eating imperfectly prepared fish." *Mr. Cornish* in his observations on the nature of the food of the inhabitants of Southern India, 1864, remarked fish is often moreover used in a semi-putrid state, from bad salting, and in this condition it undoubtedly predisposes to disease" (2nd Ed., p. 48). I think the foregoing extract sufficient to show that compelling a population to eat rotten fish may be a rather impolitic act.

and sparingly doles out the decomposing mass as a luxury to his frugal household throughout the year" (Orissa, II. p. 160). The foregoing must drive one to the conclusion that in districts where salt is cheap the people do eat salt-fish : where that condiment is dear, so that a sufficient amount cannot be afforded to adequately preserve it, the people of even the higher classes do not use it to any great extent : but where the cost of salt is prohibitory, salt-fish is totally neglected. Now we arrive at the question, is salt-fish used in Bengal? (Orissa has been alluded to). All sorts of reasons are given why salt-fish is not eaten in Bengal to any extent, "that it is not very palatable" (p. cxxiii), or that "the people are not skilled in salting fish." If, however, we cross over from Bengal into Burma, where salt instead of being Rs. 5 a maund sinks sometimes to 8 annas, we find that salt-fish is invariably consumed in the form of nga-pee by the indigenous population. I think the foregoing shows that in almost every district where salt is sufficiently cheap to allow of its being freely used in curing fish that the article is universally consumed. Where salt is moderately taxed, but salt-earth may be employed free of duty, there is a large consumption, especially amongst the poorer classes. But when the price of salt renders it unattainable by the fish-curers there is no demand for salt-fish.

LXVI. *How are the markets supplied with fish?*
The productiveness of the sea, and the harvest gathered from it by the fishermen, are two very different and distinct things. In *Sind* the coast markets appear to be well supplied, so much so, that the average exports have reached Rs. 45,889 during the last five years. In *Bombay* the local wants of the coast markets appear to be supplied to a great extent, especially at certain seasons, but very little is cured and transported inland, because "home-cured fish cannot now compete with that cured at Goa and elsewhere, where a salt excise does not exist" (p. xi). Some, however, is dried and sent inland. Taking the average imports and exports of salt and dried fish in the port of Bombay for the last five years, we find (p. xxv) the average excess of imports over exports has amounted to Rs. 73,658 annually. At *Ratnagiri* it is reported (p. xxv) that "any increase of the supply of fish would be a boon to the poor inland. At present there is no doubt that the fish imported to those places are most

The local demand for fish close to the sea is generally well met during the calm months of the year, but where salt is dear the fish is often sold and eaten putrid.

imperfectly cured, and if any means can be devised to preserve them better, a more abundant and more wholesome sustenance would be obtainable by people who are at present in a wretched state of destitution." In the *Madras Presidency*, as a rule, the supply of fish in the local coast markets is equal to the demand throughout the fine weather season, but in many places it does not appear equal to supplying inland markets. In *South Canara* (p. xliiii) the Collector observes that "great quantities of sea-fish are carried inland by boats up the many rivers which exist, or being taken by coolies to places as far as ten miles from the sea, and there exchanged for grain and condiments; these are sometimes fresh, but more frequently besmeared with salt-earth, or a slight amount of Government salt. But the salt is either so lost in mud or so expensive that very little is used, and the fish are necessarily so partially cured that in most cases putrefaction has set in before they are consumed. Thus, a wholesome food is made an unwholesome one, and the consumers' demand for it is doubtless much reduced thereby." The Collector of *Trichinopoly* observes—"No doubt good fish could be sold to almost any amount; and the supply is nothing equal to the demand" (p. lvii). At *Cuddalore* the Assistant Collector observes that "the demand of the local market can entirely consume all the obtainable supply." Some places inland are supplied from the coast fisheries to a certain extent: thus salted fish obtains a sale along the Wynaad range of hills where coffee estates exist. Likewise a fair amount is carried inland from the Western coast by rail, even for the purpose of supplying the Negapatam market on the eastern coast. Higher up we find some exported inland to supply Hyderabad, and also the people along the hills. In *Orissa* and *Bengal* no complaints are given of imperfectly supplied bazars, simply because from that Presidency the most incomplete returns have been received. The Officiating Collector of *Poree* says that "near the Chilka Lake the people subsist largely on fish all the year round. At a distance from the Chilka Lake fish is a rarity and a delicacy, except during the rainy season, at which period every chása plies his net, basket, or trap for small fry" (p. cxviii). At *Balasore*, where putrid fish is the common diet, dried fish are taken to the Tributary States during the winter for sale (p. cxxi). At *Bhaugulpoor* the markets, except during the height of the rains, are stated to be well supplied. In *Chittagong* only sufficient appears to be taken to supply the local markets, and at *Noakhally* that even this demand is scarcely met. In *Burma* the markets appear to be considered as well supplied with Nga-pee (see p. cxxix).

LXVII. It will be now necessary to examine in what way fish are consumed or prepared for consumption along the sea-coasts of India and Burma, and this must be considered under the following headings :—(1) fresh, how far can it be carried so as to be fit for food ; (2) dried fish, the sorts of which it is composed, and how it is prepared; and (3) lastly cured or salted fish, which again will have to be divided into that prepared with monopoly salt, whether foreign or indigenous, that in which salt or swamp earth is the condiment employed, and, finally, when sea-water alone has been used.

<small>Modes in which fish are cured or prepared for future consumption.</small>

LXVIII. *How far inland can fresh fish be conveyed in the plains of India so as to be fit for human consumption?* In the examination of this question, it must be distinctly understood that the use of salt or ice is not here adverted to. The distance sea-fish can be carried inland whilst fresh will depend upon several causes. The season may curtail this, as in the hot months putrefaction of course occurs very rapidly amongst fish packed closely in a basket and exposed to the full force of the sun. Again, fishermen do not bring their fish ashore as a rule (unless where there exist fixed engines and the tide suits) until after sunrise, and even that brought of an evening will rarely be conveyed inland by coolies, as they have a strong aversion to be out after dark. Of course facilities of carriage must have much to do with this question, as by rail they may be taken further than by any other mode ; but this is hardly a normal condition of things. They can be conveyed by boats some distance inland in many districts where facilities of water communication exist. But the common mode is by coolies, and what with detentions at the time of starting, and the loitering along the road, I have not yet been in any place where I have seen sea-fish when carried by coolies anything like fresh above ten miles from the place of capture. The Collector of *South Canara* (p. xliii) likewise observes on sea-fish being carried by coolies to places as far as ten miles from the sea. The tehsildar of *Tenkarei* in Tinnevelly, however, states

<small>The distance inland uncured fish can be conveyed to market in the plains of India.</small>

(p. lxxxv) "that raw fish are taken as much as 30 miles inland;" but I suspect he means fish that have not been cured, and that these raw-fish have been opened, cleaned, and rubbed inside with salt, as I shall have to refer to. In the *South Arcot* Collectorate, it is stated "fish are sometimes taken six or seven miles for sale, as far as is consistent with safety to the fish" (p. xcii). In the *Madras* Collectorate, the tehsildar of *Chingleput* observes, "fish are not taken more than twenty miles for sale" (p. xciii); at *Ponnery* "fresh fish is not carried above ten miles" (p. xciii). At *Balasore*, "only six miles from the sea, the fish sold in the markets is so stale that no European would touch it, and much of it is putrid" (p. cxxi). But the question of the kind of fish carried has much to do with this subject, the immature fish as a rule becoming putrid much more rapidly than the mature. Leaving aside the cartilaginous fishes, those which are carried inland chiefly belong to the Acanthopterygian or hard-rayed order (p. cliii) and the Physostomatous order, or those forms in which as a rule a communication exists between the air-vessel and the pharynx (p. cclxxxii). Among the first, (excluding some, as the mackerel, &c.,) are found the fishes which can be safely conveyed the longest distances in a fresh state, as the different forms of sea-perches, the Polynemi, and horse-mackerel, &c., whilst the mackerel with its wide gill-openings and the Trichiurus with its thin skin very rapidly decompose. Among the Physostomi are two reasons for decomposition, first the nature of the fish, as the herrings, secondly, due to the food which they have eaten, as the siluroids; and these last very rapidly taint. I think it may be safely assumed that fresh fish as a rule cannot be conveyed inland by coolies above ten miles, so as to be fit for human consumption. But if the fish are first opened and cleaned, some salt rubbed in and care taken in conveying them as to their being properly shaded and the direct rays of the sun avoided as much as possible, they may be carried considerably further. But salt being dear, instead of a sufficiency being employed, a very slight amount is used, and putrefaction has often set in prior to their being sold for human food.

LXIX. Having adverted to how fresh fish are disposed of, we will next enquire into how sea-fish are cured along the coasts of India and Burma, commencing with *dried fish*. This can be done with the smaller and thinner species, as *Equula's*, scabbard fishes or *Trichiurus*, many of the herrings, the Bombay duck, or *Harpodon nehereus*, &c. But for the larger fishes this process is not adapted; however, slices can be cut from them and dried, a plan I found existing in Mergui with respect to sharks. Without at present referring to the salt-tax, it is very evident that its incidence must have some bearing upon this mode of preparing fish. We first observe solely sun-dried fish wholesale in the Bombay Presidency, in fact the Collector of *Tanna* remarks "fish netted in excess of the demand is, as a rule, sliced, cleaned, and dried in the sun, and then packed off mostly to Bheuridy, which is the chief entre-pôt for the sale of dried and salted fish : whether fish is dried as above in preference to its being salted is a question I have been unable to ascertain; the fishing class say that it is their custom so to treat the fish. It is very probable that it has been resorted to in the place of curing by salt consequent on the excise duty levied on salt." As we proceed down the coast and enquire whether all the surplus fish is dried as described in Bombay, we find that it is not, the people being allowed to gather the salt-earth for this purpose, and as a consequence they prepare their fish with it in preference to having it simply dried. In the *Trichinopoly* Collectorate, "on the coast, I believe that all the salting much of this so-called cured fish gets is being buried in the sea sand, and thus getting slightly briny" (p. lvii). We hear nothing respecting the desire of the fishermen to rather dry their fish than salt it, until we arrive in Bengal, and there we find drying in the sun the almost sole means employed for curing it. Leaving Bengal and examining the reports from Burma, the sun-dried fish again become scarcely even alluded to. This brings us to the following conclusion, that wherever salt is expensive the natives have a preference (? due to the cost) to sun-dried fish : that where it is cheap, this mode of preparation is but little employed.

LXX. I now arrive at a most important question for analysis, *viz.*, how are fish cured with salt? and as I previously remarked, this must be divided under the following heads. (1)—Fish cured with monopoly or excised salt, or (2) with earth or spontaneous but untaxed salt.* Before entering into the subject contained under each of the above three heads, it will be advisable to enquire *what amount of salt is necessary to properly prepare a given amount of fish?* At *Kurrachee* the Collector of the Sea Customs reports that 20lbs. of the best salt is employed

<small>How is salt-fish prepared? with the comparative amount used in curing with monopoly salt or salt-earth.</small>

---

* Soaking fish in sea-water, either partially evaporated or not so, is so very similar in its results to merely sun-dried fish that observations on it do not appear to be necessary.

for the purpose of curing a maund or 82¾lbs. of fish. The Superintendent of Sea Customs at *Tellicherry* replies "that about 28lbs. of monopoly salt are used to a maund of small fish, as mackerel, sardines, &c.; for instance, 1,000 mackerel, weighing about two bazar maunds of 32lbs. each, are cured with five measures of salt equal to 20 and odd pounds; but if salt earth is employed, two baskets full of that substance weighing about five bazar maunds are required. It would thus appear that for the purposes of trade that the best salt is used for curing fish in the proportion of one part of salt to about three of fish; but if salt-earth is made use of it requires nearly three (above two and a half) parts of salt-earth to one part of fish. However, at Guadar, where I narrowly inspected the process adopted (salt being very cheap), a considerably larger amount of this antiseptic was used than is reported to be employed in Sind and in India. In Burma and Mergui for Nga-pee made from prawns I found one ounce of salt was required to every two ounces of these crustacea.

LXXI. *How is Government or excised salt used in the* Mode in which excised salt is *curing of the sea-fish?* In *Sind*, Go-employed in curing sea-fish. vernment or excised salt is stated to be exclusively used by the fish-curers. In the *Bombay Presidency*, either excised salt is not employed for curing fish or else it is used to such a small extent that the article is of an inferior quality. In Surat it is used for preparing the Bombay ducks or Bomloes, but they require a very small amount of this antiseptic. Fish, it may observed, can be cured with salt in sufficient quantities to render the article good, wholesome, and undecomposed for a considerable length of time, and fit for importation to distant markets. But in the Bombay Presidency it unfortunately happens that the duty per maund on salt is Rs. 1-13, whilst there being no duty in contiguous foreign states, it "generally costs two annas a maund, even if as much" (p. xii). But the sea is the same; the identical varieties of fish can be obtained by the fishermen at the same cost of capture, whilst no import duty exist on its entering British territory; the result is said to be that this mode of preparing salt-fish does not pay in the Bombay Presidency. The *Deputy Commissioner* of the salt revenue in the Southern Division observes that "at the more important ports, however, a very large trade exists in salted fish, but this is almost exclusively the produce of neighbouring foreign ports (Goa, Daumaun, and Diu)" (p. xvii). The *Assistant Commissioner* observes "foreign fisheries have decided advantages over ours, as the salt employed by them generally costs two annas a maund, even if as much, whereas British excised salt could not be removed from the pans under Rs. 2 per Indian maund. Hence the foreign fisheries are able to employ salt freely and to the full extent to cure fish thoroughly. Our fishermen, on the contrary, bearing in view the comparative cost of salt to them, cannot afford to be so lavish, and consequently use just sufficient salt to preserve the fish, for a time, but not to cure it" (p. xxi). In the *Madras Presidency*, fish is prepared in places with Government salt, especially for export to Ceylon, but as I shall presently show, there is good reason for concluding that this antiseptic is not used to any very great extent for this purpose, as it makes no perceptible figure in the amount of salt disposed of. At *Cannanore* (p. lxxv) "some of the boat-owners whose means admit of storing salted fish until they can find a good market use Government salt for curing both large and small fishes intended for consumption in the hilly countries of Coorg, Wynaad, and Mysore, &c., and also for exportation to Colombo, &c." At *Tellicherry* it is only those fish "intended for exportation to Colombo that are solely cured by Government salt, as such only can be preserved for a long time, and are capable of being carried, without deterioration, to distant markets" (p. lxxvii). In the *Travancore* State, "the fishermen themselves very seldom go to the expense of salting fish. There are export agents from Ceylon and other places who buy the fish raw, or advance money for the harvest of the season. These people buy the salt for curing" (p. lii). It is evident from the returns in the appendix that Government salt is not much used for curing fish up the eastern coast of the Madras Presidency. Whilst in Bengal it is scarcely ever thus employed, in Burma it is used for preparing nga-pee. I think the foregoing extracts show that the fishermen of India scarcely ever employ Government or excised salt in the curing of fish; that when it is so employed the salt is purchased by agents or else money is advanced by them for it, and that the article is intended for export as to Ceylon, or into the hilly countries where coffee estates exist, and the coolies can afford the better and more wholesome article.

LXXII. *Are fish cured to any extent with earth or spontaneous but untaxed salt?* This enquiry is a most important one, respecting the salt-fish trade, especially as it appears that Government or excised salt is only used for this purpose for a small local and the foreign demand, unless it is cheap. Government salt is

How fish are cured with salt-earth. The cost, and what will be the result of subjecting the salt-earth to an excise.

used for this purpose in *Sind*, but salt-earth or spontaneous salt is not thus employed. In *Bombay* we are told that the use of this salt-earth is prohibited in Surat, but that in some places in the Presidency it is employed for the purpose of curing fish, although sun-drying without salt is the usual practice. At *Ratnagiri* "the police patell of the village below Zygurh fort has found it impossible to prevent the Daldis from tresspassing on an old salt-pan there, now no longer used. These people come to dig up the earth, which is impregnated with salt, for the purpose mentioned" (p. xxv). In the *Madras* Presidency we must first examine what occurs on the Western or Malabar coast,* and subsequently the Eastern or Coromandel. On the western coast an enormous increase in the trade of curing fish with salt-earth has sprung during the last few years, the origin of which requires a short explanation. In Malabar it has been ruled that the natives possess the privilege of manufacturing salt for their own use, and that such cannot be a subject of interference by the excise. It would be difficult to guage the amount of salt required for a families use, especially if they employed such an antiseptic in curing fish for their own future consumption. Although they cannot legally sell this salt, there does not appear any law which can prevent their disposing of any surplus stock of salt-fish which they may have, and thus it has come to pass that these sea-fisheries are pretty well fished, the fishermen finding a sale for his captures is pretty steady at his work, whilst their numbers have largely increased and comparatively enormous exports of salt fish to distant places has been the result, even sea ports as Negapatam being supplied with salt-fish from Malabar by rail. It must not be supposed that this salt-earth is of the best or finest quality, for the Collector reports of Malabar salt "the people dislike it, asserting that it imparts a bitter and unpleasant savour to food, and brings on that common complaint in Malabar, the itch" (p. xxxviii). But the poor of India cannot be too particular as to the *taste* of the food which they consume; with them a far more important question is the *cost*. A few figures will explain this; the cost of excised salt to the fish-curers cannot be less than 2 rupees a maund of about 82℔s. weight : cured with this description of salt one maund of salt is required to about three maunds of fish, or omitting the cost of the fish and the wastage,

---

* From Pondicherry to Cape Comorin used to be included in the *Malabar Coast*; to this day persons from those districts are termed 'Malabars,' as will be seen in the present Ceylon report.

we have four maunds weight costing 32 annas. If, however, we examine the cost of salt-fish prepared with salt-earth, we find it to be as follows : the cost of a maund of salt-earth (p. lxxv) is from two to four pies a basket, according to size, the largest of which is capable of containing "from four to 4½ maunds, the medium 3 to 3½, and the smallest 2 to 2½ bazar maunds." Computing the bazar maund at 32lbs., we find the cost to be about 4 pies per 144lbs. weight, or 2⅔ pies per Indian maund of 82lbs. It requires, however, about three times as much of this species of salt to cure fish as it does of the excised salt, consequently 4 maunds or 329lbs. of fish cured with monopoly or excised salt require from 80lbs. (Kurrachee) to 112 (Tellicherry), which at 32 annas a maund gives an outlay* of from 32 to 48 annas. Whilst about 4 maunds or 320lbs. of fish cured with salt-earth at Tellicherry require 800lbs. weight, which gives an outlay* of from two annas to two annas six pies. If, as is proposed, this salt-earth is subjected to an excise duty of Rc. 1-13 a maund, it is very easy to foretell the result; it would enhance the cost of salting fish for the poor by 1,600 per cent., and thus entirely deprive them of it, and they will have to fall back upon sun-dried fish; the fisherman's trade will be depressed, the market for his takes will be lessened, the price of fish will fall, and that because there will be no fish-curers to purchase his captures.

LXXIII. Along the eastern coast, where its salt-earth is not so available for this trade, where its collection and sale may be considered a penal offence, (and by some officers it is, by others it is not), it can hardly be asserted that the unstricted use of salt-earth is permitted. Consequently it is in the power of any petty official or even policeman, to interfere. And thus, as will be presently shown, this coast as to the productiveness of its sea-fisheries cannot be compared with Malabar, where from a single port alone, as Tellicherry, is exported more salt-fish than from the whole extent of the Coromandel coast, whilst the richness of the fisheries as sources of supply, if utilized, is about the same. In *Madura* " the use of salt-earth is prohibited (p. liii) : In *Tinnevelly* the Collector is pathetic respecting the 'poor fisherman', and begs to record his decided opinion that any interference with sea fishing is quite uncalled for. Any restriction, therefore, would affect

Result of interfering with the collection of salt-earth to both the public, the fish-curers, and the fishermen's trade.

---

* I omit calulating the increased weight due to the use of the salt-earth.

a number of persons who have no other means of earning a livelihood." Certainly restrictions on these poor people are apparently uncalled for, but the question is, do not they now exist in the most arbitrary form, and one most disastrous to the fisherman. He continues, "it is illegal to gather the spontaneous or earth-salts, and therefore illegal to use it for fish-curing." Its use is illegal in the *Trichinopoly* district (lvii). In *South Arcot* it is a penal offence to extract salt from the earth (p. lvii). In *Madras* (p. lx), in *Nellore* (p. lxi) and the *Godavery* districts (p. lxvii) and *Vizagapatam* (p. lxix) its use is permitted. But states the tehsildar of *Ongole* "it is a custom amongst some people to employ a small quantity of salt even for large fish, and lay them underground on the sea-shore for a day, and expose them to the sunshine. The poor fishermen, unable to buy salt, wash fish in salt-water, bury them in saltish earth for a day, and manage somehow or other to make them salt-fish, but this salt-fish is not only tasteless but stinking and becomes useless in a short period" (p. xciv). In the *Kistna* district salt-earth is stated "by the Collector not to be employed" (p. lxv), but the native officials distinctly deny this (p. xcviii), saying that it is used clandestinely. And in *Ganjam* the fish-curers "are not allowed to use salt-earth untaxed" (p. lxx). The Collector of *Kurnool* states that there are no restrictions against the use of earth-salt, but there is a tax of two rupees per annum on each pan (p. lxxii). In Bengal, salt-earth does not appear to be permitted for this purpose, whilst it is not employed in Burma, probably because of the cheapness of good salt.

LXXIV. Before enquiring into the exports of these two varieties of salted fish, it will be necessary to investigate *what is the comparative, economic, and sanitary quality of fish cured with good salt or salt-earth?* We have already seen the difference in the cost of preparing them is from 12 to at least 16 times as much if a proper amount of Government salt is employed, to what it is when the untaxed salt-earth is used, but that the latter article is what is mostly purchased for home consumption, probably on account of its price.

Comparitive, economic, and sanitary qualities of fish cured with good salt or salt-earth.

LXXV. First as to salt-fish prepared with good salt.

Fish cured with excised or monopoly salt.

The price of salt along the Meckran coast and Sind (8 annas a maund) permits the use of a full amount of salt in curing fish, and having no duty to pay upon the article, they can easily undersell the natives of Bombay with a duty Re. 1-13 per maund, consequently the article properly prepared is good, wholesome, and undecomposed for a considerable time. But where a heavy salt-tax exists the quality of the article is seen to change, only just sufficient saline ingredients is used to prevent immediate decomposition, rendering it available for sale but not for keeping. Thus the article, except in its superior state, is but little better than fish cured with salt-earth. As it begins to deterioate some dealers smoke it or otherwise try to render it saleable, and at once dispose of it.

LXXVI. Secondly, respecting the quality of fish pre-
<small>Quality of fish prepared with salt-earth.</small> pared with *salt-earth*. It stands to reason that if a small amount of good salt is insufficient to cure fish properly, the probabilities must be that those prepared with salt-earth can hardly be of a superior description, both due to its inferior taste and its large admixture with impurities. In *Guzerat* (p. xxi) " the curing of fish is so utterly neglected that it will not keep for any length of time. Hence a large portion of it soon becomes unfit for human consumption, and is either cast away or used as manure." The fishermen ",endeavour to sell as much of the fresh fish as they possibly can for local consumption, at exceedingly low rates, and it is only what remains on their hands unsold that they cure imperfectly and get rid of anyhow" (p. xxi). "The high duty on salt is undoubtedly a source of epidemics, and other serious illnesses induced by eating imperfectly prepared fish" observes the Collector of *Ratnagiri* (p. xxv) whether this is due to the insufficiency of salt employed or to the use of salt-earth is immaterial. In *Tinnevelly*, the Collector states "this salted fish is much affected by the 'Paraver' caste on the coast; I never tasted it. The smell was quite enough." The upper classes would prefer a superior article, "but the lower are, I fancy, well satisfied with the article they now obtain. It has one great recommendation in their eyes, it is very cheap" (p. lv). In *Trichinopoly* the salt-fish "is liable to give pain in the bowels, and itch to those not used to its consumption." This article "generally stinks pretty offensively" (p. lvii). " If really good salt-fish were made at the present price of Government salt, the product would be beyond the reach of the chief consumers" (p. lvii). Doubtless investigations in other salt-taxed districts all point to this remark being at the whole root of the matter, and the people who use the present article have to undergo a preliminary training of "pain in the bowels and itch" before their intestines can

digest it. At *Udipy*, in South Canara, " Government salt is sometimes used by the fishermen for curing fish for their own eating, but they employ salt-earth or sea-water for that which is sold" (p. lxxiii). At *Ponany*, in Malabar, "the majority of the fishermen are very poor, and can ill afford to procure salt at the present rates. Sea-water is never employed in curing fish, but salt-earth is largely so, the consequence being that fish thus salted soon becomes unwholesome. The chief cause of the frequency of cholera in this district is popularly attributed to the indiscriminate consumption of fish salted by the above means" (p. lxxiv). At *Tellicherry* it is remarked of salt-fish, that "those intended for exportation to Colombo are solely cured by Government salt, as such only can be preserved for a long time, and are capable of being carried without deterioration to distant markets" (p. lxxvii). In *Madura* "people on the coast do not relish the salt-fish cured with salt-earth" (p. lxxxiii). In *Tanjore* the fishermen assert "that if salt-earth is used, the fish soon becomes wormy and rotten" (p. lxxxix). In *South Arcot* "fish of the smallest kind, which finds no purchaser if arrived at the stage of putrefaction, is dried up in sand and sold to the poor people" (p. xcii).

LXXVII. Anything which tends to hamper a trade or occupation must of a necessity be to a greater or less extent objectionable, and I here propose to examine the following question :—*Has the salt-tax any such effect in India?* If the purchasers of salt-fish were rich and could afford giving an enhanced price without decreasing consumption, and if the fish-curers were capitalists, the result of largely increasing the cost of salt would only be to augment the price of the salt-fish. It seems from the returns that the chief consumers are of a poor class, and due to this any increase in the price of the salted article places it beyond their reach; irrespective of this, the fishermen are likewise poor, and cannot afford the outlay which would be necessary to purchase any large amount of monoply salt.

The incidence of the salt-tax directly on the fish curers' occupation.

Salt tax continued, and how the price of salt re-acts on the Sind fisheries.

LXXVIII. The incidence of the salt-tax on the fish-curers has thus been adverted to in the different reports from the various Presidencies. In *Sind* the price of the best salt is one rupee a maund, sometimes less. The Commissioner observes respecting salt-fish " there is a good trade quite equal to the demand, and the fishermen are well off, what more can be desired? If we now examine the *exports* of dried and salted fish from this province for the last 20 years, we find the trade to have been as follows:—

5 years ending	1857-58,	value in Rupees	...	...	84,723
5 ,, ,,	1862-63	,, ,,	...	...	1,30,644
5 ,, ,,	1867-68	,, ,,	...	...	1,87,252
5 ,, ,,	1872-73	,, ,,	...	...	2,29,449

This remarkable increase, wherein we see that the exports are nearly treble now to what they were in the first five years, is worthy of examination, because about the same amount is now realised as tax, rent, or license from the fishermen, as was in the earlier period under review. The price of salt continues much the same. The number of fishermen are said to have continued stationary, and for this stimulus to the trade we must enquire further away. The first notable augmentation of exports appears occurred in 1860-61, in that year the duty on salt was raised in Bombay from 16 to 20 annas a maund of 82⅔℔s. avoirdupois: the second spurt of the export trade in Sind was in 1864-65, in 1865 the duty on salt in Bombay was again increased from 20 to 24 annas a maund: whilst none of these duties affected the salt trade of Sind.

LXXIX. Has any effect been appreciable in the Bombay Presidency in the trade of salt-fish due to the salt-tax? The following are the amounts of duty and the quantity of salt and dried-fish imported, for such years as have been furnished:—

Effects of the salt-tax in the Bombay Presidency on the fisherman's trade.

	Duty per maund.			Value of fish imported in.
	Rs.	As.	P.	Rs.
From 1852 to 1858-59...	1	0	0	P
1860-61 it became	1	4	0	P
1865 ...	1	8	0	P
1866-67 ...	1	8	0	24,499
1868-69 ...	...	...	...	1,16,246
1869-70 ...	1	13	0	91,222
1870-71 ...	1	13	0	89,899
1871-72 ...	1	13	0	64,439
1872-73 ...	1	13	0	82,019

The Collector of the salt revenue in the Northern Division observes that there can be no doubt that the salt-tax,* combined with the repeal of the duty on imports of

---

* In Hubsan it is stated that salt is given to the fishermen at ⅔ of the rate it is supplied to the public (p. xxvi), yet it is observed that "no Government salt whatever is used for curing fish." At *Cambay* formerly the fishermen "were allowed salt free of duty (p. xxvii.)

salt-fish (1867) from foreign ports, has acted most prejudicially to the interests of the British fishermen, and has deprived them to some extent of their hereditary occupation. Home-cured fish cannot now compete with that cured at Goa and elsewhere, where a salt excise does not exist, and the trade with its profits has now passed almost entirely into the hands of foreigners. Near Bombay a very small amount is said to be salted, but some is now dried. Mr. Pratt states that the fishermen only provide a sufficiency for local consumption : that in olden times salt was allowed free of duty for this purpose. It may be generally asserted that unless the fishermen can smuggle salt from the contiguous Native States, where no salt excise obtains, or are able to collect salt-earth surreptitiously, the fish-curers' trade in salt fish (exclusive of dried fish) is almost extinct.

LXXX. The most complete figures and answers have been received from the Madras Presidency, and it will first be advisable to tabulate all such as have come to hand, premising that they only refer to imports and exports by sea.

Returns from the Madras Presidency most complete.

LXXXI. *Imports and exports of salt and dried fish by sea on the Western Coast.*

Imports and exports of salt and dried fish by sea on the Western Coast.

YEARS.	IMPORTS BY SEA—VALUE IN RUPEES.				EXPORTS B SEA —VALUE IN RUPEES			
	South Canara.	Malabar.	Cochin.	Travancore.	South Canara.	Malabar.	Cochin. in cwt.	Travancore.
1853-54	?	?	?	1,008	?	?	?	28,401
1854-55	?	4,170	?	912	?	18,293	?	30,094
1855-56	?	7,887	?	1,171	?	27,687	?	44,474
1856-57	?	10,465	?	1,223	?	62,387	?	31,755
1857-58	?	8,623	?	1,498	?	42,033	?	36,826
1858-59	?	8,014	?	1,516	?	30,870	?	29,193
1859-60	?	2,487	?	2,281	?	45,246	?	43,944
1860-61	16,918	3,611	?	3,290	562	69,170	?	97,764
1861-62	16,443	5,432	?	4,266	1,034	62,140	?	70,614
1862-63	26,310	7,886	?	4,886	14,445	46,303	?	1,08,339
1863-64	24,156	5,573	?	212	10,574	87,345	...	1,08,340
1864-65	42,007	3,457	?	379	30,366	94,981	329	1,18,806
1865-66	19,061	4,749	?	0	8,750	84,882	112	1,42,264
1866-67	36,366	13,787	?	...	11,249	95,158	339	1,46,409
1867-68	45,734	25,673	?	...	8,757	1,19,708	348	1,58,378
1868-69	34,165	39,115	?	...	1,149	1,21,453	1,687	1,39,512
1869-70	34,285	23,697	?	...	20,533	2,18,573	1,639	1,19,312
1870-71	26,088	10,951	?	...	30,275	1,50,268	2,644	1,14,096
1871-72	21,478	11,442	?	...	28,452	1,72,488	2,542	1,34,895
1872-73	22,225	9,055	?	?	59,804	2,45,713	1,617	?

(58)

LXXXII. If we now examine the returns from the eastern coast of the Madras Presidency, we obtain the following results:—
Continued.

Imports and Exports of salt and dried fish on the Eastern Coast of the Madras Presidency.

Years.	Exports by sea—Value in Rupees.							Imports by sea—Value in Rupees.						
	Madras.	Ganjam.	Vizagapatam.	Godavery.	Kistna.	South Arcot.	Madura.	Madras.	Ganjam.	Vizagapatam.	Godavery.	Kistna.	South Arcot.	Madura.
1863-64 ...	291	?	?	239	0	0	1,111	13	?	?	...	?	?	0
1864-65 ...	4	65	?	894	0	...	1,080	0	?	?	...	?	?	0
1865-66 ...	602	876	?	1,485	—	...	1,495	8	?	?	669	?	?	547
1866-67 ...	80	2,282	?	2,644	65	...	1,213	10	?	?	...	?	?	992
1867-68 ...	248	608	?	1,009	396	...	864	183	?	?	60	?	?	548
1868-69 ...	3,823	1,021	?	1,360	102	...	2,667	25	?	?	...	?	?	1,085
1869-70 ...	3,277	120	?	660	316	...	7,852	329	?	?	248	?	?	694
1870-71 ...	3,060	81	?	939	52	494	6,064	83	?	?	...	?	?	722
1871-72 ...	2,046	860	?	946	96	...	4,926	85	?	?	...	?	?	3,070
1872-73 ...	239	409	20	2,654	116	90	997	13	?	?	...	?	?	5,137

(59)

The first thing that arrests one's attention is the great difference in the amount of export trade in salt-fish between what exists on the eastern from what we perceive on the western coast of the Madras Presidency. If we divide the years under examination into quinquennial periods, we see the following respecting the exports, taking the value in rupees:—

Exports of Salt-fish by sea.

Five years ending.	Western coast; value in rupees.	Eastern coast; value in rupees.
1857-58	3,21,950	?
1862-63	6,28,624	?
1867-68	11,18,991	17,531
1872-73	15,75,651*	45,137

LXXXIII. It is exceedingly curious to observe how this trade appears to be flourishing in western India and languishing in eastern.† But prior to investigating the cause, it will be as well to enquire whether in those districts where the vast increase in the salting of fish is found to be going on there is a corresponding sale of monopoly salt.

Fish-curers' trade flourishing in western, languishing in eastern coast of the Madras Presidency.

Exports by sea of salt-fish from western India, and the sales of salt, the latter having but little connection with the former.

LXXXIV. *Return showing the yearly exports of salt-fish by sea, and the annual sales of salt:—*

YEARS.	SOUTH CANARA.‡		CANNANORE.		TELLICHERRY.		COCHIN.		TRAVANCORE.	
	Fish exported; value in rupees.	Salt sold, in maunds.	Fish exported; value in rupees.	Salt sold, in maunds.	Fish exported; value in rupees.	Salt sold, in maunds.	Fish exported, in cwt.	Salt sold, in maunds.	Fish exported, in cwt.	Salt sold, in maunds.
1863-64 ...	10,574	191,002	960	11,653	14,594	72,505	...	249,206	54,167	479,062
1864-65 ...	30,366	168,279	2,194	7,932	15,047	57,516	329	160,147	60,122	477,750
1865-66 ...	8,750	184,174	111	9,856	1,941	62,135	112	210,970	70,505	461,400
1866-67 ...	11,249	151,113	120	9,728	18,253	57,381	310	34,428	73,039	463,560
1867-68 ...	8,757	174,629	3,035	8,741	20,119	56,502	348	88,006	77,642	470,760
1868-69 ...	1,149	176,465	5,202	9,045	43,195	63,340	1,637	81,639	69,614	492,000
1869-70 ...	20,533	147,173	43,405	8,807	59,395	72,610	1,639	83,199	59,325	490,920
1870-71 ...	39,275	136,967	14,702	7,933	53,019	57,624	2,644	96,349	55,691	497,010
1871-72 ...	28,452	177,482	6,951	12,009	53,404	86,674	2,542	90,674	67,336	487,260
1872-73 ...	59,804	135,839	9,511	6,985	84,296	77,332	1,617	89,105	......	......

* Return from Travancore for 1872-73 not received : taking an average of five years it would be Rs. 1,33,237 to be added, which would raise the exports to Rs. 1780,888.

† The only district on the eastern coast where anything like an export trade in salt-fish is shown is in that of Madras, and there only for four years. However, the people may collect salt-earth without paying duty.

‡ South Canara being contiguous to foreign states, where there is no salt excise, is of course one reason why the present salt-earth fish trade should not flourish there to the extent it does in Malabar.

I think the above table clearly shows that the annual exports of salt-fish in western India have very little, if any, connection with the amount of salt sold. If one looks at the state of Cochin, we see that the sale of salt in ten years, owing to the augmented rates, has reduced consumption by two-thirds, whereas it is only since this diminished demand commenced that the export of salt-fish has sprung up. If we turn to the report of the native official from the contiguous British district of Chowghawt, he informs us that the sale of Government salt has decreased, although in *one year*, 1872, the exports of salt-fish amounted to Rs. 10,674 worth; whereas in *five years* ending 1863-64, the whole value exported was only Rs. 463.

LXXXV. The reason for this is easily found : the people may gather salt-earth for this purpose, and its value a maund is something very small. Thus the fish-curer can embark in this trade without any great capital being necessary : this induces a larger demand for fish, and consequently the fishermen are thriving, and an immense amount of animal food is used by man that otherwise would be wasted. That this is a cause is shown by examining the eastern coast, where this 'prescriptive right' to gather salt is not recognised. The tehsildar of *Pattukottai* (p. lxxxix) observes that the practice of salting fish must be said to be increasing, considering that the price of fish to be cured which formerly cost one rupee has now been reduced to ten or twelve annas, and that fish-curers offer only this reduced rate. The reason of this reduced value is not due to increased prosperity of the fishing castes, for they are evidently in a depressed condition, but to the price of salt being such that a reduction in cost must take place somewhere or the price of the cured fish will place it beyond the reach of the consumer. As the price of salt remains the same, the curers can only afford a diminished rate for the fish, and that appears to me to be the true reason why their value is less than it was a few years since.

Why augmented salt-fish curing has not increased the sale of Government salt in western India.

CAUSE OF DEPRESSION IN THE SEA-FISHERIES.

LXXXVI. The foregoing, I think, lead to the following conclusions : (1) That the fisheries are in a very depressed state wherever salt is expensive or the use of untaxed salt-earth prohibited as Bombay, the Coromandel coast of Madras and Bengal.

Cause of present depressed state of some of the sea-fisheries.

(2) That the fisheries are flourishing where salt is cheap, or the use of untaxed salt-earth permitted, as Sind, western coast of Madras and Burma. (3) That good, which means Government or monopoly salt, is hardly purchased for the purpose of curing fish eaten by the largest majority of its consumers, because it is sixteen times as dear to use this antiseptic as it is to employ salt-earth. (4) That fish is extensively cured with salt-earth where permitted ; such a food, however, is a fruitful source of disease ; the article will not keep for any very lengthened period, but it is however much preferred to simply sun-dried fish. (5) That when the use of salt-earth is prohibited, the fish-curers must do one of three things, merely dry their fish in the sun, give up their trade, or purchase monopoly salt, whilst should they do the latter, they can only keep down the price of their article by reducing the cost of the raw fish and using a minimum amount of salt.

LXXXVII. This brings us to why it is the salt-tax not only is affecting the health of the people, preventing the export inland of salt-fish, but also ruining the fisherman's occupation. The fisherman works simply for the local market, he has no capital to embark in salting fish. Now a local fish market consists of a demand for the fresh fish for local consumption, and what ought to be a larger sale for salting and the inland markets where the demand is unlimited. If you cut off the latter, these people are left to subsist on the mere local demand, and it appears clear that by prohibiting the use of cheap local salt you cut most efficaciously at the very root of the fisherman's occupation ; you check demand as well as the inland consumption, and " the idleness and apathy of these drunken classes of people " is not due to their own fault, but to the little sympathy they have received and the absence of any investigation into the cause of their impoverished state. I tried to direct attention to these questions in 1865 in my " Fishes of Malabar," but until now have been unsuccessful. On May 14th 1873, however, *Madras Revenue Board* observed that the fishermen " number throughout the Madras Presidency 394,735 persons ;" that " the answers elicited by the questions put by Dr. Day, and the injunction contained in G. O. dated 22nd April 1873, No. 424, have directed the attention of the Board to the subject of the influence of the salt duties on the trade of fish-curing, and they see reason to think that a great practical hardship exists, which they would advocate

How the salt-tax affects the health of the people, and ruins both the fish-curers trade directly and the fishermen indirectly.

immediate endeavours to alleviate." They go on to say that the modes now employed to cure fish "may possibly have a direct tendency to disseminate disease."

LXXXVIII. There are, however, two other subjects relating to salt which have a bearing on this question: the *first* is that salt-fish is permitted to be imported duty free in British ports; the *second*, that no excise exists in contiguous foreign states. It is considered by most of the reporters that the abolition of the import duty of 7½ per cent. on foreign cured fish in 1867 has had a disastrous effect on the sea-fisheries of Bombay. One official, however, disputes this; it will be therefore necessary to investigate the subject. Such a trade as salting fish would necessitate a considerable capital in salt-taxed districts, if it is to be carried on in anything like an extensive scale, so as to benefit the local fishermen by affording them a market for their captures, or the consumers inland by supplying a recognised demand. Likewise in such a trade a considerable margin for profit and loss must be left, because the articles not only are perishable, but they have generally to be conveyed some distance by sea or else inland. The excise on salt is 29 annas per 82℔s. in Bombay, 32 annas per 82℔s. in Madras, and 48 annas per 82℔s. in Bengal. That fishermen can personally invest in salt so heavily taxed for fish-curing, except on a very limited scale, is manifestly impossible; the returns would not come in sufficiently rapidly, especially as they would have to borrow the necessary capital at exorbitant rates of interest in order to carry on such a trade. As the *Collector of Tanna* most justly remarks, "the fishing class, like many others in this country, look alone to immediate returns, obtained by treating of produce in as rude and primitive manner as possible, rather than prospective increased gains by laying out of capital and resorting to improved methods of preserving" (p. xxiii). The *Acting Sub-Collector of Arcot* observes that "capital and enterprise are required to stimulate a waning trade, and curing fish is not the sort of business men who are making money would be likely to embark in," yet, observing the trade is waning, he does not see "why salt-fish should be exempted from a heavy duty" (p. lix). It has already been remarked that export agents in Travancore either buy the salt or advance money for the carrying on of this trade (p. lii). In the *Malabar* Collectorate "the salt is generally supplied by the merchants who export

[marginal note: Salt-fish is imported duty free into India and this depreciates the trade of those who employ Government excised salt.]

the salted-fish (p. lxxv), or the traders " obtain advances of money from the Colombo Merchants " (p. lxxviii). In *Madura* " trade in salt-fish is looked down upon by the capitalists along the coast as being of an inferior description" (p. lxxxiii). In *Nellore* the fishermen "receive money in advance from the fish traders that come down from Golakonda and other places, and thus salt fish. It appears that they would themselves commence to trade in such fish if the price of salt is low" (p. xcvii). Without continuing extracts from the Appendix, it must be clear that at the present price of salt in India in taxed localities neither fishermen nor the generality of the residents can afford to embark in this trade, and that when such is carried on and expensive salt being employed, either traders advance money for the purpose, or else purchase the salt. Where, however, as in Bengal, the price of salt, or rather the duty, is 48 annas per 82tbs., it would appear impossible that any one would embark in such a trade, and there we find only sun-dried fish is prepared.

LXXXIX. If we now revert to the western coast, where the great trade in curing fish is carried on, we do not find that any large amount of excised salt is used, whilst probably the best article comes from contiguous foreign ports as Diu, &c. It may be urged that traders will in time advance money in the British possessions, and from Sind to Mergui the well-exhausted platitude of leaving the trade to private enterprise and the laws of supply and demand is reiterated. It appears marvellous that it cannot be understood by some of these reporters that if you tax salt to that extent that it is rendered virtually inaccessible to the fish-curers, or if it is employed that the article becomes too expensive for the poor to purchase, you are not leaving the trade to the natural laws governing such. You are placing such a prohibitory duty on it, or rather such a protective duty on salt, that the British fisherman and fish-curer has good cause to consider himself most unfairly dealt with. How can traders in western India be expected to advance money for this purpose, at least in the Bombay Presidency ? They can purchase the same article, more thoroughly prepared, consequently better flavoured, and at a very much cheaper rate in the contiguous foreign states. They have no import duty to pay on taking it into British possessions, and a dealer who under such circumstances made advances in British territory would be somewhat wanting in a due appreciation of

Why under present circumstances it does not pay to enter largely on the fish-curers' trade in the Bombay Presidency, &c.

his own interests. If a trader finds good salt at two annas a maund in one locality, and 32 or 48 annas for the same quantity in a neighbouring one, with no other differences in the cost of capturing the fish or the description of the supply, it appears most probable that he will purchase in the cheapest market. Leaving theories aside, thus it has resulted that the fish-curers' trade in Bombay is a thing of the past. " Home-cured fish cannot now compete with that cured at Goa and elsewhere, where a salt excise does not exist, and the trade with its profits has now passed almost entirely into the hands of foreigners."

XC. There are certain *minor obstructions* to the due development of this trade which require attention being drawn to. It is generally held, and I believe very correctly so, that all taxes on fishing nets or on sea-fishing are prejudicial. When salt is cheap perhaps they are not so injurious,* and the use of fixed engines in such district, or where the demand for fish is very great, may be a fit subject for raising revenue from. But surely such cannot hold good in the Bombay, Madras, or Bengal Presidencies, as without being of much pecuniary benefit to Government, they are certainly acting injuriously on the fishing interests.

<small>Minor obstructions to the fisherman's trade.</small>

XCI. In the *Bombay* Presidency in the *Junjura* district the fishermen observe. "The expense of placing fishing stakes has increased, and we do not use the 'wandope,' or large fishing net on four fishing stakes, because a tax of Rs. 6 has been imposed on its use. We have not used this net for ten years; we seldom now use any net but the 'Boorkea wole' in deep water (a net fastened by anchors under the surface of the water), for which we have only to pay Rs. 3 per annum" (p. xxvi). In the *Madras Presidency* at *Cannanore* the '*Rajah's cat*' appears to be exercising a deleterious influence on one branch at least of the fishing, *viz.*, that for sharks. It appears that in olden times one fish daily was taken from each boat as a perquisite for the Rajah's, cat or the 'Poocha meen' 'cat-fish' collection. The cats apparently have not

<small>Local taxes continued; the 'Rajah's cat,' 'curry fish,' &c.</small>

<small>* At Gwadar, where salt is cheap, the contractor for the sea-fisheries receives 1-10th of the captures, and irrespective of this there is a small export duty, still it pays to send fish to the Indian market. In the time of the Dutch at Cochin each fisherman paid 8lbs. of fish daily to the Governor. The Rajah of Travancore about the same time claimed from the fishermen 10 chukrams (a small coin 28¼ to the rupee) and three fish annually from the Quilon fishermen, because they sometimes used their nets opposite his territory. As soon as he annexed Culli-quilon a tax was annually levied of 30 chukrams for every large net and 15 for each small one, and this was the origin of the so-called poll tax.</small>

augmented so much as the fishing-boats, so this has been commuted into a money payment of two pies a day on each successful boat. In addition to this, the Rajah annually levies a tax of Rs. 2-4 on every boat and net. Half of the sharks' fins are also claimed by the Rajah's 'Poocha meen' contractor. "The fishermen very seldom turn their attention to shark-fishing, as they seem to think it not remunerative in Cannanore, owing to their having to yield one-half to the Rajah's 'Poocha meen' right" (p. lxxvi). In *Madura*, at *Ramnad*, "the trade in salt-fish is hampered to a certain extent by the interference of the villagers, who act the part of brokers between the buyer and seller. The sales are required to be made through the villagers or their representative and none else: a breach of this custom is followed by annoyance to both the buyer and seller, who therefore seek the favour of the villagers, though such occasions a little pecuniary loss" (p. lxxxii). Besides the foregoing, the zemindars of Ramnad make a collection from fishing villages "called 'karry min' or 'curry fish:' it is levied upon each boat* returning from the sea with fish. This is supposed to be fish supplied to the kitchen of the zemindar's palace" (p. lxxxiii). In *Burma* the taxes on sea-fishing are high, and as they are about being enquired into, observations appear to be unnecessary, except to observe that such ought to fall upon the fixed engines and not on the movable ones.

SEA-FISHERIES OF GREAT BRITAIN.

XCII. Having been directed to include observations on the sea-fishery report of Great Britain, I proceed to do so here, premising that I cannot see any analogy between the two cases. The salt-tax in India appears to be one, if not the chief, cause of the fisheries of India being so little worked,—such a tax does not exist in Great Britain. In India the chief subject of investigation is how to augment the working of the sea-fisheries; in Great Britain one of the main objects of the enquiry was to see if they were being overworked.

The sea-fisheries of Great Britain.

XCIII. The sea-fisheries of Great Britain have formed the subject of enquiry by a Royal Commission, the result of which was

Enquiry into the fisheries of Great Britain.

---

* Sir J. Emerson Tennent in Ceylon remarked that when the Government tax on fisheries was removed such had a disastrous effect, and that in some places the fishermen voluntarily made the rent over to the Roman Catholic churches. I see, however, that the following occurs in the report of the sea-fisheries of that country: "The Sinhalese will only give the share or the Roman Catholic churches, because the rent is sold, and if the share is not paid, the iltes of the Church are refused" (p. cxiv).

submitted in 1866.* On examining this report I have excluded a vast amount of subject matter that has no bearing on Indian fisheries, as were such included it would swell my report without any commensurate advantage being obtained. The first thing that arrests one's attention is the want in many parts of a definition of what is meant by *the supply of fish having increased.* In some places it is evidently intended to mean *the supply to the markets* or the amount captured, and has no reference to the *productiveness* of the fishing ground; in other parts this is not so. On the whole, however, the following conclusion was arrived at, that "the produce of the sea around our coasts bears a far higher proportion to that of the land than is generally imagined. The most frequented fishing grounds are much more prolific of food than the same extent of the richest land. Once in the year an acre of good land, carefully tilled, produces a ton of corn, or two or three cwt. of meat or cheese. The same area at the bottom of the sea, on the best fishing grounds, yields a greater weight of food to the persevering fishermen every week in the year. * * When we consider the amount of care that has been bestowed on the improvement of agriculture, the national societies which are established for promoting it, and the scientific knowledge and engineering skill which have been enlisted in its aid, it seems strange that the sea-fisheries have hitherto attracted so little of the public attention. There are few means of enterprise that present better chances of profit than our sea-fisheries, and no object of greater utility could be named than the development of enterprise, skill, and mechanical ingenuity which might be elicited by the periodical exhibitions and publications of an influential society specially devoted to the British fisheries."

XCIV. As regards weirs it is observed: "Suppose, for example, it could be shown that the weirs in Swansea Bay destroyed such an immense quantity of useless fry of flat fish as to impoverish the grounds of the trawlers in the British Channel, then the natural check could no longer be entrusted with the correction of the evil. For the weirs might well catch a sufficient quantity of fish to repay their owners, and yet furnish a very far less quantity of food to

<small>Reasons why everybody should be permitted to do just as they please.</small>

---

* The minutes of evidence extend over 1,379 pages, and comprise 61,831 questions the report over 107 pages, and its appendix 72, or a total of 1,558 pages, excluding the index!

the people in general, than would be supplied by the trawlers thrown out of employment; and, indeed, it is quite conceivable that the profits to the weir owners might be enhanced in proportion to the damage done to the trawlers. Supposing such a case as this to be clearly made out, it would be the obvious duty of the Legislature to interfere, and insist upon the removal of the weirs, or upon their being so altered as to be incompetent to retain and destroy the fry." In short, it was proposed "that all Acts of Parliament which profess to regulate or restrict the mode of fishing pursued in the open sea be repealed, and that unrestricted freedom of fishing be permitted hereafter," and that an identical course be adopted as regards in-shore fishing, except in certain localities as a matter of policy. All complaints were considered to be unfounded, and the best plan in future would be to let every one do as he liked,

OBJECTIONS TO ANY REMEDIAL MEASURES.

XCV. Although the impoverished state of the fishermen *Objections to any remedial measures in Bombay.* in some districts is admitted, there are some local officials who would be apathetic, leaving matters as they are and hope for better times. Still this may be due to their attention not having been sufficiently directed to the subject. However, the following opinions have been advanced by some who reside in the most impoverished districts. In *Bombay*, as at *Broach*, "that the salting of sea-fish not being a trade for the most part systematically followed, but resorted to chiefly when, owing to a large capture the supply becomes in excess of the local demand, no plan for its increase is likely to be very successful" (p. xxii). In *Kaira*, that of fish "both the supply and demand are too limited to make it worth while to risk the great increase of smuggling which would be the first result of the concession"* (p. xxiii). At *Ahmedabad*, "that in the Gogo talooka, the amount of fish caught is trifling, while in Dhundhooka it may almost be said to be *nil*. I only know of one man in Gogo who even professes to get a living by fishing; and so far as I can give an opinion, I should say that any arrangements to facilitate salting would be likely to meet with very little success in these parts"

---

* The concession referred to is this: whether if large enclosures were made near favourable localities (away from large towns), and where fish could be salted, would the following be impracticable? that salt might be sold inside them, at rates just remunerative, for the *bond fide* salting of fish.

(p. xxiii). In *Tanna*, "that it is very questionable whether the fisheries would yield greater takes than now made : if not, the simple result would be an increase of salt-cured fish with an equal decrease of sun-dried fish. I cannot concur in thinking that it is in any way the duty of the State to take action in forcing a particular article of commerce into the markets, for there can be little doubt, if really much sought for, it will be supplied irrespective of excise duty on salt. If the excise duty be taken off for the encouragement of one branch of industry, it will be necessary to make a like concession in regard to other industries requiring salt for manufacturing and other like purposes, such as chemical works, &c. I am though of opinion, as regards this district, the question is one not calling for immediate attention or adoption, as the take of fish is not so very extensive as to call for immediate action on the part of Government. Presuming, however, that it is advisable to give an increased impetus to this industry, it is, I think, very undesirable that measures should be taken thereto, until such time as some effective law for the protection of the various fisheries be passed ; for to stimulate the industry by every possible means in the power of Government without passing of laws regulating modes and seasons, &c., of fisheries, would be tantamount to holding out inducements for the destruction of the fisheries" (p. xxiv). He also objects to salting enclosures, &c. (see note, p. 68), on account of the cost to the State.

XCVI. In *Madras* the Collector of *Malabar* is of opi-

Objections to any remedial measures in Madras.

nion that the proposition to form enclosures (see note p. 68) " is not a practicable one in this district" (p. xiv). In *Travancore* the Dewan considers that such "would lead to great abuse. Salt would be purchased more for the purpose of smuggling than curing fish, as the returns of the illicit trade are sure to be quicker and more profitable. Besides, as fishing in Travancore is carried on all along the sea-board, the number of enclosures will have to be multiplied to nearly the number of fishing villages. Else the fish will be spoiled before it arrives at the curing enclosure, and the cost of transit has to be weighed against the reduced price of salt" (p. lii). In *Tinnevelly* "the fishermen of the coast are a very miserable lot of people and excessively poor. * * I beg to record my decided opinion that any interference with sea-fishing is quite uncalled. * * Surely Government interference is unnecessary, and the employment upon which so large a number of people are engaged should not be restricted." He also observes that prohibitions exist against using salt-earth, and that the high price of salt renders it prohibitory to the fish-curers (p. lv). In *Tanjore*, that if enclosures for salting fish, &c. (see note p. 68) were made, smuggling would result, the salt revenue being "far too important a matter to be interfered with on considerations of promoting particular objects, such as trade in salt-fish and the like" (p. lv). In *South Arcot*, the Acting Sub-Collector remarks respecting enclosures (note, p. 68) : " I do not think the proposition is advisable, nor do I see why salt-fish should be exempted from a heavy duty" (p. lix). The Acting Collector of *Nellore* observes, " I cannot think that much sympathy ought to be wasted on the fishermen, for they are an independent, careless, and drunken class of men" (p. lxiii). The Deputy Collector in charge of the salt remarks respecting the enclosures (note, p. 68): "Two points of doubt suggest themselves in connection with this proposition—(1) who is to sell salt in these enclosures ? The Government who hold the monopoly, or the ordinary bazarmen and dealers who now buy from Government and retail to the people ? *(2)* What is meant by 'rates just remunerative ?' If it is the Government who is to sell, and 'rates just remunerative' implies at a rate other than the fixed monopoly price, just sufficient to cover all expenses, in other words, at the prime cost of the salt to the Government, I certainly think the measure would be inadvisable. It would be the same thing as a reduction in the monopoly price for the purpose of augmenting the trade in salt-fish. * * It would, moreover, I have no doubt, lead to frauds and smuggling in various ways" (p. lxiv). In the *Kistna* district the Officiating Collector is opposed to enclosure, (note p. 68) as " the greatest number of fishing villages within a radius of 20 miles of any of them (salt-pans) is but 40, and I do not think the enclosure would pay its own expenses of carriage of salt, police guard, and of people to keep it clean" (p. lxv). In *Vizagapatam* the Collector replies. " Regarding definite proposals for dealing with salt-water fisheries, I have the honor to state my opinion that for such fisheries no regulations are called for, and no interference with the present fishermen can be otherwise than injurious." If salting depôts are established, he does not believe they will be frequented by fishermen, who are accustomed to salt their fish at their own doors, and would not be willing to carry a perishable article to a long distance; also that smuggling would be a result (p. lxix). Amongst the replies from the native officials, the Superintendent of Customs at *Tellicherry* observes of the proposition respecting enclosures (note, p. 68). "Fish are generally cured at the place where the fishermen reside, and this their fishing village is adjacent to the principal quarter of the town. To make large enclosures as proposed, away from this place, and there to carry on fish-curing operations, salt being sold inside the enclosure, would not only cause the greatest hardship to the persons employed in the trade, but could only be carried out at enormous cost to Government, for the land is private property, nor is the plan feasible in the village itself, as many non-fishermen reside within its limits" (p. lxxvii). In *Madura* the Deputy Collector of the Salt Department, Ramnad, observes respecting salt enclosures (note, p. 68), that although practicable, "this measure will necessarily involve the interference of Government servants, more or less, with the operation of curing fish. It must at least be carefully watched for the protection of the salt revenue, that the salt sold from the Government depôt is carried into the enclosure and used for the special purpose for which it had been sold. This interference will be unpopular, as it must naturally be for some time at least, with the tradesmen, who will, notwithstanding any amount of assurance, suspect the motives of Government, and the stimulus which it is hoped to give to the trade will tend to hamper it. I do not, therefore, consider the measure proposed by Dr. Day advisable, and it is my opinion that the trade should be left to be regulated by laws governing the demand and supply of other articles of food" (p. lxxxii). In Tanjore the Tehsildar of *Myaveram* observes respecting the proposal regarding enclosures (note, p. 68): "But it is impossible to have any enclosures made for the following reasons: the fish captured in the sea and brought on shore are not afterwards carried by the men. The women only carry them. If the enclosures were within a mile or two, those that are brought ashore within 4 or 5 P. M. can be carried there by the women before 5 or 6 or 7 P. M. Fish are captured and brought on shore even after 6 P. M. up to 10 or 12 o'clock in the night. In such cases it will be impossible for the women to take them to the enclosures, although they are very near. If the captures are not taken to the enclosures till the ten morning, they will become spoiled and useless" (p. lxxxviii).

XCVII. In *Bengal* the Collector of *Balasore* states
<small>Objections to any remedial</small> that "no Government interference is
<small>measures in Bengal.</small> required. * * The people of this district do not salt their fish: they dry it in the sun and eat it when it is quite putrid. They like in this way, and there is no reason why they should be interfered with" (p. cxxi). Mr. Verner, the Joint Magistrate of the 24-*Pergunnahs*, remarks respecting enclosures (note, p. 68), that "the proposition is neither advisable nor practicable. The enclosures and necessary establishment would cost much. Reduced prices can only mean removal or reduction of duty; and in this case a considerable establishment would be required to prevent the salt being smuggled away for other purposes. The only form the proposition could take, not to be utterly visionary, would be a lease to a company. The company would have to meet the cost of the establishment appointed by Government for supervision, and would have to pay largely for the concession of reduced duty; such a company would pre-suppose consumers of salt-fish" (p. cxxii). At *Chittagong* the proposal respecting enclosures (note p. 68) is not considered advisable. If the demand for salt-fish were to increase, the supply would increase in the same proportion (p. cxxiv).

XCVIII. If the foregoing answers are examined, they
<small>Objections examined in detail.</small> may be divided under the following heads: (1) that *sympathy ought not to be wasted on fishermen, for they are an independent, careless, and drunken set of men*. By careless and independent is here probably meant 'idle,' but the cause of such idleness is consequent on two reasons, *first*, that due to the incidence of the salt-tax only a local demand for fish exists; *secondly*, that were they to exert themselves they would overstock the market, the result of which would be diminished prices. As to drunkenness, it is evident all over the world that fishermen, who have so much to do with the water, do indulge in a drop now and then. But to refuse justice to a class who, in the Madras Presidency alone number nearly 400,000 persons, for this cause would appear to be harsh measure, if not to the men at least to their families. (2) That *it is not the duty of Government to foster the fish trade*. Surely if Government, by taxing salt and forbidding the use of salt earth untaxed, have unwittingly ruined a trade and decreased by such means the food of the people inland to a great extent, it is a subject which it becomes their duty to look to.

(3) That *the supply will come with the demand*. No doubt such is the case, but, as I have shown, the salt laws have stopped the demand, and by raising the price of this article have placed fish salted with excised salt beyond the reach of the poorer and labouring classes. (4) That *there are neither demands for fish nor are they captured*. Doubtless the fishermen, finding the laws so ruinous to their occupation, have taken to other pursuits, and this to the great loss of the empire at large, because the fish are not caught, the harvest of the sea is not being collected. (5) That *nothing need be done, as the local markets are well supplied with fresh fish*. Such an argument presupposes that the fish are intended solely for the use of those who live in the vicinity of the sea, and that a deaf ear should be turned to the inland demands for salt fish. (6) That *the people have become used to eat putrid fish, and no reason exists why they should be interfered with*. This humane view omits any consideration of the effect of such diet on the health of the population,—a subject generally considered worthy of some little consideration. But it unfortunately is only the poorer classes who are thus directly affected. (7) That *there are no fishermen*. This argument resembles one advanced in Southern India that, it was useless to make roads, as there were no carts; the roads, however, were made, and so were the carts. Remove the burdens from the fishing trade, and in time it will revive. (8) That *any interference must be injurious*. It is the removal of interference I propose; it is the relief from present burdens I advocate. (9) That *nothing should be commenced until laws have been passed for the regulation of the fisheries, or an increase of fishing will lead to the ruin of the sea-fisheries*. This view is, however, so utterly opposed to what is known of sea-fisheries elsewhere; laws on such, restricting the occupation of sea-fishermen, except as regards fixed engines, have scarcely anywhere proved anything but vexatious. Whilst with the enormous space of the Indian seas, with many almost uninhabited shores, banks, and islands, there is no probability of any injury being effected by the fishermen. They may scare away large fish, and only take the little ones, but at present these suffice for all demands, and could a larger amount be sold with advantage, the fisherman would soon, were the present interference with his occupation remedied, ply his occupation in the deeper water, as we see in Sind and Malabar. Respect-

ing the incidence of the salt laws on the fish-curers (10), one official *would place a heavy tax on salt-fish.* It may be presumed he is utterly ignorant of the subject on which he is giving such illogical advice. The present tax on salt being prohibitory of its use, how could such succeed? Perhaps he means, if he really means anything, that he would place his heavy tax on salt-fish prepared by salt-earth. The result would be, its cost would at once place it beyond the reach of the consumer. With respect to the proposition of selling salt inside enclosures at a cheap rate to fish-curers, simply to enable them to exercise their trade, the following objections have been raised (11) *that the erection of such enclosures would be expensive.* In certain places such may be the case, but not in the majority of localities, as will be presently referred to. (12) *That the cost of maintaining them and paying the necessary Police guard must be great.* Surely in most localities the civil officers, who are usually so adroit at administration in all its forms, could conceive some plan perhaps on the data of that which formerly obtained in England, whereby salt was permitted for this purpose tax free. In fact this appears to have been formerly the case in portions of the Bombay Presidency. (13) *That the plan is an impracticable and visionary one.* This opinion, however, is borne down by facts of what has been done elsewhere. (14) *That it will lead to the smuggling of salt.* In many places there are headmen to the fishermen caste; could not they be utilised for this purpose and made to a certain extent responsible? (15). *That Government supervision is disliked.* So it is everywhere; but when a trade can only be carried on profitably whilst such a supervision exists, it is extraordinary the vast amount of such that will be submitted to. (16) *That the fishermen cure fish at their own homes, and will not like to send them to any enclosures; that as the women carry them after they have arrived on shore, they could not go there at night, so all brought on shore after* 10 *p. m. will be spoiled.* The proposition has evidently been misunderstood; it is not that they *are* to salt fish in these enclosures, but that if they choose to do so they *may*, and will only there get good salt cheap for this purpose. From the answers given this would have to be fully explained, or else the native officials will perhaps grievously err,*

---

* I saw last year in a Madras newspaper that rewards in one district had been offered for *venomous snakes*. The first day numbers were brought, but the native official decided that *before he paid they must be skinned*, the skins, I conclude, being his voucher. It is needless to say he at once stopped the slaughter of these noxious reptiles.

and cause a dissatisfaction, but misunderstanding the plan. No interference whatever with their present mode of curing fish ought to be permitted, but an offer made of cheap and good salt if they wish it. Probably at first such would not be accepted, but it would be in time.

REMEDIAL MEASURES PROPOSED.

XCIX. As it is evident that the sea-fishermen are in an impoverished condition almost throughout the coasts of India: that the exceptions are, wherever salt is cheap, as in Sind or Burma; or in places where salt-earth may be collected free of duty and used for this purpose, as in Malabar, &c.; or a large local demand exists, as in the town of Bombay or Madras, Now, it has been considered that it would be advisable were such misery to be alleviated, such poverty remedied, especially if such can be done without entailing any cost on the State, irrespective of the utility of bringing from the ocean for the use of inland people a good supply of animal food. That this utilization of the products of the ocean is practicable, and that without any great amount of difficulty, has been allowed by many officials, and that the admitted hardship is susceptible of partial or complete relief, is shown from the opinions of some experienced civilians, whose answers are quoted.

Remedial measures proposed.

C. In the *Bombay* Presidency the Collector of *salt-revenue* considers the British and foreign fishermen should be placed on an equality, which he considers might be effected by the imposition of an import duty of 10 per cent., *ad-valorem*, on foreign fish (p. xvii). *Mr. Pratt* observes that "the fishermen on the coast away from large towns are probably too poor to provide large enough boats and suitable nets for deep sea-fisheries, and this may be a formidable hindrance to large captures of fish. It is possible that they might be induced to extend their operations if they received such advances as would enable them to provide better boats and nets. * * With adequate encouragement in the shape of advances for the purchase of nets or boats, it might be practicable to persuade men not now engaged as fishers to qualify to engage in that pursuit. * * There is but little demand for the inferior article produced by curing with salt-earth, and it is probable that the demand for salted fish would improve, if salt could be made available for curing

Remedial measures proposed in Bombay.

fish (within large enclosures or otherwise) at rates so cheap as would permit of salt being invariably used instead of salt-earth in the curing of fish," (p. xviii). The Collector of *Surat* remarks that the fishermen in this district are not possessed of sufficient capital to profit by the measures proposed (enclosures, &c.) (p. xxii). In *Broach* the Collector observes "that the proposition referred to in his paragraph 7 (of forming enclosures, &c.,) appears feasible, and is worth trying as an experiment" (p. xxii). The Acting Collector of *Kaira*, respecting the great decrease of fish during late years, remarks—" Whether this is due to the sea having receded from the north coast of the Gulf, or to a decreasing trade in salt-fish, owing to the increased price of salt, I am not able to state, but I have no doubt that the adoption of Dr. Day's proposal to reduce the price of salt used for fish-curing would give a stimulus to the trade" (p. xxiii). The Collector of *Tanna* says, "I quite concur in the opinions expressed, that reduction of the monopoly price of salt is the one plan by which the trade in salt-fish can be augmented, provided the classes engaged in fishery can be brought to see that curing by salt is better and more remunerative than drying" (p. xxiii). The Supernumerary Assistant Collector of *Ratnagiri* says, "the proposition mentioned in paragraph 7 (forming salting enclosures, &c.,) is advisable and practicable in these districts. * * * The police patell of the village below the Zygurh fort has found it impossible to prevent the Daldis from trespassing on an old salt pan there, now no longer used. These people come to dig up the earth, which is impregnated with salt, for the purpose mentioned. I believe that this place might be chosen and a similar one at the entrance of the creek at Ratnagiri, where a licensed sale of salt for curing fish on the spot might take place. Means could be found to prevent the withdrawal of salt, and as the spot chosen would be close to the mouth of the creek, a ready way would be at hand to dispose of the fish. The plan suggested by Dr. Day cannot be too strongly recommended" (pp. xxv and xxvi). The Collector of *Kanara* observes that "salting of sea-fish could be increased, it is generally believed, if salt were cheaper. The plan of enclosures in certain localities for salting fish, where salt could be procured at a cheap rate, might be tried; but great care to prevent the salt being removed and otherwise used would be necessary" (p. xxvi). Amongst the natives the headmen of the Kolies in the *Junjura district* report that "the salting of fish could be increased if we could get salt at a cheaper rate; now it does not pay us to salt all the fish we catch, so we dry in the sun a portion, and sell as much of the fresh fish as we can. We should not like to have to go any distance to cure our fish; it would be more trouble than it would be worth" (p. xxvi).

CI. In the *Madras Presidency*, the Board of Revenue state, the average imports from foreign ports for the last 5 years have averaged Rs. 47,520 annually, whilst the number of the fishermen population in the presidency is 394,735, and that to develop a trade in salt-fish "which may be of a permanent benefit to the inland as well as to the sea-board population, the Board believe no other condition is requisite than the relief of the fish-curers from the burden of the tax on salt. The Board see reason to hope that no loss need result to the revenue derived from salt by the adoption of a system of issuing salt to the fish-curers at or about cost price, under limits and securities to be hereafter determined." *Mr. H. S. Thomas, the Collector of South Canara*, observes that "with other articles that yielded less than Rs. 1,000 a year in customs dues, salt-fish were at the last settlement of the tariff placed on the free list. Consequently Indian salters who have to buy their salt at monopoly price compete at a heavy disadvantage with salt-fish, which is imported from countries where salt is free.* If the import duty on salt-fish were re-imposed at such a rate as would bring the salt consumed in their preparation up to the monopoly price prevailing in this presidency, then Indian salters would, without undue protection, be placed on an equal footing with the foreigners, and it might be expected that, where fish were sufficiently plentiful, they would find it worth while to purchase more salt for their curing, and so we should find our returns in increased consumption of salt, and the small customs receipt on imported salt-fish would be no proof that the customs duty was not a wholesome and remunerative tax, though it was an apparently trivial one in direct proceed, for it would be wholesome in that it restored

<small>Remedial measures proposed in Madras.</small>

---

* Salt-fish is now imported into Malabar and the Western coast of India from Arabia and the Persian Gulf. Mr. Thomas observes, B.C. 523, Egypt paid a tribute of fish to Persia. Doubtless they were not carried putrid, as that condition is, so far as I know, merely relished at Balasore, consequently they must have been cured by some antiseptic, as salt. Such a cumbersome article as a tribute would hardly have been selected, if the fisheries of Persia were in a healthy condition. They must have been intended for inland consumption; perhaps their fresh-water fisheries had been over-worked, as in many parts of India at the present day; and the Persians considered the food of the people a subject worthy of consideration.

for fair competition the equal balance disturbed by the presence of a monopoly price, remunerative in that it encouraged in direct profits in the more extended use of monopoly salt in Indian curing. In connection with my proposal, therefore, I would suggest a re-imposition of a customs duty on salt-fish." Respecting whether the proposition of forming enclosures in suitable places, wherein fish could be salted and salt sold at reduced rates, he replies, "yes; I think it would; and have been at much pains to get men to undertake the enterprise, and have written to the Madras Board of Revenue suggesting in detail a trial with salt sold at Re. 1 a maund, or half the monopoly price, and have made choice of a fit man to give the question a fair trial" (p. xli). Mr. Thomas's plan is detailed at pp. xli, xlii, and xliii. In *Tinnevelly*, the Collector considers that the way to increase the trade in salting fish would be "by reducing the monoply price of salt" (p. lv). In *Tanjore*, the Deputy Collector in charge of the salt department reports that "the salting of sea-fish might be increased by removing the restrictions which at present exist regarding the earth and spontaneously-produced salt." As to whether it would be practicable to erect large enclosures in suitable places wherein fish might be salted, and the salt sold at a reduced price, he continues "yes; it is practicable in this district on the great salt swamp near point Calamere, provided the restrictions referred to in regard to using salt-earth were removed inside the particular locale or enclosure" (p. lvi). The Collector of *Trichinopoly* remarks that "if the monopoly price of salt were low, I fancy the trade and quality of the commodity itself would increase, but if really good salt-fish were made at the present price of Government salt, the product would be beyond the reach of its chief consumers" (p. lvii). In *Arcot* "doubtless the salt-fish trade might be increased and improved under systematic encouragement and arrangement if such could be applied" (p. lvii). The Acting Sub-Collector observes, respecting enclosures, &c., "one such enclosure might perhaps be tried under the immediate supervision of the Salt Deputy Collector. It would cost little or nothing to try the experiment" (p. lix). In *Nellore* "the salting of sea-fish might be increased, and would certainly be improved by cheaper salt. * * The poorer fishermen would be glad to salt fish if Government salt were within their means. The proposition in paragraph 7 (see note p. 68) I have carefully considered. In so far as those who now use no salt at all would then use the cheap salt, the scheme would not lessen our revenue to any great extent. Two of my tehsildars suggest that the salt which is rejected at the pans and destroyed might be utilized for this purpose. This salt is rejected because it crumbles and does not remain in a crystallized form, but it is quite good enough for salting fish. The sale of this condemned salt for this purpose would not interfere with our monopoly prices, nor would it diminish our stores of good salt, and it would benefit the manufacturing ryots, whose salt is often condemned because of the result of bad weather, and not through any carelessness of theirs. The cost price of good salt which is paid to the ryots is only Rs. 10 a garce " (p. lxii). The Collector of the *Kistna District* considers of the salt-fish trade that " it might be increased by reducing the monopoly price of salt" (p. lxv). Amongst the *native officials*, the Superintendent of Sea Customs, Cannanore, believes "that the salting of sea-fish could be increased if the monopoly price of the Government salt be reduced; but this can only be effected by a license system restricting the sale to a certain extent to the fishermen alone for the *bonâ fide* use of salting fish" (p. lxxiv). The Deputy Collector of the salt department, *Ramnad*, observes respecting the proposition in paragraph 7, (note p. 68) " the measure is practicable, and the most convenient points of the coast to form the enclosures are the salt stations. The sale of salt at a reduced price within the enclosures may not only suppress the use of illicit salt, but also improve the quality of the salt-fish, which from the high price of salt is cured with other ingredients, as salt-earth, &c. " (p. lxxxii.) In *Tinnevelly*, the tehsildar of *Ottapidaram* states " the salting of fish would be increased if the price of salt becomes cheaper" (p. lxxxiv). In the *Tanjore* Collectorate the Tehsildar of *Tritrapundi* considers that the salting of fish could be increased by reducing the price of salt;that the erection of enclosures, wherein fish might be salted and the salt sold at a reduced price, would be advisable (p. lxxxvii). The tehsildar of Thealli likewise considers " that the salting of sea-fish could be increased by a reduction in the price of salt; erecting enclosures within which fish might be salted is not impracticable" (p. lxxxviii). The tehsildar of *Pattukottai* reports that " those who are engaged in salting and curing fish purchased from fishermen complain that their trade suffers much from the time the price of salt has been enhanced, and offer only 8 or 12 annas for fish formerly worth one rupee. Under these circumstances, I consider it advisable that the price of salt should be reduced as a stimulus for fish curing. * * Certain restrictions should be made, such as that salt should be sold at the reduced rate only in the places where the fish are cured, and that salt bought there should not be taken out or used for any other purpose" (p. lxxxix). In *South Arcot* the tehsildar states " people will use fish salted in any way; they have no idea of what is called high salt; the only consideration to them is the cost" (p. xcii). Another tehsildar considers that "the salting of fish might possibly be increased if the price of salt were reduced for the purpose and the forming of enclosures would be practicable" (p. xcii). In the *Nellore* Collectorate the salt superintendent of Sunnapugunta is of opinion that there has been a decrease in salting fish of late years, because of the rise in the selling price of salt, which has made it more difficult for the poor to procure the article and cure the fish properly (p. xcv). The tehsildar of *Handukur* observes that " the cost of salt precludes the fishermen from salting fish; it appears that they would themselves commence the trade if the price of salt is low" (p. xcvii). In the *Kistna District* the sea customs house superintendent at *Bandar* considers that "the proposition in paragraph 7 (note, p. 68) would be practicable" (p. xcviii). The Assistant Superintendent of Customs at *Vizagapatam and Kottapalem* ports think it would be good and practicable" (p. xcviii). The Superintendent at *Ipurupalem* likewise considers such will increase the amount cured (p. xcix).

CII. From *Bengal*, Dr. *W. W. Hunter* observes in

Remedial measures proposed in Bengal.

Orissa respecting assisting the fishermen with cheap salt to cure fish and the difficulties such would entail, that " Government has to consider whether it is not worth while to encounter and overcome these difficulties, rather than to continue to deprive the often famine-stricken population of the delta of a great staple article of its natural food " (p. cxviii). The Collector of Chittagong observes respecting salting fish and there being no export of it, "the fishermen object on the score of the high price of salt, and even were this objection removed, the Collector doubts whether the fishermen would be inclined to come and prepare their fish for market within the proposed enclosures" (p. cxxiv).

CIII. From the preceding answers we may thus divide

Analysis of remedial measures proposed.

the propositions advanced by various officials respecting how to ameliorate the present condition of the fisherman and encourage him to make larger captures, whereby not only the people in the vicinity of the sea, but the inland markets, may be supplied with wholesome salt-fish to meet the present demand; (1) *Reduce the price of salt to the fish-curer.* This proposition is most improtant, and appears to be at the root of the whole of the present depressed state of the fisheries. But there are a few considerations that must be carefully considered. The people of India who consume prepared fish are very poor, and the inferior article they now obtain is either dried without salt, or untaxed salt-earth is used. This latter substance costs about 2 annas to cure a maund of fish, whilst in the Madras Presidency the duty alone on the amount of salt necessary to cure the same amount is 32 annas. But this latter article will be far superior, more wholesome, keep longer, and bear transport. Great care will be necessary to enquire whether, even were salt sold at cost price and the use of salt-earth prohibited, such might not raise the prime cost of the article higher than the poorer classes could afford to give. Bad as is the present quality of earth-salt-cured fish, the sun-dried is far worse and more unwholesome. (2) *That Government salt should be sold at cost price to fish-curers* inside enclosures for the bonâ fide salting of fish.* The remarks made above will be worth consideration respecting this plan. This scheme is merely founded upon the modes pursued in Europe in bygone years, and apparently existed in portions of *Bombay*, for in *Cambay* it is said the fish-curers used to obtain their salt duty free. (3) *That the import duty on foreign salt-fish should be increased, so as to bring up the charges to a level with those incurred by fish-curers who use excised salt.* After having very carefully considered this proposition, I am inclined to doubt the benefit which will accrue to the poor. Fish in India for sale is only prepared with good salt when for the richer classes or export to distant markets. Raising the import duty on this article will have no bearing whatever on the present cheap salt-fish, but only on the first-class article.† (4) *That the restrictions on the use of salt-earth for fish-curing be removed.* This proposition in reality strikes at the very root of the evil. Were salt-earth for sale at its cost price for fish-curing within these enclosures, and also some good salt at cost price, this

---

* Not necessarily fishermen.

† In the above remarks the question is only considered with reference to its general bearing, and leaving salt to the fish-curers at its present price. Placing a prohibitory duty on foreign salt-fish will, I should think, benefit very few traders.

would give a great impetus to the trade so soon as the natural suspicion of the people had been allayed. (5) *Where salt is manufactured, it is suggested that the condemned salt should not be destroyed, but sold at cost price inside these enclosures.* Of course the cost price of such would have to be a little less than the best salt, or the manufacturing ryots would be careless respecting the quality of what they made, if the same price were given for good or bad. (6) It has been proposed to give advances for the purpose of enabling fishermen to buy nets and boats. This State assistance to fisheries, however, has not answered elsewhere, and I do not see any reason for believing that it would be of any permanent benefit in India. It might give a temporary stimulus to the fishermen, but such would soon subside, unless Government gave relief from the present cause of the depression, namely, the price of salt. That done, the trade will doubtless recover in time; it only needs fair play to become a source of profit to the fishermen and fish-curers, as well as affording an inexhaustible supply of good animal food to the inland markets.

RESULT OF THIS INQUIRY.

CIV. Unless we have a clear conception of the evil, it is useless attempting remedial measures. I therefore propose to briefly recapitulate the result of this enquiry. (1) It is impossible to deny that wherever a good demand exists for fish, either for the purpose of meeting the local consumption or material for the fish-curer, that the fishermen are in a prosperous state. (2) That wherever salt is dear the fish-curers' trade is either diminished or entirely destroyed (unless in the vicinity of large towns), and that as a consequence the fishermen are in a great state of destitution, both owing to small local demand for their captures as well as the decreased price of such in the market. (3) That certain local taxes exist on the fishermen's trade which are objectionable, as 'Rajahs cat' at Tellicherry, as 'the Rajah's or zemindar's 'curry fish' in Ramnad : as all fishing in the sea with nets which are not fixed engines ought to be free to all, especially if salt is dear. (4) That one cause depressing the fish-curers' trade, who use excised salt in Bombay and portions of South Canara, is the vicinity of foreign ports, where salt is sold at about 2 annas per 80 lbs. weight (instead of about 32 annas, as in the British territory), and no duty exists against their freely importing into British ports fish cured there.

l

(5) That salt-fish is of two descriptions, *first*, that cured with Government salt for the rich, for the sick and for export to distant markets, as it keeps well, is wholesome, and nourishing, the cost of the salt employed being equal to about $1\frac{1}{3}$ pie to 1 lb. of fish.* *Secondly*, that cured with untaxed salt-earth for the poor, which does not keep well, and is considered as either pre-disposing to or a direct cause of disease, the cost of the salt employed being equal to $\frac{1}{10}$th of a pie to 1 lb. of fish. (6) That in many places where the salt excise is strictly enforced, the poor have to consume their fish putrid, or simply bury it along the sea-shore or soak it in partially evaporated sea-water and then dry it in the sun, these plans being reported as fruitful causes of disease. (7) That there is no diminution of fish in the sea appears clear, because wherever there is a large and steady demand for them, such appears to be invariably met, and it is merely in those localities where no steady demand exists that a decrease is reported. (8) That there may be a temporary absence of migratory fish, as the mackerel and sardines, that such may extend over several successive seasons, but proof is still wanting to definitely conclude what such absence is due to. In like manner, predaceous fishes, which follow and prey upon these shoals of migratory ones, will necessarily be absent at these periods, because they are pursuing their prey. This, however, can hardly be considered a diminution of these fishes, but only their temporary absence. (9) There is, however, in certain places a local decrease of sea-fish due to local causes, and the result of how the fishermen carry on their captures confining their labours to capturing the fry and small fish along the shore and in the estuaries; but as these smaller fish were the baits which induced the larger marine forms to come in, the fishermen's labours remove this lure; they disturb this water and the larger and more valuable species seek their prey further out to sea. This, however, is only one of the effects of dear salt, as the smaller fish can be dried and cured with a minimum amount of this antiseptic quite insufficient to preserve the larger and more nourishing species.

REMEDIAL MEASURES.

CV. If the foregoing deductions from personal investigations and the reports of numerous local officials are correct, the conclusion cannot be avoided that the chief cause of the

Remedial measures.

---

* These figures are approximative, deduced from the Kurrachee and Tellicherry returns, and likewise from Malabar reports.

present depressed condition of the fisherman's and fish-curer's trade is to be sought in the incidence of the salt-tax, and that those who deprecate "any interference with the fishermen who are so very poor," or in a "miserable state of destitution," appear to be unaware of the state of the case. One cannot suppose such advisers to be oblivious of the distresses of those amongst whom they reside, or would really wish to feed the poor on putrid fish, considering the realization of the salt revenue as of immeasurably more importance than the lives, health, and comfort of their fellow creatures. Assisting the fishermen with pecuniary advances for the purpose of purchasing boats and nets, as proposed, is scarcely to be expected to be sufficient in itself to place the sea fishermen's and fish-curers' trade in a healthy state; and if it is in a healthy condition, such advances would not be required. Seeing that it is only in the heavily salt-taxed districts, or in the vicinity of a contiguous foreign state where salt is untaxed, that the fishermen are badly off and the fish-curers' trade may be considered a thing of the past, one feels justified in laying the blame on the incidence of the salt-tax. It is evident in portions of the Bombay Presidency that this was foreseen, for the fish-curers used to have salt to carry on their trade duty free, a remission which appears to have been withdrawn, whilst the salt-fish trade has also disappeared. Thus expensive salt is beyond the reach of the majority of fish-curers; it is ruinous to their trade, and whilst diminishing or destroying that, the demand for fish decreases, and the fisherman becomes involved in the common ruin. The evil, however, does not cease here; it compels the poor of many districts to either entirely go without fish at their meals, consume them putrid, merely sun-dried, or salted with such impure salt as to predispose to or even induce disease. This being the case near the sea-coast is much more apparent inland, where the fresh-water fisheries, from the reckless manner in which they are being poached, evidently in places show signs of exhaustion. To give an impetus to the fisherman's calling, to raise him from a miserable state of poverty to comparative comfort by creating a demand for fish, and which would aid in supplying the inland markets at no expense to Government, would surely be both a humane as well as a politic action, even should such entail some little additional trouble on the local civil officers. As the root of this evil is evidently the price of salt, so this of itself is the first consideration. Of course, as in Burma, there may exist a fisherman population who do not care to risk fishing far from land, but these are for local more than for general considerations.

CVI. With reference to the incidence of the salt-tax there are two main questions for consideration; *the first* is relieving those who cure fish with Government or excised salt from unfair competition with the foreigner, who can now prepare his fish with the same quality of salt at from 2 to 4 annas a maund whilst the Indo-British curer rarely gives less than 32 annas for the same quantity in Bombay, 38 in Madras, and from 54 to 80 in Bengal. This may be met to a certain extent by an import duty of 10 per cent. *ad valorem* on foreign cured fish, but even then it must be evident that the foreigner has an advantage, as a much higher duty than 10 per cent. would be necessary to raise his expenditure from say 4 annas a maund for salt up to from 32 to 80 for the same quantity. But this will only afford relief to those merchants who now use Government salt, and this is employed merely for the richer classes, the sick, and export to Ceylon or inland, as to the Wynaad and Hyderabad. If the proposed augmentation of the salt-tax in Sind is carried out, this tax must also partially assist the Sind fish-curers to compete with the foreigner in the Bombay market.

<small>Incidence of the salt tax.</small>

CVII. But it appears to me that there is a very much more important question involved than simply assisting those few fish-curers who salt fish for the richer classes or more distant markets. The general health of the people seems worthy of consideration, and there are three causes at least (irrespective of the state of the fishermen) that call aloud for remedial measures. (1) The inland districts are not now supplied with salt-fish from the maritime ones, because the article, if prepared with Government salt, is too expensive to be within the compass of the means of the poorer classes, who would be its chief consumers. (2) Owing to the inability to purchase salt, semi or quite putrid fish is being consumed by the general population, either in a dried or partially raw state, in certain districts. (3) That the use of fish cured with salt-earth is productive of disease, and is only purchased because of its cheap cost. If the price of the article is raised much above what it stands at now, it will be placed entirely beyond the reach of its chief consumers; and the problem to be solved is this, how can good salt-fish be sold at its present price? This I fear is impossible, but certain remedial measures might be tried, and

<small>Curing fish with salt-earth most important to the poor.</small>

though I here propose pointing them out, I must leave the civil department, in whose province it lies, to initiate the details.

CVIII. Admitting dear salt is the root of the evil, we

Selling salt at or about prime cost to the fish-curers.

arrive at the question how could its expense be lowered to the fish-curers? There are two articles now to be investigated, pure salt and salt-earth. If Government would sell, in a way to be hereafter detailed, salt at prime cost to fish-curers *bonâ fide* for the curing of fish, the result would be a diminution of at least from 50 to 75 per cent. This would not only cheapen the article now manufactured, but give an impetus in time to the fisherman's trade. As the fish-curer found that due to this concession, and a duty on the foreign article, he could with a profit increase his trade in the British territory, doubtless it would induce many to carry it on. Wherever this occurs, a steady demand for fish must spring up, and the fishermen's trade revive. Not only this, but the article will be within the means of a larger number of people; and if the use of salt-earth is also permitted in the same place, there cannot be a doubt but that some of this cheaper Government salt will be mixed with it, and the salt-fish as a result will be cured in a better manner.

CIX. If likewise the collection and use of salt-earth for

The collection of salt-earth for fish-curers used inside enclosures to be permitted.

this trade is permitted, that alone will give a great stimulus to fish-curing, and those who now have either to go without cured fish or eat the abominable substance sold in Bombay, the Coromandel Coast, and throughout lower Bengal, would procure a more saline article instead of a thoroughly putrid one. Not one ounce of monopoly salt is now used, and it will only be by degrees that the employment of this saline will be introduced; and as the trade goes on it is to be hoped that the same benefits would result as have been perceptible of late years on the Malabar Coast.

CX. The mode in which I proposed that salt should be

Salt disposed of inside enclosures.

sold to the fish-curers was as follows, by the erection of enclosures wherein the fish should be salted, and not removed until cured. If this were done and a memorandum kept of the salt conveyed inside these places, and the amount of salt-fish weighed as it was taken away, a single glance at the figures would show if any smuggling were taking place; and I would draw attention to the circumstance that in many places headmen of the fishermen exist in every village with a superintending headman over several villages. Probably the civil authorities might advantageously employ their services. Of course in such a plan it must be distinctly laid down that no interference with the fish-curers will be permitted. If they do not wish to use cheap salt inside these enclosures, they may go on as they always have done : if one person commences and it pays, I have no doubt but others will speadily follow, but nothing must be done to cause trouble or suspicion to these people.

CXI. I would strongly urge that in salt-taxed districts the fish-curers be permitted to collect salt-earth for use in their trade duty free, provided they employed it within enclosures as proposed. Such would be no loss to Government, but it would permit this ingredient to be employed openly instead of as now by stealth, rendering the fish-curer almost at the mercy of every petty official.

The collection of salt-earth from this purpose no loss to the Government revenue.

CXII. Relieve the fish-curer of his present burdens, and he will be able to purchase fresh fish and cure it in such a manner that he can provide a wholesome article to distant inland markets at a price which the consumer can afford to pay. By cheapening salt, some at least can be added to the salt-earth now used for the local supply, so that the article will keep longer and be more wholesome to the local poor. Permit the use of salt-earth in salt-taxed districts, and instead of putrid fish being eaten as a rule, disseminating disease, some saline can be employed which certainly will be found beneficial to the health of the poorer classes.

Result of cheapening salt to the trade of salt-fish.

APPENDICES.

BELUCHISTAN.

1. In commencing the appendices to this report on the sea-fisheries, I have considered it expedient to give a short account of personal investigations made at *Gwádar* on the Meckran Coast, as the trade in salt-fish may be said to have almost sprung up since the establishment of a telegraph station and a British Political Officer at that port. From this place a considerable amount of salt-fish is exported to Bombay and elsewhere. Prior to 1867, fish imported into Bombay paid a duty of 7½ per cent., whilst the price of salt has been considerably raised in the British territory by an increase in the duty levied upon it.

Sea-fishery at Gwádar: how it has sprung up of late years. Import duty on salt-fish into India abolished; salt tax in India increased.

2. At Gwádar the take of fish and the exports had been as follows during 10 months in 1871,—the contractor receiving one-tenth of the captures, and there being a small export duty:—

Gwádar, its fisheries; mode of preparing the captures.

	Captured.	Exported.
Pish-hul ...	90,000	80,000
Seir-fish (*Cybium*) ...	3,000	2,200
Coompa (*Pristipoma hasta*) ...	8,000	8,000
Mush-hul ...	95,000	95,000
Pulla (*Clupea*) ...	25,000	23,000
Soh-lee (*Sciæna axillaris*) ...	73,000	71,000
Kur (,, *diacanthus*) ...	8,500	2,500
Soh-ru (*Pagrus spinifer*) ...	12,000	8,000
Teg-gu-lum (*Stromateus*, or *Pomfret*) ...	10,000	10,000
Sah-rum (*Chorinemus*) ...	21,000	10,000
Suk-kun (*Sciæna semiluctuosa*) ...	825	
Gulloo ...	3,000	1,000

The salt used for curing fish is here of a very superior quality, and a camel-load of about 280 ℔s. weight costs about Re. 1-8-0, or one-seventh of what it does in many maritime districts in Bombay and Madras. The following is the method followed in curing fish in this place:—Salt or brine pits are formed in the ground under a light but high shed, so that, although the direct rays of the sun are not allowed to fall on the fish, any breeze that exists blows freely through the building; the whole of this is surrounded by a light wooden enclosure. The pits are usually about six feet square and three deep, but some are shallower; they are puddled with a tenacious blue mud obtained from a neighbouring hill, and this entirely prevents the water from soaking into the soil. Over this blue mud is spread a clean bamboo mat and a layer of salt. The fish, having been split open and cleaned, are placed as a second flat layer; then some salt, next another layer of fish, some more salt, and lastly sea water just sufficient to cover them is added: the brine was said not to be used twice. The fins of sharks and fish-maws or sounds are treated in the same way. After two or three days they are removed, re-salted, and packed in large heaps, as the people of the country will not purchase them if too dry: these heaps are either matted or thatched over. Sharks' fins are not, as in India, simply cut off and dried in the sun, but a considerable portion of the muscles along their bases are also removed, and the whole cured in the salt-pits as already described: 120 kupputs weight, each of which equals 336 ℔s., had been exported to Bombay, or a total of 360 cwt. in 10 months in 1871. Fish-maws or sounds, known as rough isinglass, is also largely collected; it is chiefly obtained from the air-vessel of the *Sciœna axillaris* and *S. diacanthus*, their value at the customs house being estimated at Rs. 65 a hundred. The fishes that are exported to India form a great part of the merchandise which is sent, and from its sale, materials for making nets, and country cloth, &c., are purchased. In touching at some other places along this coast, the same prosperity in the fishing classes and trade was apparent, and which was not only supporting thousands of fishermen and their families, providing for their daily wants, but also enabling them to obtain by its produce many articles of necessity, and even luxury, from the British possessions and other places. In no locality was a decrease of fishermen complained of; on the contrary, they were reported to be largely increasing and in a thriving state. The excellent and cheap salt permits any amount of fish to be cured; this ability to preserve large quantities creates a demand for fish from the fishermen, whilst the quality of the salt-fish is such that it may be exported to distant markets without becoming putrid like salt-fish which is merely cured with salt-earth. From the foregoing it would appear that cheap salt is a proper stimulus to improving the condition of the fisherman by giving him a market for any amount of fish he can capture, and the fish-curer an article capable of being exported with profit to the salt-taxed districts of Western India.

SIND.

Sea-fisheries of Sind.

3. The province of Sind does not possess an extensive sea-coast; but due to the river Indus opening by several mouths along its border, a great amount of food suitable to both large and small fish is carried down to the ocean. As a consequence, it is a favorite resort for both predaceous and other marine species, some of which come in vast shoals for breeding purposes. Irrespective of this, salt exists in enormous deposits, and as yet the Government have not imposed upon it the amount of taxation that obtains in the rest of British India.

Favorable prospects of the fishermen.

4. Thus everything most favorable for the sea-fisherman's occupation exists—an inexhaustible wealth of fish, and cheap salt. Irrespective of this—for without a market the trade would be unremunerative—salt-fish may be exported into India duty-free: there an unlimited market is to be found, provided the article be good, and the native of Hindustan has to purchase his salt at double or treble the price at which it is sold in Sind. And thus the Sindee in good salt-fish can undersell the Hindu in his own market, as I shall subsequently have to explain.

5. The following figures show the exports and imports :—

Return showing the quantity and value of fish-oil, dried and salted fish, and fish-maws and shark-fins, exported from Sind for the last twenty years.

YEARS.	Fish-oil.		Dried and salted fish.		Fish-maws and shark-fins.		GRAND TOTAL.
	Gallons.	Value, Rs.	Cwts.	Value, Rs.	Cwts.	Value, Rs.	Value, Rs.
1853-54	1,493	859	...	10,750	1,389	32,766	44,375
1854-55	1,181	630	...	11,525	369	10,767	22,922
1855-56	945	510	...	25,027	645	25,495	51,032
1856-57	...	...	...	21,537	572	20,466	42,003
1857-58	510	500	...	15,884	476	12,659	29,043
1858-59	25	22	...	17,376	384	12,012	29,410
1859-60	100	2,197	...	24,684	392	13,717	40,598
1860-61	1,556	1,274	...	39,629	554	17,214	58,117
1861-62	...	...	...	33,291	342	13,690	46,981
1862-63	84	86	...	15,664	422	19,236	34,986
1863-64	67	60	...	26,316	363	15,863	42,239
1864-65	...	...	...	54,518	819	32,775	87,293
1865-66	...	...	...	37,742	692	25,595	63,337
1866-67	2,474	2,365	...	29,976	786	40,979	73,320
1867-68	3,299	3,549	...	38,700	906	54,263	96,512
1868-69	1,231	1,483	...	78,963	1,920	95,750	1,76,196
1869-70	7,633	9,317	...	34,402	1,284	69,680	1,13,399
1870-71	4,572	5,361	...	46,346	1,514	73,024	1,24,731
1871-72	4,670	4,949	...	29,030	1,258	76,869	1,10,848
1872-73	6,309	7,527	...	40,708	1,315	55,440	1,03,675
TOTAL	36,149	40,689	...	6,32,068	16,402	7,18,260	13,91,017

If we divide the foregoing into periods of five years, we obtain the following respecting the value of the *dried, and salt fish exported* :—

			Value in Rupees.
5 years ending 1857-58	...	...	... 84,723
5 „ 1862-63	...	...	... 130,644
5 „ 1867-68	...	...	... 187,252
5 „ 1872-73	...	...	... 229,449

6. In fact, the great stimulus to the trade appears to have occurred during the last five years under review, which commenced when Government took off the duty on imported salt-fish. If we look at the yearly figures what do they demonstrate? The first great increase in the trade was in 1860-61; in that year the duty on salt in Bombay was raised from Re. 1 to Re. 1-4 a maund; the next great rise was in 1864-65; in 1865 the duty was again raised in Bombay to Re. 1-8 a maund,—these duties not affecting the fishermen of Sind.

7. *Could the salting of sea-fish be increased in Sind?* The Commissioner appears to consider no necessity exists for any extension, the supply being equal to the demand. But the subject has to be looked at on a broader basis; the supply in India *is not equal to the demand.* The native official considers the practice of salting fish has remained stationary. But it appears from the sea customs house returns that the trade on the contrary has largely increased of late years, and, as it is yearly expanding it may well be left alone, provided the price of salt is not increased.

Salting of fish increasing.

8. Government salt is said to be exclusively used for curing fish. Its price is stated to be about one rupee a maund, but it varies, and is sometimes even less.

Government salt exclusively employed: price Re. 1 a maund.

9. Although the Tehsildar gives instances in which the fishermen's privileges have been curtailed of late years, investigations lead to an entirely opposite conclusion. They have been relieved of certain cesses, and instead of the fisheries having been let by auction to a contractor, who was allowed to tax them largely, they now merely pay a yearly license on their boats by the ton; otherwise they are perfectly unrestricted in the modes and times of fishing.

Fishermen's operations unrestricted.

10. The fishermen have certain headmen, whose duties appear to be to settle questions of caste and sundry other matters of a trifling nature, and to conduct the religious ceremonies connected with marriages and funerals.

Headmen of caste.

11. *How are the fishermen supplied with boats and nets?* From the answer furnished by the Tehsildar it would appear that the old ruinous plan is still in force,—a plan which involves the fisherman so completely that, having once been entangled by these money-lenders, he can rarely, if ever, disentangle himself. Without possessing any capital, he looks round for some one to lend it him for the purpose of obtaining the implements of his craft, *viz.,* nets and boats. The exorbitant rate of interest charged in Karáchi appears to be this: He borrows the money for the purchase of the

Ruinous mode in which the fishermen provide themselves with nets and boats.

materials, giving a bond to hand over the whole of his captures to the bond-holder at half the ruling market rates. Thus every means of subsistence is taken away; the bond-holder now supplies him at usurious rates with all the necessaries of life, and the fisherman in reality is his slave. This unfortunate state of affairs was reported upon years since, and to give relief to this class of people, cesses and other taxes bearing upon them were removed; but according to the Tehsildar's statement, they have not benefitted therefrom.

12. As the captures are much larger than formerly, whilst the fishermen are said to have decreased, to whom must this increase of wealth have gone? If the Tehsildar's statement is correct, an increase of captures effected by a decreased number of persons must have the effect of increasing the emoluments of the fishermen, or if all their captures go to their bond-holders, they may obtain the benefit. The Tehsildar states that "some of the fishermen have given up the business, owing to their inability to procure nets, boats, &c., which are very costly, and also owing to the low rate at which they have to sell the fish to the persons from whom they borrow money." Figures seem to show that if such is the case, they must be most apathetic in looking after their own interests, as an enormous increase of receipts goes now to the fishermen craft above what they received in former years. Whilst their taxes have decreased, the price of salt has not been raised, and a splendid market in India has been opened to them.

The fishermen represented by the Tehsildar as being in a very impoverished condition.

13. But the question arises *is such a fact?* I cannot help surmising that it is not. On personal enquiry I could not hear of such a miserable condition of the fishermen at Karáchi. Lastly, the Commissioner observes that "the fishermen are well off," and that the salt-fish trade "is in a natural and healthy state now and needs no assistance; it is best left alone." In this conclusion, I cannot resist coinciding, and to likewise remark that, with salt at one rupee or less a maund, the fishermen of Sind have enormous inducements offered them to supply the salt-fish market of Bombay.

Reasons for doubting the correctness of the Tehsildar's opinion.

14. The sea-fishermen of Sind, besides fishing in the deep sea and in-shore, engage as lascars in coasting vessels, collect jungle from within tidal influence, the bark of which is useful for preserving timber for building purposes, and the leaves as food for camels and horned beasts. It appears that during the time of the Amirs, and up to 1846, the fisheries were farmed out, and the contractor was allowed to exact his share either in money or kind. The following were the percentage cesses which the Karáchi contractor was entitled to:—one-sixth of all fish taken inside the harbour or within Manora Point (the farthest point of the mainland abutting on the harbour): one-fourth of those netted or taken by hook and line outside the harbour, provided they were less than sixty, but if over that number then one-eighth: on every boat-load taken outside the harbour, five fish were selected by the contractor: he also received one-sixth of the fish-sounds extracted, and a little over 10 per cent. of sharks' fins. On every boat-load of fish taken

Taxes which formerly affected the sea-fishermen.

near Kiamari, one-sixth : on fish exported by sea Rs. 12-11-1½ per cent.,—if to Daraja, Sháhbandar or Sonmiani, Rs. 13-10-9 per cent., if to any inland village Rs. 14-10-4½ : on fish-sounds conveyed inland Rs. 13-8-8 per cent : on fish caught in nets, the Government share of some kinds was at the rate of twenty-two rupees a hundred : the same of some captured by hooks and lines : of others, only sixteen rupees a hundred : whilst of a few kinds the share was recovered at the market rates. Other cesses also affected the fishermen, as head-money on lascars, taxes on charcoal and on wood, &c.* The contract sold annually for between four and seven thousand rupees. First, the extra cesses were discontinued : in 1845 the contract system was abolished, and for the following reasons :—up to the period just mentioned the fishermen were all more or less involved in pecuniary embarrassment, and entirely in the hands of bania creditors, by whom not only their boats and nets, but even the necessaries of life were supplied in consideration of the profits of their labors being relinquished to them, the share of the contractor being first deducted.†

15. On August 23rd, 1851, the *Deputy Collector* of Customs

Annual licensing of boats followed by a return to the auction system.

observed that, on the abolition of the Mirs taxation, annual licenses were adopted, which had produced the following sums :—

	Rs.	A.	P.
1846	673	0	0
1847	3,255	8	9
1848	2,995	9	9
1849	3,445	8	0
1850	3,473	5	0
1851	757	1	0

and he proposed, in consequence of the small amount realised, that the fisheries should be sold by auction to a contractor. To this the Commissioner in Sind (Sir B. Frere) gave his approval, and that year (1851) Rs. 5,250 were offered and accepted. This plan was persisted in until 1856, when the late General Jacob, whilst Acting Commissioner (September 1856) instructed the Deputy Collector of Customs to introduce a system of licensing each fishing boat,—a plan apparently not approved of by the Deputy Collector, as he never carried it out, adhering to the auction system. However, a year later, the Commissioner directed that the following plan should commence, and it has been in force since 1858 :—

16. The present mode in which the fishermen are taxed is by

Present plan of levying a license of Rs. 5 a ton annually.

levying a license of Rs. 5 a ton on fishing boats, and this has realised in the five years ending 1870-71 Rs. 13,596-7-5, or at about the rate of Rs. 2,719-4-8 per annum. Thus in six years ending 1851 under the annual licensing Rs. 2,433 per annum was averaged, or about Rs. 286 less than the average amount now obtained. The following are the scales of licenses now in force :—Canoe or toney Rs. 3 a year : a butel of 1¼ tons Rs. 5 : from 1½ to 1¾ tons Rs. 7-8 : from 2 to 2¼ tons Rs. 10 : from 2½ to 2¾ Rs. 12-8 : from 3 to 4 tons Rs. 15 :

---

* Probably it extended to grass cut in the salt marshes, but I omitted to take a note of it.—F. D.

† According to the Tehsildar, they are in much the same condition now, the contractor taking all their captures at half the market rates

from 4¼ to 5 tons Rs. 20: from 5¼ and upwards Rs. 5 a ton. Fractions less than half not charged for; exceeding half charged for as a ton.

17. The amount of fish on the sea coast of Sind are as numerous

<small>Sea-fish very abundant.</small>

as they are off Beluchistan; the supply in fact is inexhaustible. Even the shad termed here the pulla has the Indus up which to ascend to breed,—a river fished with the greatest difficulty and not spanned by any works of irrigation, and they are taken in enormous numbers during the cold season. I now propose adverting to the manner in which these fisheries are worked, how the captures are disposed of, and the way the fish or portions of them are prepared in the bazar. First, as already observed, the fisherman has to obtain a license for his boat, which pays according to its tonnage, leaving him at liberty to employ it in any way, in any place, or at any hour he deems most profitable to himself. During the south-west monsoon, outside fishing is only practicable during any breaks in the weather, the best months for the fish and fishermen being those of the cold season. The larger sea-going boats are mostly engaged in capturing sharks, skates, saw-fishes, and large *Sciænas*, the three first being valuable for their fins, which are exported to Bombay for the China market; the last for its air-vessel, termed also 'fish-maws' or 'fish-sounds,' likewise an article of commerce with China due to the quantity of isinglass it contains. These fish are taken, as a rule, by means of large nets, and not by *baits* as is the custom in Malabar. Having obtained a boat-load, they make for the shore, not always returning direct for Karachi, but preparing the fins and oil at the nearest suitable spot on the coast. I measured a shark captured in one of these nets; it was upwards of twenty feet in length, and demonstrated the great strength of the twine which has to be employed for this purpose. The *Sciænas* are also large fish, rarely under 15 or 20 lbs. weight; their chief value is their air-vessel, which is worth from 8 to 12 annas when removed from the fish and simply dried in the sun; of course wholesale, the price would not be quite so much. The flesh of these last fish is somewhat coarse, and not held in much estimation. Moderately-sized boats are also used in the open sea, but this is more for line or hand-fishing, the most killing baits being prawns, which are netted with small meshed cast nets in the inner harbour. In this manner most of the edible fish which take baits are procured: only two hooks are usually attached to each line: a suitable bank having been selected, the boat is anchored, and four or five lines are employed, one to each man. The fish taken a few miles out, of course, vary with the season of the year, but may be considered to average from 1 ℔. to 2 ℔s. each. The booroo, *Otolithus argenteus*, and small *Sciænas* were very numerous when I was out fishing in the month of January; the nearer inland the smaller the size of the fish, which likewise are generally of an inferior description as food. In some boats the fish are cut open, cleaned and salted as taken; in others this process is entirely carried out on shore.

18. The fish in 1871 were all or almost all brought to the fish

<small>Mode of disposing of the captured fish.</small>

market outside the town and there sold by auction. They arrive by camel-loads, donkey-loads, or cooly-loads, and as they come are sold in lots by the Mukaddams who charge one pice or three pice on each rupee realized, and also take a little of the fish which they sometimes sell. The Tehsildar observes (1873, p. ix) that now the fishermen have all the fish, except that the person from whom they borrow the money to supply themselves with boats and nets puts into the bond that they, the 'banias,' are to have all the fish landed at half the market rates.

19. The fish having been sold by auction, and either salted or dried, and the air-vessels removed from the sorts which possess them, and the fins from others, a few points still remain for explanation respecting the fish-maws or sounds, and also regarding the fish-oil. The principal fish from which the fish-maws or rough isinglass is produced are the sea perches, *Serrani*, as the gussir, *S. lanceolatus*, attaining a large size ; the dangara or begti, *Lates calcarifer*; the booroo and souwah amongst the *Sciænas*, and the roeballs or *Polynemi*. Here the air-vessels are not extracted from the cat-fishes or *Siluroids*, which family are, however, comparatively rare. This accounts for the shape of the 'sounds' from Karáchi and the Meckran Coast, differing, as a rule, from those obtained from Malabar ; for the air-vessels of these Sind fishes are, when dried, of a long oval-purse-shaped or hour-glass form, and no marks of having been roughly torn away from the back-bone. On the contrary, the Malabar ones, which are mostly from *Siluroids*, are rounded, and have a rent on one side where they have been roughly detached from the vertebral column to which they are firmly adherent. The air-vessels are removed from the fish by the purchaser immediately he becomes its possessor, as it is very essential that they should be prepared as quickly as possible ; they are taken out with a fair amount of care and subsequently dried in the sun.

How the fish are prepared.

20. The fish-oil at Karáchi is obtained entirely from the livers of the sharks and their allies, the saw-fishes, rays, and skates. Very little care is taken in its preparation, and it is of the same foul character as elsewhere. The oil sardines, *Clupea Neohowii*, are plentiful at times at Karáchi and along the sea-coast, certainly as high as Gwádar, but they are not employed for the purpose of obtaining oil from : their arrival and departure are stated to be too capricious for its being worth while to construct nets for their capture, whilst often they are usually present only at such periods as they are unsuited for this manufacture.

Fish-oils.

21. Fish are either dried in the sun, or else having been salted are subsequently dried. The fins are generally first removed from the back and abdomen, as well as the head, if small ; they are then split open, salt rubbed in, and the sun does the rest. I could not hear that anywhere in or near Karáchi, the same pains were taken in this manufacture as at Gwádar.

Salted-fish.

22. I obtained about 190 species of fish when at this port. It is rather remarkable that in the markets there was an almost entire absence of the pomfrets, *Stromateus*, although *S. atous*, and *S. niger* abound in the sea. Horse mackerels *Caranx*, roeballs *Polynemi*, and cat-fishes *Siluridæ*, were also by no means numerous. On the other hand, there was a great abundance of *Sea Perches*, *Serrani*, of *Pristipomatidæ* and *Sciænidæ*, not only in species but in numbers composing each species. *Gobies*, more especially of

Varieties of fish.

smaller varieties, were numerous; they abounded in the muddy estuaries within tidal influence, running about on the soft mud as the tide ebbs, in order to obtain any little insect or animal substance that may be moving about, but diving down out of sight on the approach of anything which betokened danger. *Blennies* also were common on the rocks in the harbour, where they resided in large basins which were covered by every tide. Soles, *Pleuronectidæ*, were both numerous and attaining a very large size. Sharks, saw-fishes, rays and skates, as already observed, were brought largely to the markets for the purpose of obtaining their fins or converting their livers into oil.

23. The *Commissioner of Sind* (January 1st, 1873) replies that breeding and immature fish are not destroyed in the tidal estuaries, so far as he knows. He does not see that any increase to the trade of salt-fish is wanted. "There is a good trade quite equal to the demand, and the fishermen are well off, what more can be desired." As regards whether the proposition in paragraph 7 (of forming large enclosures in suitable places, wherein fish might be salted, and salt sold at a lower rate for this sole purpose) is advisable or practicable, he continues—"No, there is no reason whatever for fostering the trade by exceptional premiums.* It is in a natural and healthy state now and needs no assistance. It is best left alone."

Opinion of the Commissioner of Sind.

24. From the *Native Officials* the following has been sent. The *Tehsildar of Karáchi* replies that the selling price of salt is Re. 1 a maund, that Government salt is employed for curing fish, but salt-earth and sea-water are not. The practice of salting fish has remained stationary of late years. "In old times the fishermen had the privilege to tunur grass which grew in the salt marshes about Karáchi and its suburbs. The fishermen themselves used to sell the grass or to charge the persons who cut it at the rate of one anna per bundle. The money derived from this source amounted to about Rs. 400 or 500 a year. About 10 years ago this privilege of the fishermen caste was taken away by Captain Giles, the Port Officer, and Mr. Price, the Superintendent of the harbour works, as stated by them.† There are four divisions in the fishermen caste, *viz.*, Karáchi, Lara, Bundri, and Wungra, each division having its own head. This headmanship is hereditary; moreover, their headmen have been recognised as such in the sanads granted them by Sir Bartle Frere, Commissioner in Sind; the duty of the headman is to settle caste and sundry other matters of a trifling nature, and to conduct the religious ceremonies connected with marriages and deaths. On marriage occasions, the headman receives *lungis*, varying in value according to the circumstances of the parties undergoing such ceremonies. The emolument to the headman from this source of income is so small that it is not worth noticing. It will not be improper to observe that the practice to give a fish to the headman of the fisherman's own division, on his return from fishing, has been discontinued since last year. Formerly the fishermen of this place fished

Opinion of a Tehsildar.

---

* Salt is not taxed in Sind as elsewhere in India, thus the fish-curers have exceptional benefits.

† Instead of the fishermen in Karáchi having been deprived of privileges, they have had many conferred on them. (see p. VI ante.)

within certain limits; no outsider was allowed to fish within these limits. The fishermen, therefore, complain that they have been deprived of this right or privilege by the introduction of the license system which was ordered to be carried out by the then Commissioner in Sind, Sir Bartle Frere.* The fishermen borrow money from merchants and others to buy boats and nets. A net (description not recorded) costs about Rs. 400 or 500. The old net being worn out every year, a new one is generally made. A boat costs about Rs. 1,000, and generally lasts for some years. The fishermen sell all the fish which they get to the persons from whom they borrow money for the purchase of boats and nets at half the ruling rates in payment of the sum borrowed, there being a clause to that effect in the bond; for instance, if a fish is worth 1 rupee, the fishermen give it to them for 8 annas. The fishermen have decreased, some have died of sickness, such as cholera, &c., and some have given up the business, owing to their inability to procure nets, boats, &c., which are very costly, and also owing to the low rate at which they have to sell the fish to the persons from whom they borrow money."

25. The following shows the weight of *salt-fish* carried inland *by rail* from Karachi:—

Amount of fish carried inland by rail.

	Mds.	Seers.
From January 1st to December 31st 1872	274	18
Do. do. June 1873	156	14

* This complaint is merely that now any one may take out a fishing license, the contractor having been done away with at the fishermen's request.

BOMBAY.

26. The sea-board of the Presidency of Bombay is of considerable extent, and, owing to certain physical circumstances, well adapted for fisheries, except during the south-west monsoon. The Gulf of Cambay affords a moderately quiet expanse of water to some sorts of the larger marine fish which resort there to obtain food and a quieter locality than the open sea. Whilst along the coast, especially in its southern portion, numerous creeks penetrate miles inland and along the entire length of some districts.

Sea fisheries of the Bombay Presidency.

27. *Is the supply of sea-fish equal to the demand?* If we merely examine this question with regard to the local wants of the inhabitants who live in the vicinity of the sea, it is generally considered to be so. But if we take a more comprehensive view, and enquire whether the sea furnishes a good supply of fish equal to the requirements of the people in the inland markets? the answer most unquestionably is, that it does not. An unlimited inland market exists, and this is moderately or badly supplied with salt or dried-fish in one of the following ways—(1) imported from countries or districts where the tax on salt does not prevent the pure article from being used for this purpose : (2) salt-fish of a very inferior quality, preserved by the salt-earth which is collected by stealth or which fish-curers are permitted to employ in some few districts, or (3) simply the fish dried in the sun.

The supply of fish only equal to the demand along the sea-coast districts.

28. *What is the state of the salt-fish trade locally, has it increased, decreased, or remained stationary?* The Collector of Salt Revenue, Northern Division, observes that there can be no doubt that the salt tax, combined with the repeal of the duty on imports of salt-fish from foreign parts, has acted most prejudicially to the interests of the British fishermen, and has deprived them to some extent of their hereditary occupation. Home-cured fish cannot now compete with that cured at Goa and elsewhere, where a salt excise does not exist, and the trade with its profits has now passed almost entirely into the hands of foreigners. Near Bombay a very small amount of fish is said to be salted, but some is sun-dried. Mr. Pratt states that the fishermen only provide a sufficiency for local consumption ; in olden times salt was allowed free of duty for this purpose. From the Guzerat District is the same account, except that Bombay Ducks or Bomloes are sun-dried, and other fish are very imperfectly cured with salt : or some take salt to sea and cure what they capture, but only in small quantities, and of an inferior quality. In Surat, we are informed

The local salt-fish trade has decreased or entirely ceased, the article being merely sun-dried.

that the rise in the price of salt has not had a restrictive effect on the trade, which has increased rather than otherwise of late years; the fish which are principally exported being the 'Bomloes,' which are dried in the sun with very little salting. At Kaira, there is a decreasing trade, the reasons of which are doubtful. In Tanna fish are sun-dried, but the Collector has been unable to ascertain whether fish is now dried in preference to being salted, or whether it has been resorted to consequent on the excise duty on salt. At Ratnagiri, that the high price of salt is a source of epidemics induced by eating imperfectly-prepared fish. From the opinions expressed by the foregoing European officials, we observe that (1) considers the trade has increased, (2) that it has decreased, whilst the remaining (3) appear to think that either fish is merely sun-dried or insufficiently salted, and the trade apparently has almost gone into the hands of foreigners, who have their salt without paying duty, and are permitted to import the salted article free into the British possessions.

29. The following are the native opinions given upon this question :—

Four-fifths of the native officials consider that the local trade has decreased.

The headmen of the Kolies in Jungura consider that the practice of salting fish has decreased. In Broach that it is stationary; in Kaira that it has decreased; the same in Ratnagiri and Kanara. Thus, out of five answers, one considers the practice of salting fish is stationary, the other four that it has decreased.

30. *What is the selling price of salt?* is an important question in this trade. It appears that formerly fishermen, in some districts at least, were permitted to have salt duty free, evidently because this occupation cannot be carried on with much pecuniary benefit, if dear salt is employed. The following are the amounts of duty which have been levied for 20 years ending 1871-72 :—

The Government and the retail price of salt in the various districts.

From 1852-53 to 1858-59 the duty per maund was	Re.	1	0	0
In 1860-61 it became	...	...	...	„	1	4	0
In 1865 it was raised to	...	...	...	„	1	8	0
In 1869-70 an augmentation again took place to	...	„	1	13	0

On turning to the reported selling price to the public, we find it given as follows :—In the Concan Districts it varies from Rs. 2 to Rs. 3-8 a maund, or, as observed in Guzerat, British excised salt cannot be obtained from the pans under Rs. 2 an Indian maund, including the cost of removal at 1 anna. In Surat it is from Re. 1 to Re. 1-4 per Surat maund.* In Hubsan it is stated that salt is given the fishermen at two-thirds the rate it is sold to the public. In Kaira Rs. 2-8; in Kanara Rs. 3 a maund. In short, the cost of salt is at the place of manufacture :—Re. 1-13 the excise, plus the price of the salt, whilst the carriage of the article and the profits of the retail dealers have subsequently to be added. It is therefore very improbable that fish-curers who have to pay this large sum could compete in the markets with the neighbouring foreign curers, who have salt which "generally costs 2 annas a maund, even if as much."

---

* A Surat maund is from 37¼ to 38lbs. avoirdupois, whilst an Indian maund is 82⅞lbs. avoirdupois.

31. *Is Government or rather excised salt used for curing fish?*
Mr. Pratt remarks that it is not, as a rule,
Government salt not employ- employed : in Surat solely excised salt is stated
ed by the fish-curers.
to be used, but 'Bomloes' are the fish principally exported, and they are dried in the sun with very little salting ; in Tanna it is not used, neither at Ratnagiri, whilst the Salt Deputy Commissioner of the Southern Division observes that on his tour he has "not unfrequently fallen in with vessels returning from sea with cargoes of fish cured with salt obtained in the Goa territory." The *native officials* remark that in Hubsan "no Government salt whatever is used for curing fish;" in Broach and Kaira that it is.

32. *Is salt-earth or sea-water used for curing fish, or are they simply dried?* Mr. Pratt states that both
Sea-water and salt-earth used, sea-water and salt-earth, the latter not large-
also drying fish in the sun.
ly, are in some places employed for the purpose of curing fish; also the same reply comes from Ratnagiri and Kanara. The use of this article is prohibited in Surat and some other places, whilst sun-drying, without salt, appears to be the usual practice, or, as remarked by one set of native officials, we sell all we can fresh, and dry the remainder in the sun.

33. *What is the quality of Indian prepared fish, cured or dried in British territory?* I may here remark
Four descriptions of Indian- upon there being four main descriptions of
cured fish—(1) with good salt, (2)
salt-earth, (3) sea-water, (4) sun- fish prepared for commerce—(1) salted with
dried.
good materials; (2) cured with salt-earth; (3) prepared with sea-water; (4) simply dried in the sun. Each of these requires a few words of explanation, as the varieties are infinite.

34. *First*, as to those descriptions salted with good materials, or salt of the best quality. This may be used
Fish cured with good salt.
in sufficient quantity to render the article good, wholesome, and undecomposed for a considerable length of time, and fit for export to distant markets. But when at places contiguous to the British possessions the identical varieties of fish can be obtained by the fishermen at the same cost of capture, with salt at two annas a maund, and no import duty to pay on taking it into British territory, it does appear sanguine to hope that the Indian fish-curer with salt at as many or more rupees a maund as the foreigner pays annas, can enter into any competition. The result is that either he must retire from the trade or put less salt into his article; some have followed one course, some the other. Consequently it is to be anticipated that sufficient saline ingredients are only used in the British cured salt-fish to prevent its immediate decomposition, so as to render it just available for sale, but not for keeping. As it gets bad, some dealers smoke it; but without entering into details I will quote the remark of the Collector of Ratnagiri—"The high duty on salt is undoubtedly a source of epidemics and other serious illnesses induced by eating imperfectly-prepared fish."

35. *Secondly*, respecting the quality of fish prepared with salt-earth. It stands to reason that if a small
Fish cured in the British amount of good salt is not sufficient to cure
territory with salt-earth.
fish well, that the probabilities must be that those prepared with the salt-earth can hardly be of a superior description.

Its quality is reputed to be very inferior, but, as I have not personally examined it in the Bombay Presidency, I leave my remarks on this subject to a future occasion under the head of the Madras Presidency.

36. *Thirdly*, as to the quality of fish prepared with sea-water. This

Fish prepared with sea-water. may be done in two ways,—evaporating the water to a certain consistence, and then using it for curing fish, or the fish may be simply dipped in the sea, or buried in some place close to the sea, and subsequently dried in the sun. Of course, it is only small fishes for which this is adapted, but with such and also with some larger but very thin ones as the scabbard-fish, *Trichiurus* and *Equulas*, this is very extensively carried on. As long as the weather is dry, they may be kept for some time, but as the monsoon sets in and the atmosphere becomes saturated with moisture, they rapidly decompose, but are still used as food by the lowest classes.

37. *Fourthly*, as to the quality of the dried fish. This, of course,

Sun-dried fish. is the same as that of the last-referred-to description, with this difference that no salt whatever is used.

38. *Has the present rate of duty and the removal of the import duty*

The present rate of duty on salt, perhaps, also combined with the abolition of the import duty on foreign-cured fish, has apparently ruined the trade in most parts of this Presidency. *on salt-fish had any depressing effect on this trade?* The previous observations will show that the answers received, as a rule, seem to demonstrate that it has, but one officer thinks that it has not. It will therefore be necessary to briefly investigate this point. It must be admitted that to engage in an extensive trade some amount of capital is necessary, more especially when such concerns perishable articles. Fishermen, always improvident, are usually considered as amongst the most indigent of the people ; their average earnings in sea-fishing are computed in several places on the Madras coast at eight annas a day when at work, but it is only at certain seasons that they are so, and even at this time of the year storms and other causes may prevent their having daily occupation. But admit from Rs. 10 to 15 a month as their earnings, how can they invest in salt at two or three rupees a maund for salting fish ! the returns would not come in rapidly enough, and they would have to borrow money at exorbitant rates of interest to carry on their trade ; whilst their articles being perishable, they are liable to lose their whole consignment. But it may be urged that traders will advance the money for the purchase of salt, and the expansion of the trade may well be left to private enterprise. Unfortunately this sanguine view of looking at the prospects of an impoverished class is hardly consonant with the facts brought to light. Traders can purchase the same article, more thoroughly preserved, consequently better flavored, and at a cheaper rate in the contiguous foreign States, whilst, since 1867, the import duty of $7\frac{1}{2}$ per cent. has been removed ; thus any dealer who made advances in the British territory would be somewhat wanting in due appreciation of his own interests. If a trader finds salt two annas a maund at one locality, and in a neighbouring one two or three Rs., and no other differences in the cost of capture of fish, or the description of the supply, it appears probable that he will import to Bombay and elsewhere from the cheapest places. Thus the want of capital amongst the fishermen, and the enhanced price of salt (2 or 3 rupees a maund), appear to be the reason why foreign fish cured in contiguous ports with salt at 2 annas a maund, and imported without paying duty, has ruined the occupation of the fishermen on the coasts of Bombay, and the trade and its profits have now passed almost entirely into the hands of foreigners. As a natural result now, the fishermen, who do not reside near large towns, prefer going as sailors in coasting vessels to following their hereditary trade.

39. *Have the sea fishermen increased, decreased, or remained stationary?* From the replies received from two European and five native officials we obtain the following. Both the Europeans consider that the fishermen have decreased of late years; three of the native officials are of the same opinion, but in Ratnagiri and Kanara an increase is reported. In short, it may be taken as a fact that, with decreasing trade, the fishermen have taken to other pursuits, as sailors, &c., but where salt-earth can be employed for curing fish without any duty being imposed, a stationary or increased state exists.

Five-sevenths of the reporters consider the fishermen have decreased of late years.

40. *The foregoing seem to carry the following conclusions:* (1) That the supply of fresh fish on the sea coast is generally equal to the demand; (2) that the amount of salt-fish cured is insufficient for the local inland markets; (3) that the practice of curing fish has largely decreased in the British territory; (4) that the present price of salt renders it impossible for the British fish-curer to compete with fish salted in contiguous foreign territories with untaxed salt, and imported duty free into India; (5) that excised salt is rarely employed for fish-curing; (6) that salt-earth is; (7) also sea-water; (8) but chiefly the captures are merely sun-dried; (9) that the quality of the article prepared in the British possessions is inferior to that in the neighbouring States; (10) that the present rate of duty is an almost insuperable obstacle to the curing of fish; and (11) lastly, that the fishermen are decreasing.

Conclusions as to reports from the Bombay Presidency.

41. *We now come to what remedies have been or are proposed to ameliorate the present state of ruin to many of the Indian fishermen?* which must be considered under each different proposition that has been made.

Respecting remedial remedies which have been proposed.

42. The Collector of the salt revenue, considering the repeal of the import duty of 7½ per cent. in 1867 as one if not the chief cause of the present deterioration of the sea fisheries, suggests *an import duty of 10 per cent. ad valorem on foreign fish.* Before deciding this to be the best course to pursue, two considerations arise: first, what relief will this afford the British Indian fish-curer? and secondly, what effect will it have on the consumers of salt-fish? (1) The cost of capture and carriage to the port of trade will be about equal to both parties, and it only remains to consider, will a 10 per cent. duty place the two on the same level? This must most decidedly be answered in the negative, the one who purchases his salt at 2 annas a maund and has a 10 per cent. duty will be in a far better position than the other who pays from 32 to 46 annas a maund. Another difficulty also arises in the British territory—from where is the fisherman to obtain the money to

Imposing an import duty of 10 per cent. ad valorem. The proposition examined.

cure a remunerative cargo of salt-fish if he has to pay so highly for his salt? If he had his salt at 2 annas a maund and no duty on his fish, he could compete with the foreigner, a course it appears to be considered could not be permitted, due to the smuggling that such would occasion. However, such a tax would be a little boon to the fishermen, and perhaps induce some traders to make advances on this article in British territory. (2) What effect would such a duty have upon the consumers of salt-fish? The first result of course must be to raise the price of the foreign or superior salt-fish from 10 to 20 per cent. Then arises the question, might not the consequence be that the poorer classes will be compelled to purchase the cheaper but inferior article produced in British territory, and may not this be very prejudicial to health? It has also been proposed *that Government should give advances to enable the fishermen to purchase boats and nets.* But this mode of stimulating the fishing trade will no more succeed than did the giving of bounties increase the real fishermen in Great Britain. If the trade itself is on a healthy footing, the fishermen are best left to their own resources. The question here is, have not Government put such a weight on their own people that they cannot compete with foreigners? Have they not, by taxing our salt to the extent they have, and by permitting the salt-fish from foreign states to be imported free of duty, virtually created a protective duty not in favor of their own fishermen but in that of the foreigner?

43. It was suggested whether, if enclosures were made in favorable localities, away from large towns where the demand for the fresh article absorbs most of the captures, and inside which places fish could be salted, would the following be impracticable? *That salt should be sold inside them at rates just remunerative for the bonâ fide salting of fish.* This scheme which resembles that pursued in years gone by of letting the fishermen have salt duty-free, meets with opposition on several grounds. That it would be expensive or impracticable, as special establishments would be necessary to supervise them, a necessity of keeping up considerable supplies of duty-free salt, and the cost of conveying it to its destination, whilst some of the fishermen observe they should not approve of having to salt fish anywhere except where they do at present. On the other hand, at Broach, it is considered feasible and worth trying as an experiment; in Ratnagiri, that it is advisable and practicable; and in Kanara it might be tried, but with great care to prevent smuggling.

Whether salt might not be sold cheap inside enclosures erected in suitable places for the bonâ fide curing of fish.

44. Lastly, we come to the following opinion, that " to stimulate the industry by every possible means in the power of Government, without passing of laws regulating modes and seasons, &c., of fisheries, would be tantamount to holding out inducements for the destruction of the fisheries." As the Collector of this very district strongly deprecated protecting the fresh-water fisheries that are being ruthlessly destroyed by every means that man can employ, without paying any rent or being subject to any regulation, it appears surprising that he should be averse

Opinion that if anything is done to stimulate fishing in the sea, such would be tantamount to holding out inducements for the destruction of the fisheries.

to any attempts being made to reap the rich harvest in the sea off his own coast. It appears to me that sea and fresh-water fisheries in India require exactly opposite treatment: we ought to help ourselves to some of the inexhaustible wealth of the former, whilst we should prevent the annihilation of the fishes in the latter. In the fresh-waters, fishermen, when permitted, can destroy the breeding fish; in the open sea, it will be many generations before such a result can accrue. But this question is discussed at considerable length in preceding pages.

45. *The Collector of Salt Revenue* (March 15th 1873) submits extracts from reports by Mr. Deputy Commissioner Taylor and Mr. Assistant Commissioner Faulkner on the Sea Fisheries of the Southern and Northern Divisions of this Presidency—" It appears that the demands for fresh fish are amply met everywhere along the coast, but there can be no doubt that the salt tax, combined with the repeal of the duty on imports of salt-fish from foreign ports, has acted most prejudicially to the interests of the British fishermen, and has deprived them to some extent of their hereditary occupation. Home-cured fish cannot now compete with fish cured in Goa, Damaun and other places where there is no salt excise, and the trade and its profits have now passed almost entirely into the hands of foreigners. The adoption of the measures proposed in paragraph 7 of Dr. Day's letter (the forming of enclosures, &c.) would, no doubt, relieve British fishermen, but I doubt whether it would much increase the trade, as salt-fish is already so cheap as to be within the reach of the poorest classes. The construction of enclosures proposed would be expensive, and special establishments would be necessary to supervise them. I think that the expansion of the trade might well be left to private enterprise if British and foreign fishermen were placed on an equality, which might be effected by the imposition of an import duty of 10 per cent. *ad valorem* on foreign fish." *The Deputy Commissioner of the Southern Division* (January 25th 1873) replies—" At all the sea ports on this coast, fish could be easily captured in larger quantities than are ordinarily netted, but, as a rule, more fish is taken than can find a ready sale at local markets. In the neighbourhood of Bombay, where the consumption is considerable, fish is captured in large quantities, the markets being chiefly supplied from Mahim, Worlee, and other large fishing villages situated on the sea-shore on the west coast of the island of Salsette, and also to some extent from Trombay, Oorun, and Panwell in more immediate vicinity to the harbour of Bombay, but at none of these places is fish salted for exportation or even for local consumption, excepting perhaps in very small quantities. A considerable trade in dried fish is carried on at all the ports on this coast, but the consumption is restricted to places not remotely situated from the seacoast. At the more important ports, however, a very large trade exists in salted fish, but this is almost exclusively the produce of neighbouring foreign ports (Goa, Damaun, and Diu) where, owing to the cheapness of salt, and the fact also of its being more easily obtained, greater inducement is presented to the curing of fish than in the British territory. On my tours along the sea coast in the Southern Concan, I have not unfrequently fallen in with vessels returning from sea with cargoes of fish cured with salt obtained in the Goa territory. These vessels for

months together are employed on the above work, leaving Ratnagiri and other places at the opening of the fair season, and returning with their cargo for disposal before the setting in of the south-west monsoon. It seems most desirable that greater inducements should be held out to encourage in our own territory the curing of fish for transport into the interior. *Mr. Assistant Commissioner Pratt*, in reference to this subject, observes that—" the difficulties which stand in the way of promoting the above object may probably be the following :—(1) Fishermen on the coast away from large towns are probably too poor to provide large enough boats and suitable nets for deep-sea fisheries, and this may be a formidable hindrance to large captures of fish. It is possible that they might be induced to extend their operations if they received such advances as would enable them to provide better boats and nets. (2) At present no larger number of men are engaged on fisheries than are required to provide an amount of fish sufficient for local consumption. Large masses of the population, however, are during a greater part of each year in more or less distress, and would probably be thankful to be put in the way of earning a livelihood. With adequate encouragement in the shape of advances for the purchase of nets or boats, it might be practicable to persuade men not now employed as fishers to qualify to engage in that pursuit. (3) Salted fish would only find a ready sale in the interior, if well-cured fish could be put cheaply into the markets. There is but little demand for the inferior article produced by curing with salt-earth, and it is probable that the demand for salted fish would improve, if salt could be made available for curing fish (within large enclosures or otherwise) at rates so cheap as would permit of salt being invariably used instead of salt earth in the curing of fish." Mr. Pratt is further of opinion that the cheapening of salt for the curing of fish would not alone be a sufficient stimulus to the fisheries. A stimulus, he thinks, seems to be chiefly required in the direction of encouraging people to make large captures of fish, and he sees no way of promoting this important object except by seeking to attract to the employment of fishing (by a judicious system of advances) larger numbers than now follow that pursuit. " As a rule, Government salt is not used for the curing of the fish captured anywhere on this coast. Fishermen object to pay duty on the salt used for such-purposes, and naturally therefore resort to places where their wants can be more easily supplied. Prior to the year 1867 there used to be a levy of 7½ per cent. customs duty on foreign salt-fish : now it is allowed free import, so that fish cured in British territory cannot possibly compete with that imported from foreign markets. Hence it is reasonable to suppose that, of late years, there has been a decrease in the number of sea fishermen compared with former years. It only remains for me to add that there is too much reason to believe that the fishermen on the coast are insensible to the improvidence of destroying breeding and immature fish in tidal estuaries." " In the Concan Districts the selling price of salt varies from Rs. 2 to 3½ per maund. Government salt (*i. e.*, excised salt) is not as a rule used in the curing of fish on this coast. Both sea-water and salt-earth (the latter not largely) are in some places employed for the purpose of curing fish. The practice of curing fish has to a great extent diminished, owing partly to the falling off in the amount of fish usually captured, and also the duty charged on salt in British territory. Excepting that salt was formerly allowed free, there appears no ground for supposing that fishermen possessed in old times any particular privileges. There appear to be no headmen among fishermen, nor is the pursuit of fishing confined to any one particular caste. No one possesses any right as regard sea fisheries on this coast. The fishermen supply themselves with boats and nets. On the whole, it is supposed that the number of sea fishermen have decreased, but correct data upon this point are not available." *The Assistant Commissioner in charge of the sea-shore salt-works and ports of Guzerat* (February 15th 1873) replies—"I beg to submit the following information which I have been able to glean, regarding the 'fisheries' between Damaun and Surat, the only portion of the sea-coast within my range that I have been able to visit since taking charge. It was my intention to have withheld this report for some time longer, hoping in the meanwhile to extend my tour further north into the Gulf of Cambay, and include the results of further inquiry in this letter; but as it is overdue, and has been called for, I must confine this notice within the geographical limits above specified. The 'fisheries in these parts may be divided into four distinct classes : *first,* local fisheries ; *secondly,* sea fisheries ; *thirdly,* monsoon fisheries; and *fourthly,* foreign fisheries. (1) Every town of importance along the coast has got its 'local fishery,' which supplies fresh fish for daily consumption to the inhabitants. The positions of the fishery grounds are generally at the entrance of, or up a river, and sometimes in the open sea, but in such cases they are not far removed from the shore. They are invariably so situated as regards distance that the fish can be brought into market and disposed of while it is in a perfectly fresh and sound state. It very seldom happens that the supply exceeds the demand, except perhaps under unusual circumstances, but it may be said, as a rule, that the demand is greater than the supply, and this condition strongly favors the idea that fishing is not looked upon as either a popular or profitable occupation. The following are the principal local fisheries between Surat and Damaun which supply the markets indicated. Although fish is caught and consumed all along the coast, the 'local fisheries' mainly owe their existence to the demand for fish by Christians, Parsees, and Mahomedans, who never follow the occupation of fishermen, and whose circumstances are generally such as to enable them to pay for the indulgence of using fresh fish, which, as a rule, is much cheaper than meat or poultry." Then follows a list of the local fisheries and the vernacular names of a few of the fish captured in them. "The supply of the varieties of fish are not equal throughout the year, but are dependent upon certain seasons when they are more or less plentiful. Some of them, however, are comparatively rare, but, as they are met with, they have been placed in the list. Others again are caught in such large quantities as to exceed the demand for them in a fresh state, when they are either dried or salted, and retained for subsequent use. 'Bomloes,' or Bombay ducks *(Harpodon nehereus)* and 'Bhing' come under this category. The former are simply dried in the sun, but the latter and all other kinds of fish require the use of salt to cure them. The class of boats employed on these fisheries are either 'Muchwas' or 'Cottias' (canoes), ranging in capacity from two to five candies each and are lateen-rigged. They are open-decked, and are unsuited for deep-sea fishing. The fishermen purchase their boats, but make the sails, rigging, and all other gear necessary to work them. The fishing nets are made of hemp and cotton (twine and thread), and are dyed with shemby, babool, and barks of other trees to preserve them. They are of such forms and sizes as are adapted to the requirements of each locality and the description of fish to be caught. They are generally attached to stakes, imbedded in the ground, which are so placed, either in rivers or by the sea-shore, as to catch the full effect of the tide, and thus trap the fish that are carried by the current. But hand-fishing with round and oblong nets is not uncommon, especially in shallow waters and the banks of rivers. The latter are generally trailed against the current by two men or women, one of whom has a hold of each extremity. As a rule, the fishermen make their own nets and dye them. They also manufacture twine and thread from the raw materials (hemp and cotton) of which the nets are made." The different kinds of nets used in the 'local fisheries' are known by the following names :—' Chog' or ' Choga,' round hand-nets, leaded at the bottom, made of cotton thread; Golwa or large nets attached to stakes) they are of a conical shape made, as are the four following, of hempen twine) ;'Kundaru' and'Mag,' both oblong-shaped hand nets used in catching mullet; 'Wadee' the same, but for taking prawns; ' Pandia,' a small trailing net used in ponds; and ' Murree'. There are no restrictions, tolls or supervision on the local fishing trades. " Fishermen are, as a rule, poor. The precariousness of their occupation and the uncertain profits derived from it often compel them to accept service as sailors in coasting vessels, laborers, and, in fact, anything that will ensure them a steady and certain means of living. These remarks apply to the whole body of fishermen engaged in the various kinds of fisheries spoken of in this letter. It need, however, be scarcely pointed out that every fisherman does not possess a boat or net, and it is chiefly those who are dependent on their labor for a livelihood—and they are by far the greatest number— that seek employment in other ways". " (2) The ' sea-fisheries' are not extensive; as far as I can learn, only about 200 vessels compose their fleet, but the number depends on the coasting trade, such having the preference as being most remunerative. The carrying capacity of each vessel ranges from 20 to 40 candies. They remain out for about a week, but the actual period of their stay is regulated by the success they meet with. They take small quantities of salt to cure the fish they catch. The fish thus obtained is taken to port and sold to dealers, who dispose of it for local and inland consumption." " The quantity probably thus brought is not considerable, and certainly of a quality and description to meet with favour only from the poorest class of natives. The fishing season lasts for 10 or 12 weeks between January and March, and the fishing grounds are most of them south of Demaun, extending as far as Bassim in the Tanna Collectorate. (3) ' Monsoon fishing' is carried on at the mouths of the rivers during July and August. It is computed that at least 500 vessels engage in this occupation annually, at considerable risk to both life and property. The inducement is the certainty of obtaining large hauls of the Bhing fish, which is then very abundant, especially in the Nurbudda, Taptee, and Midhola estuaries, and which is in great request for the roes that are obtained from them. The fish is salted on board, and finds ready sale at every port or place the vessel puts into, for by a preconcerted plan dealers are always in waiting to purchase the whole stock. The vessels employed are of the same class as those engaged in the sea-fisheries, ranging from 20 to 40 candies each. They take advantage of breaks in the weather to start for their destinations, and so calculate their movements as to be able to make some particular port or place before they are overtaken by a storm or foul weather. They never cast anchor at sea, and are careful to keep as close in-shore as possible. The run from one place to another generally ensures a good haul of fish. This they dispose of and then prepare to start again on the next favorable opportunity. 'Surat fish-roe' has quite a wide-spread reputation. It is esteemed a great delicacy, and is in demand by all classes of the fish-eating community, especially the Parsees. There is no doubt that if the Bhing fishery was properly fostered and conducted, it would prove a source of considerable emolument to those engaged in it, as the supply appears unlimited, and the demand would certainly keep pace with the supply. But, so far as I am able to learn, the catching of this fish is conducted in a mere helter-skelter manner, and the curing of the fish is so utterly neglected that it will not keep for any length of time. Hence a large portion of it soon becomes unfit for human consumption, and is either cast away or used as manure; and here I would mention an important fact in connection with this fishery as operating against success and extension, viz., that the fishermen employed in it endeavour to sell as much of the fresh fish as they possibly can, for local consumption, at exceedingly low rates, and it is only what remains on their hands unsold that they cure imperfectly and get rid of anyhow. They do not seem to be alive to the importance of the fishery and the advantages that might be derived from it if systematically conducted and proper attention paid to the curing of the fish. They evidently look upon the time spent in catching the fish as affording them occupation and the means of subsistence at a period of the year when they cannot pursue their ordinary calling, or obtain employment as sailors, laborers, &c. This fishery, if properly fostered and conducted, might, in my opinion, be developed to such an extent as to make it an important item of commerce and means of industry. (4) Regarding the 'foreign fisheries' of Demaun, Diu, Jaffiabad, Cutch, &c., I am not in a position to afford any information beyond stating the fact that salted fish, chiefly Pomfrets, Seir fish and Bomloes, are largely imported from these places into many of the ports of Guzerat and the Concan, and find its way beyond the Ghâts and other distant places. Care seems to be taken in curing the fish, as it is better flavoured, keeps longer, and is consequently preferred to all other kinds of preserved fish to be met with in the markets. Probably the simple secret of this superiority lies in the circumstance that good unadulterated salt is used in curing the fish. In this respect the foreign fisheries have decided advantage over ours, as the salt employed by them generally costs 2 annas a maund, even if as much, whereas British excised salt could not be removed from the pans under Rs. 2* per Indian maund. Hence the foreign fisheries are able to employ salt freely and to the full extent required to cure fish thoroughly. Our fishermen, on the contrary, bearing in view the comparative cost of salt to them, cannot afford to be so lavish and consequently use just

---

* Including cost of removal, 1 anna.

sufficient salt to preserve the fish for a time, but not to cure it. Little or no refuse fish is used for manuring purposes. Sharks' fins and fish maws or sounds seem to be unknown in this part of Guzerat, but they form an important item of traffic between the ports in the Arabian and Persian Gulfs, the African coast and Bombay, whence they are chiefly exported to China." " I may briefly say (1) that breeding and immature fish are indiscriminately caught with others; (2) that the salting of the 'Bhing fish' and roe could be both increased and improved (see ante); (3) that the proposition of forming enclosures, &c., does not seem to me to be either advisable or practicable, unless at considerable expense to Government and risk to the revenue." *The Assistant Political Agent, Junjura* (October 29th, 1872), observes that the questions on the sea-fisheries have been put to the headmen, whose answers are believed to be correct and are appended. *The Collector of Surat* (October 23rd, 1872) replies—"Breeding fish are found only in the month of June, and then they are not caught in great numbers. The fishermen, however, do not scruple to take such miniature or breeding fish as they can get. The rise in the price of salt has not had a restrictive effect on the trade, which has increased rather than otherwise of late years. The fish which are principally exported are 'Bomloes', which are dried in the sun with very little salting. The fishermen in this district are not possessed of sufficient capital to profit by the measures proposed (enclosures, &c.). I doubt whether any reduction in the price of salt would have a great effect upon the trade, and the plan suggested must necessarily afford increased facilities for defrauding the salt revenue. The price of salt in this district is from Re. 1 to Re. 1-4 per Surat maund. The salt used for salting fish is purchased from sellers, or from the Government ' Agurs.' No salt-earth is used for salting fish, but sea-water is employed in rare cases. There has been a slight increase in the practice of salting fish on the whole. The fishermen have had no peculiar privileges. There are no headmen of the fishing castes ; no one claims any right in respect to the sea-fisheries. They buy their nets or make them themselves, and purchase boats if required. Sea-fishermen appear on the whole to have decreased.

46. *The Collector of Broach* (January 14th 1873) answers "that breeding and immature fish are destroyed in considerable numbers in the tidal rivers of this district. That the salting of sea-fish not being a trade for the most part systematically followed, but resorted to chiefly when, owing to a large capture, the supply becomes in excess of the local demand, no plan for its increase is likely to be very successful. That the proposition referred to by Dr. Day in his paragraph 7 (of forming enclosures, &c.) appears feasible and is worth trying as an experiment." *Mr. Whitworth, the Supernumerary Assistant Collector*, suggests that the experiment should be made "on the Hansot coast, near the village of Kutpor, where, by the closing of the salt-works, numbers of persons are put out of employment and are ready for a new profession."

Opinions of the European officials of Broach.

47. *The Acting Collector of Kaira* (December 7th, 1872) observes that as regards sea-fisheries "there are none worth the name in this collectorate; except at Cambay and Baroda, the demand for fish is very small. The fish-consuming population of Guzerat is very limited. I have ascer-

Opinions of European official in Kaira.

xxiii

tained also from Cambay and from the villages on the tidal portion of the Mhye that the supply of sea-fish appears to have greatly decreased of late years. Whether this is due to the sea having receded from the north coast of the Gulf, or to the decreasing trade in salt-fish, owing to the increased price of salt, I am not able to state, but I have no doubt that the adoption of Dr. Day's proposal to reduce the price of salt used for fish-curing would give a stimulus to the trade. At the same time I do not urge the adoption of the suggestion in this collectorate. Both supply and demand are too limited to make it worth while to risk *the great increase of smuggling*, which would be the first result of the concession, and for which the vicinity of foreign territory offers *here* peculiar facilities."

48. *The Supernumerary Assistant Collector, Ahmenabad* (November 29th, 1872) observes that "in the Gogo talooka the amount of fish caught is trifling, while in Dhundhooka it may almost be said to be *nil*. I only know of one man in Gogo who even professes to get a living by fishing; and so far as I can give an opinion I should say that any arrangements to facilitate salting would be likely to meet with very little success in these parts." *The Acting Collector* continues (December 3rd, 1872)—"There is but one fixed net on the Gogo coast; it catches but little, and the amount of fish caught by temporary and hand-nets is less. No facilities for salting would, he believes, increase the number of nets or fishermen."

Opinions of European officials in Ahmenabad.

49. *The Collector of Tanna* (March 31st, 1873) reports—"From enquiries made throughout the district it does not appear that breeding and immature fish are destroyed to any extent in tidal estuaries. It is only at certain seasons of the year that fish enter the tidal estuaries for breeding purposes, *viz*., chiefly during the monsoon months, when there is a considerable influx of fresh water in the estuaries. In many of the tidal streams stakes are placed down and nets cast; little injury though, it is said, is caused by such fishing to fry, the meshes of the nets used being, as a rule, sufficiently large to prevent their being caught. Some slight increase to salting of fish might, there can be no doubt, be effected. It is though very questionable whether any very great or perceptible increase could be brought about. In many localities the take of fish exceeds very considerably the local demand. Fish netted in excess of the demand is, as a rule, sliced, cleaned, and dried in the sun, and then packed off mostly to Bheuridy, which is the chief entrepôt for the sale of dried and salted fish; whether fish is dried as above in preference to its being salted, is a question I have been unable to ascertain; the fishing class say that it is their custom to so treat the fish. It is very probable that it has been resorted to in the place of curing by salt, consequent on the excise duty levied on salt. If so, it is though I consider very doubtful, whether salting in the place of drying would be resorted to, even if the duty were lessened, as the fishing class, like many others in this country, look alone to immediate returns, attained by treating of produce in as crude and primitive a manner as possible, rather than prospective increased gains by laying out of capital and resorting to improved methods of preserving. I quite concur in the

Opinion of the Collector of Tanna.

opinions expressed, that reduction of the monopoly price of salt is the one plan by which the trade in salt-fish can be augmented, provided the classes engaged in fishery can be brought to see that curing by salt is better and more remunerative than drying. On the other hand, it is very questionable whether the fisheries would yield greater takes than now made; if not, the simple result would be an increase of salt-cured fish, and an equal decrease of sun-dried fish. The proposition made as to the sale of salt within enclosures at low rates would, I think, be found to be in practice impracticable, as, to carry out the suggestion, it would necessitate the retention of very considerable supplies of duty-free salt for curing purposes, over, it may be, extensive areas in each district. Such salt would, I presume, have to be conveyed to the enclosures at Government cost, and, until consumed, have to be under the supervision of Government officers, all which must entail cost on the State, unless all such expenditure be recovered from the fish-curers. On the other hand, the cost at which salt would be supplied would have to be fixed sufficiently low as to induce the fishermen to use it in the place of drying as now: unless this be so, and it can be proved that salted fish fetches higher and more remunerative rates than dried fish, there is, I anticipate, little prospect of the enclosures being extensively patronised. Throughout, Dr. Day, I gather, assumes that there is a demand for salted fish in preference to sun-dried. I am though, I must say, somewhat doubtful as to whether this be the case; assuming though that it is, I cannot concur in thinking that it is in any way the duty of the State to take action in forcing a particular article of commerce into the markets, for there can be little doubt, if really much sought for, it will be supplied irrespective of excise duty on salt. If the excise duty be taken off for the encouragement of one branch of industry, it will be necessary to make a like concession in regard to other industries requiring salt for manufacturing and other like purposes, such as chemical works, &c. That the question of fish-curing by salt should engage the attention of the Inspector General of Fisheries is apparent. I am though of opinion, as regards this district, the question is one not calling for immediate attention, or adoption, as the take of fish is not so very extensive as to call for immediate action on the part of Government. Presuming, however, that it be advisable to give an increased impetus to this industry, it is, I think, very undesirable that measures should be taken thereto, until such time as some effective law for the protection of the various fisheries be passed, * for to stimulate the industry by every possible means .in the power of Government, without passing of laws regulating modes and seasons, &c., of fisheries, would be tantamount to holding out inducements for the destruction of the fisheries."

50. The *Collector of Colaba* replies (23rd March 1872) that

Opinion of the Collector of Colaba.
"owing to the flatness of the country, the tides run so far up the different creeks; the fish in them are of a coarse kind, though sometimes of a very large size."

---

* NOTE,—6th January 1872. The *Collector of Tanna*, in respect to fresh-water fisheries, observed: "no distinction is made between breeding fish and others, whilst the young are also caught. There is no restriction whatever; none is recommended."

The following statement shows the imports and exports of fish dried and salted at the Port of Bombay since 1866-67 :—

Years.	Imports.	Exports.
1866-67	24,499	17,450
1867-68	not available.	
1868-69	116,246	13,558
1869-70	91,222	10,386
1870-71	89,809	23,755
1871-72	64,439	10,938
1872-73	82,019	16,894

51. *The Collector of Ratnagiri* (25th January 1873) observes:

Opinion of the Collector, &c., of Ratnagiri.

" There are several fish-markets in this zillah, such as at Kharepatam, Mhaprul, and in fact almost all the large villages on the tidal rivers. The cultivators at these generally barter grain or wood for salt-fish. The high duty on salt is undoubtedly a source of epidemics and other serious illnesses, induced by eating imperfectly prepared fish. It would be a great boon to the poorest classes were Government to reduce the tax on salt, so as to enable fish to be properly salted. " " Besides the sea fishery are those up the numerous creeks or arms of the sea which penetrate inland for several miles along the whole length of the coast district. Far inland, in depths too shallow for navigation, at low tides, and during the monsoon, some parts of these creeks contain fresh water." *The Supernumerary Assistant Collector* (4th January 1873) reports that breeding and immature fish are destroyed to a considerable extent. The proposition mentioned in paragraph 7 (forming salting enclosures, &c.,) is advisable and practicable in these districts. At the inland harbours of Ibrahimputan and Sungmeshwur, there is a weekly bazar of fish, which is exchanged for wood and grain. Any increase of the supply of fish would be a boon to the people inland. At present there is no doubt that the fish imported to those places are most imperfectly cured, and if any means can be devised to preserve them better, a more abundant and more wholesome sustenance would be obtainable by people who are at present in a wretched state of destitution. The Mamlutdar states that sea-water and not salt-earth is used for curing fish, but I know of one place at least where salt earth is employed. The police patell of the village below the Jygurh fort has found it impossible to prevent the Daldis from trespassing upon an old salt-pan there, now no longer used. These people come to dig up the earth, which is impregnated with salt, for the purpose mentioned. I believe that this place might be chosen and a similar one at the entrance of the creek at Ratnagiri, where a licensed sale of salt for curing fish on the spot might take place. Means could be found to prevent withdrawal of salt, and as the spots chosen would be close to the mouth of the creek, a ready way would be at hand to dispose of the fish. The plan suggested by Dr. Day cannot be too strongly recommended."

52. "The *Collector of Kanara* (February 17th, 1873,) replies that breeding and immature fish are destroyed daily to a great extent." " Salting of sea-fish could be increased, it is generally believed, if salt was cheaper. The plan of enclosures in certain localities for salting fish, where salt could be procured at a cheap rate, might be tried, but great care to prevent the salt being removed and otherwise used would be necessary."

<small>Opinion of the Collector of Kanara.</small>

53. The following are the replies from the native officials :—In the *Junjura district* the *headmen of the Kolies* report—"Fish are caught the whole year round; doubtless many are destroyed which are immature. Sometimes, not often, from some cause utterly unknown to us, the fish die in the creeks; we suppose from bad water. The salting of fish could be increased if we could get salt at a cheaper rate; now it does not pay us to salt all the fish we catch, so we dry in the sun a portion, and sell as much of the fresh fish as we can. We should not like to have to go any distance to cure our fish; it would be more trouble than it would be worth. In Hubsan salt is sold at one maund a rupee to all classes but Kolies; we, the Kolies, get one maund and a half the rupee. No Government salt whatever is used for curing fish, neither is salt-earth nor sea-water. The practice of salting fish 'has decreased in quantity,' because there are many less fish caught; because the expense of placing fishing stakes has increased, and we do not use the 'wandope' or large fishing net on four fishing stakes, because a tax of six rupees has been imposed on its use. We have not used this net for ten years; we seldom use now any net but the 'boorkea wole' in deep water (a net fastened by anchors under the surface of the water) for which we have only to pay Rs. 3 per annum." "In old times we had certain privileges which we have not now" (what they were is not stated). Regarding whether they have any hereditary headmen, they continue—" Yes, we, the Sir Patell and Chogala of Kolies, have Yakoob Khan's sunnud authorising us and our heirs to exercise all the authority and duties of the Sir Patell and Chogala of Kolies: all sircar's orders, &c., must pass through our hands; all disputes, &c., amongst Kolies of Hubsan are decided by us alone, and we both receive kucqs from the sircar and all Kolies, and the Kolies of Hubsan acknowledge our rights." * No one claims any rights over the sea-fisheries. The fishermen supply themselves with boats and nets; six or ten club together, build a boat, make a net, and divide the produce into shares. The sea fishermen have decreased, because many of them now go in merchant vessels.

<small>Opinion of natives respecting the fish and fisheries.</small>

---

* "This is quite true. I have seen the sunnud, a very old one indeed, conferring on these men's ancestors absolute authority over all Kolies, and defining their kucqs, &c." *Note by the Acting Collector.*

54. From the *Broach Collectorate* the following replies have been received from the *native officials*. The four *Mamlutdars* answer that "the price of Government salt in the different talookas is as under:—

Opinions of the native officials of Broach.

Broach	...	...	...	Rs. 1 4 0	per Surat maund.	
Ahmode	...	...	...	„ 1 8 0	„	„
Jambusar	...	...	...	„ 1 4 0	„	„
Anklesar	...	...	...	„ 1 8 0	„	„

It is employed for curing fish, but neither salt-earth nor sea-water are used for the purpose. The practice of salting fish has neither increased nor decreased of late years. In old times the fishermen-caste had no privileges which they do not now possess. At Broach, Jambusar, and Hansot this caste has patels at its head. Their appointment is not hereditary, but is made by their own community, and all the members have a voice in it. Their principal duty is to settle caste disputes. They are not paid for their services. The right to sea-fishing does not vest in any particular person. The fishermen make nets themselves, and sometimes purchase them from their caste-men. They have generally boats of their own, and those persons who have no boats get them on hire from others. The number of fishermen appears to have decreased.

55. From the *Kaira Collectorate* replies from the *Mamlutdar of Borsud* have been forwarded:—A considerable number of breeding and immature fish are destroyed during inundations in the monsoon. The fishing has greatly fallen off of late years. The supply of fish is now too scanty to render the adoption of Dr. Day's suggestion necessary. The cost of salt is Rs. 2-8 per Bengal maund. Government salt is used in curing fish, but neither salt-earth nor sea-water is. The practice of salting fish has decreased of late years. The fishermen apparently have all the privileges they formerly possessed, except that at Cambay they were allowed salt free of duty. They have a headman called 'patel.' The office is not hereditary, but the appointment is made by the 'Zumat.' He resides at Sarode in Broach, and a few acres of land are allowed him by the Thakore of that place. His duty is to settle all disputes, &c., amongst the fishermen. No one claims any rights in respect to the sea-fisheries. Boats are little used; the nets are made by the fishermen themselves. The fishermen have decreased with decreasing trade.

Opinion of native official in the Kaira Collectorate.

56. From the *Ratnagiri Collectorate*, the following reply has been received from the *Mamlutdar*:—"The price of salt is Rs. 2 a maund in the town; it is not used for curing fish. Sea-water is employed for this purpose. The practice of salting fish has decreased within the last fifteen years in consequence of the increase in the price of salt. In the fishing castes there are persons distinguished by the names of 'hodekur' and 'patel,' though these titles find no place in Government records of any sort; the duties of these persons seem to be those of mediating between parties engaged in any small disputes, of which there are a good number. There do not seem to be any emoluments attached to the office of 'hodekur,' though the latter was exempted from the poll tax, according to information drawn from the dufters of the office." No claims are put forward to the sea-fisheries. "In the creeks, posts are

Opinion of native official at Ratnagiri.

placed to serve as boundaries for the exclusive right of fishing in particular parts thereof." The fishermen supply themselves with boats and nets. "The sea-fishermen have decidedly increased in numbers."

57. From *Kanara* the following answers have been received from the Tehsildars :—The selling price of salt is Rs. 3 a maund, and it is not used for curing fish, but salt-earth is. The practice of salting fish has much decreased; in olden times the fishermen had no privileges they do not now possess. There are headmen whose duties are confined to settling caste disputes, for which they have no regular emoluments. No one claims any rights in respect to the sea-fisheries. The fishermen build their own boats with timber purchased from Government or elsewhere: they also make their own nets. The number of fishermen have increased.

<small>Opinion of the native officials of Kanara.</small>

58. In examining the returns of the *Great Indian Peninsula Railway* as to the amount of fish carried inland from Bombay or contiguous stations, we obtain the following :—

<small>Salt-fish carried inland by the Bombay railway.</small>

			Maunds.
1868 first half-year	...	...	4,572
,, second ,,	...	...	1,870
1869 first ,,	...	...	1,931
,, second ,,	...	...	3,782
1870 first ,,	...	...	8,245
,, second ,,	...	...	8,021
1871 first ,,	...	...	9,638
,, second ,,	...	...	6,861
1872 first ,,	...	...	10,998

MADRAS.

Sea-fisheries of the Madras Presidency: those on the Eastern or Coromandel Coast differ from those on the West or Malabar side.

59. The seaboard of this Presidency is about of equal extent to those of Bombay and Bengal, whilst being, as a rule, better populated, its marine fisheries are of very considerable consequence. It may be divided into two distinct portions,—its Western or Malabar, and its Eastern or Coromandel Coasts. The Malabar Coast receives the full force of the south-west monsoon, commencing about the end of May or commencement of June, and continuing for the next three months. Its fishes differ generally from those of the Eastern Coast, as siluroids, clupeoids, and sharks are very abundant. On the Coromandel Coast, there is a large variety of sea perches and horse mackerels.

The supply of fish, as a rule, is locally sufficient during the fine weather.

60. *Is the supply of fish equal to the demand?* As a rule, it is so locally during the fine weather season, but in many places it does not appear to be equal to supplying inland markets. In fact, the same reply may be given here as has already been given in the Bombay report.

61. *Have the fish increased, decreased, or remained stationary?* Some officials report one thing, some another;

It is difficult, if not impossible, to decide whether the supply of marine fish is the same as in former years, because certain species migrate and often remain away for several consecutive seasons.

but it must be borne in mind that many of the replies refer to the vast shoals of fish, as the sardines, whose advent and disappearance is most uncertain. Again, the predaceous fish which feed on them, following these shoals, are of course absent, should their food have migrated as evidently they migrate with them. Thus, a very large source of supply becomes suddenly stopped, or rather does not arrive when expected, and it is thought that the supply is decreasing. Another reason, which seems to affect this answer, is whether the fishermen are, or are not, capturing the harvest of the sea which comes almost to their doors?

62. *Are the fish being captured?* is then the next point for consideration. If a large local market exists for fresh fish, and any amount captured will find a sale, it appears that the supply fairly, often entirely, meets the demand. Then the fishermen go further out to sea, because they are aware that they will be able to dispose of their captures. Thus in Malabar, we find the

If there is a sufficient market, the fishermen capture the deep sea-fish; if the market is bad, they fish along the shore for small fish, and scare away the larger sorts that are coming for their prey.

export alone of salt-fish, which averaged Rs. 38,054 annually in five years ending 1858-59, had augmented to Rs. 1,81,699 annually in five years ending 1872-73, showing increased demand met by increased labour. On the Coromandel Coast, on the contrary, one hears of no increase of captures, and in many stations a decrease of fish is reported; but I do not think this latter statement can be accepted. It appears due (owing to causes presently to be explained) to no market existing which would repay the fishermen to increase his means of capture. If he has no market, he naturally only plies his trade along the sea-shore, where, with less expense, he is able to capture sufficient for the local demand, whilst the small sorts taken can be dried or cured with care. Disturbing the shore and taking the small fish, he captures the food which previously decoyed the large ones in, and thus he scares away the supply of the more important sorts, and asserts the fish to have decreased.* This, however, does not signify to him so long as the market is supplied, and sufficient obtained for his family requirements.

63. Respecting the effect of the *salt-tax* upon the trade of salting fish, a few considerations have first to be noted. In the Madras Presidency, on October 4th, 1869, the monopoly price of salt was raised from Re. 1-11 to Rs. 2 a maund, at which rate it still continues. During the succeeding nine months consumption was checked, but in 1870 an increase became again apparent. At a first glance the raising the price of salt might be thought to have an effect on the salt-fish trade, and so it would, were Government salt employed for this purpose. But if salt-earth is collected untaxed, might not the raising the selling price of salt cause the poorer classes to fall back on salt-fish imperfectly cured with salt-earth?

Salt-tax in the Madras Presidency: when it may not directly affect the curing of fish.

64. *Is it permitted to gather salt-earth untaxed for the purpose of curing fish?* It appears that opinions differ in different collectorates upon this head : thus in twelve districts the following orders are in force as seen from the *answers of the European officials.* Its collection is *prohibited* in six, viz., Madura, Tinnivelly, Tanjore, Trichinopoly, Ganjam, and South Arcot. Whilst it is *permitted* in the other six, viz., South Canara, Malabar, Madras, Nellore, Kistna, and the Godavery.

Salt-earth permitted to be collected free for fish-curing in some districts.

65. If we examine the returns received from the *native officials* of those districts *where the use of salt-earth is prohibited,* we find the following statements. In the six collectorates where the use of salt-earth is prohibited, 30 native officials state what they consider occurs in their districts:—

How fish are cured in districts where salt-earth is prohibited.

11 assert Government salt alone is used.
4 „ „ „ and sea-water are used.
2 „ salt-earth alone is used.
10 „ Government salt and salt-earth are used.
1 „ sea-water only is used.
1 „ fish are only dried.
1 „ they are only imported.

How fish are cured in those districts where the use of salt-earth is permitted, and effect of the salt-tax on this trade.

66. In those collectorates wherein the *use of salt-earth is permitted, we have* 26 *answers from native officials:—*

9 assert Government salt alone is used.
3 „ that salt-earth alone is used.
16 „ „ „ „ and Government salt are used.
1 „ fish are merely dried.

* If the Collectorate of Malabar, with a sea-coast hardly one-fifth of that of Coromandel, can *export* yearly upwards of Rs. 38,000 worth of salted-fish by sea, one would imagine that some specific cause must be in operation which prevents any or a very slight quantity being cured on the latter salt-taxed coast, or obliges the dwellers there to import from Malabar, where salt-earth is used untaxed for this purpose.

xxxi

Here, however, I would observe that the various reporters may be referring to different articles. The great place to which salt-fish is exported from Malabar appears to be Colombo, and the sea customs house officer at Tellicherry asserts that what is intended for this market is solely cured by Government salt, or they would not keep : he gives the value of these exports from Tellicherry for the last five years at Rs. 58,460-8-1 annually. Increased steam communication between the two places has done much for this trade, and also due to a bank having been opened, and where the fish-curers can obtain advances upon the security of their bills of lading. But it does not appear that everywhere solely Government salt is used for this purpose : thus from Cannanore the value of these exports for the last five years are given at Rs. 13,954 annually, and the sea customs house officer remarks, it is used to a very limited extent and that for curing large fish : in this case the salt is generally supplied by the merchants who export the salted-fish to Colombo, Tuticorin, Negapatam, &c. At Ponany the sale of Government salt is said to have decreased, but Rs. 10,674 worth of salt-fish were exported in the *one year*, 1872-73. whereas in *five years* ending 1863-64, the export reached Rs. 463 worth only; here salt-earth must consequently be the substance employed. Fish that has to be stored for some time before being exported either inland to the hilly districts or to Ceylon, has to be cured with good salt. At Calicut, the difference in the price of the Indian mackerel cured with Government salt or salt-earth I found to be eight annas a thousand. The richer classes of natives and fishermen for their own family use are said to invariably have fish cured with Government salt in some parts of Malabar. But for the poorer classes the article cured with salt-earth is evidently what they generally consume. The Collector reports that the people so dislike the native salt* " that all who can afford the expense prefer Bombay salt at eight times the price, and if an excise of Re. 1-13 a maund were added to its present market value (about four annas) Malabar salt would be utterly unsaleable." Now, it is largely collected, the pans are described as existing in thousands, and the substance is very largely used for the curing of fish. If the duty will entirely stop the manufacture, such must be a very clear proof of the exceedingly small profits these poor people are making on their salt-fish, and how the proposed excise will probably act in a most deleterious manner on the occupation of curing fish with salt-earth. The trade that will be left can only be that for the rich and the foreign market, even much of this, I anticipate being destroyed, because my conv.ction is that the amount of monopoly salt reported in places to be u¹sed for this purpose is overstated. The salt-earth is purified sometimes, or perhaps mixed with a certain proportion of the best salt to improve its quality. Annual returns of the sale of monopoly salt will show this. Whether any concession could be made respecting the sale of salt to the fish-curers that would obviate this is a consideration, for if one great

---

* I have not considered it necessary in this place to advert to the opinions of those who consider that natives relish badly-cured salt-fish more than they would do a well salted article ; the fact being the poor people cannot afford to buy the expensively prepared salt-fish, so perforce have to be *contented* with the inferior, however detrimental to health such may be.

branch of this trade suffers, the fisherman's work is decreased; if the market for his captures diminishes, those who follow this occupation must become less, and then a vast amount of animal food now going to nourish the poorer classes will be left uncaught and consequently useless to man. The salt-fish trade is now pretty fairly destroyed in most places along the Coromandel Coast, a diminution in actual food supply has occurred, and with the proposed introduction of a salt excise into Malabar and Canara, the same results are not unlikely to result from similar causes. It must be remembered that every year vessels arrive in Malabar from the Persian Gulf and Arabain Coast bringing cargoes of well-salted fish. This untaxed article finds a ready sale, but fishermen on the Coromandel Coast have no such foreign opposition. Irrespective of this, in some of the Portuguese settlements in Western India, good salt may be purchased at two or three annas a maund, and I cannot quite think that fishermen in neighbouring British ports will be able to compete if they have to give some 1,500 per cent. more for the identical article.

67. The returns received respecting the present state of the sea-fisheries in such places as the salt-tax exists appear to show that it is only in South Canara, Malabar, and Travancore that the fish trade is doing well, and some very interesting facts have to come to light. In 1868 it was decided that the manufacture of salt in Malabar and Canara was not illegal. This was originally intended to be merely for home consumption, but, of course, advantage was at once taken to employ it for the curing of fish. How much goes to the hill ranges of the Wynaad is not known, but the rail from Beypore carries inland on an average upwards of 2,000 tons annually; it even imports it into Negapatam on the eastern from the western coast. If the salt tax is detrimental to the salt-fish trade, the raising the price of salt ought to be productive of good to the fish-curers of Malabar and Canara, and the Native State of Travancore, because they being able to collect salt free of duty, could undersell traders elsewhere who would have to purchase monopoly salt. This is just what has occurred. In Malabar, in five years ending 1863-64, the exports of salt-fish were Rs. 3,01,204; but in five years ending 1872-73, they had augmented to Rs. 9,08,495. In Travancore the exports in five years ending 1862-63 averaged 41,320½ cwts. annually; but in 1867-68 they had risen to 67,103 cwts. Out of 30 native officials who have reported on the state of the salt-fish trade, three give an increase (in one of these, however, it seems he is mistaken); nine give a stationary condition, and 18 a decrease. As this subject is fully considered in my report, I will only conclude this paragraph by observing that it seems pretty clear that the salt-tax has a most important bearing on the salt-fish trade of Malabar. Where the salt-tax, pure and simple, exists, the fish-curers, as a rule, cannot buy the salt, and must borrow money for the purpose at exorbitant rates, or give up the trade. Where salt is cheap, fish-curing flourishes and fishermen thrive, the two results, I believe, being solely connected with the incidence of the salt-tax.

How the salt-tax acts? Has it improved the prospect of fishermen where salt my be collected free of duty, and ruined it of those who have to buy monopoly salt?

xxxiii

68. *As to whose the fisheries are?* The sea ones appear to exclusively belong to Government, but in some places headmen, as will be explained, claim or obtain certain rights; even in one locality (Bandaṛ in the Kistna District) they are said to have monopolised these fisheries to themselves.

Sea fisheries Government property.

69. *Who are the fishermen?* These may be briefly divided into (1) those who fish in the sea, and (2) those who employ their time in the estuaries, backwaters, and such like places. Of the *first* division many are still to be found near large stations, as Madras, or along the Malabar Coast, but where their trade has languished, they seem to be emigrating or taking to a sea-faring life. Of the *second* division thus described by the Madras Revenue Board: "the estuaries, however, are fished by a distinct class, who have most probably no other support." It is very difficult to obtain any definite information as to who these persons really were; of course some are, as explained by the Collector of Madras, mere idlers, "no man willingly accepting work if he thinks he has a chance of catching a few fish; it leads to the waters being perpetually disturbed and to the fish being frightened away, as no party now intends himself to preserve the fish." Again, there appear to be others whose ancestors from immemorial ages have had something to do with fishing; that they never went to sea for the purpose appears clear, but what their exact work was I have not been able to ascertain. From various enquiries, however, it appears far from probable that, besides fishing in such localities, they were the sellers of all captured fish and the curers of those brought in from the sea; and it is far from probable that their present degraded or impoverished condition is due to the incidence of the salt-tax having destroyed their legitimate occupation. Sixty-three native officials have remarked upon the present state of the fishermen :—On the *Malabar Coast* out of eight answers, six assert they are largely increasing; one that they are stationary, due to a local cause; and one that they are decreasing at Cannanore, evidently owing to the taxation they are subject to. On the *Coromandel Coast*, out of 55 answers, 22 assert they have increased or slightly so, or on the whole; 15 that they continue stationary, and 18 that they have decreased : evidently they are in a much worse condition on the eastern than on the western side of the Presidency.

The fishermen: (1) those who ply their trade in the sea, but owing to decreased trade have to go as sailors, &c., elsewhere; and (2) the estuary fishermen, who very probably in old times were the sellers and curers of fish, but whose occupation has been ruined by the salt-tax.

70. In the Madras Presidency the *fishermen have customs* of a *pariarchal nature,* which are more strictly observed upon the east coast than on the west; but even on the former they appear to be falling, at least in places, into disuse. There seem to be three different grades of headmen; of the highest or *priestly* which number two at least on the east coast, one at Madras, the other in Cuddalore. The Madras one appears to hold religious sway over the fishermen up the coast, settles quarrels respecting the observances of caste and customs, and is the ultimate referee for disputes. He receives certain emoluments from the hereditary headman. When he travels, great respect is shown him, and apparently he has no expenses. The *hereditary* head-

Patriarchal customs among fishermen.

E

man holds sway over one or more villages; should he die without heirs, an election ensues amongst all the people under the jurisdiction, or else one is selected from amongst the fishermen by the local authorities. He is the representative of the people, settling domestic and caste disputes, but in most places has to earn his own livelihood by fishing. In one locality, however, in the Kistna district, it is stated (correctly?) that the headman has possessed himself of the sea-fisheries as his by right of his office. He presides at all bridal and religious ceremonies, taking the first place, being helped to betel-nut before every one else, whilst he generally receives some fee for marriages, but not for funerals. Thus it is observed in one talook that he is paid three annas for putting on the bridegroom's turban; in another district certain lands are said to be attached to his office. He appears in some places to be exempted from work; should a vessel strand or be in distress off his district, he orders out all the able-bodied men and gives help, or should Government direct men to be collected, it is his duty to do so. In some cases he levies small fines, but the uses to which they are put are not everywhere the same; sometimes he has them personally, in other localities half goes to the priestly headman, or else he has a percentage, and the remainder goes into a fund to be used at certain festivals. In some places a small fee is levied on nets or boats used in fishing, or a percentage of fish is paid him. Lastly, they have their *elective* headman, who is appointed by the votes of the people of a village or hamlet; he decides minor matters, determines the labor to be done, fixes the rate of wages, &c., and obtains some few fish for his pains. On the Western or Malabar Coast, in many towns, the fishermen have no headmen, and in some localities the Native Christians are largely employed in this trade. There are, however, to be found places wherein the same patriarchal customs prevail as on the Coromandel Coast. Thus, in South Canara there is a *priestly headman* who appears to have spiritual control over a large district. He frames rules in regard to caste matters, and imposes penalties on those who infringe them. The fishermen raise money amongst themselves from which he receives a portion of his emoluments. His office is hereditary. There are also *hereditary headmen* of one or more towns, and *elective headmen* of each village, these last only holding office during the will of the people.

71. *As to whether breeding fish or fry are destroyed?* It appears to be almost universally admitted that they are, more especially in estuaries and rivers, up which the young ascend to procure food and shelter. This is effected by nets, bamboo screens, or fixed engines of many descriptions, and are divisible into (1) those which are fixed and (2) those which are moveable.

Breeding fish and fry destroyed.

72. *Fixed engines* for capturing marine fishes are either stake nets parallel or nearly so to the shore, placed in bays or other favorable localities, or if in rivers or backwaters, they may partially or entirely impede the passage of fish. Those most adapted to fishing purposes have a fair-sized mesh and do not entirely impede any waterway; but when they have small meshes and go quite across a piece of water, permitting feeding fish to enter with the tide, but cutting off the return of the most

Fixed engines for capturing fish, some legitimate, others most unfair; and destruction sometimes wasteful.

minute, are strongly to be deprecated. I would draw especial attention to my remarks (at para. 91) on the young fish being sold in the markets of Madras: captured legitimately along with other and larger fish, and having become entangled in the nets, cannot be avoided, but the practice of destroying their fry for sale as food, does not commend itself to one's ideas of economy. One way is very much the same as that employed in irrigated fields to take the fresh-water fish; the owners of swamps cut open the banks of rivers so as to allow water to flow in with the full tide, and as soon as it is thought a fair amount of fish have entered, a cruise is placed in the opening, and as the water recedes, everything is captured. Were the interstices of these traps of fair size, so much injury would not be inflicted, but they capture the most minute fish, and these little ones are generally left to perish in the mud, not being considered worth the trouble of *collecting*. Large weirs are constructed on the same principle, entirely spanning creeks. There are many other forms of fixed engines, as the dip-net, so extensively used along the Malabar Coast; but these, provided the mesh is of fair size, can hardly do much injury to the fisheries.

73. *Moveable engines* for capturing fish are of diverse forms, from the casting net to the large seines. In some of the latter minute meshes are employed. As a rule, when the mesh is of a fair size, nets are a legitimate way of taking the fish, and should not be interfered with; the fishermen who employ them in the sea are the best judges of the size of the mesh they find most appropriate for their trade. Shoals of mackerel or sardines appear at times off Malabar, as already remarked upon, and the mesh suited for one species is not well adapted for taking the other.

<small>Moveable engines for capturing sea-fish.</small>

74. The replies from the Madras Presidency and personal investigations seem to show, (1) that the sea is rich in fish; (2) that the supply, as a rule, is locally sufficient during the fine weather; (3) that the fisheries are practically inexhaustible; (4) that on the east coast, except near large stations, these fisheries are hardly worked; (5) that on the west coast these fisheries are fairly fished with an increasing trade; (6) that it is not apathy on the part of the fisherman to which is due the non-capture of the fish, but the want of a market; (7) that when the local demand is not large, the fishermen only capture those easily taken in shore fish, and in doing so drive away the more valuable kinds; (8) that in some districts salt-earth may be collected for curing fish, in others it may not; (9) that, as a rule, Government or monopoly salt is only used for the larger and better class of fish; (10) that a large salt-fish trade exists in Malabar and South Canara, where the people have the right of collecting salt duty free; (11) that the fisheries are public property; (12) that the fishermen are in a very impoverished state on the Coromandel Coast, but comparatively well off on the western side of the Presidency; (13) that in places certain customs of a patriarchal nature exist amongst the fishermen; (14) that breeding fish and their fry are destroyed when they can be taken; (15) that fixed engines for the capture of marine and estuary fish exist everywhere, sometimes spanning whole piece of water, and there the fry are massacred in a wasteful manner.

<small>Conclusions.</small>

xxxvi

75. *What remedies have been proposed in order to increase the supply*
Remedies which have been *of sea-fish and make them more available for the*
alluded to. *inland market?* Eighteen European officials
reply to this question : one knows of no remedy: one would give systematic
encouragement and arrangement to the trade, but omits to mention what
such should be : four consider that a reduction in the monopoly price of salt
would effect this; in which nine others concur: two would be apathetic,
one of whom remarks that he does not think any special stimulus is
necessary : one would allow the use of salt earth for this purpose, provided
the salt revenue were not affected: nine give their opinions on a proposition
I submitted, to which I will allude in the next paragraph. Much of the
philanthropy so freely volunteered respecting the poor fishermen as regard
the fresh-water fisheries, appears to have vanished when the subject of
sea-fisheries is mentioned. It seems to have been considered by several
that only the local fish trade in their own districts requires looking after,
the more general question of an adequate inland supply being placed
quite in the background, and, lastly, fears as to the salt-tax being evaded,
&c., which will presently be adverted to. Respecting philanthropy here
as regards helping the sea-fishermen to capture more fish so as to augment
the food supply,* one official observes, " I cannot think much sympathy
ought to be wasted on the fishermen, for they are an independent,
careless and drunken set of men." Another remarks, " I do not see why
salt-fish should be exempted from a heavy duty, whilst other articles prepared with salt have to pay."

76. The proposition circulated was whether, if enclosures were
made near favorable localities away from large
Answers to a proposition as to towns where a sufficient demand for fresh fish
whether salt might not be sold might be anticipated, it might or might
cheap to fish-curers under certain
restrictions. not be advisable to sell salt inside to be used
for the *bonâ fide* salting of fish, at a rate just
sufficient to cover the expense? Nine have replied to this question : five consider such a proposition as unadvisable or impracticable on account of the
fear of its increasing the amount of salt smuggling : two suggest it might
be tried : one that it is practicable, but not advisable, and one that it is
both practicable and advisable.

77. *The Madras Revenue Board* (May 28th, 1869,) observe that
" the estuaries, however, are fished by a disObservations of the Revenue
Board on the fishermen, and a tinct class, who have, most probably, no
proposal to extend the excise other support; and if these waters are farmed
on salt to Malabar, and South by outsiders, the class of fishermen, who are
Canara.
already anything but well off, will suffer.
The Board would, therefore, suggest that either the headmen of these fishing
villages should hold the fisheries at a small royalty, and under some such
rules as are proposed by Dr. Day, or that protective rules should be
embodied in a brief Fishery Act. With regard to the proposed restriction
in the use of nets having meshes below a certain size, it appears to the

---

* Since this question has been brought to notice from the circular of last year, I
understand (June 6th, 1873,) that the Madras Revenue " Board have called upon all Collectors of Maritime districts to report what measures can be adopted to obviate the hardship
of fishermen being unable to obtain salt for fish-curing except at monopoly rates."

Board impossible to forbid (as proposed by Dr. Day*) the use of a mesh of less than four inches in circumference, for they believe that whiting of the average size and all the smaller mullet could pass through meshes of that size, to say nothing of prawns, immense numbers of which are daily consumed in Madras and elsewhere, and which are taken in fine-meshed casting nets." I have been furnished by the Revenue Board with a most interesting report by *Mr. Pennington,* (September 22nd, 1871,) respecting the salt-tax from which I make the following extracts: The Board "are of opinion that the Madras Salt Excise Act should be brought into operation in the districts of Malabar and South Canara *only* in the first instance; but that no time should be lost in putting it in force there. The inexpediency of treating these two districts exceptionally, as regards salt supply, has been so frequently admitted by Government that it is unnecessary to discuss the question† further: but since it became generally known (in 1868) that the manufacture of salt in Malabar and Canara is not illegal, there has been (in the former district at any rate) a very considerable increase in this private manufacture‡, and the question has become more pressing. This fact was duly brought to notice, but * * the Government declared the absolute right of the people to manufacture and use such salt as they required for their own purposes, and observed that it was only intended to prohibit the 'sale of salt to others than Government.' Government proceeded to say that any such interference with immemorial custom in Malabar and Canara would be tantamount to a breach of faith, and (in para. 23) remarked that 'the proper remedy is, undoubtedly, the enactment of the proposed excise law, which will tax *all* the salt produced.' * * Mr. Ballard (August 27th, 1868,) pointed out that it was 'carried on with little or no system, wherever there is a backwater or a salt-marsh, without even regular preparation of the ground, and without any intimation to the authorities either of place or time of manufacture or disposal of the article.' He added that it would be utterly impossible for him to effectually guard against the sale of this salt to other than officers of Government, unless a preventive establishment was allowed him on a scale which could never be sanctioned, and which would be 'corrupt and inquisitorial to a degree.' He speaks of the manufacture being

---

* I think my remarks have been a little misunderstood; provided it is considered I proposed no nets with a mesh of less than four inches in circumference be employed within tidal reach. In the report, I conclude, referred to (see Proceedings, Madras Government, April 30th, 1869), I observed as regards the propositions, exclusive of rent, "I have not entered into the question of the sea and coast-fisheries of the Madras Presidency, as I am not quite certain whether it was intended I should confine my investigations solely to the rivers." * * "The sea-fisheries appear to require as much stimulus as the freshwater ones do conservation." * * "Where means of salting fish exist, and the trade is not carried on, it appears that a stimulus might be advantageous." I then considered that the poverty of the sea-fishermen might be due to revenue regulations, and the incidence of the salt-tax, but I required more proof before stating such a belief.

† Surely the interest of "the class of fishermen" has been entirely lost sight of. Should the Board's views be carried out, the salt-fish trade of Malabar must rapidly sink to a similar state as that on the Coromandel Coast.

‡ See the returns of annual exports of salt-fish from Malabar, which tend to show that the increase in the manufacture of salt must be great at least for this purpose. If the Madras Government adhere to their decision, "that no interference with the customary manufacture of salt for private use," will be allowed, of course fish-curers may collect it for such. But this is difficult to be reconciled with their order that, "*all* salt manufactured will be taxed."

'spread over *thousands* of petty pans; these pans are mere pits, the work of a few moments with a 'mometty,' and concludes that the Excise Act will necessarily interfere with the rights of land-owners as much as the measures he had proposed, because, of course, the Collector would only grant licenses to such persons as would manufacture so largely that the duty leviable at their pans would be sufficient at least to pay for the expense of collecting it. All other manufacture must be summarily suppressed; and it is difficult to see how this should not include by far the greatest part of the 'thousands of petty pans' referred to by the Collector." "The present price of Madras salt at Coimbatore is about Rs. 2-8 a maund, and it probably could not be sold at Beypore for much less than Rs. 2-12; Bombay salt, which is popularly considered to be much *stronger*, can be imported and sold after payment of the full duty for about Rs. 2-12 to Rs. 3." "It is evident that the importers of English salt will soon find it to their advantage to land their cargoes at Calicut in preference to Madras, as the normal price of the salt there is likely to be nearly 50 per cent. higher than in the latter place."

78. The *Collector of Malabar* (March 6th, 1872,) addressed the Revenue Board on this subject, observing that—" It may perhaps be deemed almost superfluous to remark that if the privilege of manufacturing salt for home use is to remain unaffected by the Excise Act, the introduction of that Act is a matter of little moment. The existing law, as at present understood, allows no more, and every grain of salt made in the district is so made, under the pretence of being required for home consumption. It is because this privilege affords an effectual cover for making salt for *sale* that a change in the existing law is desirable. It must not be supposed that an excise of Re. 1-13-0 a maund is compatible with the manufacture of Malabar salt for the market. The people dislike it, asserting that it imparts a bitter and unpleasant savour to food, and brings on that common complaint in Malabar, the itch. Their aversion to it is so strong, that all who can afford the expense, prefer Bombay salt, at eight times the price, and if an excise of Re. 1-13-0 a maund were added to its present market value (about four annas) Malabar salt would be utterly unsaleable. If, therefore, Section 11 of the Excise Act be taken as legalizing the levy of the excise on all salt made, the immediate effect of the Act will be to put a stop to the manufacture. With regard to the petty pans now in use, it is impossible to give more than a guess at their number. It was estimated in 1858 that there were 1,800 acres of land in the district on which salt might be produced; but the fact that a vigorous attempt at manufacture was made on some of what were supposed the choicest spots, and that the resulting salt 'scarcely exceeded a sample in quantity,' shows that the estimate is not to be trusted. I think, however, that the extent of land on which the manufacture of salt is actually going on would not be overstated at 600 acres. The pans, which are sub-divided into partitions like a chess board, average about four feet square; and there may be from one to a score of them on an acre of ground, the number depending on the soil and the state of the weather. I think it would be below the mark to take the number of petty pans in use at 3,000. Of these, at least two-thirds are in the neighbourhood of Chow-

marginal note: Collector of Malabar on the salt excise.

ghat, a fact to be accounted for not only by the facilities afforded by the soil and nature of the place, but also because it was there that in 1867 a notoriously ineffectual attempt was made to stop the manufacture. The salt-makers are coolies who pay the owners of the land a share of the salt produced, generally fixed at one-third. At Chowghat, where three or four hundred persons are engaged in the manufacture under the very nose of the Salt Superintendent, you may see on the far side of the marsh several persons carrying umbrellas. These are the overseers before whom the land-owner's share is measured out. At the sight of a European they vanish, and the salt-makers hurry off in all directions, showing that they are still uneasy as to the lawfulness of their calling. Any few of them who may be intercepted declare that they don't know to whom the land belongs, and that they are only engaged in making a handful of salt to season their rice. In reality, after the clay filter in which the salt-earth is dissolved, the basins into which the brine runs, and the little clay-lined pans in which the brine is evaporated, are completed, the amount of salt which one man can make in a day is not less than a maund. An exact description of the way in which salt is produced in Malabar, with a sketch in illustration, is to be found on pages 91 and 92 of the "selections from the records of the Madras Government, No. XVI, Memorandum of Salt," published in 1855, where the Collector of Bellary describes the process of manufacture in that district. The only differences are that here the hollow mounds are only about 18 inches high; the reservoir is a hole in the ground lined with clay about the size of an ordinary basin, and the brine, when poured into the pans, takes less than a day to evaporate. I would respectfully urge that the new system should not be introduced until the burst of the monsoon lends us its aid by obliterating the works and putting a natural end to the manufacture for six full months. This would give the salt-makers time to become reconciled to the fact that they will next year have to pay the excise or seek other employment. The number of persons who have learned to look to the manufacture for their daily food is now very considerable, and they will not be driven away to other and less profitable occupations till many of them have been fined and imprisoned, and much discontent and misery will be the result. When the monsoon has stopped the manufacture, it will be comparatively easy to prevent its recommencing. I think the question of compensation may depend, to a great extent, on the sagacity of the people; but it will never be necessary to award compensation unless a license to manufacture is refused, and I should propose to grant a license to any land-holder applying for one. I believe no license will be applied for when the right to manufacture carries with it the obligation of paying the excise, but should licenses be applied for, the work of supervision will not be difficult, as, in most cases, the Taluq establishment is near. The question as to the necessary preventive force can be decided only after licenses are applied for. With regard to the withdrawal of Government from the Import trade, the question in no way depends on the introduction of the Excise Act. I have no doubt that merchants will be found to undertake the trade. The only obstacle to as free a supply of salt as of grain, or any other necessary of life, lies in the indispensable precaution taken at Bombay of making the exporters find security for the amount of duty, besides taking from them a deposit of four annas a maund. This suffices to prevent any but a wealthy merchant from engaging in the trade. I doubt not that many will seek to avoid the necessity of finding security by importing salt from Arabia. In that case the importer brings with him no reliable document to 'show the amount of his cargo, and it rests with the Customs House officials whether all the cargo shall be brought to account. To obviate the effects of this, it will be necessary to prescribe the importation of foreign salt, except salt imported from England to ports where there is a European officer, *viz.*, Calicut and Tellicherry. Bombay salt can be imported at each of the six ports where a salt store now exists. Salt so imported will, of course, be bonded, and the duty paid as it is taken from the warehouse. The rooms in our present sheds will serve as warehouses, and no doubt will readily be rented. I feel bound, in spite of all the correspondence that has already passed, again to urge the impolicy of throwing open the trade suddenly, instead of, in the first instance, promoting a growing private trade by an import duty so fixed as to admit a margin of profit to the private trader beyond what he gets by supplying Government. The present difference between the duty and selling price is only three annas a maund, whereas if the merchant sells the salt to us, and then again buys it from us wholesale, he gets seven annas a maund for putting it into the golah, and 5 per cent. discount for taking it out again, we bearing, in addition, the cost of warehousing and weighing. If we abolish the discount, and admit salt at a duty of Re. 1-6-0, a steady private trade will begin, and in a comparatively short time, we shall be able to withdraw in full confidence of the withdrawal being permanent. Otherwise the necessity of re-introducing a Government trade appears highly probable, and it is needless to point out that confidence in the security of private trade from Government competition having been once shaken, its establishment on a firm basis will be indefinitely postponed."

79. Having, for the purpose of continuing this investigation of the bearing of the salt tax on the sea fisheries, applied to the Revenue Board for a return of the annual export and import of salt-fish for fifteen years, I received the following answer from the *Secretary*: "I find that I cannot give you a return showing the annual exports and imports of salt-fish from each collectorate for the last fifteen years. Previous to 1867-68 our accounts show 'salted provisions' in the aggregate. (*) From 1867-68 they show 'provisions' without any particulars. We *might* be able to extract the information *for the last five years only* from the Collectors' monthly reports, if we have plenty of time and a few extra clerks, but I am not sure even of this." Unable to obtain this most necessary and important return from the Revenue Board, the Collectors were applied to, and their various answers are given in the Appendix, from which it appears that a longer period than five years might be furnished.

Application for returns of exports and imports of salt-fish, &c.

(*) Hams, bacon, and tongues are, I believe, salted to a small extent in Madras for home consumption, but I think only fish for export; consequently, one would imagine the returns spoken of, but not furnished, would give the required information in the gross, if not in detail.

80. *Sir Arthur Cotton*, K. C. S. I., observed (1867): "I should suppose that the injury to the coast-fisheries must be very great, now that seven of the principal rivers on the east coast, *viz.*, the Coleroon or Cauvery, the Pennair in South Arcot, the Pellaur, the Pennair in Nellore, the Kistna, the Godavery and the Mahanuddy are thus barred."

Injury to coast-fisheries occasioned by irrigation weirs.

81. *The Collector of South Canara* (March 5th, 1868,) remarked regarding the improvement of the sea-fisheries: "It may also be expected to be a source of indirect profit in increasing the expenditure of salt sold at monopoly price, and in swelling our exports, for fish in different shapes are already exported both by sea and inland. With other articles that yielded less than Rs. 1,000 a year in Customs dues, salt-fish were at the last settlement of the tariff placed on the free list. Consequently, Indian salters, who have to buy their salt at monopoly price, compete at a heavy disadvantage with salt-fish which is imported from countries where salt is free. If the import duty on salt-fish were re-imposed at such a rate as would bring the salt consumed in their preparation up to the monopoly price prevailing in this presidency, then Indian salters would, without undue protection, be placed on an equal footing with the foreigners, and it might be expected that, where fish were sufficiently plentiful, they would find it worth while to purchase more salt for their curing, and so we should find our returns in increased consumption of salt, and the small Customs receipts on imported salt-fish would be no proof that the Customs duty was not a wholesome and remunerative tax, though it was apparently a trivial one in direct proceeds; for it would be wholesome in that it restored for fair competition the equal balance disturbed by the presence of a monopoly price, remunerative in that it encouraged indirect profits in the more extended use of monopoly salt in Indian curing. In connection with my proposal, therefore, I would suggest a re-imposition of a Customs duty on salt-fish." Subsequently, he (H. S. Thomas, Esq.,) replied (March 31st, 1873,) that breeding and immature fish, " are diligently destroyed to the fullest extent possible." The salting of sea-fish would be increased, " as suggested in my piscicultural report of 1870, or as suggested by you, or by a combination of both plans." Respecting whether the proposition of forming enclosures in suitable places wherein fish could be salted and salt sold at reduced rates, he replies: "Yes, I think it would, and have been at much pains to get men to undertake the enterprise, and have written to the Madras Board of Revenue, suggesting in detail a trial with salt sold at Re. 1 a maund, or half the monopoly price, and have made choice of a fit man to give the question a fair trial." The letter adverted to (dated April 3rd, 1873,) is as follows: "In accordance with the suggestion contained in paragraph 7 of Dr. Day's letter, enclosed in the Proceedings of Government, dated 2nd September 1872, No. 1274, I invited tenders to cure fish with salt given at half the monopoly price. I first applied myself to the fishermen, and tried my best to induce them or their headmen to undertake the enterprise; but they are an improvident race, always poor, because always drinking their surplus cash, and the irrational nature of the objections raised showed very clearly

Opinion of the Collector of South Canara.

F

that it was vain to hope to persuade them to diverge ever so slightly from their few stereotyped habits of thought and life. I then addressed myself with no better success to the Mussalman dealers in imported salt-fish. I found also that Hindu merchants held back from connection with what they considered a demeaning trade. I then fell back on the Christian part of the community, whether Protestant or Roman Catholic, and the result, after some enquiry, was the enclosed petition and the promise of two more from others of the same class. The petitioner is prepared to accept the terms I offered, which were—

I.—That he shall receive as much Government salt as he wants at half price.

II.—That he shall pay cash, or give security for all he receives.

III.—That he shall receive it inside a guarded enclosure within which he is to use it, taking out the salted fish only, and under inspection.

IV.—That the enclosure and the guard are to be provided by Government, and the curing buildings within it to be his own erection on our site and liable to removal on requisition.

"The petitioner adds one stipulation, namely, that the use of salt-earth shall be prohibited, for he thinks he cannot pay the half price for salt, and thoroughly cure fish in competition against those who half-cure with earth-salt for which they pay nothing. The petitioner wished at first to have the monopoly of curing for a certain period, but I objected to this on the ground that it would practically give him the power of compelling the fishermen to sell their surplus fish to him at any price he chose to dictate. The arrangement, if confirmed, will stand therefore as follows : petitioner may perhaps secure the monopoly to himself for a time by fair dealing, that is, by offering the fishermen a fair market value for their surplus, by curing well, and by selling the cured fish at fairly remunerative rates, and yet not exorbitant prices. As soon as he fails in any of these points he will be liable to the competition on the same terms within a contiguous enclosure of any one else who may thereby be led to think he can do better. It will thus become his interest to treat both the fishermen and the public well. The petitioner has the advantage of being able to provide ample good security, and sale agency at Mercara, as long as his honesty in dealing is to the satisfaction of the head of the Basel Evangelical Mission, of whose congregation he is a member. Whether the arrangement will be profitable to Government is the other side of the question. Salt costs Government from the contractor eight annas a maund of Bombay salt, or five annas four pie of Goa salt. Sold to the petitioner at half the monopoly price, or Re. 1 a maund, the margin for Government will be eight annas or ten annas eight pie a maund, and out of this will have to be paid the guard and the carriage to the enclosure, and the erection of the enclosure. For the enclosure, a high bamboo railing will suffice, and that may be very cheaply run up. The carriage will be a mere trifle, for the site selected is Sultans' battery, so as to be inoffensive to the town and near the fishermen, and one or two boats a month will carry all that is wanted along the estuary between it and the salt kotaurs, and this may be borne either by the Government or the petitioner, as thought fit. The guard need not be extensive, no great amount of salt being exposed at a time, and the only thing to be guarded against being, the carrying of salt out past the gate by any of the employed laborers with a view to selling it at a profit. If the amount of fish cured be small, Government will scarcely be remunerated, but if it be large, there may not only be no loss, but a decided profit. Though the amount cured may perhaps be small at first, there is every reason to anticipate that a man who desires to live by it entirely, and has no lack of means, will endeavour to make it as extensive as he can. I think the enterprise should be given a trial, for even if it yield no profit to Government it will still benefit the fishermen, and the curers, and the inland public, which last will thus have an additional food not previously within their reach, and this without any injury to the salt revenue. It is quite possible, moreover, that the demands in the interior may be found sufficiently large to develope an extensive trade; and in such case, the Government will be the gainers by so much more of their salt being sold at a price, which, though not the monopoly price, is still a profitable price. I do not anticipate that the use of salt-fish in curries, can ever be so general as injuriously to affect the ordinary salt sales. If it does, the time for raising the price of salt to the curer will have arrived. The illicit salt question, which includes the use of salt-earth has been dealt with at length by my predecessor, and a decision thereon is needed, as much in the interests of the salt revenues generally, as in the decision of the present matter. I have the honor to request, therefore, that the Board will be pleased to instruct me how I may deal with the users of salt-earth, and what answer I am to make to the application enclosed."* In 1870 Mr. H. S. Thomas, in his report on the pisciculture of South Canara, made some interesting observations on the sea-fisheries of that district, considering them most fruitful and important. Its sea-board extends over 120 miles, besides possessing 404 miles of estuaries. Great quantities of sea-fish are carried inland by boats up the many rivers which exist, or being taken by coolies to places as far as 10 miles from the sea, are there exchanged for grain and condiments; these are sometimes fresh, but more frequently besmeared with salt-earth or a slight amount of Government salt. But the salt is either so lost in mud, or so expensive, that very little is used, and the fish are necessarily so partially cured, that in most cases putrefaction has set in before they are consumed. Thus, a wholesome food is made an unwholesome one, and the consumers demand for it is doubtless much reduced thereby. So long as the salt monopoly is maintained at its present repressive figure, so long will the sea-fisheries of India labor under this disadvantage; while the price of this necessary of life is so high, that the poor cannot afford to cansume with their daily food enough even for the maintenance of health, it is not to be expected that they can indulge in fish cured with this expensive luxury. As long also as fish thoroughly salted in countries where salt is untaxed, is imported into India free of customs duty, so long will the Indian curer be unable to compete therewith. In the sea and estuaries, as in the rivers, numbers of immature fish are destroyed in small meshed nets. But here also, as in rivers, the same difficulty exists in simultaneously permitting the capture of small sorts of fish, and protecting the young of larger

---

* When going through some parts of this district, sun-dried and salt-fish which had been imported from the Arabian Coast were seen being sold in the various bazars.

descriptions. Nets are hung continuously day and night across the larger half of tideways, and fish entering the rivers, and passing and re-passing with each tide are thus intercepted; but as the centre, which is deep and required for the shipping, is necessarily left open, the obstruction is not complete. There are marshes by the river side that are flooded by every high tide; the fry of the sea-fish frequenting the estuaries, are in the habit of coasting along the very edge of the rivers and running into all shallow places. When the tide rises over these marshes, the fry go in with it, probably finding more insect food amongst the swamp grass and on the freshly inundated land. But when they think to return with the ebbing tide, they are met by long lines of close wattle and fine leaf basket work, that allows the water to pass, but not the fry. At every tide in the daytime the fry are thus waylaid and then left high and dry, thickly strewn in long lines, whence they are carried away in basket loads. The mullet suffer much in this way. They are a desirable sea-fish, and the wholesale destruction of their fry in this manner should be prevented. It is pointed out that shrimps are the natural food of the larger fishes, and only a luxury to man, and in order to decrease such destruction on account of luxury, a tax on shrimp nets is proposed; the same remarks are almost equally applicable to prawns. The protection of the smaller sorts of fish and the fry of sea-fish, by prohibiting meshes less than three inches in circumference, is also calculated to increase the food, and consequently the numbers and size of the larger sea-fish, sardines (*Clupea Neohowii*), used to visit this coast in such numbers that they were the ordinary diet of convicts, and the cocoa-nut trees were manured with them. For nine years, however, they had disappeared, and only stray individuals were caught at intervals; but this year they have reappeared in shoals, and have for the last eight months been daily captured in such numbers and with such ease, that all the consumption of the large town of Mangalore and of the interior could not raise the price above three pies, say a farthing and a half a hundred. A canoe full has been caught in an hour and a half or two hours with a single casting net; and the rocks in the neighbourhood are completely covered with these small fish, besmeared with salt-mud and exposed to dry in the sun. As was to be expected, the porpoise, the firm fleshed seir, *Cybium*, and other large fish have followed in their wake, and sea-fish of all sorts have this year abounded. But rays, hammer headed, and other sharks are generally plentiful on this coast. There is no lack of the delicate pomfret, *Stromateus*, mullet, *Mugilidæ*, and Indian whiting, *Sillago*.

xlv

The following statement shows the value of salted-fish, shark-fins, maws and oil imported and exported from 1860-61 to 1872-73:—

OFFICIAL YEAR.	IMPORTS.				EXPORTS.			
	Fish, salted.	Shark-fins.	Fish-maws.	Fish-oil.	Fish, salted.	Shark-fins.	Fish-maws.	Fish-oil.
	Rs.	Rs.	Rs.	Rs.	Rs.	Rs.	Rs.	Rs.
1860-61	16,918	...	...	...	562	1,305	763	5,126
1861-62	16,443	...	...	45	1,034	1,685	270	4,363
1862-63	26,310	...	...	56	a 14,445	2,350	263	480
1863-64	24,156	435	37	1,720	a 10,574	4,702	2,498	2,920
1864-65	42,007	140	...	850	a 30,366	7,961	2,988	1,566
1865-66	19,061	170	62	...	a 8,750	6,666	1,161	...
		Fish-maws and shark-fins. d				Fish-maws and shark-fins. d		
1866-67	36,636	..	345	1,461	a 11,249	...	7,763	225
1867-68	b 45,734	...	...	565	a 8,757	...	5,928	62
1868-69	34,165	...	...	660	1,149	...	8,022	205
1869-70	34,285	...	1,372		c 20,533	...	8,477	135
		Shark-fins and fish-maws. e				Shark-fins and fish-maws. e		
1870-71	26,088	200	...	...	c 30,275	5,726	5,493	1,762
1871-72	21,478	985	100	...	c 28,452	5,886	1,450	1,177
1872-73	22,225	255	170	13	c 59,804	7,865	4,106	634

(a). Part of the imports re-exported to Malabar.
(b). The Import duty on salt-fish was removed by Act XVII of 1867.
(c). The work chiefly of Bombay fishermen, who catch off our coasts, and some of whom cure with smuggled salt.
(d). Up to February or March 1866.
(e). Up to September 1869, and October 1870.

82. *The Acting Collector of Malabar* (April 30th, 1873,) considers that, "breeding and immature fish are not destroyed in the tidal estuaries to any great extent. The salting of fish could be increased if the price of salt were lowered. The present selling price of Government salt is not prohibitive of its employment on fish-salting, but there can be no question, that in this district, wherever salt-earth is easily procurable (as at Tellicherry), fish-salting flourishes. This, of course, is not the sole reason for the flourishing state of the Tellicherry fishery, but it has a good deal to do with it." "Colombo is supplied to a very considerable extent with salt-fish from Tellicherry. The Singhalese frequently come up on fishing excursions in their peculiar outrigger boats, as far as the Sacrifice rock, which lies a few miles off the coast, midway between Tellicherry and Calicut." The proposition to form enclosures, wherein fish may be salted, and the salt sold at a reduced rate, "is not a practicable one in this district." The *Collector* of the district informed me that an old custom, law, or regulation exists; due to which the people are permitted to collect salt-earth, or use salt prepared by themselves from the sea

Opinion, &c., of the Collector of Malabar.

water, and that without any hindrance from the revenue authorities who look upon it as a prescriptive right. He also considered that if this were not permitted, the salting of fish would be decreased. I now propose offering a few remarks upon what I personally witnessed or information which I obtained at Calicut, except with reference to fish-oils, which will be treated subsequently in a paper by itself. Sharks are captured by means of baits, putrid beef or porpoise flesh being employed: large pieces are buried one or two days prior to their being used as baits. The hook is attached by a chain to the line whilst the fishing is carried on in the deep sea from 4 A. M. until sunset, but never at night time. The fishermen assert that the darker it is the better, and moonlight nights are good for this species of game; however, they prefer daylight. The oil sardines breed in June and July, subsequent to which period they commence to become fat, and from August to November are suited for the manufacture of fish-oil; at the commencement of the year they are moderately lean again. The natives of this coast assert that if eaten when very fat, and in large quantities, they are liable to occasion sickness, but subsequently, or at the commencement of the year, any amount may be consumed without poisonous results. A boatload of sardines computed to contain 14,000 at the time the fish are not fat enough to be used for the manufacture of oil, is worth from 14 to 15 annas, provided the fish are common. When oil is present the price rises to five or six rupees. If it is intended to use the bodies of these fish during the oil season, as well as extract the livers, the following course is pursued:—The heads are cut off, the abdomen opened, cleaned, prepared with salt, and then dried in the sun. If the monopoly salt is employed, the article is worth 12 annas a maund: if salt-earth 8 annas a maund; from this it must be apparent how small a margin the trade has, if it is wished to prepare the superior article, only 4 annas in $82\frac{2}{3}$ ℔ of salt-fish. Now, four annas would not purchase above 4 to $4\frac{1}{2}$ ℔ of good salt, irrespective of which, the profit has to be made. Most of the deep sea-fish appear to breed between the October monsoon, and the following March, for which purpose they come towards the land. Certainly some of the *Serrani*, *Pristipomas*, and *Therapons* breed early in the north-east monsoon, their young being common in Malabar in October. Fishes here are dried as well as salted; if the former, it is done on the sand, but not on mats; if the latter, they are laid on mats. The Indian mackerel is taken here and off Cochin in enormous quantities; from four to five thousand are considered a boatload valued at from Rs. 10 to 15, sold fresh; but they must be obtained cheaper than this by the fish-curers, because if salted with monopoly salt, their value is only about two rupees a thousand, whilst if earth-salt is employed, it is one rupee eight annas for the same quantity. Oil is not obtained from the livers of these fish. The following fish are mostly those which are salted in this district: roe-ball *Polynemus*, siluroid, catfishes, scir-fishes, and pomfret. At the town of *Cochin*, I observed that (1872) the number of standing dip nets, has considerably increased during the last ten years. Also along the backwater, that the weirs have augmented, but these last belong to the native States of Cochin or Travancore. Since the foregoing was written, the *Collector* of Malabar has obligingly furnished me with the following returns, to which I have added those prepared for me, whilst Mr. Ballard was Collector, and published in the ' Fishes of Malabar.'

xlvii

Annual imports of salt-fish into Malabar.

Year.	Cannanore. Value Rs.	Tellicherry. Value Rs.	Kalai. Value Rs.	Badagara. Value Rs.	Quilandi. Value Rs.	Calicut. Value Rs.	Beypore. Value Rs.	Tanur. Value Rs.	Ponani. Value Rs.	Chowghat. Value Rs.	Cochin. Value Rs.	Total. Value Rs.
1854-55	268	304	...	...	...	160	...	...	113	...	3,325	4,170
1855-56	620	291	...	55	...	264	...	...	70	...	5,687	7,887
1856-57	1,354	1,211	434	500	...	2,453	...	...	...	...	4,513	10,465
1857-58	107	199	210	269	6	393	...	...	110	...	7,335	8,623
1858-59	374	130	...	...	...	582	...	...	143	...	6,779	8,014
1859-60	330	169	...	119	...	897	...	...	10	...	960	2,487
1860-61	161	150	...	...	...	158	...	2	...	...	1,803	3,611
1861-62	8	...	...	...	...	36	...	...	1,359	...	3,709	5,432
1862-63	348	3,246	...	...	88	681	226	295	1,384	...	2,090	7,886
1863-64	747	921	...	...	368	498	450	...	1,210	...	891	5,573
1864-65	...	...	78	...	55	210	249	...	1,700	...	2,152	3,457
1865-66	500	...	...	...	532	...	...	424	...	...	...	4,749
1866-67	938	...	...	...	...	1,055	2,359	...	788	...	4,908	13,787
1867-68	1,256	5,665	...	...	112	3,179	4,304	315	1,085	...	12,669	25,673
1868-69	4,157	1,115	...	...	72	3,668	10,282	...	125	...	15,146	39,115
1869-70	4,074	708	...	...	103	425	12,398	...	680	...	4,902	23,697
1870-71	1,625	3,698	...	...	200	1,086	2,075	...	210	...	4,449	10,961
1871-72	40	...	...	...	300	2,000	1,949	...	...	...	765	11,442
1872-73	1,545	879	...	148	701	1,065	1,688	...	...	...	3,529	9,055

The subject of these imports is difficult of analysis, because large quantities of this is merely the fact that the fish has been shipped from one port and landed at another. Thus Beypore used to import none; now that little place is the terminus of the Madras Railway, which carries this exceedingly foul-smelling article far inland.

Annual exports of salt-fish from Malabar.

Year.	Cannanore. Value Rs.	Tellicherry. Value Rs.	Kalai. Value Rs.	Badagara. Value Rs.	Quilandi. Value Rs.	Calicut. Value Rs.	Beypore. Value Rs.	Tanur. Value Rs.	Ponani. Value Rs.	Chowghat. Value Rs.	Cochin. Value Rs.	Total. Value Rs.
1854-55	11	1,902	545	...	358	632	...	902	900	...	14,443	18,293
1855-56	96	1,581	58	54	162	118	...	...	...	...	25,618	27,687
1856-57	134	5,515	5,500	168	350	887	...	985	232	...	49,128	62,387
1857-58	1,784	5,680	44	430	120	215	...	247	...	...	33,474	42,033
1858-59	2,707	3,963	...	61	360	318	...	434	...	...	32,012	39,870
1859-60	1,083	6,744	3,352	68	1,742	840	...	12	31	...	32,374	46,246
1860-61	818	10,769	4,780	96	4,750	501	...	961	...	...	40,713	69,170
1861-62	4,942	11,647	3,773	2,472	2,983	1,002	...	...	...	...	34,411	62,140
1862-63	97	7,267	1,878	964	16,418	784	...	10	...	...	28,626	46,303
1863-64	960	14,694	2,622	4,230	16,263	688	...	23	...	...	48,060	87,315
1864-65	2,194	15,047	3,694	2,297	15,263	3,275	...	18	...	...	53,203	94,981
1865-66	111	1,941	129	268	15,278	2,765	605	...	...	...	63,895	84,882
1866-67	120	18,263	165	...	8,074	10,150	...	30	...	...	58,376	95,158
1867-68	3,035	20,119	715	906	3,497	215	...	1,226	...	...	90,695	1,19,708
1868-69	5,202	23,355	695	...	5,245	3,862	...	75	...	...	83,019	1,21,453
1869-70	43,406	88,661	3,680	8,691	21,694	4,900	...	...	...	...	97,012	2,18,573
1870-71	14,702	40,999	3,836	10,475	23,677	7,705	1,125	161	...	...	40,598	1,50,208
1871-72	6,951	47,489	...	17,513	43,469	4,270	20	1,500	...	...	51,346	1,72,488
1872-73	9,611	71,828	2,802	7,278	78,365	8,621	1,275	3,642	10,674	2,775	49,452	2,45,713

The foregoing figures show that in Malabar, where the use of salt-earth for fish-curing is permitted free of duty, and with a tax insisted on elsewhere, the fishermen are doing duty a most wonderfully improving trade commencing in the year 1863-64, when the salt tax was increased and made applicable to the Native States of Travancore and Cochin.

5 years ending 1858-59, exports to the value of ... Rs. 1,90,270
5   ,,   ,,   1863-64,   ,,   ,,   ...   ,, 3,01,204
4   ,,   ,,   1867-68, not received.
5   ,,   ,,   1872-73, exports   ...   ,, 9,08,495

Value of Fish Maws and Fins imported into Malabar.

Years.	Cannanore.		Tellicherry.		Kallai.		Badagara.		Cuilandi.		Calicut.		Beypore.		Tanur.		Ponani.		Chowghat.		Cochin.		Total.	
	Maws.	Fins.	Maws.	Fins.	Maws.	Fins.	Maws.	Fins.	Maws.	Fins.	Maws.	Fins.	Maws.	Fins.	Maws.	Fins.	Maws.	Fins.	Maws.	Fins.	Maws.	Fins.	Maws.	Fins.
1868-69	3,612	...	...	...	...	...	...	...	...	...	...	...	...	...	...	...	...	...	...	...	11,414	...	...	...
1869-70	...	15	...	...	...	...	...	...	...	...	...	...	...	...	...	...	...	...	...	...	16,351	780	...	...
1870-71	1,840	...	...	50	...	...	...	...	...	...	985	...	...	...	...	...	...	...	...	...	16,792	420	...	...
1871-72	360	...	...	50	...	...	...	...	...	...	500	100	...	...	...	...	...	...	...	...	19,300	...	...	...
1872-73	1,613	...	66	...	...	...	...	...	...	...	...	...	...	...	...	...	...	...	...	...	11,827	...	...	...

The imports of fish maws and shark-fins are only seen in the large shipping ports, or where there is a direct or coast traffic with Bombay or Europe, as some of this rough isinglass is apparently shipped to the latter part of the world, and all does not find its way as it did fifty years since to China.

Weight and Value of Fish Maws and Fins exported from Malabar.

Years	Cannanore.		Tellicherry.		Kallai.		Badagara.		Quilandi.		Calicut.		Beypore.		Tanur.		Ponani.		Chowghat.		Cochin.		Total.	
	Cwt.	Value Rs.	Cwt.	Value Rs.	Cwt.	Value Rs.	Cwt.	Value Rs.	Cwt.	Value Rs.	Cwt.	Value Rs.	Cwt.	Value Rs.	Cwt.	Value Rs.	Cwt.	Value Rs.	Cwt.	Value Rs.	Cwt.	Value Rs.	Cwt.	Value Rs.
1854-55	...	...	74	1,319	...	...	3	61	2	20	34	306	...	...	1	24	...	...	14	190	5	289	133	2,259
1855-56	1	7	65	418	...	...	...	...	...	...	47	395	...	...	1	27	...	...	7	259	12	444	123	1,430
1856-57	...	...	40	500	...	...	8	37	...	...	37	319	...	...	14	103	9	128	5	207	7	335	126	1,919
1857-58	...	...	35	362	...	...	...	...	...	...	61	465	...	...	14	64	1	8	2	23	7	363	130	1,366
1858-59	...	...	55	999	...	...	...	...	...	...	44	466	...	...	...	...	1	0	1	16	5	233	106	1,729
1859-60	...	...	137	1,265	...	...	5	21	...	...	90	464	...	...	...	...	...	...	1	70	20	773	253	2,578
1860-61	...	...	169	3,430	...	...	22	693	...	...	155	5,117	...	...	20	152	...	...	15	892	18	931	421	11,215
1861-62	...	...	233	3,278	...	...	...	...	3	23	36	1,767	...	...	...	...	26	1,394	1	19	11	658	310	7,049
1862-63	...	...	215	10,771	...	...	25	318	...	...	103	5,172	...	...	21	996	...	...	...	...	23	1,133	386	18,390
1863-64	...	...	220	11,057	...	...	1	30	21	1,045	146	7,292	...	...	18	984	...	...	...	...	12	530	418	20,958
1864-65																								
1865-66	No returns received.																							
1866-67																								
1867-68																								
1868-69	14	761	285	7,948	...	...	12	217	6	94	163	4,245	...	...	...	...	...	...	3	68	...	...	503	16,329
1869-70	4	157	611	22,075	...	...	9	294	...	...	215	7,292	...	...	...	...	...	...	23	540	...	...	862	30,328
1870-71	26	1,395	410	16,103	...	...	...	...	7	211	130	3,786	35	...	...	...	1	19	35	535	...	...	647	21,556
1871-72	14	882	569	17,670	...	...	93	2,278	63	1,186	401	9,409	...	...	...	...	...	...	30	395	9	70	1,180	32,093
1872-73	72	2,632	930	30,174	...	...	29	686	...	...	397	9,365	60	1,200	...	...	...	...	9	198	59	1,906	1,556	45,956

As it has been shown that the amount of fish captured for salting has largely increased of late years, it also appears that the quantity of fish-sounds and shark-fins have augmented. This latter productive industry being greatly dependant upon that of the fish curer, because capturing fish for their sounds or fins solely, would not be a profitable occupation.

83. The returns of the Madras Railway, for the last few years, showing the tonnage of salt-fish carried inland from the western coast, have been obligingly furnished by the Traffic Manager, and are as follows:—

Salt-fish carried by the railway inland.

1865	... ...	1,641 tons.
1866	... ...	1,465 ,,
1867	... ...	2,068 ,,
1868	... ...	2,565 ,,
1869	... ...	1,863 ,,
1870	... ...	2,604 ,,
1871	... ...	2,274 ,,

The *Acting Traffic Manager*, June 24th, 1873, declines furnishing a copy of the return for 1872, observing that "all statistics compiled by this Company are sent to the Consulting Engineer to Government, to where I beg to refer you for any you may require." Time has not admitted of this further reference.

84. The *British Resident of Travancore and Cochin* (April 17th, 1873) observes with reference to the first of these Native States—"The report from the Dewan of Travancore will follow when it has been received." I extract the following figures up to 1862-63 from my work on the "Fishes of Malabar." The figures given subsequent to that date are from three returns furnished me by the Resident on November 6th, 1868, and August 1873.

Return from Travancore.

Imports and Exports of Salt-fish into Travancore.

Years.	Imported.		Exported.	
	Quantity in cwt.	Value in rupees.	Quantity in cwt.	Value in rupees.
1853-54	958¼	1,008	26,982	28,401
1854-55	867¼	912	28,590	30,094
1855-56	1,113¾	1,171	42,251¼	44,474
1856-57	1,172¼	1,223	30,213½	31,755
1857-58	1,424¼	1,498	34,984	36,826
1858-59	1,440¼	1,516	27,730½	29,193
1859-60	2,167½	2,281	39,116¾	43,944
1860-61	2,962	3,290	49,665½	97,764
1861-62	4,012¼	4,266	35,923½	70,614
1862-63	4,395¼	4,886	54,166¾	1,08,339
1863-64	200½	Not given	54,167	1,08,340
1864-65	357	...	60,122	1,18,806
1865-66	...	...	70,505	1,42,264
1866-67	...	...	73,039¾	1,46,409
1867-68	...	...	77,682	1,58,378
1868-69	...	...	69,614	1,39,512
1869-70	...	...	59,325	1,19,312
1870-71	...	...	55,691	1,14,096
1871-72	...	...	67,336	1,34,895

The foregoing shows that the amount of salt-fish, imported from the British possessions into this Native State, was considerable up to 1862-63, since which period the trade has been ruined. Now it will be perceived that Travancore, instead of being a large importer, has become, on the contrary, an energetic exporter. The use of salt-earth used to be, and I conclude still is, allowed for salting fish. An export duty of about two annas a cwt. also appears to be in existence. The foregoing table affords some interesting subjects for analysis. Up to the year 1867-68 the *imports*—

in five years ending 1857-58 averaged cwt. 1,107 yearly,
   ,,   ,,   ,,   1862-63   ,,   ,,   2,995¼ ,,
   ,,   ,,   ,,   1867-68   ,,   ,,   151¼   ,,

The extinction of the import trade commenced after 1862-63, and if we turn to the duty on salt, we find it was raised two annas a maund in 1861-62. That year's supply of salt-fish having been exported from the British possessions to the Native State, the trade now languished for two years and then entirely ceased. If, on the other hand, we examine the export returns from Travancore, we shall see just the reverse. Up to the year 1867-68, the *exports*, much of which go to Ceylon—

in five years ending 1857-58 averaged cwt. 32,604 yearly,
   ,,   ,,   ,,   1862-63   ,,   ,,   41,320½ ,,
   ,,   ,,   ,,   1867-68   ,,   ,,   67,103   ,,
in four years ending 1871-72   ,,   ,,   62,991½ ,,

During the early portion of this period the price of black salt was Rs. 77 a garce, or about 10 annas a maund. About 1863-64 the British Government entered into a treaty which binds the Rajahs of Travancore and Cochin to raise the selling price of salt to what obtains in Madras, or upwards of 100 per cent.

The amount of salt sold has been as follows in Indian Maunds:—

1862-63	...	460,450	1867-68	...	4,70,760
1863-64	...	479,062	1868-69	...	4,92,000
1864-65	...	477,750	1869 70	...	4,90,920
1865-66	...	461,400	1870-71	...	4,97,040
1866-67	...	463,560	1871-72	...	4,87,260

The system of enclosures, observes *the Dewan* (July 1873), "would lead to great abuse. Salt would be purchased more for the purpose of smuggling than for curing fish, as the returns of the illicit trade are sure to be quicker and more profitable. Besides, as fishing in Travancore is carried on all along the seaboard, the numbers of enclosures will have to be multiplied to nearly the number of fishing villages. Else the fish will be spoiled before it arrives at the curing enclosure, and the cost of transit has to be weighed against the reduced price of salt. It may be well to note here that the fishermen themselves very seldom go to the expense of salting fish. There are export agents from Ceylon and other places who buy the fish raw or advance money for the harvest of the season. These people buy the salt for curing."

85. The *Dewan of Cochin* (April 3rd, 1873) replied that it is presumable that breeding and immature fish are to some extent destroyed, but there is no data to form a judgment. In some places up the backwaters, Cocculus Indicus and other drugs fatal to fish are employed for

Replies from the Dewan of Cochin.

their capture, and it is probable that by this means many fish are killed, besides those actually taken. As regards the trade in salt-fish it " is carried on to a considerable extent, and facilities already exist for indefinitely increasing it. Should any practical suggestions be offered on the subject, the sirkar will be prepared to consider them."

The exports of salt and dried fish from the two ports of Cochin have been as follows :—

	cwt.		cwt.
1863	...	1868 ...	1,687
1864	329	1869 ...	1,630
1865	112	1870 ...	2,644
1866	339	1871 ...	2,542
1867	348	1872 ...	1,617

The selling price of salt is now two rupees or 32 annas a maund, whilst in 10 years ending 1863-64, the average price had been slightly above 11 annas a maund.

The amount of salt sold has been as follows in Indian Maunds :—

1862-63	...	249,206	1867-68	...	81,639
1863-64	...	166,147	1868-69	...	83,199
1864-65	...	210,970	1869-70	...	96,349
1865-66	...	34,428	1870-71	...	90,674
1866-67	...	88,006	1871-72	...	89,105

Returns of the exports in salt-fish alone are not available, consequently the effects of this rise of price on the trade of salt-fish cannot be observed. However, along the Malabar Coast, there is another consideration, *viz.*, that a great quantity of the salted and cured fish is prepared from sardines, *Clupea Neohowii, C. & V.*, and fishes of the shark family. These sardines are very uncertain as to their appearance; in some years they arrive in enormous quantities, in other seasons they are almost entirely absent, whilst the presence of the sharks depends upon these shoals of small fish upon which they prey.

86. *The Officiating Collector of Madura* (December 18th, 1869) observed that "the seaboard in the district belongs to the Zemindars of Ramnad and Sivagungah, two permanently settled estates, and we have therefore nothing to do with the salt-water fisheries. The use of salt-earth is prohibited to the fish-curers of the district; the quality of the salted-fish is good generally, but the natives would prefer it of a superior description; knows of no way in which the trade in salting fish could be augmented. Subsequently (June 1st, 1873) the same officer furnished me with the following return :—

Opinion, &c., of the Collector of Madura.

liv

Return showing the Exports and Imports of Salt Fish, Fish Fins, and Fish Maws in the Ports of the Madura District from 1863-64 to 1872-73.

YEARS.	FISH MAWS.						FISH FINS.								SALT FISH.						
	Import.			Export.			Import.				Export.				Import.				Export.		
	Quantity.			Quantity.		Value.	Quantity.			Value.	Quantity.			Value.	Quantity.			Value.	Quantity.		Value.
	cwt.	qrs.	lbs.	cwt.	qrs. lbs.	Rs.	cwt.	qrs.	lbs.	Rs.	cwt.	qrs.	lbs.	Rs.	cwt.	qrs.	lbs.	Rs.	cwt.	qrs.	lbs.	Rs.
1863-64	...	...	...	...	... ...	...	2	2	14	52	9	2	19	213	...	...	...	...	241	...	...	1,111
1864-65	...	...	...	2	... ...	109	39	...	...	780	14	...	...	290	...	...	...	...	311	2	...	1,090
1865-66	...	...	...	...	... ...	...	40	3	...	935	9	1	...	185	92	1	14	547	333	1	14	1,485
1866-67	...	...	...	...	... ...	...	21	3	14	438	27	1	...	545	236	1	...	992	243	2	...	1,213
1867-68	...	...	...	...	... ...	...	28	...	...	560	23	1	18	468	250	...	...	548	168	1	...	864
1868-69	...	...	...	...	... ...	...	31	...	14	622	21	...	...	430	325	...	...	1,085	761	...	...	2,667
1869-70	2	...	...	...	... ...	...	11	...	...	230	36	3	...	751	278	...	...	694	2,109	2	...	7,862
1870-71	...	...	21	...	... ...	...	34	2	14	608	20	...	...	380	243	...	...	723	1,799	...	...	6,064
1871-72	...	...	15½	...	... ...	7	53	2	15	1,054	38	2	...	830	1,057	...	...	3,070	1,094	...	...	4,926
1872-73	9	1	17½	...	... ...	471	26	...	9	500	23	1	14	467	1,984	2	14	5,137	231	1	...	997

87. *The Acting Collector of Tinnevelly* (November 5th, 1869) reported that, as a rule, the fishermen of the coast are a very miserable lot of people and excessively poor; the way in which they now work is by a system of advances from their 'Chummaties' or headmen, a few of whom reside in each village, and supply nets, lines, boats, &c., for the use of which a certain share (one-third) of all the fish caught is taken by the Chummaty. Sea-fishing is the daily employment of a large number of the inhabitants living on the sea coast; these men have certain contracts to supply fish with headmen of the Paraver (fisher) caste, distinct from the Chummaties. Subsequently (February 22nd, 1870) the Acting Collector continued that he was averse to letting out the right of fishing in the rivers and backwaters of the coast by public auction "for the reason that it would give the renter a certain power over a poor class who now earn their living as fishers of these creeks and backwaters." "The effect of giving any one person the right of fishing the mouths of the Tambrapurni and Vypar, and the creeks and backwaters along the coast, would be certain injury to the poor fishermen who with their forefathers have been accustomed to gain their living from these waters from time immemorial." "I beg to record my decided opinion that any interference with sea-fishing is quite uncalled for. Sea-fishing is the daily employment of a large number of the inhabitants living on the sea coast. Any restriction, therefore, would affect a number of persons who have no other means of earning a livelihood. These men have certain contracts to supply fish with headmen of the Paraver (fisher) caste (distinct from the Chummaties.) Fish are dried in the sun, salted, and all those not sent inland for consumption are exported to Ceylon. The supply of fish is large and not decreasing, and while such is the case, surely Government interference is unnecessary, and the employment upon which so large a number of people are engaged should not be restricted." It is illegal to gather the spontaneous or earth salts, and therefore illegal to use it for fish-curing, but no doubt they do, as the high price of salt prohibits their using it in the quantities necessary to cure the amount of fish caught on the coast: it answers very well for this purpose. "This salted-fish is much affected by the 'Paraver' caste on the coast. I never tasted it. The smell was quite enough." The upper classes would prefer a superior article, "but the lower are, I fancy, well satisfied with the article they now obtain. It has one great recommendation in their eyes; it is very cheap." The way to increase the trade in salting fish would be "by reducing the monopoly price of salt."

Replies from the Collector of Tinnevelly.

88. *The Collector of Tanjore* (January 21st, 1873) replied that all fish are destroyed with the exception of very small ones which cannot be entrapped in the net. The salting of sea-fish might be increased by reducing the price of salt. The forming of large enclosures near favorable localities wherein fish could be salted, and the salt provided at reduced rates, would be "unadvisable, as it would be obviously impossible to see that the salt sold was actually employed in salting fish alone, and the salt revenue is far too important a matter to be interfered

Opinion of European officials of Tanjore.

with on considerations of promoting particular objects, such as trade in salt-fish, and the like. The present salt revenue of Tanjore alone is fourteen lacs of rupees, and a revenue more easily collected, and at the same time attended with so little actual pressure or popular murmurs, can hardly be imagined. Any measure, therefore, calculated to affect this important branch of the public revenue should not in my opinion be adopted unless after the most careful consideration, and *after* it has been shown to be urgently needed. The answers of some of the native officials evince an incorrect comprehension in some respects of the questions put to them, but they nevertheless give all required information in respect of the mode in which the salting of sea-fish is carried on, and the present state, as compared with the past, of the trade in salt-fish, as well as the fishermen classes. Taken on the whole, these answers do not show that the trade of salting sea-fish is materially affected by the price of salt." Subsequently (March 14th, 1873) he continued—" I agree with the Acting Sub-Collector in considering the proposition in paragraph 7 of Dr. Day's letter (respecting forming enclosures for salting fish in, and where the salt would be disposed of below the monopoly price) unadvisable. As pointed out by Mr. Happell, it would be obviously impossible to see that the salt sold was actually employed in salting-fish alone, and the salt revenue is far too important a matter to be interfered with on considerations of promoting particular objects, such as trade in salt-fish and the like." "The answers of some of the native officials evince an incorrect comprehension in some respects of the questions put to them, but they nevertheless give all required information, &c." *The Deputy Collector in charge of the Salt Department* answers that breeding and immature fish are destroyed in tidal estuaries both by mankind and larger fish which feed upon them; the latter mode of destruction cannot well be obviated, being a principle of life and nature's diction; but the former is very objectionable on account of the small size of the mesh of nets employed, and might be put a stop to by legislation. The salting of sea-fish might be increased by removing the restrictions which at present exist regarding the earth—and spontaneously produced salt, but which would detrimentally affect the salt revenue to a certain extent. As to whether it would be practicable to erect large enclosures in suitable places wherein fish might be salted, and the salt sold at a reduced price, he continues, yes, it is practicable in this district on the great salt swamp near Point Calamere, provided the restrictions referred to in regard to using salt-earth were removed inside the particular locale or inclosure. *The Sub-Collector* replied that the meshes of the nets are very small, and fishing is carried on all the year round. It is, therefore, a matter of course that both breeding and immature fish are destroyed in large quantities, but no apparent diminution is observable. "In this district fishermen's hamlets are generally situated at a few miles distance from large towns or numerous populous villages, so that, generally speaking, the greater part of the fish caught in the sea are consumed fresh, and only small quantities salted." Respecting whether forming large enclosures in suitable places and selling salt inside at reduced prices for curing fish is advisable, he replies " it would be practicable enough, but certainly not advisable, as it would be nearly impossible to see that the salt sold was actually employed in salting fish alone."

89. The *Collector of Trichinopoly* (October 12th, 1869) replied that the fish-curers "are not allowed to use salt-earth, taxed or untaxed; but owing to the high price of Government salt, especially in the inland talooks, I have no doubt that this earth-salt is surreptitiously used pretty largely. On the coast I believe that all the salting, much of this so-called cured-fish gets, is being buried in the sea sand and thus getting slightly briny. It is illegal to manufacture earth-salt; therefore, none ought to be produced for curing." The salt-fish " is liable to give pain in the bowels, and itch to those not used to its consumption. As the whole subject of earth-salt is kept dark, being illegal, I cannot say how they do or might employ it when curing fish." The salted-fish " generally stinks pretty offensively." As to whether the natives would prefer a superior article were such obtainable, he replies, " no doubt they would. A few old hands might be so wedded to former usage as to prefer the present nauseous article, in the same way as there are still found elderly people in England, who object to the use of hip and sponge baths." The purchases of Government salt for fish-curing are probably but trifling. " If the monopoly price of salt were low, I fancy the trade and quality of the commodity itself would increase, but if really good salt-fish were made at the present price of Government salt, the product would be beyond the reach of its chief consumers. No doubt good fish could be sold to almost any amount, and the supply is nothing equal to the demand."

Opinion of the Collector of Trichinopoly.

90. *The Collector of South Arcot* (May 28th, 1869) replied that " there is no express rule either prohibiting the use of salt-earth for curing salt-fish or allowing its use, and I do not believe that salt-earth is generally used, since the extracting of salt from the earth* is a task of labor and somewhat expensive, besides being a penal offence. There is very little salt-earth available near the coast, where the curing of fish is mostly carried on, and sea-salt is more readily procurable. Where salt-earth is found in any quantity, there is the least quantity of fish to be had. The salt-earth must first be purified before using for curing purposes." Natives would prefer a superior quality of salt-fish to that now sold were such obtainable. " In some places there is a ready market for fresh fish for salting, and where the curing is carried on, Government salt is of course purchased for the purpose; the extent I cannot tell, but it will be regulated by the trade, and I am not of opinion from my enquiries that the trade is anything very great in my district." " Doubtless the salt-fish trade might be increased and improved under systematic encouragement and arrangement, if such could be applied."

Replies from European officials of South Arcot.

---

* If preparing this substance is a *penal offence*, one would have imagined that a *rule prohibiting its use* is in existence.

Statement shewing the Export and Import of the undermentioned articles by Sea and Land from and into the Ports of South Arcot during the last ten years.

YEARS	BY SEA							BY LAND								
	Exports							Exports				Imports				
	Salt fish		Fish fins		Fish maws			Salt fish		Salt fins		Fish fins		Fish maws		
	Quantity	Value	Quantity	Value	Quantity	Value		Quantity	Value	Quantity	Value	Quantity	Value	Quantity	Value	
	Cwts. lbs.	Rs.	Cwts. lbs.	Rs.	Cwt. lbs.	Rs		Cwts. lbs.	Rs.	Cwts. lbs.	Rs.	Cwts. lbs.	Rs.	Cwts. lbs.	Rs.	
1863-64	... ...	...	... ...	...	... ...	...		511 101	1,490	416 96	991	... ...	...	... ...	...	
1864-65	... ...	...	2 ...	100	... ...	...		392 78½	1,678	1,107 84½	2,337	... ...	...	... ...	131	
1865-66	... ...	...	... 35	66	... ...	...		964 107	3,276	1,129 101	2,390	... ...	...	... ...	23	
1866-67	... ...	...	1 ...	...	... ...	...		783 34½	2,468	1,474 59½	2,993	... ...	...	... ...	79	
1867-68	... ...	...	... ...	18	... ...	...		129 78½	1,057	237 91	478	... ...	...	... ...	...	
1868-69	... ...	...	6 19	68	... ...	...		... ...	...	... ...	...	... ...	6	... ...	...	
1869-70	... ...	...	8 ...	241	... ...	...		... ...	...	... ...	...	... ...	...	... ...	...	
1870-71	118 ...	434	11 56	361	45 ...	11		... ...	...	... ...	...	... ...	...	... ...	14	
1871-72	... ...	...	13 28	326	— ...	...		... ...	...	... ...	...	... ...	...	... ...	...	
1872-73	40 56	90	14 20	20	56 ...	13		... ...	...	... ...	...	... ...	...	... ...	...	
TOTAL	158 56	524	55 ...	1,300	101 ...	24		2,721 80	9,949	4,626 94½	9,959	... ...	6	... ...	245	

The quantity in these cases cannot be ascertained.

The Superintendent of Sea Customs, Porto Novo, reports that the records in his office up to 1867-68 having been destroyed, the required information cannot be given. The exports entered in this statement for those years are therefore those of Cuddalore only.

The Assistant Collector (December 2nd, 1872) observed that fishermen destroy everything in the shape of a fish that comes into their power. Does not think that the salting of fish could be increased in the Cuddalore district, inasmuch as the demand of the local market can entirely consume all the attainable supply. *The Acting Sub-Collector of South Arcot* (January 15th, 1873) reported that fishes of all size are indiscriminately taken, but that as regards forming large enclosures in favorable localities, wherein fish could be salted, and the salt provided at reduced rates, if the suggestion were carried into effect, the salting of fish could be increased except in the Cuddalore taluq, where the take of fish barely meets the demand. " I do not think the proposition is advisable, nor do I see why salt-fish should be exempted from a heavy duty, whilst other articles prepared with salt have to pay; capital and enterprise are required to stimulate a waning trade, and curing fish is not the sort of business men who are making money would be likely to embark in. The scheme would be practicable only if the enclosures were within the police guards at the salt pans; nobody could be trusted to take away the salt; one such enclosure might perhaps be tried under the immediate supervision on the Salt Deputy Collector. It would cost little or nothing to try the experiment, and six months' trial would be worth many sheets of opinion." *The Collector* (March 1st, 1873) remarked that breeding and immature fish are destroyed in the tidal estuaries. Fishermen are not provident as a class, and literally all is fish that comes to their net. What they do not use they leave where they leave their nets. The salting of sea-fish might possibly be increased if the price of salt were reduced for the purpose, but not probably to any great extent : as a rule, the demand of the local market absorbs the supply afforded by the local take, and even the advantage offered by the reduced price of salt would hardly stimulate fishermen to employ the extra exertion necessary to increase the latter. The making of large enclosures near favourable localities inside which fish could be salted, and where salt might be sold at reduced rates, would be practicable, but with the employment of an additional preventive force to check smuggling. It would hardly be advisable as being unlikely to produce any measure of public good for the reasons already given. *The Assistant Collector* (Dec. 2nd, 1872) replies—fishermen destroy everything in the shape of fish which comes into their power. Does not think the salting of fish could be increased in his (the *Cuddalore*) district, inasmuch as the demand of the local market can entirely consume all the attainable supply. At the end of the Sub-Collector's letter referred to, follow the ensuing answers from native officials, but, as they are unsigned and not docketed, they appear to come from the same official. The cost of salt is Rs. 2 a maund, Government salt is employed for curing fish, but earth salt is not much used. The practice of salting fish has decreased of late years in all places except Chellumbrum. The fishermen have all the privileges they formerly possessed. There are headmen "Nattamakaran" amongst the fishing castes; the office is hereditary. His dutes are to regulate the ceremonies attendant on marriages as well as funerals, and to dispose of such disputes as arise among his caste people. When a fine is levied a small share goes to him, and the remainder merges into a fund reserved for the performance of certain festivals among the caste. In marriages he is paid a fee ranging from 4 annas to Rs. 2. No one claims any right respecting the sea-fisheries. The men use nets and boats of their own. Except in the Cuddalore talook, there is an increase in the number of sea-fishermen.

91. The *Acting Collector* of *Madras* (June 23rd, 1870) replied—

Opinions of European officials of Madras.
"fish is not cured by salt-earth in the interior of the district. In the coast villages the practice does obtain to a certain extent, and no tax is levied from the people." There are no restrictions against its use. "Salt-earth is used in its natural state for curing fish in the district. I am not aware of any process of purification being gone through before the earth is used for curing fish." The quality of the salt-fish is good, and it is largely consumed. A better quality, however, would be preferred provided it was not too expensive. Government salt is purchased to a large extent in Pulicat, Coromandel, Sadras, Covelong, and other villages on the sea coast for salting fish, and large quantities are cured for the purpose of trade. "I do not think any special stimulus on the part of Government is necessary to augment the trade of salting fish." The next day (June 24th, 1870) he continued,—"The salt-water fisheries in this district have never been rented out. I find, on referring to the records of the office, that the question of letting by public auction the right of fishing in the various backwaters in the district formed the subject of a correspondence between the Government and the Collector in June last, in consequence of a reference made by Mr. Fraser, the Executive Engineer, regarding the right of fishing in the backwaters at Ennore and Kattivakum. On that occasion Mr. Fraser advocated the disposal by public auction of the right of fishing in all the backwaters of the district. I enclose a copy of Mr. Jones' letter to Government, and beg to express my concurrence in the opinion therein expressed. The salt-water fishery appears to be at present monopolized by a few well-to-do fishermen, who employ the poorer members of the same caste to catch fish for them on daily wages. Under the existing system fishing in the backwaters is carried on most recklessly, and to an unlimited extent, and in consequence there is a wholesale destruction of the young fish. The only way of putting a stop to this, is renting out the right of fishing under conditions similar to those recommended in the case of fresh water fisheries." The *Collector's* letter (June 14th, 1869) referred to contains the following :—" I must say I do not see why the fisheries of all the backwaters and canals should not be made to produce a revenue which would go a long way in supplementing the very inadequate local funds of this district." "I have no hesitation in recommending the renting out the fisheries of all backwaters, more especially as, so far as I can see, the fishery now being in common leads to a great deal of idleness, no man willingly accepting work if he thinks he has a chance of catching a few fish; it leads also to the waters being perpetually disturbed, and to the fish being frightened away, as no party now intends himself to preserve the fish." [The several fish and meat markets in the town of Madras are worth a visit by any one desirous of personally ascertaining the comparative supply of the former article to the demands for it. This cannot, however, be carried out without some personal

discomfort; the buildings are very inferior though they are better kept than in Calcutta, but cannot be compared with the well ordered Bombay markets. There are different periods at which these places should be examined. The supply is most abundant when the sea is smooth, and the fishermen able to ply his trade. At such periods (except under very exceptional circumstances), I do not think the supply will be found to exceed the demand. I admit that it is easy for those who are well off, and possess sufficient means to purchase or procure fish at almost any time, to consider the supply cannot be insufficient as they never have to go without. But this is not the question, they are as units to the general population, and it is concerning the latter I mostly allude to. For every single enquirer after butcher's meat, 30 to 40 persons will be seen looking after fish, and this being so when the supply is most abundant, the difference is still more apparent at times when such is comparatively inadequate to the demand. Every one appears to be elbowing and pushing their neighbours as to which can first get access to the stalls, whilst the vendors are careful not to lose the chance of enhancing the price to the utmost limits they can obtain. When good sea-fish are obtainable, small fresh water species, unless they are air-breathers, are not much appreciated, but even in the markets of Madras a vendor of small estuary and sea-fishes will almost invariably be found, the sea-supply being generally insufficient, the late comers have to purchase his inferior commodity, another reason also doubtless being that they are cheap. I closely examined some of their baskets, in fact I purchased many as they were brought, and in them found the following:—little specimens of *Ambassis*; the small fry often not one inch in length, of fishes which when adult attain 20℔s., 30℔s. or more in weight, as *Serranus lanceolatus, Pristipoma hasta*; or those which also are the young of good table fish which grow to 10 ℔s. or 15 ℔s. when adults, as *Mesoprion fulviflamma, M. Johnii* and *M. chirtah*, but which averaged 30 to one ounce in weight, and a large number are left in spirit in the Madras museum for the inspection of those who are sceptical on these points. The chief localities from which they are obtained are the Adyar and Coum rivers, where indiscriminate fishing is permitted, with nets having any size of mesh. I was informed that about 400 persons were employed netting in these two pieces of water, which communicate with one another and with the sea, and up which small fry ascend for the purpose of obtaining security, rest and food. Exclusive of crustacea, these immature fish form the majority of those the prohibition of whose capture is deprecated. These observations are not the result of hearsay evidence, but of personal observation and I think that those who deny such destruction takes place, would do well to personally inspect the quality of the fish which are there slaughtered. Salt and dried fish are also sold in these markets, but are not in such request as the fresh.]

92. The *Acting Collector of Nellore* (October 26th, 1869) replied that there is no tax levied on salt-earth, the quality of which is inferior, whilst the fish-curers may purify it if they like. Rich and poor alike consume the salt-fish which is of an inferior description, and they would probably prefer it of a better quality. Government salt is purchased to some extent for curing fish. The trade in salt-fish is purely local, and is not likely to be augmented. *The Collector of Nellore* (April 9th, 1873) observed, that the

Replies of Nellore officials.

price of salt sold wholesale is Rs. 2 a maund, whilst the bazar retail rate is somewhat more. Government salt is used in curing fish where there is a market in which a proper price can be obtained. Thus the Pulicat Lake fishermen use Government salt for the fish which they send to Madras, and the fishermen in the north of the district use Government salt for the fish which they sell to traders and contractors from the Nizam's dominions and other inland territory. Small fish are dried in the sun. Earth salt and sea water are used considerably, but fish thus cured is a very inferior article, and does not find its way far inland. Only a conjecture can be made as to whether the practice of salting fish has increased or decreased of late years. Some officials say that the enhanced price of salt caused a decrease; others say that there has been an increased demand for salt-fish of late. The fishermen of this coast have parted with no privileges. Each village or Polliem on the coast has its hereditary headman, who was originally selected by the fishermen themselves. His only emolument is the exemption from personal labor attached to his office. His duties are to attend festivities in the village, to procure labor, &c. There are hereditary headman who exercise priestly authority all along the coast. They settle disputes and receive fees on occasions of marriages, &c. There are two of these priests in Triharikota, but they are themselves subordinate to a headman in Madras, to whom they pay a portion of their fees. This Madras headman occasionally visits the villages on the coast, when he travels in a palanquin borne by the fishermen. But I am not aware that there is any trace in this district of any system of advances made by the headmen to the poorer fishermen. No one claims any rights as regards sea-fisheries, but different villages are extremely tenacious of particular local limits, within which they claim exclusive rights of fishery. The fishermen purchase their own boats and nets, which are often pledged to the contractor or soucar who advanced the purchase money. The number of fishermen has remained stationary. Breeding and immature fish are destroyed in tidal estuaries to a considerable extent. From personal observation I know that the fishermen do not return into the water the innumerable small fry that are caught along with the large fish in the draw-nets, but I do not think it is worth while to attempt to change this practice, because any interference on our part would be vexatious and in all probability useless, and also because the quantity of fry thus destroyed is trifling in comparison with that destroyed by birds and fish, and occasionally to a very great extent by the drying up of estuaries in the hot weather. The salting of sea-fish might be increased and would certainly be improved by cheaper salt. In the north of the district, those only can afford to purchase salt who have obtained advances from traders who carry fish inland or ship it to Madras. The poorer fishermen would be glad to salt fish also if Government salt were within their means. The proposal in paragraph 7 I have carefully considered. In so far as those who now use no salt at all would then use the cheap salt, the scheme would not lessen our salt revenue to any great extent. Two of my tehsildars suggest that the salt which is rejected at the pans and destroyed, might be utilized for this purpose. This salt is rejected because it crumbles and does not remain in a crystallized form, but it is quite good enough for salting fish. The sale of this condemned salt for this purpose would not interfere with our monopoly prices, nor would it diminish our stores of good salt, and it would benefit the manufacturing ryots, whose salt is often condemned because of the results of bad weather, and not through any carelessness of theirs. The cost price of good salt, which is paid to the ryots, is only Rs. 10 per garce. The 'rates just remunerative' suggested by Dr. Day would be very much below monopoly price. The great obstacle to this scheme is the difficulty and expense of preventing smuggling. The preventive service on which the police are now employed at the pans, is most unpopular in the force, and it is to be feared, that in spite of all our watchfulness it is not always effective. If it is believed that this experiment may establish an additional source of food for the population, it may be worth while to undertake it; but it will certainly increase the difficulties of the Superintendent of Police, and may impair our salt revenue. I cannot think that much sympathy ought to be wasted on the fishermen, for they are an independent, careless, and drunken class of men, and their gains are not dependent upon fluctuations of the season to such an extent as are those of the agricultural ryots. Subsequently (June 14th, 1873) he observed that "there have been no exports or imports of fish by sea from or to this district during the last ten years. There is a large fish trade with Madras by means of the Pulicat Lake and East Coast Canal, but we have no returns of this trade." *The Deputy Collector in charge of the Salt Department* (November 23rd, 1872) replies that the number of native officials consulted on the coast (see returns from native officials) is 18, extending along a length of seaboard of some 170 miles. There is no doubt that breeding and immature fish are destroyed in the tidal estuaries of this district and in the Pulicat Lake, which is also tidal, possessing as it does one or two perennial outlets to the sea, but no reliable estimate, even approximately, can be formed of the extent of such destruction. From enquiries I have made along the length of the coast above referred to, the returns varied as to the quantities destroyed by man. Besides these there are no doubt other means of destruction such as the eating up and consumption of small and immature fish by the larger species, and full grown ones, by sea gulls and other birds, &c., a large number also die and putrify annually in the swamps and other shoals in the hot weather when the bars close, and many of the smaller estuaries and creeks cease to be tidal, and become perfectly dry. No idea can be given of the quantities lost by these means, but it is evident that the destruction one way and the other is by no means little. Nevertheless it may, I think, be safely assumed that no deficiency of fish exists, for not only do local wants seem to be adequately met, but large quantities are salted and sent from certain places on the coast by bandy into the interior, and from the shores of the Pulicat lake and canal by boat into Madras. The only mode that occurs to me by which the salting of sea-fish could be increased in this district would be, I think, to make the use of salt-earth and sea-water in curing fish penal. The practice of using salt-earth and sea-water for the purpose is now very generally and most freely carried on, and were it put a stop to under certain penalties similar to those prescribed for the illicit manufacture and smuggling of salt, the immediate result would no doubt be to increase the use of salt. The quality of the salted fish would thus be much improved. It would be more tasty and wholesome, and fetch a better price, and our salt revenue would, doubtless, improve by creating a demand for salt for a purpose to which it is now very seldom applied; such a measure would, however, in my opinion, have an injurious tendency otherwise. By raising the price of salted fish, it would make it less accessible for purchase to the poorer classes, and by diminishing consumption, and consequently the demand, it would further affect the chief livelihood of the fishermen on the coast, and decrease their already insufficient means of subsistence. The gain, if any, to Government in the increase of salt revenue would thus be at the expense of the poorer classes, and I should be loth to advocate the measure; moreover, if less fish is sought for and less fish salted, there will be little or no demand for salt for the purpose, and the question of gain to Government, thus viewed, becomes altogether chimerical. The proposition in paragraph 7 of Dr. Day's letter, as now put, stands thus :— whether, if large enclosures were made near favorable localities (away from large towns) where fish could be salted, it is desirable or practicable to sell salt inside those enclosures at rates just remunerative for the *bona fide* salting of fish. I do not think the adoption of this proposition in this district is either advisable or desirable. It may not be impracticable, but this also is open to question. Two points of doubt suggest themselves in connection with this proposition. (1) Who is to sell the salt in these enclosures? The Government who hold the monopoly, or the ordinary bazarman and dealers who now buy from Government and retail to the people? (2) What is meant by 'rates just remunerative'? If it is the Government who is to sell, and 'rates just remunerative' implies at a rate other than the fixed monopoly price, just sufficient to cover all expenses, in other words, at the prime cost of the salt to the Government, I certainly think the measure would be inadvisable. It would be the same thing as a reduction in the monopoly price for the purpose of augmenting the trade in salt-fish, and Dr. Day himself is of opinion, "this being generally effected is very improbable." It would moreover, I have no doubt, lead to frauds and smuggling in various ways. The enclosures could not be close to, or supervised by, the existing agency at our present sale depôts, for not only are these not favorable localities for the salting of fish, but there would be other objections to their use as such. If, on the other hand, it is the ordinary bazarmen and dealers that are to sell within the enclosures, why not their present places of vend in the public bazar and resort of each village do as well. We have no complaint against their rates, which it must be presumed are now just sufficiently remunerative to give them a small profit of a pie or two per seer. *Cui bono* then Government interference to make them sell in certain places at certain rates. We may be assured that where a demand exists, a market will establish itself, and it is useless to set up markets and then try to create a demand.

93. *The Officiating Collector of the* Kistna district (November 10th, 1869) reported that " sea-fish are caught in large quantities both for home and inland consumption, and cured with salt before exportation. Fishermen extract oil from certain sea-fish in small

Opinion of the Collector of the Kistna district.

quantities for lighting their huts, but not for sale; and I am not aware that any oil is extracted for medicinal purposes in this district." Salt-earth is stated not to be employed. The "fishermen say that they always use salt for curing fish. It is, however, said that salt-earth is also used, but very rarely, by the poorer classes of fishermen for curing very small fish of inferior kinds. There are no restrictions against salt-earth being employed; it is procured directly after the marshy soil absorbs the periodical flow of sea-water. It may be purified before being used, but it is generally applied here in its uncleaned state, and washed off in sea-water; after twelve hours the fish are dried. It is tasteful, wholesome, and can be preserved for several years; there is a considerable trade in this article with the interior. Fish cured with good salt is of a very good quality and is freely used as diet by native physicians." The quality is good; the trade might be increased by reducing the price of salt required *bonâ fide* for curing fish. It is stated that the fishermen of one village prevent those of the neighbouring village from catching fish within the limits of their own village; if this practice be put a stop to, the trade may increase. The district is well supplied, and much is exported. He continues (June 12th, 1873), that all kinds of fish that can be caught in the tidal estuaries of the district are captured; that "the only method (that occurs to me) of increasing the salting of sea-fish would be to increase the number of fishermen. There are not many on the coast of this district, and I believe all fish (except the small quantity that is eaten fresh) that are caught are salted. I do not know of fish being thrown away, or being used as manure. At the same time, the coast from north of Bander to Nizampatam, being for the most part mud, is not by any means favorable for sea-fish, and it may be that all the fish that are to be had are caught, and that more fishermen could not find support. Certainly there are no such hauls of fish made here as I have seen on the Western Coast, or even up in the Vizagapatam and Ganjam districts, while the fishing villages are much smaller, less largely populated, and much further apart here. I do not think the proposition in paragraph 7 would be advisable here; the salt-pans close to the coast are China Ganjam, Nizampatam, and Manginapudi. The greatest number of fishing villages, within a radius of 20 miles of any of them, is but 40, and I do not think the enclosure would pay its own expenses of carriage of salt, police guard, and of people to keep it clean. An experiment might, however, be easily made at China Ganjam."

Statement showing the Salt-fish, &c., exported by Sea from the District of Kistna for the last ten years.

Official year.	Articles.	Whither exported.	Quantity.				Declared real value.		
			Tons.	Cwts.	Qrs.	℔s.	Rs.	A.	P.
1863-64	Fish-maws	Madras	1	17	0	8	1,853	11	3
	Sharks' fins	Ditto	0	1	0	8	20	0	0
			1	18	0	16	1,873	11	3
1864-65	Fish-maws	Ditto	0	10	3	25	548	10	3
	Sharks' fins	Ditto	0	4	1	1	68	4	0
			0	15	0	26	616	14	3
1865-66	Fish-maws	Ditto	0	13	1	9	666	8	5
	Sharks' fins	Ditto	0	4	0	0	65	0	0
			0	17	1	9	731	8	5
1866-67	Salt-fish	Rangoon	0	5	2	0	65	0	0
	Fish-maws	Madras	1	3	1	3	1,161	13	8
	Sharks' fins	Ditto	0	3	1	5	60	0	0
			1	12	0	8	1,286	13	8
1867-68	Salt-fish	Rangoon	3	3	2	0	396	0	0
	Fish-maws	Madras	0	9	2	27	542	4	9
	Sharks' fins	Ditto	0	2	2	15	48	0	0
			3	15	3	14	986	4	9
1868-69	Salt-fish	Rangoon, &c.	0	9	0	0	102	0	0
	Fish-maws	Madras	0	11	3	24	436	0	0
	Sharks' fins	Ditto	0	2	0	22	40	0	0
			1	3	0	18	578	0	0
1869-70	Salt-fish	Rangoon, &c.	0	17	2	0	316	0	0
	Fish-maws	Madras	0	0	0	24	12	0	0
			0	17	2	24	328	0	0
1870-71	Salt-fish	Rangoon, &c.	0	1	3	0	52	0	0
	Fish-maws	Madras	0	15	1	26	775	0	0
	Sharks' fins	Ditto	0	3	2	2	64	0	0
			1	0	3	0	891	0	0
1871-72	Salt-fish	Rangoon, &c.	0	2	3	0	96	0	0
	Fish-maws	Madras	0	5	1	6	287	8	0
	Sharks' fins	Ditto	0	6	0	16	112	0	0
			0	14	0	22	495	8	0

lxvii

Official year.	Articles.	Whither exported.	Quantity.				Declared real value.		
			Tons.	Cwts.	Qrs.	lbs.	Rs.	A.	P.
1872-73	Salt-fish	Rangoon, &c.	0	7	2	0	116	0	0
	Fish-maws	Madras	0	18	2	19	510	0	0
	Sharks' fins	Ditto	0	8	3	22	72	0	0
			1	10	0	13	698	0	0
		Grand Total	14	4	2	10	8,485	12	4
	Salt-fish	......	5	7	2	0	1,143	0	0
	Fish-maws	......	7	6	0	3	6,793	8	4
	Sharks' fins	......	1	11	0	7	549	4	0
			14	4	2	10	8,485	12	4
	Madras.								
	Salt-fish	......	0	7	3	0	101	0	0
	Fish-maws	......	7	6	0	3	6,793	8	4
	Sharks' fins	......	1	11	0	7	549	4	0
			9	4	3	10	7,443	12	4
	Rangoon.								
	Salt-fish	......	4	19	1	0	1,036	0	0
	Vizagapatam.								
	Salt-fish	......	0	0	2	0	6	0	0
		GRAND TOTAL	14	4	2	10	8,485	12	4

94. *The Collector of the Godavery district* (October 28th, 1869) observes that the use of salt-earth for fish-curing is not interfered with. It is employed either purified or not so. The quality of the salt-fish is tolerably good; much is exported to Madras and Hyderabad. A superior quality would be preferred were such obtainable. Government salt is never used for this purpose. The markets are fully supplied. [When I was at *Coconada* in 1868, there were about 40 fishermen; the quality of their salted-fish was very inferior; they used salt-earth and sea-water. When the former was employed they put small fish into it, left them there for 24 hours, and subsequently dried them in the sun. They considered monopoly salt as too expensive for it to be profitable to them to use it in curing fish.]

Opinion of the Collector of the Godavery District.

Statement showing the quantity and value of Exports and Imports of Salt-fish, &c., from and into the ports of the Godavery District from 1863-64 to 1872-73.

lxviii

EXPORTS.

Names of Articles.	1863-64.		1864-65.		1865-66.		1866-67.		1867-68.		1868-69.		1869-70.		1870-71.		1871-72.		1872-73.	
	Quantity. cwt. lbs.	Value. Rs.	Quantity. cwt. lbs.	Value. Rs.	Quantity. cwt. lbs.	Value. Rs.	Quantity. cwt. lbs.	Value. Rs.	Quantity. cwt. lbs.	Value. Rs.	Quantity. cwt. lbs.	Value. Rs.	Quantity. cwt. lbs.	Value. Rs.	Quantity. cwt. lbs.	Value. Rs.	Quantity. cwt. lbs.	Value. Rs.	Quantity. cwt. lbs.	Value. Rs.
Salt-fish	58	294	185 49	894	209 69	1,485	445 42	2,644	164 9	1,009	155 93	1,360	187 97	600	98 85	999	178 76	940	245 10	2,654
Fish-fins	... 34 107	467	16 15	326	4 28	71	1 38	30	... ...	...	17 105	680	... ...	...	... ...	...	... 82	12	... ...	...
Fish-maws	... ...	75 33	2 105	71	2 102	43	1 68	59	2 40	32	6 66	90	... ...	...	1 ...	...	1 62	24

IMPORTS.

Names of Articles.	1863-64.		1864-65.		1865-66.		1866-67.		1867-68.		1868-69.		1869-70.		1870-71.		1871-72.		1872-73.	
	Quantity. cwt. lbs.	Value. Rs.	Quantity. cwt. lbs.	Value. Rs.	Quantity. cwt. lbs.	Value. Rs.	Quantity. cwt. lbs.	Value. Rs.	Quantity. cwt. lbs.	Value. Rs.	Quantity. cwt. lbs.	Value. Rs.	Quantity. cwt. lbs.	Value. Rs.	Quantity. cwt. lbs.	Value. Rs.	Quantity. cwt. lbs.	Value. Rs.	Quantity. cwt. lbs.	Value. Rs.
Salt-fish	...	...	...	...	252 90	969	...	...	19 8	60	...	...	36	240	...	...	...	...	...	...
Fish-fins	...	...	...	...	...	...	...	...	...	...	...	...	...	1	...	...	...	...	...	...
Fish-maws	... 61½	27	...	...	...	...	...	...	...	...	...	...	...	...	...	...	...	...	...	...

95. *The Collector of Vizagapatam* (July 17th, 1869) observed that the fish-curers in his district are allowed to use salt-earth untaxed and without restriction. If purified before being used, the purifier becomes subject to punishment for a breach of the salt laws. The trade of salting would only be augmented by lowering largely the price of salt, which on other grounds cannot be anticipated. Subsequently (September 14th, 1869) he continued :—" Regarding definite proposals for dealing with salt-water fisheries, I have the honor to state my opinion that for such fisheries no regulations are called for, and no interference with the present fishermen can be otherwise than injurious. The tidal backwaters do not require preservation to ensure a plentiful supply of fish, and any revenue that may be raised by farming them to the present fishermen is too insignificant to be worth collecting, while its exaction would be felt as a hardship by a poor and hard-working class of men." The *Collector* also observed that until the abolition of the moturfa, a tax was levied upon fishermen, which might perhaps be called a rent for the right of fishing, but which was in fact a capitation tax on the castes that exercised the fishermen's profession. At the permanent settlement the amount of this tax was set down at Rs. 3,463-2-9, but this sum included the tax on sea fishermen, from whom by far the greater portion of it was collected, as appears from the incidence of the tax falling almost entirely on the estates on the coast. At present no taxes are reported to especially affect these people. *The Acting Collector* (November 9th, 1872) reported that he had no reason to suppose that the demand for salt-fish in his district exceeded the supply. Large quantities are sent inland and no rise in the price had been heard of. If salting depôts were established, he does not believe they will be frequented by fishermen, who are accustomed to salt their fish at their own doors, and would not be willing to carry a perishable article to a distance. Such depôts would lead to salt smuggling, a traffic which it is already sufficiently difficult to keep down in the district.

<small>Opinions of European officials in the Vizagapatam district.</small>

Return showing the Exports and Imports in the year 1872-73.

	Exports.				Imports.		
Where to	Description.	Amount.		Value.	Description.	Quantity.	
		cwt.	lbs.	Rs.		cwt.	lbs.
Calcutta ...	,,	1	52	20	...	0	52
Moulmein ...	,,	14	0	23	...		
Madras ...	,,	4	0	56	...	...	
Gopalpore ...	,,	4	44	60			
Madras ...	fish-maws	0	44	30	...	...	

96. *The Acting Collector of Ganjam* (June 14th, 1870) reported that the fish-curers in his district are not allowed to use salt-earth untaxed, and that when used it must be prepared first; that the quality of salted-fish is rather good, although they would prefer a superior article if they could obtain it cheap. Government salt is not purchased to any great extent for salting fish. [When I was at *Gopaulpur* in 1868, sea-water was being used for fish-curing. I was informed that during the preceding few years, shoals of small fish about the size of sardines had annually appeared about November. They were only employed as an article of diet; oil was not prepared from them. At Ganjam I also enquired into the fisheries; salt appeared to be obtained with great facility, and it was surmised that much of the rejected article was used by fish-curers. Salt-fish at both these places was exchanged with the Brinjaris for turmeric and other articles.]

Reply of the Collector of Ganjam.

lxxi

Statement showing the annual Exports of Salt-fish and fish-fins from ports in the Ganjam Collectorate for ten years from 1863 to 1873.

Years.	GOPALPORE.							BARWAH.							CALINGAPATAM.							TOTAL.						
	Quantity.			Value.				Quantity.			Value.				Quantity.			Value.				Quantity.			Value.			
	cwt.	qr.	lbs.	Rs.	A.	P.		cwt.	qr.	lbs.	Rs.	A.	P.		cwt.	qr.	lbs.	Rs.	A.	P.		cwt.	qr.	lbs.	Rs.	A.	P.
Salt-fish.																												
1863	...	...	...	...	...	...		...	...	...	...	...	...		32	0	16	65	0	0		32	0	16	65	0	0
1864	...	...	...	...	...	...		...	...	...	...	...	...		30	3	0	76	0	0		312	3	12	876	0	0
1865	10	2	24	243	0	0		262	3	12	800	0	0		55	3	3	364	0	0		538	3	27	2,282	0	0
1866	29	3	24	205	0	0		472	2	0	1,675	0	0		58	0	9	343	0	0		88	0	6	608	0	0
1867	1	1	6	35	0	0		...	...	...	...	...	...		24	2	0	57	0	0		235	1	24	1,021	0	0
1868	1	0	13	10	0	0		209	1	19	929	0	0		28	1	0	110	0	0		29	0	13	120	0	0
1869	1	0	24	24	0	0		...	...	...	...	4	...		1	3	13	57	0	0		11	1	9	81	0	0
1870	9	0	24	15	0	0		100	3	20	800	0	0		13	1	16	45	0	0		119	2	4	860	0	0
1871	5	0	...	...	...	...		69	2	26	339	4	0		15	2	16	70	0	0		85	1	14	409	4	0
1872	...	...	...	...	...	...		...	...	...	...	12	0		...	...	...	...	...	...		...	...	12	...	12	0
TOTAL ...	57	2	2	592	0	0		1,135	1	21	4,514	0	0		259	2	17	1,187	0	0		1,452	2	12	6,353	0	0
Fish-fins.																												
1863	...	...	...	...	...	...		...	...	...	...	...	...		...	...	...	...	...	...		...	...	...	...	...	...
1864	...	...	...	...	...	...		...	...	...	...	...	...		...	...	...	...	...	...		...	...	...	...	...	...
1865	...	...	...	...	...	...		...	...	...	...	...	...		...	...	...	...	...	...		...	...	...	...	...	...
1866	...	...	...	...	...	...		...	...	...	...	...	...		...	...	...	...	...	...		...	...	...	...	...	...
1867	...	...	...	...	0	0		...	...	...	...	...	...		...	...	...	...	...	...		...	...	...	12	0	0
1868	1	1	24	12	0	0		...	...	...	...	...	...		...	...	...	...	...	...		1	1	24	200	0	0
1869	4	0	2	200	0	0		...	...	...	...	...	...		...	...	...	...	...	...		4	0	2	227	0	0
1870	4	0	18	227	0	0		...	...	...	...	...	...		...	...	...	...	...	...		4	0	18	450	0	0
1871	7	0	20	450	0	0		...	...	...	...	...	...		...	...	...	...	...	...		7	0	20	200	0	0
1872	5	2	5	200	0	0		...	...	...	...	...	...		...	...	...	...	...	...		5	2	5	...	0	0
TOTAL ...	22	1	13	1,080	0	0		...	...	...	...	...	...		...	...	...	...	...	...		22	1	13	1,089	0	0

97. *The Collector of Bellary* (June 17th, 1869) observes:—"I believe the trade of salt-fish in this district, at diminished rates, might be enormously increased."

Observation of the Collector of Bellary.

98. *The Collector of Kurnool* (November 6th, 1869) states "that there is no extra tax for using earth-salt for curing fish; on earth-salt there is a tax of two rupees per annum on each pan." There are no restrictions against the use of earth-salt. The natives would prefer a superior quality of salt-fish, but all depends of course on the price. The supply appears to be equal to the demand.

Replies of the Collector of Kurnool.

99. From the *South Canara Collectorate,* the following reports have been received from the *native officials.* The *Tehsildar of Mangalore* observed (in 1872) that the supply of fish in the market is not always equal to the demand; about 75 per cent. of the people eat it. Small dried fish are generally sold at 12 annas a maund, and large ones at Rs. 12 per 100. Fish are speared by torch light. He now (1873) replies that the average selling price of salt in the bazar is Rs. 2-4; that Government salt is used for curing fish; sea-water is not employed for this purpose, but salt-earth is. The practice of salting fish has remained stationary. Fishermen do not appear to have had in old times any privileges which they do not now possess. There are headmen called Guricars among Mogers, who form one of the fishing castes; his office is hereditary. His duties are to make enquiries regarding the observance of the caste rules by the members belonging to his caste. He is entitled to get the usual honors and betel-nut, &c., on occasions of marriage and such ceremonies, but derives no other emoluments. No one claims any rights respecting the sea-fisheries. Fishermen purchase boats and also obtain them on hire. They likewise buy nets as well as manufacture them; sea-fishermen appear to have increased. *The Tehsildar of Kasargod* observed (in 1872) that during the rainy season people usually eat cured fish; that locally prepared is too small to meet the demand, so some is imported from Arabia and elsewhere. About 85 per cent. of the people eat fish; that the average retail price of salt is Rs. 2-6 a maund. Monopoly salt and salt-earth are both employed for curing fish, but sea-water is not used. The practice of salting fish has decreased of late years. The fishermen now possess the privilege of using salt-earth as they like, the Government having placed no restrictions on the use of the same, and of freely catching fish. They did not possess any peculiar privileges in old times. There are no headmen of the fishing caste; the fishermen act as they choose; no one claims any rights in regard to the sea-fisheries. They buy boats and sometimes nets, but usually manufacture these last themselves. The sea-fishermen have increased of late years. *The Tehsildar of Udipy* replied (in 1872) that the markets are well supplied; 68 per cent. of the people eat fish. Sea-fish which ascend rivers are trapped; also they are speared or killed with knives at night time, when they are attracted to the surface of the water by the light of torches. He now (1873) observes that the retail price of salt averages the following:—Goa salt one anna, Bombay salt one anna two pies a seer;

Opinions of native officials in South Canara.

this would be Rs. 2-8 or Rs. 2-10 a maund. Government salt is sometimes used for curing fish prepared for their own eating, but they employ salt-earth or sea-water for that which is sold. On the whole, the practice of salting fish is said to have increased of late years. The fishermen formerly had no privileges they do not possess still. Mogers and Kharves are the only two fishing castes who follow it as an employment. But the Kharves are only in limited numbers; they have a common place of residence which is styled a Kéri (a row of houses). For each such Kéri there is a headman called Guricar, who investigates caste matters amongst all the people who reside there. The office is hereditary; no emoluments accrue to it. In a similar way, Mogers live in groups of houses which are termed Patna (town). For every such Patna there is a headman. He is termed Guricar, and he decides all questions of caste. It does not appear that he derives any emoluments, except that he is entitled on auspicious or inauspicious occasions to precedence in receiving betel-nut. His office is not hereditary, and his election or removal depends on the will of the people. Over the Mogers of all the Collectorate (except Kundapur) there is a spiritual preceptor named Mangal Pujary; he resides at a place called Benne Kudru, near Barkur. His duties are to frame rules in regard to caste matters, to see if the people conform to them or not, and to impose penalties on those who infringe them, &c. The people of the caste raise money for him; his office appears to be hereditary. No one claims any rights respecting the sea-fisheries. Fishermen procure boats and manufacture their own nets; their number according to the last census appears to have increased 15 per cent. over the former one. *The Tehsildar of Kundapur* replied (1872) that besides the local fishermen, Daljis come in the hot season from Bhankot, Ratnagiri, and other places to the north, take fish, cure them, and carry them away for sale. The markets on the coast are generally well supplied. At markets inland the price is double or treble what obtains in Kundapur, owing to the cost of carriage. Usually paddy or rice is given in exchange for fish. About 85 per cent. of the people are consumers of fish. Subsequently (1872) he states that salt is retailed at Rs. 2-8 a maund. Government salt is mostly used for curing fish; poor people sometimes employ salt-earth for this purpose, but not sea-water. The practice of salting fish remains stationary. The fishermen in olden times had no privileges they do not now possess. There are no headmen of the fishing caste. All fishermen believe they have a right to the sea and river fisheries; those which are well-to-do have their own boats or hire them; nets they make themselves from hemp they grow or purchase; these people appear to have increased of late years. *The Tehsildar of Uppinangadi* replied (1872) that at his inland station dried fish are brought from Mangalore and elsewhere for sale at 1 anna 4 pies a seer. Forty per cent. of the people are fish-eaters; the supply of fresh fish is inadequate to meet the local demands.

100. From the *Malabar Collectorate* the following answers have been received from the *native officials*. *The Tehsildar and the Superintendent of Sea Customs, Cochin* (March 13th, 1873) reply that breeding and immature fish are not destroyed to any great extent. That forming enclosure, &c. (as suggested), is not feasible, the area of British

Opinions of native officials in the Malabar Collectorate.

Cochin being very limited. The price of salt is Rs. 2 a maund, and it is used for curing fish, but salt-earth or sea-water is not so employed. The practice of salting fish has decreased owing to a scarcity of fish of late years. The native christians who engage in sea-fishing here are not of the fisherman caste in the proper acceptation of the term; they had therefore no peculiar privileges which they do not now possess, nor have they any headman. No one claims any rights as regards the sea-fisheries. About 30 of the fishermen here possess boats and nets; the owner of a boat has generally sufficient members in his family to man the boat. Where they fall short of the required number of hands, neighbours make up the deficiency, the latter get an equal share of each of the fish captured. The number of fishermen has remained stationary. *The Superintendent of Sea Customs at Ponany* (March 24th, 1873) replies that, during the breeding season, which embraces the months of January and February, wholesale destruction of breeding and immature fish is caused by fishermen in tidal estuaries. The meshes of their nets are so small that even the smallest fish cannot escape. The salting of sea-fish cannot be increased at Ponany, as there are only a certain number of persons who understand the curing of fish. These persons undertake to salt as much as they are capable of. The proposition in paragraph 7 (respecting enclosures) would be very advisable if the privilege of manufacturing salt-earth is removed; unless this is done, the reduction in the price of Government salt would be of no material advantage, for, as long as salt-earth is allowed to be manufactured, its illicit sale in the market cannot possibly be prevented. The monopoly price of salt is Rs. 2 an Indian maund; owing to its high price only a limited amount is used for curing fish. The majority of the fishermen here are very poor, and can ill afford to procure it at the present rates. Sea-water is never employed in curing fish, but salt-earth is largely so, the consequence being that fish thus salted soon becomes unwholesome. The chief cause of the frequency of cholera in this district is popularly attributed to the indiscriminate consumption of fish salted by the above means. The manufacture of salt-earth is openly carried out to a great extent in all the villages of this talook. In my visits to the sub-ports, I have come across salt pans in many places, and have seen people manufacturing there. The sale of Government salt in this district has of late decreased, owing to large quantities of salt-earth having found its way into the market. Unless the privilege of manufacturing salt-earth is restricted, it is useless to reduce the price of Government salt. The practice of salting fish appears to have remainéd stationary of late years. The fishermen had no peculiar privileges in olden times which they do not now possess. There are no headmen amongst the fishing castes here. No one claims any right as regards the sea-fisheries. The majority of the fishermen have their own boats and nets, while others of more limited means obtain them on hire. There is an annual increase in the number of fishermen, but no data are available from which the actual increase in each year can be ascertained. *The Superintendent of Sea Customs, Cannanore* (March 27th, 1873) answers that all fishes, whether breeding or immature, are destroyed without distinction, and to a large extent. That the salting of sea-fish could be increased if the monopoly price of the Government salt be reduced; that this can only be effected by the introduction of a license system, restricting the sale, to a certain extent, to the fishermen alone for the *bond fide* use of salting fish.

Statement showing the annual sale of monopoly salt in the Cannanore District for the last ten years ending 1872-73.

Years.	Sold at Cannanore.		Total Sale.	
	Maunds.	Seers.	Maunds.	Seers.
1863-64	11,653	17	48,634	17
1864-65	7,932	0	54,718	0
1865-66	9,856	0	54,649	0
1866-67	9,728	0	60,937	0
1867-68	8,721	0	39,467	0
1868-69	9,045	0	58,076	0
1869-70	8,807	0	63,569	0
1870-71	7,932	0	61,233	0
1871-72	12,008	0	66,848	0
1872-73	6,985	0	58,303	0

That the proposition in paragraph 7 (respecting enclosures) would be practicable if a large establishment can be provided to check frauds and deceits. The retail price of Government salt is from Rs. 2-4 to Rs. 2-6 a maund; it is used to a very limited extent for curing large fish; in this case, the salt is generally supplied by the merchants who export the salted fish to Colombo, Tuticorin, Negapatam, &c. Some of the boat-owners, whose means admit of storing salted fish until they can find a good market, use Government salt for curing both large and small fishes intended for consumption in the hilly countries of Coorg, Wynaad and Mysore, &c., and also for exportation to Colombo, &c. The fish so stored is generally purchased from them by petty merchants coming from those places. Sea-water is not, but salt-earth is, used for curing fish; this earth is procured from marshy lands. The women of the fishermen caste collect the earth, paying the proprietors of such lands at the rate of 2 to 4 pies a basket according to their size, the largest of which is capable of containing from 4 to $4\frac{1}{2}$, the medium 3 to $3\frac{1}{2}$, and the smallest 2 to $2\frac{1}{2}$ bazar maunds. The earth is collected during the dry weather, stored in large quantities and used in its impure state. The fishes are cut open, the entrails removed, salt-earth rubbed in, and then they are thrown in a tub, and left there until the following day, when they are washed in the sea-water and exposed to the sun until well dried. The practice of salting fish has increased of late years, due to the increased demand from Colombo and other places to which they are exported. The fishermen still possess all the privileges they ever had; they have no headmen; the owner of boats and nets are considered to be chiefs of their parties. The Rajah of Cannanore claims a right in respect to the sea-fishery. This is called the 'Poocha Meen' or 'fish for the Rajah's cat.' In former days one fish was given from each boat out of their daily captures, but now or of late years it has been commuted into a money payment of 2 pies a day on each successful boat. In addition to this the boat and net-owners pay the Rajah a tax of Rs. 2-4 annually on every boat and net. The fishermen, boat and net-owners, in Cannanore, are the 'Mukuwars,' a low class of Hindús: the 'Collakars' or 'native christians,' and the 'Moplahs.' The Rajah lets the 'Poocha Meen' right on contract, renewable annually; from this contract the first-named class of fishermen, boat and net-owners, are excepted; and from the others the contractor collects daily the money due on account of the right in question. The Mukuwars pay annually a lump sum of Rs. 70 for their boats and nets, and also for the 'Poocha Meen.' The fishermen are supplied with boats and nets and other requisites for fishery by the owners of the boats and nets, who also advance them a certain sum of money (charging them no interest for the same) to ensure their services. The money thus advanced is not deducted from their daily labour. It is generally refunded by them, should they be unwilling to work for the party advancing the money; and in some instances in case of death, inability to work from extreme old age or infirmity, or in cases of desertion, the money advanced is a loss to the boat-owner. The owners remain on shore while the fishermen go out, and on their return to the shore the owners of the boats and nets sell their captures. Such as remain unsold are taken by the boat-owners (for salting purposes) at the average rate at which the portion sold realized, and the sale being thus completed they divide the proceeds equally between themselves, *viz.*, one-half to the owners of the boats and nets, and the other half to the fishermen; but should the latter prove unsuccessful and capture only sufficient to realize their expenses for the day, the boat and net-owners surrender their share in favour of the fishermen. The sea-fishermen have decreased of late years in consequence of the extinction of some families from cholera and other diseases, and also from poverty. Sharks' fins are cured, and a moiety of the value thereof goes to the Rajah's 'Poocha Meen' contractor. Fish-maws are also cured. Fish-oil is extracted in small quantities for the use of boats, as such is considered to prove a good preservative. The fish-maws and sharks' fins are only cured to a small extent and exported to Bombay and Calcutta. Deep sea fishing is resorted to at this place during the seir-fish season, from November to January. The fishermen very seldom turn their attention to shark-fishing, as they seem to think it is not remunerative in Cannanore owing to their having to yield one-half to the Rajah's 'Poocha Meen' right. Ceylon fishermen resort to the Malabar Coast for fishing during the seir-fish season, but do not do much. There are at present 72 boats at Cannanore and 24 at Highicode. *The Superintend-*

ent of the Sea Customs at Tellicherry (April 9th, 1873) replies that breeding and immature fish are destroyed in the tidal estuaries of this place to some extent. At high tide many young fish and a small number of breeding ones at the estuaries of the Darmapatam and Mahe rivers enter into the marshy grounds along their sides, where they are either netted, or caught without the use of any apparatus, when the water recedes at low tide, and the fish are left on the surface of the ground. As these rivers seldom, during the hot season, overflow their banks at high tides, the fishermen, in order to get the fish into the low marshy grounds enclosed by them for this purpose, often cut open the banks and thus enable the water to flow in, carrying fish with it into these enclosed spaces. As soon as a good supply of fish is collected, that part of the bank left open is closed by a valve made of split bamboos, generally so contrived as to let the water flow back through its interstices and at the same time bar the egress of the fish. When the water has so receded, the fishermen simply have the trouble of picking up the fish. A good deal is however wasted, inasmuch as only such fish as are capable of being used are taken, while the rest, generally very small ones, are left to perish. This is the process obtaining in the hot season of the year, viz., from January to May. During the monsoons the Darmapatam river often overflows its banks where these latter are low, and water is thereby collected in various places in the marshes for a long time, and from such localities the fishermen catch the fish by means of hand nets. The quantity of fish so obtained is not so great as that captured in the hot season. It appears that as these estuaries are often covered by the tides, many young and some large fish enter, the latter for breeding purposes, and are often taken. Sometimes during the hot season of the year, boat-loads of fish, including many breeding ones, are known to have been taken. For the purpose of increasing the salting of sea-fish, no better plan than a reduction in the price of salt, and the exemption of the boats and nets and huts of the poor men among the fishermen caste from the local municipal tax, can be proposed. Fish are generally cured at the place where the fishermen reside, and this, their fishing village, is adjacent to the principal quarter of the town. To make large enclosures as proposed, away from this place, and there to carry on fish-curing operations, salt being sold inside the enclosure, would not only cause the greatest hardship to persons employed in the trade, but could be carried out only at an enormous cost to Government, for the land is private property; nor is the plan feasible at the village itself, as many non-fishermen reside within its limits. The proposition is neither advisable nor practicable in this place. Salt is sold wholesale in the bazar at Rs. 2-2 or Rs. 2-3 a maund, and retail at Rs. 2-6 or Rs. 2-7 a maund; it is used to a great extent for curing fish, a little more than half the quantity of salt-fish manufactured here being fish so cured. Those intended for exportation to Colombo are solely cured by Government salt, as such only can be preserved for a long time and are capable of being carried, without deterioration, to distant markets. Fish principally cured with this salt are mackerel, cat fishes, seir-fish, pomfrets, sharks, &c. Of these kinds of fish, mackerel and cat fishes are cured to a very great extent and exported to Colombo, as well as to some parts of the Tinnevelly district. Salt-earth is, but sea-water is not, employed for curing fish. Sea-water was formerly used for the purpose, but that practice was stopped by the interference of the local authorities, as it was apprehended that fish so cured would bring on disease. Salt-earth is extensively used for this purpose; it is obtained from the swamps here, and is used by being rubbed in its impure state into the fish. The owners of swamps cut open the banks of the rivers along which they are situated, thus allowing a sufficient quantity of salt-water to flow into the enclosed portion of the swampy ground, and settle there for some days. Fishermen are then allowed to scrape up the upper crust for a consideration of 8 pies a basket-load of 3 country maunds. The principal fish which are selected for being cured with salt-earth are mackerel, sardines, eels, cat fish, &c. A large quantity of this kind of salt-fish is carried inland to the Wynaad and Mysore. A good lot of the same, intended for Palghaut, Coimbatore, &c., is taken from here by sea to Beypore, and from thence, the same carried to its destination by rail. It is, however, inferior in quality to the salt-fish cured with Government salt, cannot be preserved for any length of time, and is sold at a cheaper rate. A large quantity of this is yearly exported to the Tinnevelly ports. The following return of the export trade of this article for the last five years will show that it is on the increase. The practice of salting fish was formerly confined to the fishermen castes only, but of late years fish-curing operations are performed by Moplahs and Kallakars (Syrian christians who came from Quilon and settled here for the purpose). Increased steam communication, and the establishment of the Oriental Bank at this place, have augmented the trade, as the salt-fish traders here now export their cargo to Colombo by steamer, and can have money advanced to them from the Bank on the security of their bills of lading. The demand for this article in Ceylon is so great that the traders here often obtain advances of money from the Colombo merchants. The fishermen do not seem to have any privileges. They have, however, by consent, made certain rules which are strictly observed. The most noticeable amongst these rules is the right of the first discoverer, among a lot fishing together, of a shoal of fish. In this case, the man who first saw the fish is allowed to capture them without hindrance from the others, even though at the time when the fish were discovered he was not prepared to launch his net. Should any disputes arise on this head, the matter is laid before certain wealthy men of their own caste, whose decision is final. In reality there are no recognised headmen among the fishermen here, but the owners of the boats and nets have certain respect paid them among this class of people. Certain of the wealthy among them hear all complaints arising between themselves, and settle all caste disputes, &c. In some cases these arbitrators are remunerated for their trouble, but no fixed compensation is given them. It depends on the importance of each case. Their meetings for purposes of arbitration are held in a house built by subscription for this purpose, their assemblies being generally held at night. It is a noticeable fact that so thorough is their belief in the integrity of their arbitrators, that their decisions are always implicitly submitted to. No one claims any rights as regards the sea-fisheries. The rich fishermen here are the boat and net-owners; they do not go to sea themselves, but supply the poor among them with fishing apparatus, and pay them besides for working for them. Such contracts are often reduced to writing. The remuneration is half the supply of fish captured, the other half going to the boat-owner, unless the take is very small, when all goes to the fishermen. About 200 boats from this port go to sea, besides those which come here from neighbouring places for fishing and to sell their captures in this town. The sea fishermen have of late years very largely increased; every year people from Chowghaut and Quilon come and settle here for fishing and fish-curing.

Statement showing the description and value of dried Salt-fish, &c., exported from Tellicherry during the last 5 years.*

YEARS.	Dried Salt-Fish.					Dried Sharks' Fins.							Fish-Maws or Sounds.							Total Value Exported.		
	Number of bundles.	Value.				Weight or quantity.			Value.				Weight or quantity.			Value.					
		Rs.	A.	P.		cwt.	qrs.	lbs.	Rs.	A.	P.		cwt.	qrs.	lbs.	Rs.	A.	P.	Rs.	A.	P.
1868-69	11,227	43,195	6	1		177	0	25¼	6,034	0	9		107	3	12⅔	1,914	11	9	51,144	2	7
1869-70	13,591	58,395	11	3		451	1	7½	17,829	5	5		169	2	23	4,245	15	2	80,470	15	10
1870-71	9,016	53,019	8	0		213	1	10¼	8,217	11	6		197	0	8¼	6,885	13	3	68,123	0	9
1871-72	4,521	53,404	9	4		282	3	20¼	11,122	0	0		276	3	8¾	6,747	1	0	71,273	10	4
1872-73	28,253	84,296	8	9		598	3	9¾	22,405	13	0		340	3	1¾	7,780	14	10	1,14,483	4	7

* These figures do not agree with the Collector's returns, paragraph 66.

But of the foregoing the following was exported to Colombo :—

		Rs.	A.	P.
1868-69 to the value of	...	13,294	15	3
1869-70 ditto	...	20,908	7	3
1870-71 ditto	...	25,655	8	0
1871-72 ditto	...	37,653	9	8
1872-73 ditto	...	64,018	8	9

In the year 1864 (see *Fishes of Malabar*) I obtained from the Collector a return showing the exports of salt-fish from this same port, and the gradual rise in the amount has been most remarkable :—

Export of salt-fish for five years ending 1858-59 ... 18,621
Ditto ditto ditto 1863-64 ... 50,004
Ditto ditto ditto 1872-73 ... 292,311

Statement showing the annual sale of monopoly salt at Tellicherry for the last ten years ending 1872-73.

	Amount of salt sold in		Value in		
	Maunds.	Srs.	Rs.	As.	P.
1863-64 ...	72,505	0	1,03,851	14	4
1864-65 ...	57,516	0	83,535	14	0
1865-66 ...	62,135	0	91,662	9	6
1866-67 ...	57,381	0	93,453	3	0
1867-68 ...	56,502	4	93,006	14	8
1868-69 ...	63,340	24	1,02,538	13	0
1869-70 ...	72,616	26	1,29,615	15	4
1870-71 ...	57,624	0	1,06,171	7	2
1871-72 ...	88,674	0	1,67,261	4	6
1872-73 ...	77,332	10	1,46,327	8	11

In the work referred to (page 15), I observed that "in British Malabar the natives are permitted to collect salt-earth or impure salt from the ground, and employ it for the purpose of curing fish, as no duty is levied on it, and no fiscal impediment exists in the way of their using it, so that for this purpose it is almost exclusively employed, and manufactured salt used to a very limited extent. It may, perhaps, be a fact that in the British territory, where salt-earth can be obtained duty free, the 'salt monopoly' does not directly affect the fish trade, or rather its *price*, irrespective of *quality*. But it is open to question as to the quantity of saline matter which would be contained in a given quantity of the two kinds of salt, and whether, if the manufacturer could procure the better commodity at a cheaper rate than at present, he would not cure his fish in a superior manner. In such a case they would keep better, be more wholesome for the consumer, or could be conveyed to much greater distances inland." It is to be regretted that up to this time no returns from any of the Tehsildars have been received, whether they are in inland talooks or those bordering on the sea. The *Superintendent* further observes (July 30th, 1873) that about 28℔s. of monopoly salt are used to a maund of small fish, as mackerel, sardines ; for instance, 1,000 mackerel, weighing about 2 bazar maunds of 32 ℔s. each, are cured with 5 measures of salt equal to 20 and odd pounds ; but if salt-earth is employed 2 basketsful of that substance, weighing about 5 bazar maunds, are required. From enquiries now made, it appears that mackerel and other small fish are cured with salt-earth to a very great extent. But to cure an Indian maund of large fish, such as sharks, seir fish, large cat fishes, &c., the above quantity of salt is not sufficient ; thus 82 ℔s. of monopoly salt is used for curing 100 seir fish weighing about 160 ℔s. This last species of fish are now often cured with monopoly salt alone.

101. From the *Madura Collectorate*, the following answers have been received from the *native officials :—The Tehsildar of the Madura talook* (1869) says that the salt-fish imported is generally considered good, but that a larger supply would meet with a sale. *The Tehsildar of the Tiroomungalum talook* (1869) replies that the supply is equal to the demand. *The Deputy Collector of the Salt Department, Ramnad* (March 31st, 1873), observes, first, with reference to forming enclosures for salting fish, &c. : " The measure is practicable, and the convenient points on the coast to form the enclosures are the salt stations. The sale of salt at a reduced price within the enclosure may not only suppress the use of illicit salt, but also improve the quality of the salt-fish, which from the high price of salt is cured with other ingredients as salt-earth, &c. But this measure will necessarily involve the interference of Government servants, more or less, with the operation of curing fish. It must at least be carefully watched for the protection of the salt revenue, that the salt sold from the Government depôt is carried into the enclosure and used for the special purpose for which it had been sold. This interference will be unpopular, as it must naturally be, for some time at least with the tradesmen, who will, notwithstanding any amount of assurance, suspect the motives of Government, and the stimulus which it is hoped to give to the trade will tend to hamper it. I do not therefore consider the measure proposed by Dr. Day advisable, and it is my opinion that the trade should be left to be regulated by the laws governing the demand and supply of other articles of food. There are not estuaries of any importance in this district, except that formed at the mouth of the river Vigay, near Attenkarai. This abounds with breeding fish, which are indiscriminately destroyed between February and September. The mouths of the minor streams are dry, except during freshes ; any small fish which run up the streams during the tides are caught by the villagers. The trade in salt-fish is progressive in this district. The tradesmen, I may say, live from hand to mouth, but the readiness with which the salted fish is taken off their hands has induced many along the coast, chiefly Labbai Mussalmans, who carried on other occupations, to take up this trade. The trade in salt-fish is hampered to a certain extent by the interference of the villagers, who act the part of brokers between the buyer and seller. The sales are required to be made through the villagers or their representative

and none else; a breach of this custom is followed by annoyance to both the buyer and seller, who therefore seek the favor of the villagers, though such occasions a little pecuniary loss. Trade in salt-fish is looked down upon by the capitalists along the coast as being of an inferior description. If their minds were disabused of this prejudice, and they would invest money in the trade, it would be likely still more to improve. The retail Government price of salt is Rs. 2 a maund of 3,200 tolas, but the market retail rate varies according to the distance of the market from the Government depôt. Government salt is supposed to be used for curing fish. But fishing villages are the readiest markets for illicit salt; fishermen also use clandestinely salt found on the islands of the southern part of the coast of this district. Salt-earth is also largely used, collected from the islands; and fish so cured is, I am informed, much liked by the Singhalese, and is therefore exported to Ceylon. People on the coast do not relish the salt-fish cured with salt-earth; sea-water is not employed here for curing fish. The practice of salting fish has of late years increased, but no statistics are available; from the fact, however, that a greater number of men along the coast have now betaken themselves to sea-fishing, and that an improved trade in salt-fish is carried on, it is obvious that the practice has increased. The fishermen had no privileges formerly which they do not now possess. The fishing population now includes principally the following castes:—Mussalmans or Labbai, Karayan, Pallavarayan, Pavavan, and Kadayan. They have each their caste headman in the village. He is the representative of the community, and his duties are confined to the settlement of domestic or caste differences. His office is hereditary and supposed to be honorary, but he derives his own share of fines and penalties levied for breach of caste or social rules; this share is regulated by custom. On the coast of this district, except the right of chank fishery, which belongs to the zemindars of Ramnad and Thevagunga, that of other sea-fisheries is claimed by none. In fishing villages near the head quarters of the zemindars of Ramnad a contribution called 'karry min' or 'curry fish' is levied upon each boat returning from the sea with fish; this is supposed to be fish supplied to the kitchen of the zemindar's palace. Fishermen are supplied with nets by the better-to-do of them, called 'Sammanothy.' The fish caught is divided equally between the owner of the boat and the fishermen, but the amount is regulated by circumstances. The sea-fishermen have increased on the whole, as the lower castes on the coast have taken to it. The aboriginal fishermen castes, Paravars and Karayans, have decreased, many of their families having emigrated to other parts of the coast; a great part of the Parava population have given up fishing and betaken themselves to sea-faring. The Pallavarayan and Kadayar castes have remained stationary."

102. From the *Tinnevelly* Collectorate the following answers have been received. In the Tinnevelly talook (March 1870) there is no sea-board, and it is observed that the fresh-water fishes are dried but not salted. The present quality of salt-fish supplied from Travancore is preferred to that from the coast; an improved quality would find a sale; the supply is insufficient for the towns and more important villages. The *Tehsildar* of *the Sattur talook* reports that the salt-fish imported from the coasts of

Opinion of native officials in the Tinnevelly Collectorate.

Tuticorin and Vypar are not considered to be so good as the people desire to have an improved quality would meet with a sale, the supply is not equal to the demand. *The Tehsildar of Ottapidaram* replies that in his talook there are forty fishing villages, having an average population of 234, and that the fishermen have no other employment. Their castes are Paravars, Valiars, Shanars, Mussalmans and Singhalese. Those which fish in the backwaters are mostly the poorer members of the classes eating fish. The Shanars, Paravars, Maravars, Mussalmans and Pariahs salt fish which they also purchase for sale. Fish are taken long distances for sale, as Madura, Dindigul, &c.; the price is from " 4 pies to 8 annas for countable fish; for non-countable, *i. e.*, (very small) fish 6 to 12 annas per thoolam, and for fish which are brought from a very long distance in the sea by Singhalees, and which are of a very superior quality, 4 annas to 4 rupees; but the last mentioned are not purchased in considerable quantities by contractors." Their average daily earnings are 2 annas a day all the year round, excluding costs. There are no taxes on fishermen. The local market is sufficiently supplied, but more are not taken than are sold, and perhaps if more fish were taken they could be sold. Fish for eating is preferred fresh. The fish in the sea have decreased; the fishing population has remained stationary; 7 boats, 73 vallums, and 88 catamarans go to sea, whilst there are two more nets than there were. Very small fish are taken in great quantities. The best fishing months in the sea are from October to January. Fish are salted to a great extent with purchased salt, but not with salt earth, which last, however, is to be obtained in brooks at Vipar and other villages adjacent to the sea. The present quality of the salt-fish is considered to be good; an improved quality at present prices could be sold in larger quantities. All classes who consume fish will eat it salt; and the salting of fish could be increased if the price of salt becomes cheaper. Salt-fish is exported. Smoking fish or preserving their roes is unknown, but the maws or sounds are prepared. Fish oil is manufactured from livers only. Fish are used as charms. The *Tehsildar of Mangunery* states that in his whole talook there are 197 fishing villages, containing 1,16,100 occupants in all; besides fishing they have other employments. The Paravars, Muckuvars, and Mussalmans fish in the sea, and those castes which capture fish also salt them. Fish are taken long distances for sale, and the contractors generally allow the workers ¼ of the captures, the average daily earnings being from 1 to 6 annas. Some markets are fully, others insufficiently, supplied. Fresh fish is preferred to the salted article; 315 fishing boats go to sea, their numbers and those of the nets having remained stationary of late years. Very small fish are taken to a certain extent. The best months for fishing in the sea are from April to August. Fish are salted to a great extent with purchased salt, whilst "the salt earth in the inland parts is not used in curing fish; but in the localities bordering on the sea it is used for curing fish along with salt." Salt earth is not sold, and salt is not procured from the sea-water. The present quality of salt-fish is considered to be good, but an improved one at the present price would be preferred. Only sea-fish is salted, and no further improvement in augmenting the trade can be suggested; some is exported. Roes are salted, and fish dried but never smoked, whilst the sounds are not preserved.

Fish liver oil is prepared. The *Tehsildar of Tenkarei* replies that the same castes as reported in the last talook fish in the sea, &c.; that raw-fish are taken as much as 30 miles inland, but salt-fish are to a longer distance than this,—even to foreign districts. The remuneration paid by contractors to fishermen is $\frac{1}{3}$ share if large, $\frac{1}{2}$ if they are small, and their average daily earnings vary from 2 to 8 annas. The local markets are fully supplied; fresh fish are preferred to the salted ones. The fishing population has decreased of late years; about 200 boats go to sea; these and the nets have also decreased. The best fishing months in the sea are from September to February; only a small amount of fish are salted, for which purpose the salt is bought; salt earth is also used, being obtained from the land nearest the sea, but is not sold. Salt is not obtained from sea-water. Fresh-fish is preferred to the salted, but the quality of the latter is considered to be good, but a better one would be preferred. "The salting of fish can be increased, if the sale of raw-fish be stopped." Salt-fish is exported to Ceylon and other places. Fish are not smoked, but certain ones are dried, whilst roes are salted, and the fins and sounds preserved. Damaged and salted-fish are used as manure for cocoanut trees. Oil is prepared from the livers of some species.

103. *In the Tanjore Collectorate* the following replies have been forwarded from the native officials. The *Tehsildar of Negapatam* (January 1873) observes that salt is sold by Government at 2 rupees a maund, and by retailers in the bazars at 2 rupees 6 annas or 7 annas. Government salt is used in small quantities for curing fish, as are also both salt-earth and sea-water. The practice of salting fish must be said to have increased of late years in this part, owing to the railway* as also fairs recently introduced in some places, at which salt-fish is largely purchased. The fishing classes still possess all the privileges they have ever enjoyed. Pattanavans, Carayrs, and Sambadavars had, till the close of 1870, the privilege of catching fish within the distance of a mile in the estuaries of tidal rivers, without paying any impost to Government. In pursuance of orders they were then let out, at which the fishermen complain as an infringement of their immemorial rights and customs. There is at Negapatam one man styled 'nambian' as the head of the Pattanavans of the fishermen castes inhabiting the villages on the coast between Cuddalore and Vedaraniem. His place of residence is also called 'nambian cooppam'; his office is hereditary, and on his death all the fishermen unite together, and appoint his heir as their headman. His main occupation is to settle disputes arising amongst the Pattanavans. Sometimes he uses a net of his own, and employs coolies who catch and sell fish for him. He goes in a palanquin to the villages inhabited by the said Pattanavans to enquire into matters of custom; the villagers come in advance to meet him, and present their respects to him, and conduct him to the village. During his sojourn in a village his expenses are borne by the fishermen

* On receiving this statement I wrote to the *Traffic Manager* of the railway, who states (June 9th 1873) that "during the last 12 months, the weight of salt-fish received at stations on this Company's line from the western coast was 288 tons. During the same period the quantity forwarded from the eastern coast (Negapatam station) was 16$\frac{1}{4}$ tons." Thus the increase of salted fish is due to its being imported from a district where the use of salt-earth is freely permitted for this purpose, and the Tehsildar's opinion is erroneous.

of that village; he enquires into the offences committed, punishes the offenders and collects the fines, &c. If a large net is nearly ready to be used, a present of 7 pon's, an ancient silver coin, and a cloth is made to the headman, who gives a chit for the same, and it is only thereupon that the net is worked. If a marriage takes place in a house, it is not to be performed without a present of 200 betel-nuts, and as many leaves, and of 2 fanams (5 annas) being first made to the headman. In the event of a marriage being celebrated in the village where the 'nambian' resides, rice and vegetables are to be presented to him besides the aforementioned presents; the fishermen who live in the same village as the 'nambian' are in the habit of giving him fish for his diet as 'Valaikari.' These are his emoluments, in addition to the income derived from using his own net; a document to the above effect, executed by all the people of the fishermen caste to the nambian in old days, is still in his possession. The present nambian being a minor, the terms of the document are not strictly carried out. Pattanavans and Carayrs, who catch sea-fish, assert that they possess from time immemorial an executive right to the sea-fisheries; but their assertion is not supported by any record, nor does it appear that this alleged privilege has ever been the subject of contention between them and any other people. The fishermen make their own nets. Boats are not used in these parts for sea-fisheries, but the fishermen purchase catamarans at their own cost. The number of fishermen remains stationary. *The Superintendent of Sea Customs, Muthipettah*, replies that the selling price of salt is Rs. 2-3-7, and that Government salt is not used for curing fish. As regards whether salt-earth or sea-water is employed for this purpose, he continues, " it appears no, but fishermen in some places buy large fish for cash, and cure them as follows. They first split them, put salt into the split, and bury them for some time in places contiguous to the sea, and then disinterring them, wash them in salt-water, and lastly dry them up. As regards small fish, the process of opening is simpler: it is by spreading them in the said places and exposing them to the sun. The practice of salting fish appears to have decreased of late years in these parts, the profit of fish-curers having, it is stated, become very limited. The existing privileges seem to have been also enjoyed in old times. But it is said that, in the estuaries of the sea into which the Coriar and Pamaniyar empty themselves, the fishermen possessed, till so late as 10 or 12 years ago, the privilege of capturing fish as far as sea-water runs in the said estuary, without any let or hindrance; but that, as the fishery in the said estuary has subsequently been rented out by Government, that privilege has since been disallowed. [This probably refers to the year 1870, as explained by the *Tehsildar of Negapatnam*.] The fishermen of this place are Sanagars, who are divided into three factions, *viz.*, Periakatchi, Sinnakatchi, and Nadukatchi, each of which party having a headman of its own styled 'Marakayar.' This headmanship is hereditary. The headman of each faction settles disputes as to relationship, &c., arising amongst the Sanagars, and takes precedence by the chief men on the occasions of marriage and the like ceremonies; he, however, receives no kind of income from fishermen as emolument for his post. No one claims any right as regards the sea-fisheries. The fishermen procure small boats and nets at their own cost; but those who have neither the one nor the other join those who have them, and go along with them for fishing; the income derived from the fish captured is divided into as many shares as there are men engaged in the job, with 1½ extra share (one for the boat, and half for the net), thus the share of each man being equivalent to that allotted to the boat. The number of sea fishermen have decreased, because sea-fish are not caught now to the same extent as formerly. *The Tehsildar of Tanjore* replies that the monopoly price of salt at the Government depôts throughout the district is 2 rupees per Indian maund, the retail selling price in the market is Rs. 2-3 in the villages adjacent to the salt depôt, and Rs. 2-6 in places more remote. For the purpose of curing fish Government salt is not used to any extent, but they are chiefly prepared with salt-earth. The practice of salting fish has decreased of late years. In olden times the fishermen possessed no peculiar privileges. There are headmen of the fishermen castes. There are certain degrees of headmen; the highest is styled Nambiar or Puttum Kattigal, who has authority over a number of fishing villages along the coast, and whose word is supreme; the office is hereditary and carries with it emoluments in the shape of a percentage upon the fish captured. The duties of the office, like most hereditary ones, are light, chiefly of a patriarchal nature, consisting of the settlement of disputes amongst themselves, attending the celebration of marriages, &c. The Nambiar or Puttum Kattigal has the privilege of receiving the first betel-nut amongst this class of men. The next degree of headman is called Nattamaikkaran, who is the recognised head of the village, and whose duties and emoluments are similar to those of Nambiar, though on a small scale, whilst the office is not hereditary. No one claims any rights as regards the sea-fisheries. As a rule, the fishermen purchase boats and logs for catamarans (rafts), and make their own nets. The number of sea fishermen have slightly increased. *The Tehsildar of Tritrapundi* replies that, all fish are destroyed, with the exception of very small ones which cannot be entrapped in the nets. That the salting of fish could be increased by reducing the price of salt. That the erection of enclosures wherein fish might be salted and salt sold at a reduced price would be advisable. The selling price of Government salt is 2 rupees per maund, but the merchants who buy the salt at this rate sell it at Rs. 2-4-2 a maund, whereby obtaining a profit of four annas (6 pence) on every maund. Government salt is used for curing fish, but only after it has been purchased by merchants. Salt-earth is used to a small extent, according to the Deputy Tehsildar's statement. In other places neither salt-earth nor sea-water is used for this purpose. The practice of salting fish has decreased. It does not appear that in old times the fishermen had any privileges they do not now possess. *The Deputy Tehsildar* writes as follows :—"There are headmen of the fishing castes ; their post is hereditary, and when all the heirs are extinct it is bestowed on a competent man selected for the purpose by the residents of each hamlet or street. Each headman determines the labour, &c., to be performed by men subject to his jurisdiction, fixes the rate of wages, &c., and gets such work done by them. He obtains for his services as much as each of those working under his control receives as his wages, and another extra share as a special remuneration for his headmanship. He also settles ordinary disputes regarding caste and custom that may arise amongst those classes. His emoluments cannot be ascertained otherwise than in the aforesaid manner." *The Tehsildar* continues, that no one claims any right respecting the sea fisheries. The fishermen procure nets or boats at their own expense, and those who cannot afford to do so get a loan thereof, while some join those who are possessed of nets or boats in catching fish. In the latter case fish caught are divided into three parts, of which two form the share of the owners of nets or boats, the third part going to those who actually catch them. The fishermen as to numbers have remained stationary in Topputurai and thereabouts, but decreased in Mutupettai. *The Tehsildar of Myaveram* states that breeding and immature fish are being destroyed. The salting of sea-fish could be increased were all the large ones thus cured. But it is impossible to have any enclosures made for the following reasons :—The fish captured in the sea and brought on shore are not afterwards carried by the men. The women only carry them. If the enclosures were within a mile or two, those that are brought ashore within 4 or 5 P. M. can be carried there by the women before 5, 6, or 7 P. M. Fish are captured and brought on shore even after 6 P. M. up to 10 or 12 o'clock in the night. In such cases, it will be impossible for the women to take them to the enclosures, although they are very near. If the captures are not taken to the enclosures till the next morning, they will become spoiled and useless. All the captures which are not sold fresh should be split and salted so soon as they have been brought on shore. All the fishermen have not got nets; some eight or ten persons engage themselves as coolies under a net-holder. The net-holder may deliver his whole share at the enclosures for being salted, but the coolies will suffer loss if they are also to deliver up their shares, because they would not get a good price ; every coolie carries at once his share to other places and by selling them for grain or cash, earns his livelihood. They can get a better price if they go to other places; even in the locality where fish are captured, the price is not stationary. If the enclosures are made, an arrangement may be made that salt may be sold there at rates remunerative for the exclusive purpose of salting fish. The monopoly price of salt is Rs. 2 a maund; it is used for curing fish. Salt-earth is not used, but sea-water is employed for the purpose. Those that are captured in the evenings are split and thrown into a tub filled with salt-water, and allowed to remain there the whole night. Next morning they are taken out and dried. The practice of salting fish has decreased for the last ten years, because a large amount of the fish is sold fresh. The fishing castes never had any peculiar privileges. There are headmen called Nattameigars ; the post is hereditary. The headman employs coolies of his own for fishing. On the occasion of marriages and funeral ceremonies, the fishermen caste people act up to his orders. On marriages he has 8 annas to 1 rupee, according to their ability. This is all his income ; he gets nothing for funeral ceremonies. No one claims any right in respect to sea-fisheries ; the fishermen supply themselves with their own nets and boats; their numbers have increased of late years. *The Tehsildar of Thealli* considers that breeding and immature fish are not destroyed. The salting of sea-fish could be increased by a reduction in the price of salt. Erecting enclosures within which fish might be salted is not impracticable. Provision should be made for bringing to the enclosures all the fish that are to be salted, for buying salt just enough for curing them, and for removing them afterwards out of the enclosures in a cured state. The price of salt is Rs. 2-3-5 per maund. Government salt, as well as salt illicitly manufactured from earth, and spontaneous salt collected by stealth, are used for curing fish. Salt-earth or sea-water is not employed for the purpose. The practice of salting fish has decreased of late years, as the captures have become less in quantity. The fishermen have all the privileges they ever possessed. There is one headman for each "Cuppam" or small village on the sea-coast; the rank is hereditary. His duties are to settle the disputes amongst them regarding their caste, and to be the chief for carrying out marriage as well as funeral ceremonies. His emoluments are, that he receives from each family a fee at the rate of eight annas for each marriage. He also fishes for his own livelihood. No one claims any right as regards the sea-fisheries. The fishermen supply their own boats and nets; they have increased of late years. *The Tehsildar of Pattukottai* replies that the retail price of salt is Rs. 2-4 a maund; that it is used for curing fish, the process being as follows: A long split with a knife is made on the back of the fish from his head to his tail; salt is put into it; then it is closed, and the fish buried in a pit for some days. By so doing the salt preserves the fish from putrefaction and draws out the fluid matter which is in the fish. The fish-curers, however, assert that they don't use sea-water for the purpose, and plead that if sea-water is poured into the split, and the latter is folded up, the water, incapable of remaining long in the split, not only comes out but also has not the power of drawing out the fluid matter in the fish : so the fish consequently gets putrified by either kind of fluid, emits stench and becomes thus spoiled. As for using salt-earth they urge the same plea, and add that if salt-earth is used, the fish soon become wormy and rotten; but this I think is said through fear, since I learn, on enquiry, that they nevertheless do use salt-earth in curing fish. Those who are engaged in salting and curing fish purchased from fishermen complain that their trade suffers much from the time the price of salt has been enhanced, and offer only eight or twelve annas for the fish formerly worth one rupee. Under these circumstances, I consider it advisable that the price of salt should be reduced as a stimulus for fish-curing. A suggestion is made for salting and curing fish in an extensive place outside large towns, and with this view for selling salt in that place at a cheap rate. There is no objection to this being done, provided there be a sea for fishing in the vicinity of such town, or so near it that the fish caught may be taken there as soon as possible. This precaution is necessary in order to provide against the contingency that if fish have to be taken to that place from a long distance for being cured, they will be spoiled unless salted, some kinds within twelve hours, and others within twenty-four hours, from their being taken from the water. In case of reducing the price of salt for the sole purpose of encouraging fish-curing, certain restrictions should be made, such as that salt should be sold at the reduced rate only in the place where fish are cured, and that salt brought there should not be taken out or used for any other purpose. The practice of salting fish must be said to be increasing, considering that the price of fish to be cured, which formerly cost one rupee, has now been reduced to ten or twelve annas, and that fish-curers offer only this reduced rate; and seeing also that large quantities of salt-fish are daily imported by rail. The fishermen formerly had no privileges they do not now possess; no one claims any right respecting the sea-fisheries. Boats are bought or manufactured by the fishermen themselves at their own expense. Nets are either bought or made of flax or cotton-thread; the former material being purchased in Calamere Point and Vedenaniam, and the latter in the local markets. Formerly the only classes of fishermen were (1) Sambadavers, and (2) Karaiyais, whereas now it appears that certain other classes such as Kadayars, Sanagais and Valayars are also engaged in fishing, and that the Sambadavers remain stationary. *The Superintendent of Sea Customs, Negapatam*, states that salt is sold by Government at Re. 1-13 a maund, whilst the retail price is Re. 1 per five marcals of 10½ seers. (Tanjore marcal is supposed to be equal to 3 Bengal seers.) (The Collector notes a marcal is 25½ tolahs of rice). Government salt is used for curing fish, which are first washed in the sea-water, and then Government salt is employed for curing them. The practice of salting fish has decreased this year; the fishermen still possess all the privileges they formerly had. Amongst the Pattanaver caste each suburb has one or two headmen, who are appointed by the villagers, and whose appointment is hereditary. These headmen procure at their own cost catamarans and nets in like manner with others, and earn their livelihood thereby. Pattaneivars claim an exclusive right in respect to the sea-fisheries: but there is no objection to others also catching fish in the sea. The fishermen weave their own nets, as the practice of fishing in boats does not obtain here. Trees called Malaivembu are brought from Colombo, made into catamarans, and then used for fishing. The number of fishermen have remained stationary.

xci

Return showing the quantity and value of Imports and Exports by sea of salted fish, &c., in the district of Tanjore from the year 1863-64 to 1872-73.

YEAR.	IMPORTS.								EXPORTS.							
	FISH MAWS.		SHARK FINS.			SALTED FISH.			FISH MAWS.			SHARK FINS.			SALTED FISH.	
	Quantity.	Value.	Quantity.		Value.	Quantity.		Value.	Quantity.		Value.	Quantity.		Value.	Quantity.	Value.
	Cwt. lbs.	Rs.	Cwt. lbs.		Rs.	Cwt.	lbs.	Rs.	Cwt.	lbs.	Rs.	Cwt.	lbs.	Rs.	Cwt. / No. lbs.	Rs.
1863-64	... ...	...	... 84		15	404	...	1,304	81	17	4,058	100	12	4,075	542 Cwt. ...	1,578
1864-65	... ...	...	3 14		62	379	58	1,454	136	14	5,944	...	...	...	284 No. 899,900 / 7	1,907 / 1,373
1865-66	... ...	...	... ...		...	374	28	828	1	0	50	114	98	4,604	473 No. 2,906,458 / 70	6,103 / 1,440
1866-67	1 70	81	12 14		242	645	63	2,339	...	...	...	116	...	3,681	585 No. 142,000 / 28	331 / 2,957
1867-68	... ...	...	12 84		255	537	63	1,935	11	14	556	75	77	2,724	481 No. 10,000 / ...	224 / 3,135
1868-69	... ...	...	26 46		521	343 No.	49 650	1,492 13	34	7	1,692	241	7	8,478	403 81	2,286
1869-70	... ...	...	19 70		383	228 No.	56 1,000	787 43	10	...	240	42	70	1,089	391 56	6,952
1870-71	... ...	...	14 56		290	602 No.	98 7,300	2,260 110	41	56	1,885	167	...	6,244	525 No. ... / 16,000	2,540 / 32
1871-72	... ...	...	8 0		160	563 No.	56 5,157	1,495 97	22	0	490	70	56	2,110	172 70	1,068
1872-73	... ...	...	... ...		...	...	...	...	...	...	...	43	...	1,106	211 No. ... / 46,500	952 / 180

104. From the *South Arcot Collectorate* the following replies have been received from the *native officials*. The Tehsildar (July 31st, 1869) replies that fish-curers are not allowed to use salt-earth untaxed; if they were, it would better first to have it manufactured into salt. People will use fish salted in any way; they have no idea of what is called "high salt"; the only consideration to them is the cost. Men of the Carriar, Patnaver and Pullie castes fish in the sea, whilst it is solely Pullies who do so in the back-waters; they are known by the name of Shemdavers. Fish are sometimes taken six or seven miles for sale, as far as is consistent with safety to the fish. The sea-fishermen on the coast earn between ten and twelve annas a day. The fish on the coast are said to have diminished, and the cause is attributed to the bad seasons we have had of late. The fish appear to have receded into the deep sea. Any amount of fish captured would meet with a sale; everything is captured as all find purchasers. Fish are salted to a great extent with salt stated to be purchased from the Government depôts, and to a small extent perhaps it is. But my impression is that salt is obtained in good quantities by lixiviating salt-earth which abounds in the marshes and back-waters on the coast. This is got for nothing, and its manufacture costs equally nothing. Sea-water is not used. Salt-fish is not exported. None are smoked: the process observed here is kippering, *i. e.*, to rip open the fish, clear it of the intestines, and, having incised the fleshy part, to apply salt and leave it to dry. In reply to "are fish dried"? he says "I do not understand this, if it is meant to ascertain whether fish are dried for sale independent of salting, I answer not to any extent. Sea-fish, never. Fresh-water fish of the smallest kind, which finds no purchaser, if arrived at the stage of putrefaction, is dried up in sand and sold to poor people who are unable to purchase the larger fish; otherwise salt-fish is essentially dried fish." Neither fish fins nor sounds are preserved. *The Tehsildar of Cuddalore* (1872) replies that salt is sold at Rs. 2 a maund, and is used for curing fish. Salt-earth and sea water are also employed for this purpose. The practice of salting fish has decreased of late years. The fishermen caste have no particular privileges; they have headmen; the post is hereditary. They attend marriages and other ceremonies occurring in the caste, and distribute betel-nut to the people on the occasion, for which they receive from 4 annas to 2 rupees, according to the circumstances of the parties. No one claims any rights in regard to the sea-fisheries. Fishermen supply themselves with their boats and nets: these men have been on the decrease of late years. *The Tehsildar of* ? (locality not stated) replies that fishermen are not provident as a class, and capture everything. What they do not use, they leave where they leave their nets. The salting of fish might possibly be increased if the price of salt were reduced for the purpose, but not to any extent. As a rule, the take is absorbed by the local market. The proposition in paragraph 7 (respecting enclosures) would be practicable, but with the employment of an additional preventive force to check smuggling. It would hardly be advisable, as being unlikely to produce any measure of public good for the reasons given. The selling price of salt is Rs. 2 a maund; it is almost exclusively used for salting fish: salt-earth or sea-water is not employed with the knowledge of the

Opinions of native officials in the South Arcot Collectorate.

xciii

authorities, and probably not at all. The practice of salting fish has decreased to a slight extent, possibly owing to the price of salt. Fishermen had no privileges in old times which they do not now possess. They have an hereditary headman named "Nattamakar"; the duties are simply those relating to the social relations of caste, such as presiding at ceremonies, arbitrating in disputes. The emoluments—petty presents or fines on such occasions. No one claims any rights in respect to the sea-fisheries. The number of fishermen has increased except in the Cuddalore talook where they have decreased.

105. From the *Madras Collectorate* the following answers have been received from the *Native Officials*. The *Tehsildar of Conjaveram* (1871) replies that his talook does not extend to the sea. Fish are salted to a great extent; common salt is usually employed; only a few persons make use of salt-earth, but it is criminal to sell such. The present quality of salt-fish is considered to be good, and the supply equals the demand. *The Tehsildar of Chingleput* (1871) says that the castes which fish in the sea and back-waters are Patanavur, Pullee, Parambar, Karayar, and Jonakar. Fish is not taken more than twenty miles for sale : the local markets are sufficiently supplied. The fishing population is decreasing, but the boats and the number of nets continue much the same. The fishing months in the sea are chiefly from January to June ; it is also carried on, but on a much smaller scale, from August to November, whilst in November and December respectively it, becomes still less. Fish are salted to a great extent ; only Government salt is employed, salt-earth is not used. More salt-fish could be sold ; it is not exported ; fish are not cured by smoking, but they are dried ; roes are salted ; fins or sounds are not preserved. Fish liver oil is manufactured. *The Tehsildar of the Madurantakum Talook* (1871) observes that the local markets are sufficiently supplied. The fishing population is increasing, as are also the boats and nets. Fish are largely salted with ordinary salt, but not salt-earth or sea-water. The quality of salt-fish is good, and the supply is equal to the demand ; some is exported. Fish are neither cured by smoking nor their roes or sounds preserved : also fish are dried and fins collected, but oil is not manufactured. *The Tehsildar of Ponnery* (1871) states that fresh fish is not carried above 10 miles ; the fishing population is increasing ; the number of boats (50) continues the same, but nets have increased. Fish are salted to a great extent with Government salt and salt-earth, but sea-water is not employed for curing fish. More salt-fish could be sold. Salt-fish are exported ; fish are dried, roes salted, but fish are not cured by smoking, nor are fins or sounds preserved. Fish liver oil is manufactured. *The Tehsildar of Sydapet* (1871) reported that the local markets are sufficiently supplied ; sometimes the captures cannot be sold fresh, and then they are salted. The fishermen are increasing : the number of boats (64) have continued the same, but the nets have increased. Fish are salted to a great extent with Government salt, but neither salt-earth nor sea-water is employed for this purpose. The present quality of the salt-fish is considered to be good. In answer to the question "is there a good market for salt-fish or could more be sold ?" he replies : " High demand for salt-fish in the bazars. They could not be sold in greater quantities than at

Opinion of Native Officials in the Madras Collectorate.

xciv

present." Salt-fish is exported. Fish roes are salted, small fish dried, fins are preserved, but sounds are not so. Fish liver oil is extracted.

106. From the *Nellore Collectorate* the following replies have been received from the *native officials*, whose answers are compiled by the Collector (October 26th, 1869). The following castes generally fish in the sea: Pallevandlu, Pattapuvandlu and Chembadivandlu; they all as well as men of many other castes likewise net the back-waters and rivers. Some retail dealers do not fish, but merely purchase to re-sell: some is exported to the western districts of Kurnool, Bellary, Cuddapah, &c., and Hyderabad. The local markets are insufficiently supplied inland, so fish are brought from the coast talooks of Gudur, Kandukur and Atmakur. The rule is that the fish are hawked about and are generally exchanged for grain; 13 Tehsildars and Deputy Tehsildars give the following as the proportion of people who eat fish: one estimates them at 25 per cent., five at from 50 to 75 per cent., four at from 75 to 85 per cent., and three at from 85 to 95 per cent. In Ongole, Rapur, Nellore, and Kavali the fishing population is supposed to have decreased of late years, whilst in Gudur, Atmakur, and Kandukur an increase is reported, whilst a stationary state exists in the two remaining talooks. The fishing boats have increased in Gudur, decreased in Nellore and Ongole. Amongst the five talooks bordering on the sea, the nets have decreased in four and increased in one. The best fishing months in the sea are from January to April, inclusive. In three out of five of the maritime talooks fish are salted to a great extent: at Atmakur sea-water is used, elsewhere purchased salt is stated to be employed, and some of the Tehsildars consider salt-earth is not used for this purpose. The present quality of salt-fish is considered good. It is exported from all the coast talooks except Kavali. Fish are generally dried, fins and sounds are prepared, as well as fish-oil, and in some places roes are salted. The Tehsildar of Ongole (January 29th, 1873) replies that salt in the zillah is sold for Rs. 2 a maund and is made use of for curing fish if large, small ones being simply dried. "It is a custom amongst some people to employ a small quantity of salt even for large fish, and lay them under ground on the sea-shore for a day and expose them to the sunshine. The poor fishermen, unable to buy salt, wash fish in salt-water, bury them in saltish-earth for a day and manage somehow or other to make them kara-vallu (salt-fish), but this karavallu is not only tasteless but stinking, and becomes useless in a short period." Small fish dried and salt-fish are exported to Hyderabad, Kurnool, &c., whilst some goes to Madras. The Palle caste people, Patnapu (Tuli) or Mahomedan caste men catch salt-water fish, and these men, it appears, were formerly related to one another. But Palle people generally fish in salt rivers, and Tuli people in seas, but people of both castes give their fish to traders on contracts, or sell them personally, and use what remains for themselves. Fishermen have one religious headman to whom they give at every marriage 4 annas with $2\frac{1}{4}$ seers of rice and other grains. There is another man named priest in their caste, to whom also they give at marriages Rs. 2 with $4\frac{1}{2}$ seers of rice and other grains. The latter man (priest) has frequently to visit the coast and other places where fish are taken, but does not go with the fishermen. Their headman

in religion has to decide their disputes, and if he is unable to settle it the priest has to pass a final order. The head in religion and also the priest are hereditary officers. The Tuli caste people have lived by fishing for an indefinite period of time, and they appear to think that they have an indisputable right to fish in the sea. Most of the Tuli men are very poor, their daily earnings only just covering their expenses. Whenever they may be in need of new boats or nets, or at least of repair for their old ones, they borrow money from traders, and give them fish every now and then in satisfaction of their debts. Some fishermen also obtain money by exporting grain on their own vessels, or giving their vessels for freight. The number of fishermen in the Ongole talook is 200 or thereabouts. Fishermen in Gudur talook come with their families to the coast of Ongole about the month of January and quit the coast by the end of June, during which period they export quantities of fish to their own district, and besides take home with them the remaining lot. The Tuli fishermen are few and do not possess above 10 or 15 nets and boats. The opinion suggested to decrease the value of salt only in the case of selling to fishermen, is thought favourably of, and an encouragement to men of the Tuli caste to salt fish, yet it appears that the Government will become partial. Now in stations, officers of the Salt Department throw away some quantity of salt in the water on the ground that it is something like powder or black-coloured. The Tehsildar suggests that this would do for the fishermen, selling it at Re. 1 a maund, and prohibit it from being re-sold. The replies of the *Salt and Sea Customs officials along the coast or on the shores of the Pulicat lake* are thus given. (November 23rd, 1872.) The selling price of salt is Rs. 2 a maund. The Salt Superintendents of Kanuparti, Pakala and Gogulapalli, and the Sea Customs Superintendents of Kottapatam and Ramayapatam state that Government salt alone is used for curing fish, and that salt-earth or sea-water is never made use of. The Padarti Salt Superintendent says, Government salt is chiefly so used, but that sometimes spontaneous salt also is smuggled and applied for the purpose. At Tunnualapentah large fish are said to be cured with Government salt, and small ones with salt-earth and sea-water. At Kristnapatam also, both Government salt and sea-water or salt-earth is made use of, but fish cured with Government salt fetches a better price. The Isakapalli Salt Superintendent states, that mostly salt-earth and sea-water is made use of, and that Government salt is seldom or never known to be bought for the purpose ; and the Dugarazupatam, Tada, and Sunnapugunta Superintendents appear to be of the same opinion and express a doubt as to the quantity cured one way or the other, but they seem to think that both measures are commonly resorted to. The Salt Superintendent of Kanuparti expresses a doubt as to whether the practice of salting fish has increased or decreased of late years. The Sea Customs Superintendent of Kottapatam, and the Salt Superintendents of Padarti, Tunnualapenta, Isakapali and Sunnapugunta are of opinion that there has been a decrease of late years; the last named because of the rise in the selling price of salt, which has made it more difficult for the poor to procure the article and cure the fish properly, and the others on account of generally deficient rain-fall of late years, whereby the bars have become silted up, and many tidal creeks closed from the sea. The Gogulapalli, Dugurazupatam and Tada Salt Superintendents think that the practice has increased owing to a greater demand for salted fish, whilst the Superintendents of Kristnapatam and Pakala and the Sea Customs Superintendent of Ramayapatam are of opinion that it has remained stationary. The fishermen have now all the privileges they formerly possessed, but some are of opinion that the means of livelihood have generally decreased, as the demand for salted fish has generally diminished. Others again think, where the trade in salt-fish has increased the income of the fishermen has likewise improved. The fishing castes in this district are four, (1) Palle, (2) Tuli, (3) Patapu —and (4) Chambadi. Each caste, and in fact each village or hamlet where a number of them club together and reside (forming what is called a Palliem), has its own headman, called in some cases 'Pedda Capu' and in others ' Pedda Arkattu.' This office is hereditary, and on the failure of heirs, the community join together and select one from amongst their number to be their future headman. The duties are mostly honorary. The headman presides at all marriage and religious ceremonies, for which he receives certain 'russooms' or fees. He settles all petty quarrels and disputes in his ' Pollien ;' he is looked up to and his word obeyed with greater respect than any one else in that ' Pollien.' He has the privilege of being exempted from work. If a vessel strands or comes off his hamlet in distress, he gathers together all the able-bodied men and gives help, so likewise at any Government call he furnishes help and collects labour, but is exempt from personal work. The duties of the office are not thus defined but recognised merely by custom and long usage. The emoluments likewise are not regular but consist of contributions or fees paid by the people from long-acknowledged habits. All the Superintendents are unanimous in stating that no one claims any right to the sea-fisheries, but the fishermen of each hamlet or pollien are very tenacious as to their peculiar rights to fish within certain limits, whether in the sea, or in any tidal creek or estuary. These limits have never been defined or recognised by any authority, but have been admitted by long-established usage among the fishermen themselves from time immemorial, and if the fishermen of one hamlet are found transgressing their limits, and plying their trade within the limits of a neighbour, the result is a never-ending source of dispute and quarrel among the men of both hamlets. But the quarrel seldom goes further than themselves, and receives no countenance at the hands of any authority. For the purpose of supplying themselves with boats or nets, they purchase materials and construct them themselves; when unable themselves to make them, they employ men skilled in the work for the purpose and pay such men the ordinary rates of cooly hire. The Kanuparti Salt Superintendent states that in his division no fishing boats are used, but only nets. The Tunnualapenta Salt Superintendent says that the value of a large net for sea-fishing is 100 rupees, and of small nets used in tidal creeks and estuaries 4 to 5 rupees. The Iskapalli Superintendent thinks that the fishermen will sell a large net to any one wanting the same for 50 or 60 rupees. As regards whether the number of fishermen have increased, decreased or remained stationary, the Superintendent of Kanuparti, Padarti, Tunnualapenta and Tada and the Sea Customs Superintendent of Kottapatnam are of opinion that the number has decreased. In Tada the decrease is attributed to a falling-off in the means of livelihood. The Iskapalli Salt Superintendent also thinks there is a decrease, except in the 'Pattapu' caste of fishermen, in which caste, owing to a larger number of nets, he thinks there is an increase. The Salt Superintendent of Gogulapalli, Kristnapatam, Dugarazapatnam and Sunnapugunta, and the Sea Customs Superintendent of Ramapatam, state that the number has increased on account of increased means of livelihood. The Pakala Salt Superintendent is the only one who reports the numbers to remain stationary, but he assigns no reasons for his belief. *The Tehsildar of Striharikota* (1872) replies that in this district the breeding and immature fish are caught and destroyed by means of nets, by the class known as Pallepuvandlu, &c. The breeding months are September and October, and one-tenth of the fish caught at these times are breeding. Salt-fish is purchased by merchants and exported to Madras. If the price of salt thus used were to be reduced, the salting of fish would very likely increase. The working of the proposition (enclosures) in paragraph 7 is impracticable, because there are no salt stations in the division except beyond a distance of 10 or 15 miles. Government salt appears to be used for curing fish. Small sea-fish after being washed are exposed to the sun on the sands, but salt-earth does not appear to be used for this purpose. The practice of salting fish has remained stationary. Fishermen have the same privileges now that they formerly possessed. Amongst the fishing caste there is one Adimulam Setti at Madras, who is the principal headman amongst the fishermen of the Pattapuvandlu caste: there are two others Dalavaya Venkatraya Setti and Mantrichina Venkatraya Setti as headmen at Puliyenjeri Kuppam in Strikarikota division; this headmanship is hereditary. The one at Madras is regarded with priestly reverence, while the other two are looked to for the settlement of religious disputes. These headmen fish like others for their maintenance; at marriages a fee of 6¼ anuas is paid to the headmen, as well as fines for breach of religious rites. Half of these collections go to the headman at Madras, and the remaining portion is enjoyed by the said two headmen. But this practice is said to be gradually falling off, because some pay on the occasion, whilst others get their wants attended to without payment. No one claims any rights respecting the sea-fisheries. The fishermen Pattapuvandlu freely fish in the sea. Fishermen procure boats and nets at their own cost. During the last four years the quantity of fish taken has been less than in previous years, consequently the sea-fishermen are in a poor condition. *The Tehsildar of Handukur* (1872) observes that fishes of all sizes are captured: that the small ones are destroyed by fishermen, large fish, and birds. The fishermen residing in the villages along the sea coast in this talook are not themselves able to salt fish, as they are poor; but men of the Palle and Pattapu caste receive money in advance from the fish traders that come down from Golakonda and other places and thus salt fish. It appears that they would themselves commence to trade in such fish if the price of salt is low, and thus the work would come to a prosperous state. It would be advisable to store up the quantity of salt required for fish in extensive premises selected for the purpose with reference to the said paragraph 7, and to sell it at a lower price, and strict orders may be issued, so that the salting of the fish may be effectual in the said premises only and no salt removed from it. If they are allowed to sell the fish in any other place but in the particularized one, they would sell the salt to others at high prices. Therefore it is advisable to salt the fish in a particular place, or the plan will not prosper. It appears that some salt manufactured by the ryots is destroyed as of a bad quality, but such might be sold at a low price for this purpose. Government salt is used for curing fish, but if salt-water were employed the fish would putrify, and if salt-earth is used the fish will be destroyed by insects. The practice of salting fish has decreased, as fewer fish are captured. The fishermen have all the privileges they formerly possessed; they have a headman termed Aurikuttu; the office is hereditary; he also fishes and sells his captures like other fishermen. He investigates their religious disputes, but receives no fees. No one claims any right regarding the sea-fisheries. The fishermen provide themselves with nets and boats. Their numbers are stated to have increased.

107. From the *Kistna District*, the following replies have been received from the *officials in the Sea Customs Department*.

Opinions of native officials of the Kistna District.

The *Sea Customs Superintendent of Bandar* observes (May 1873), that breeding and immature fish are destroyed by fishermen and birds. The headmen of the villages consider themselves as privileged, being used to capture fish to the exclusion of others: the practice of salting fish has not increased. The proposition in paragraph 7 would be practicable. The price of Government salt being high, it has not been usual in this district to purchase it; salt-earth and swamps salt being procurable everywhere, fish are claudestinely cured with the same. In the Vizagapatam and other countries sea-water is used for curing fish, but not in this district. The fishermen possess the same privileges they had formerly : those who are thus referred to being privileged to catch fish are headmen. Though they are not privileged by any competent court, yet they are enjoying that privilege as hereditary through the favour of influential members for the time being. Their privileges are to catch fish and to obtain emoluments at festivals and other happy occasions. The headmen of the villages claim a right to fish in the sea; they do not allow others to fish. The fishermen are stated to have remained stationary. *The Assistant Superintendent of Customs at Vizagapatam and Kottapalem ports* answers (May 1873) that all descriptions of fish are destroyed, viz., small fish by nets of close meshes, and large fish by nets made of hemp. January, February, and March being the breeding seasons, the fish come in shore when they are netted. Persons who trade in salt-fish go in the first instance, purchase salt from the Government store, and either cure the fish on the sea-shore or at their own homes. The proposition in paragraph 7 would be good and practicable. Retail price of salt Rs. 2 and wholesale Rs. 1-14 a maund; it is used for curing fish, but salt-earth or sea-water is not for such fish as are required in the trade, but some people cure small fish by means of the salt obtained from the process of mixing the salt-earth in water clandestinely, but this cannot be proved. Owing to paucity of fish last year due to northerly winds, the amount salted is said to have decreased. Fishermen have all the privileges they ever possessed. There are headmen, and the post is hereditary; in default of any who have ceased to be headmen, others are appointed from amongst the fishermen by the authorities. Their duties are to perform acts connected with their religious duties and Government work, such as exporting and importing goods, &c. At feasts, &c., they receive at the rate of 8 or 4 annas. They are allowed to set up stake nets in rivers, and a share is allowed to them out of the fish caught by the other fishermen. Their emoluments are small, but the exact amount is not known. The number of fishermen remains stationary. *The Superintendent of Sea Customs at Ipurupalem* (May 1873) remarks that breeding and immature fish are destroyed in the tidal estuaries and in the sea; they are usually salted, but some are disposed of fresh, which the purchasers salt. As there is an estuary between China Ganjam and Peda Ganjam within my range, large quantities of fish are caught there by means of rafts and nets: elsewhere fish are not in any large quantities. If an enclosure be made near the China Ganjam Salt Depôt, and fish salted inside it with salt sold at the proposed rates, the amount cured would be increased. The retail dealers sell salt at Rs. 2-4-0 a maund. At the port of Ipurupalem it is sold at Rs. 2-8-0; it is used for curing fish, but neither salt-earth nor sea-water. Formerly, large quantities of fish were procurable at the ports of Pachamagali, Ipurupalem, Naidupalli and Motupalli. The fish are not procurable now in such large quantities. There are no headmen of the fishermen caste at Ipurupalem port; every man is independent. There is a headman at China Ganjam and Peda Ganjam. He is supreme in matters of religion and festivals, but derives no emoluments; whilst the moturpha tax existed he was exempted from it. Among the Pattapu caste people who catch fish by employing boats there is a headman who settles their family disputes and religious customs; he is annually allowed 8 annas for every boat employed in fishing. The fishermen at Ipurupalem have remained stationary, whilst those at China Ganjam, Peda Ganjam, Motupalli and Naidupali have increased. From the *Kistna District* the following returns have been received from *the other native officials*. *The Tehsildar of the Bapatla talook* (1871) replies that the fishermen both capture and salt fish, and that their numbers continue about the same, with the exception of the Tulis who have migrated here within the last ten years, and fish in the sea, consequently boats and nets have become more numerous. During the months of November and December fish are reported to be taken in small quantities, and during January, February, and March in large quantities; a considerable amount of fish is salted, for which purpose Government salt is used and not salt-earth (chowdu): however, the latter is said to be collected: sea-water is not employed. The quality of the salt-fish is reported to be good; it is exported inland but not by sea. Fish roes are dried, not salted. Fish are thus cured: as soon as caught they are opened, the entrails removed, they are washed in sea-water, then unground salt is put into them, and they are exposed to the sun. The next day more unground salt is put in, and they are again exposed to the sun's rays. On the third day, they put in a little quantity of ground salt and lay them out in the sun: on the fourth day they are again put out in the sun and then stored until sold. Sharks' fins are dried; Jonangi people come from the south and purchase them at a dear rate. Some small fish are simply dried. Shark's liver oil is manufactured; fish sounds are prepared, and it is used as a glue for repairing boxes. Subsequently (May 1873), in reply to the questions circulated in 1872, he observes: except the sorah chápa (sharks) all other fish produce roe, that of the perch is useful, while that of the other species becomes useless so soon as the fish is opened. Pundu chápa is rarely netted in the estuaries. The fishermen living along the coast from Peraly up to Peda Ganjam, fish for four miles out to sea, obtaining perches. Other species are caught along the shore in great quantities, whilst those taken in the sea are few. The fish when brought ashore are sold to merchants who salt and trade in them. Should no merchants arrive the fishermen salt the fish themselves. Though the price of salt is now high, yet the fish fit to be cured* are being salted when they are procured in sufficient quantities. By reducing the price of salt, the trade in salt-fish would not be augmented, as all the fish caught along the sea-shore and estuaries are salted. There are no considerable number of fishermen in this taluq. Fish merchants usually salt fish at their own houses, as do also the fishermen. Salt is sold at Rs. 2 a maund at the Government depôts, and in distant villages at an increase of 1 auna in the rupee. Government salt is used for curing fish both by merchants and fishermen. Were sea-water employed the fish would putrefy, not being sufficiently saline; salt-earth likewise is not used. The best fishing months are January, February and March. The practice of salting fish has remained stationary. Among the fishing castes, each village had one or two headmen; the fishermen do not understand how the headmen were formerly appointed; they do not possess any certificate or patta to show by whom they were appointed. The office is hereditary; the headmen decide family disputes, and direct the fishermen to furnish supplies to Government when required. At marriages they receive betel-nut; on festive occasions they are asked prior to relatives. During festivals of the village goddess the headmen perform the ceremony, the cost being paid by the fishermen. First the headman's sheep is sacrificed, subsequently those of other people. Any one may fish as they please in the sea or estuaries, but the Public Works Department sometimes prevent raising dams across Romparu, between Peda Ganjam and China Ganjam to catch fish, on the plea that the drainage of surplus water will be prevented from passing into the sea. Boatmen from the Nellore district, or about 20 fishermen, have emigrated to China Ganjam, and certain other fishermen from Negapatam to Perali, consequently the fishermen along the coast in this taluq have somewhat increased. *The Tehsildar of the Bunder Talook* (1869) answers that the large fish which are caught in the sea, salted and dried in the sun, are very good, whilst the smaller ones thus prepared are not so tasty. The fishermen have decreased since the cyclone, in which many were washed away; about 20 boats go to sea for fishing purposes, which as well as the nets have decreased. The best months for fishing in the sea are from December to March. Fish is not salted to any great extent. Government salt is employed for this purpose, and earth-salt is not used† in curing large fish; when used it is collected from the

* The superior kinds of fish, if of a large size, must be cured with good salt or they will decay if kept any length of time.

† See commencement of this paragraph, where this is distinctly contradicted by the Sea Customs Superintendent of this port, the meaning, I suspect, being that Government salt is used for curing the larger and better sorts of fish.

swamps. Salt is not prepared from sea-water by the fishermen; the quality of the salted fish is excellent; the quantity made continues about the same. Some is exported to Hyderabad and other countries for sale. Small fish are dried in the sun; none are smoked; fins are collected, and fish-oil is manufactured. Subsequently (May 1873) he replied that breeding and immature fish are destroyed in the tidal estuaries to any extent without distinction. Immature fish are caught by means of thick nets, the meshes being as small as Bengal gram. Fish to be salted are first split open; their intestines removed; they are then rubbed with salt and dried for a week. " I do not think that the proposition in paragraph 7 would be advisable or practicable. Government salt is used for curing fish, but neither salt-earth nor sea-water. The salting of fish increased prior to the cyclone, when a great number of fishermen perished and their nets were washed away. Whilst the moturpha tax existed, the residents of villages had the peculiar privilege of catching fish in channels and estuaries within their respective villages. After the abolition of the moturpha tax, the fishermen have not had that privilege. Now-a-days every one catches fish. There are no headmen in the fisheries. Estuaries and the mouths of the Kistna have fisheries. It has been usual for the fishermen living there to fish in them. The usage observed many years ago was that each fisherman among the villagers had a fixed number of nets. The limits within which they could catch fish were fixed. If the residents of some village had set up stake nets contrary to usage, complaints were made to the local authorities who decided them. This usage is observed in the port of Bunder, Gilakaladindi, the sea-side and estuaries of Latchimipuram and Padapatnam, &c., villages in this taluq. All fishermen have an equal privilege in regard to the sea, Kistna river, and other channel fisheries. The fishermen who decreased due to the cyclone are now stated to be increasing." The *Tehsildar of the Repalli talook* (1869) observes: Salt-fish sells at two rupees a maund; each fish weighs one quarter of a maund, and when fresh is worth four annas. The supply of fish does not equal the demand; dried fish are exported; fishermen report that they are decreasing in numbers, as are also their boats and nets. Salt-earth is said not to be used for curing fish, nor is sea-water employed. Small fish are simply dried in the sun. No fish-oil is prepared. Subsequently (May 1873) he replied that the fishermen go 1½ mile out to sea for fish which they salt: but there is only a demand for small sorts. At Nizampatam, Kottapalem, and Lankavenidibba fish of good quality are salted and exported to Hyderabad and elsewhere. The fishermen being unable to purchase salt for salting fish, take advances of money for their livelihood from fish merchants coming from Bunder, &c., to whom they deliver their captures, and the merchants have them salted by coolies employed by them. These are exported inland, there being no local demand for them. Government salt is used for curing fish, but neither salt-earth nor sea-water. For the last two years fish have been scarce. Each village has a headman who is termed Pedda Capoo: he receives four annas at a marriage. When the moturpha tax existed he was exempt; he has to obtain boats when required for Government service. When the fishermen captured fish he used to have a share. When the moturpha tax existed they were allowed palmyra trees for canoes and palmyra leaves for covering their houses without taxing them for the price of the same. Those fishermen who are unable to obtain nets, &c., employ themselves as boatmen, sailors, &c.

108. *From the Godavery Collectorate* the following returns were received from the *native officials*. The *Tehsildar of Ramachendrapur* replies (1869), that both fishermen and boatmen salt fish. " The daily earnings of those who fish in the sea will be one rupee, while the other fishermen who fish in canals, &c., earn four or two annas a day, which is not more than sufficient for their subsistence." The supply is not equal to the demand. " The fishing population has decreased in consequence of a few having resorted to Moulmein and other coasts for carrying on their trade." The best fishing months in the sea are November, December and January. Fish are salted with Government salt; salt-earth is not employed: sea-water is not used for this purpose. Salt-fish is not exported. Some fish, after they have been dried in the sun, are smoked. Fish roes are salted; small fish are dried without being salted. A few sharks' fins are preserved and sounds are collected. Fish liver oil is manufactured. The *Tehsildar of Amalapuram* answers (1869) that his district does not reach the sea coast, but that fish which remain unsold fresh are salted with Government salt, the salt-earth not being collected, nor the sea-water used. The salt-fish is considered to be good, and the supply sufficient to meet the local demand. The *Tehsildar of Narsapuram* (1869) observes that men of the Pallevallu, Bestavallu and Vaddi castes fish in the rivers and back-waters; they also sell and salt fish. The fishing population remains stationary; catamarans not boats are used for sea-fishing, and their numbers as well as those of the nets remain the same. The best months for sea-fishing are from November to February inclusive. Fish are salted to a small extent with salt purchased in the markets, but salt-earth is not employed, neither is sea-water used for this purpose. The present quality of salt-fish is considered to be good and the quantity supplied equal to the demand. The only sort of fish cured by smoking are damaged cray fish. Fish are dried and roes salted; fins and sounds are preserved to a very small extent. Fish liver oil is also prepared. The *Superintendent of Sea Customs at Narsapur* (1872) replies that the local markets are fully supplied during eight months of the year: from March to June insufficiently so. *The Tehsildar of the Peddapuram talook* (1869) reported that salt-fish is carried coast-wise for 500 or 600 miles, and in a fresh state 40 or 50. The local markets are insufficiently supplied, the fresh being preferred to the salted article. The fishing population has increased of late years; about 1,500 boats go to sea, and those as well as the nets are more numerous than formerly. The best fishing months in the sea are from December to March inclusive. Fish are salted to a great extent; for some Government salt is used, for others salt-earth, which is excavated from the swamps. Sea-water is not employed for this purpose. The quality of the salt-fish is considered good. Fish are cured by smoking or drying; the roes are salted; sharks' fins are also preserved. In the *Tanaku talook*, which does not border on the sea, salt-fish are said to be preferred to the fresh, and the supply does not equal the demand. In the *Ellore talook*, an inland one, salt-fish are said to be preferred to the fresh, but as they are dearer the poor cannot purchase them. If salt-fish were brought in enormous

numbers and their price reduced a little, every one will purchase them. The *Deputy Tehsildar of Coconada* (1869) replied that there are nine fishing villages, two of which have 200 occupants, whilst the others have from 20 to 30. The fishing population has decreased of late years; about 40 boats go to sea, and both these and the nets are less than formerly. The best fishing months in the sea are February and March. Fish are salted; salt-earth is employed, which is obtained free of cost from the swamps near the sea-shore: sea-water is also used. Fish salted during the summer are considered good, and those prepared during the cold season are not so. Salt-fish is exported. Fish are dried; prawns are smoked; roes are salted; fins and sounds are preserved. The porpoise is taken for obtaining oil from, and this oil is used for curing rheumatic pains. The tail of the ray is used for smoothening wood, and sharks' fins for colouring palanquins, &c. Fish liver oil is also manufactured. *The Deputy Tehsildar of Coconada* (1872) replied that Government salt is seldom used for curing fish, but salt-earth and sea-water are often employed for this purpose. The practice of salting fish has remained stationary. Fishermen had no particular privilege in old times which they do not now possess. There are two kinds of headmen of the fishing castes, *viz.*, Kulapedda (head of the caste) and Jattupedda (head of an assembly). The first sort of headman is hereditary, whilst the latter is conferred on some one by all the inhabitants of the village. The Kulapedda will be headman of the caste for two or three districts, and such headmen employ themselves in settling religious disputes, in conducting such public affairs as may have to be performed on behalf of fishermen, &c., and in disposing of cases such as adultery, &c., if committed in these castes. They have neither land nor other emoluments. Presents are given them at times of marriages. Inhabitants of villages claim the right of the fisheries in front of their land. The number of fishermen continues stationary. *The Deputy Tehsildar of Tuni* (1869) observed that his talook does not border on the sea, but that sea-fish are brought there daily. That fresh and salted are equally preferred; the quality of salt-fish is considered to be good. *The Deputy Tehsildar of Pittapur* (1869) reported that there are eleven fishing villages, the occupants of which number 627, and that they have other trades as well. A fisherman generally earns 2 annas a day. The Sudras prefer fish fresh, the Pariahs prefer it salted. Fishermen have not increased, because a few have embarked for Moulmein on account of famine, while some have died of cholera. About 50 fishing boats go to sea; both they and the nets have decreased. The best fishing months in the sea are from December to March inclusive. Fish, for which there is no sale whilst in their fresh state, are usually salted with salt-earth collected from tidal creeks, but not with Government salt. The salt-earth is mixed with the sea-water and is thus employed for curing fish which are subsequently dried in the sun. The quality of the salt-fish is considered to be good, and more could be sold; it is consumed by Pariahs to a great, and by Sudras to a small, extent, but is said not to be fit to be exported. Fish fins and sounds are preserved; fish liver oil is manufactured; and its gall used as medicine. Subsequently (March 26th, 1873) he continues, that the monopoly price of salt is Rs. 2 a maund in his district; that it is not used for salting fish, but both salt-earth and sea-water are employed for this purpose. The practice of salting fish remains stationary. One person acts as headman in each fishing village ; the rank is hereditary ; they both fish and act as palanquin bearers, whilst it is reported that they have lands in some villages. Persons of all villages on the seaside are privileged to use nets in certain places allotted. Fishermen make their own boats and nets; their numbers have remained stationary. *The Deputy Tehsildar of Coringa* (1869) replies that fish is exported to Moulmein and other places ; the average daily earnings of sea-fishermen are about one rupee, but of those who fish in the rivers perhaps four annas. Sufficient are taken to supply the local markets. Fresh fish are generally preferred to the salted article. Fish are salted to a great extent with Government salt: and salt-earth, which is taken from the swamps along the coast, is used for curing mackerel; sea-water is not employed. Fish are dried, roes are salted, and fins and sounds are preserved; fish liver oil is also manufactured. Subsequently (1873) he continues, salt is usually sold in this division at Rs. 2-4 a maund. It has been reported that the fishermen residing in the villages of Coringa, Tallaveru, &c., use the Government salt, and that those residing in the suburbs in the immediate vicinity of the sea, employ salt-earth for curing fish. It has also been reported that salt-earth is used for this purpose. The practice of salting fish has remained stationary. Fishermen had no peculiar privileges in old times which they do not now possess. It does not appear that there are any headmen of the fishing castes. No one claims any rights in regard to the sea-fisheries. The fishermen purchase boats and nets at their own cost. Their numbers have remained stationary. The localities where fish are caught are about 5 or 6 miles distant from the place where the fishermen reside. It will therefore be convenient to fishermen to cure fish at their homes, consequently the sale of salt at the places where fish are caught does not appear to bring about any benefit. *The Deputy Collector* (February 28th, 1873) observes that salt is sold at 3 rupees a maund in this division, and is generally used for curing fish. Salt-earth and sea-water are employed in some places for this purpose secretly, because rich men will not prefer to eat such fish. The practice of salting fish has certainly increased of late years, as salt fish is exported to Hyderabad and other places from this district in large quantities. Fishermen possess all the privileges they formerly had, besides being freed from the moturpha tax. In villages where there is a fishing caste, the fishermen themselves appoint a person to be a headman. His duties are to settle caste disputes, &c.; to visit with fines, &c., those who commit immoral crimes, such as adultery, &c. The office is hereditary indeed. No emoluments are assigned to him, but he is allowed to receive certain presents, and is much respected on the occasion of marriages, &c., of fishermen. There are many who claim rights regarding sea-fisheries opposite to their huts or places of residence ; such disputes give rise to civil actions. The fishermen generally construct boats themselves, as they are apt carpenters, while some who are ignorant of the work get them constructed by workmen. Likewise they prepare their nets. In some cases they hire boats from soucars, &c. There is an increase in the sea-fishermen as in other classes of late. This is attributed to the bounty of Providence, but not to any particular reason.

109. *From the Vizagapatam Collectorate* the following returns of native officials have been received (November 1872). The *Deputy Tehsildar of Bimlapatam* replies that salt is sold at the pans at Rs. 2 a maund, but retailed in the bazar at nearly Rs. 2½. The Government salt is used for curing large fish, and salt-earth or sea-water for the small ones. The practice of curing fish has remained stationary of late years. In old times the fishermen had no peculiar privileges they do not now possess. There are headmen of the fishing castes termed Pillaho; the office is hereditary: the headman settles all caste disputes, for which he receives a small remuneration in money or fish as a present. As regards sea-fishing, the villagers residing on the seashore consider they have a claim to cast their nets before outsiders. The number of fishermen has remained stationary; they supply themselves with boats and nets. *The Deputy Tehsildar of Vizagapatam* observes that salt is retailed in the town at Rs. 2-10-8 a maund. It is customary amongst Vaddi caste people (sawyers) and certain fishermen in the town who have money to cure big fish with Government salt, but the poorer fishermen do not employ it. Salt-earth and sea-water is used in curing fish. The practice of salting fish has neither increased nor decreased of late years. In old times the fishermen had no peculiar privileges they do not now possess: they have hereditary headmen whose duty it is to settle caste disputes; his emoluments are (1) if he goes and asks the fishermen when they catch fish they give him 2 or 4 pies' worth; (2) in marriages he receives 3 annas for putting a turband on the head of the bridegroom. No one claims any right as regards the sea fisheries. The number of fishermen has remained stationary of late years.

110. *In the Ganjam Collectorate* the following are the returns from the *native officials*. The *Tehsildar of Chicacole* (June 1870) observes that there are about 38 fishing villages in his district, averaging from 90 to 900 persons. In some the persons are also agriculturists or labourers. Those of the Meela, Wuda, and Jalary, castes fish in the sea, whilst some of the Meela, Pully, Khandra, Jelary, and Neyala castes employ themselves in the rivers and back-waters; the latter and also the Kavety and Relly castes sell and salt fish. Fish is carried long distances for sale to the various fairs and into the interior. The general scale of remuneration to fishermen on the coast is two-thirds of the fish taken by them, and one-third to the owner of the nets; this gives each man about Rs. 4 a month. For eating it is preferred fresh. The fishing population continues about the same: about 200 boats go to sea between this and Poondy: both boats and nets continue about the same. The best fishing months in the sea are from November to January inclusive. Fish are salted to a great extent; neither salt nor salt-earth is employed for this purpose, but sea-water is used in a fermented state. The quality of the salt fish is pretty good, but an improved one would be preferred, and more could be sold, as all classes who eat fresh fish will also consume the salted article. Fish, except *crustacea*, are not smoked; roes are salted, sharks' fins are dried, sounds are not preserved. Fish liver oil is extracted from sharks and skates. *The Tehsildar of Berhampore* replies that in his district there are about

cvi

sixteen fishing villages, having an average of about 219 occupants. Men of the Tero, Kevuti, Kandra, and Mila casts fish in the sea, back-waters, and rivers. Those of the Tero, Kevuti, Kandra, Mila, Reddika, Jalari, and Haddi or Pariah caste sell fish, whilst Kevuti, Mila, and Kandra men salt them, and the salt-fish is taken a long distance for sale. Fishermen working as coolies are not paid in money, but receive half the fish captured in fresh-waters, and one-third of those taken in the sea goes to the owners of the nets and two-thirds to the coolies, who earn about two annas a day. The supply is sufficient for the demand; for eating it is preferred fresh. The fishing population continues about the same; about 400 boats and 500 catamarans go to sea, which, and the nets, are about the same numbers as formerly. The four months of the cold weather are those in which most fish are taken; salting is carried on to a great extent; salt is not used, neither is salt-earth, which is obtainable from swamps and places near the Chilka lake, but cannot be taken without the circar's permission. Sea-water is used by some persons for seasoning the fish. The present quality of salt-fish is good, the supply of which does not equal the demand. Fish are dried and their roes salted; neither fish fins nor sounds are preserved. Fish liver oil is prepared; crocodiles are not in large numbers, but are found in fresh water here and there. There are some creatures like cats (otters) which destroy fish a good deal.

111. The following return* from the *Collector of South Canara* had it arrived in time should have been inserted at page xlv :—

---

* *Statement showing the quantity and value of "Fish," dried and " salted," and "Fish oil" imported into, and exported from, the Port of Bombay during the last twenty years.*

YEARS.	IMPORTS.			EXPORTS.		REMARKS.
	Fish dried and salted.	Fish oil.		Fish dried and salted.	Fish oil.	
	Rs.	Gals.	Rs.	Rs.	Gals.	Rs.	
1853-54	...	...	90,856	31,870	...	80,243	55,999	The information regarding fish, dried and salted (which is registered in value only), for the year previous to 1866-67 and for 1867-68 cannot be furnished, the same not having been separately registered.
1854-55	...	...	1,09,274	63,751	...	1,73,587	1,01,022	
1855-56	...	...	2,842	1,827	...	2,806	1,804	
1856-57	...	...	8,998	6,508	...	10,845	7,843	
1857-58	...	...	77,631	62,981	...	29,180	23,449	
1858-59	...	...	10,982	8,924	...	53,532	43,017	
1859-60	...	...	35,268	28,340	...	41,222	33,170	
1860-61	...	...	19,096	8,982	...	8,527	6,852	
1861-02	...	...	14,941	12,006	...	34,591	30,932	
1862-63	...	...	12,385	5,911	...	17,729	14,269	
1863-64	...	...	45,821	30,278	...	36,494	32,220	
1864-65	...	...	13,629	7,687	..	12,306	89,235	
1865-66	...	...	62	40	...	140	135	
1866-67	...	24,499	600	2,695	17,450	1,880	2,203	* Not separately registered.
1867-68	...	...	...	...	...	90	40	
1868-69	...	1,16,246	2,800	3,856	13,558	4,931	7,064	
1869-70	..	91,222	396	1,327	10,386	1,281	1,364	
1870-71	...	89,899	1,008	1,717	23,755	...	...	
1871-72	...	64,439	5,389	5,800	10,938	4,608	5,554	
1872-73	...	82,010	896	2,899	16,894	21,432	25,833	

cvii

Return showing the quantity and value of Government salt sold in the District of South Canara for Home (i. e., District) and Inland (i. e., Mysore and Coorg) consumption from the year 1861-62 to 1872-73.

YEAR	HOME CONSUMPTION.						INLAND CONSUMPTION.						TOTAL OF COLUMNS 2 AND 4 AND 3 AND 5.					
	Quantity.			Value.			Quantity.			Value.			Quantity.			Value.		
1	2			3			4			5			6			7		
	Mds.	Srs.	Ts.	Rs.	A.	P.	Mds.	Srs.	Ts.	Rs.	A.	P.	Mds.	Srs.	Ts.	Rs.	A.	P.
1861-62	1,34,092	25	20	1,98,025	4	8	1,26,871	27	40	1,90,101	7	5	2,60,964	12	60	3,88,126	12	1
1862-63	1,61,534	34	5	2,41,840	9	6	1,85,961	20	30	2,78,942	5	9	3,47,496	14	35	5,20,782	15	3
1863-64	1,91,002	15	45	2,85,608	14	3	2,10,382	3	5	3,15,573	2	8	4,01,384	18	50	6,01,082	0	11
1864-65	1,68,279	37	35	2,51,543	1	5	1,97,677	36	75	2,96,516	14	4	3,65,957	34	30	5,48,059	15	9
1865-66	1,84,174	9	52	2,79,900	8	9	2,10,263	19	42	3,20,117	6	0	3,94,437	29	14	6,00,017	14	9
1866-67	1,51,113	11	29	2,54,767	10	11	1,28,781	15	3	2,17,318	9	5	2,79,894	26	32	4,72,086	4	4
1867-68	1,74,629	0	0	2,93,896	6	0	1,72,785	0	0	2,91,574	11	0	3,47,414	0	0	5,85,471	1	0
1868-69	1,76,465	0	0	2,95,868	0	9	1,60,011	0	0	2,70,018	9	0	3,36,476	0	0	5,66,886	9	9
1869-70	1,47,173	0	0	2,72,279	1	8	1,22,851	0	0	2,37,643	9	0	2,70,024	0	0	5,09,922	10	8
1870-71	1,36,967	0	0	2,73,033	6	0	1,27,894	0	0	2,55,788	0	0	2,64,861	0	0	5,28,821	6	0
1871-72	1,77,482	0	0	3,52,639	1	6	1,54,977	0	0	3,09,954	0	0	3,32,459	0	0	6,62,593	1	6
1872-73	1,35,839	0	0	2,70,806	12	5	1,24,101	0	0	2,48,202	0	0	2,59,940	0	0	5,19,008	12	5

CEYLON.

112. The *Colonial Secretary, Ceylon,* observes (Sept. 20th 1872) that the selling price of salt has been thus from 1825 to 1836 : "During these 12 years the price of salt appears to have been at the rate of 18 fanams, or 2*s*. 3*d*. per 'parrah,' equal to about 2/3 of a bushel. Taking the average weight of salt at 70 ℔s. per bushel, the price of 1 lb. was $\frac{27}{70}$ of a penny, or a little more than $\frac{3}{8}$ of a penny. From 1837 to 1842 : the price during this period was at 2*s*. 4*d*. a bushel, or $\frac{2}{5}$ of a penny per ℔., which is slightly above the former rate. 1843 to 1845 : during these three years the price was 2*s*. 8*d*. a bushel, or $\frac{16}{35}$ of a penny per ℔., that is, a little above $\frac{7}{16}$ of a penny." *The Auditor General of Ceylon,* in his report on the salt revenue of Ceylon (June 22nd 1871), observes : "the Central, I need hardly say, not being a maritime province, derives its supply from the Southern and North-Western Provinces, whence also the Colombo market and the Western Province generally is furnished, there being no production or manufacture of salt along its sea-coast. The monopoly rate is 4*s*. 8*d*. per cwt., whereas that in Madras is Rs. 2 a maund = 5*s*. 5½*d*. per cwt."

Price of salt in the island of Ceylon: and table of imports of salt fish and salt.

Imports into Ceylon for 10 *years ending* 1871.

Fish dried and salted.				Salt.			
Year.	Cwts.	Qr.	℔s.	Year.	Cwts.	Qr.	℔s.
1862	61,041	3	9	1862	121	1	25
1863	60,905	0	19	1863	4,996	0	9
1864	75,248	2	3	1864	133	3	8
1865	66,970	2	6	1865	4,037	0	22
1866	70,190	2	11	1866	96	2	11
1867	71,794	3	17	1867	4,555	1	1
1868	73,293	3	22	1868	6,481	2	25
1869	75,188	3	0	1869	293	0	21
1870	76,968	1	15	1870	71	0	2
1871	78,574	3	16	1871	3,213	0	4

113. From the Report of the Commissioners appointed to enquire into the *sea-fisheries of Ceylon* I have extracted the following information, premising that the whole question of imports and exports of salt fish, dried fish, fish oils, fins, and sounds does not appear to have been examined. It would have been more interesting to have investigated the following questions :—Has a rise in the price of salt had any effect on the fishermen

Investigations into the state of the sea-fisheries of Ceylon.

and fish-curers' trade ? Has an increased amount of salt fish been imported from India as the monopoly price of salt was augmented in the island? *Sir J. Tennent* observed—" and it is a remarkable fact, as illustrative of the singular habits of these people, that each diminution of the duty (or rather a tax, or license on fishing), instead of leading to increase of the trade, and an addition to the revenue, had, in every instance, the directly contrary effect. On an average of four years from 1830 to 1833, whilst the tax was one quarter, the average amount of revenue was £7,389 per annum. From 1834 to 1837, when it was reduced to one-sixth, the average annual receipt was £6,694. And from 1837 to 1840, when the duty was one-tenth, the receipts fell off to £4,821." Now this decrease of trade may have been due to the incidence of the salt tax, there being only two ways of disposing of fish, *viz.*, fresh or salted. The reduction of captures is given as between 1834 and 1837, and it was in the latter year that the price of salt was increased. These figures are absolutely necessary before coming to any conclusion. The Commissioners consider that scarcity of fish was the cause. But whether this dearness ought to be attributed to the diminished supply and scarcity of the article is a point on which surmises only can be made. " On the western side of the island (no salt produced ?), so far as we have been able to carry our investigations, it appears that fishing operations are almost continuous, being interrupted only by the changes of the south-west and north-east monsoons. The mode of fishing during the one monsoon differs in some measure from that which is carried on during the other. * * We have now to consider the complaints which have been brought against foreign fishermen, that is, men who come here from the continent of India. Fishermen of nearly every class complain that their operations are injuriously affected by the mode of fishing adopted by these strangers, and there is reason to believe that its prosecution is not beneficial to the public. The principal objections to its use are, that the fishermen, not content with having the open sea to fish in, resort to the well-known bait grounds, disturbing the settlement of the fish there, and in capturing everything without regard to the interests of the other fishermen, that they bring up with the fish, amongst other things, a peculiarly offensive matter called múdú-hori, all which objectionable refuse is packed up in mat bags or baskets with the fish and brought on shore. Though the condition of the fish is thus much deteriorated, it is admitted that the poorer classes purchase the fish on account of its cheapness. However, it is important that there should not only be an abundant and cheap supply of fish in the market, but also that it should furnish a wholesome and nutritive food, and we are therefore of opinion that some restriction should be imposed on this style of fishing. * * After giving the best consideration to the evidence, and the law as it exists at present, we are of opinion that it is advisable to legislate for the preservation and improvement of the sea-fisheries of this island, so as to secure to the public the greatest possible advantage from the various modes of fishing. In this view, the law should provide that rules may be made; that a breach of those rules should be rendered penal; that the fish captured under such breach should be forfeited and destroyed; and that the implements of capture should be publicly burned. * * That the rules be submitted in pursuance hereof may permit the prohibition of the use of the following nets:—The *páweni* or *wali-dela*, the *súda-dela*, the *adappu vallei*,* and restriction in the use of the *kattumarang-dela* ;† and that they may also render penal the bringing ashore and sale of small and ill-conditioned fish."

114. The following are the descriptions given by *Mr. Pareira* in the report as to the *various modes* of *fishing* and nets in use. "The fishing district of Colombo may be said to extend from Mount Lavinia on the south as far as Pamunugama on the north, embracing a coast line of about twenty miles. I shall therefore confine my remarks to the fishing operations carried on along this sea border, as my connection with this part of the coast qualifies me to speak from personal knowledge. The *south-west monsoon*, which sets in about the latter end of May, and is called by fishers the Wárakan season, is marked by considerable disturbance of the currents and tides, but this disturbance soon subsides, and a few days after the break of the monsoon the sea along the western coast settles into its ordinary calm, when fishing operations, which had been temporarily suspended owing to the wind and waves, are again resumed. The fishing during this season is, however, carried on further out at sea, and is principally confined to large fish which are taken by hook and line. The *má-dela*, or seine, is never used during this season. The boat-fishing is divided into two classes, those that go out at early dawn, and at midnight. The former first proceed to some well-known *bait ground*, where, by suspending a small thread net from the boat, while at the same time still moving on, a sufficient quantity of small fish and prawns are caught to serve as bait. From these bait grounds the boats proceed a considerable distance seawards, and when arrived at the proper fishing ground, especially over sunken rocks, the boats are anchored and the fishing by hook and line commences. The fish caught in these boats are generally of the large descriptions, such as the weṭiya, tumbeya, maguruwa, laweya, kaṭtáwa, seir, &c., being mostly good table fish. The boats that start at midnight fish with artificial bait made of strips of cocoanut and the fibrous bark of the eheṭu tree. This bait, attached to a *hook and line*, is cast into the sea, while the boat is under full sail, and attracts fish of the bonito class, principally kelawalla, eṭawalla, baleya, and even seir fish, &c. The boats that start in the morning continue out at sea till about 3 or 4 o'clock, and those that start at midnight till 12 or 1 o'clock, and return on shore. There are only two descriptions of nets used during the southwest monsoon—the *móra-dela* and the *kumbuṭu-dela*. The former is made of thick twine, and, as its name implies, is intended to catch sharks, although other varieties are also caught. The latter is made of finer twine, and is intended to take any kind of small fish indiscriminately. There is a third description called the kaṭṭa-dela, also made of fine

---

* This net was introduced from India about 1861. During the ebb tide in the lake, men fasten "several nets together, and bury the lower edge below the mud: and at the flood, the upper edge is raised and propped by sticks, thus making a kind of fence. At the second ebb, these men go out and pick up the fish which have been left on the mud, having been prevented from escape by this fence of nets. Only the large fish are picked up, the smaller kinds being left to rot on the mud. The quantity so left is much greater than that which is picked up."

† Catamaran fishermen.

twine, and is intended to take larger sized fish than the kumbuṭu-dela. The móra-dela is laid overnight, and is dragged in the morning. The others are laid in the morning, and are dragged ashore during the forenoon. Each description of these nets has its own particular ground and particular hours for spreading, and the ground or time of the one cannot be interfered with by the other. As is well known, boats assist in the laying of these nets, as they proceed along the intended line, and when the whole has been paid off the boats return for line fishing, or come on shore. After a sufficient time has elapsed for the fish to get into the nets, the boats go back, and haul them on board, and return on shore with the captures. These nets have not the small bag attached; but the fish are caught in the meshes. The nets are kept straight by means of floats at the top and weights at the bottom, and the ends are marked by buoys. A floating net called the *paweni-dela*, made of hemp, is also used. One end of this net is attached to a buoy which is left at a selected spot, from whence the boat moves on across the line which fish generally take in entering the harbour, and the net is dropped and allowed to drift with the current. This net is so laid on during the night, and is hauled up the following morning. The fourth description, called the *uḍa-dela*, is used at the middle of the south-west monsoon. It is made of cotton thread, and is spread in the same manner as the má-dela. A rope attached to one end being left on shore, the net is carried out to sea by a boat, and, after a considerable circuit has been made, the rope at the other end is brought ashore, and the area so enclosed is dragged. The fish captured are generally of the smaller descriptions. The time of laying the net depends altogether on the quantity of fish present in the harbour, the operation being sometimes repeated three or four times in the day, when an abundance of fish offers inducement. Another net, the *súda-dela*, is employed at the middle of the south-west monsoon. Each of these nets is about 30 fathoms long, and sometimes as many as five or six are joined together. The net is broad enough to reach the top, where it is kept straight by floats, and at the bottom there are stones which make it rest on the ground. The net is made of thread, and the meshes are very small. This net is carried out by a boat, and when a large shoal of small fish is found, it is laid out in a circle to enclose the shoal. After this is done, a terrible splash and noise is made in the enclosed space by beating it with oars and clubs. The fish are caught in the meshes, and are generally of the sardine class. Throughout the year, *angling* from the rocks on the coast is also carried on, but the fish generally caught are small, and belong to what is commonly known as the rock-fish class. The last description is known as *kattumarang-del*. This is a recent introduction. This net is carried out into deep sea in a kattamarang or raft manned by three men, accompanied by a second raft with two men. When the fishing ground has been reached, a portion of the net is put into the second raft, and both commence paying it off from this point, proceeding in a circle, until both meet again, when they lash the rafts together, and haul up the net. All the fish enclosed are forced by the gradually diminishing circle into the sack attached to the middle of the net. All descriptions of fish, from the largest to the smallest, and even the spawn, are scraped and taken. During the *north-east monsoon*, the two descriptions of boat-fishing carried on in the south-west monsoon are practised, and the same kind of fish caught. Besides, a boat called the *Gahenang Oruwa* goes to sea, carrying nets for securing bait. These nets, which are known as *gahenang-del*, are small cast nets, which are cast in the different bait grounds (Maddes). When a sufficient quantity of bait has been secured, these boats go far out to sea, and they fish with rods. The rod is generally a bamboo, and the hook is of considerable size. The seir fish, the angilas, etawallas, kaṭṭáawas, &c., are thus easily captured. Another class of boats called *Eng-oru* go out especially to the bait ground, where, after capturing small *fish for bait*, they are bound to supply the deep-sea fishers with them. The net used by the *Eng-oru* is made of thread, is of a circular shape, and is cast from a rope attached to the middle. *Pudu-oru*, another description of boat, goes out in the mornings, proceeds to the fishing ground with an *artifical bait* made of the bark of the eheṭu plant. These boats being small do not go beyond the roadstead. The fish is caught with a hook and rod, and the description caught is the kaṭuwalla. There is night fishing in the north-east monsoon, which, unlike the night fishing of the south-west monsoon, commences in the evening, and continues till the following morning. The boats are anchored at the rocks, and fishing by line and hook commences. The fish thus caught are generally of the large size. Night fishing by torchlight is also practised during the prevalence of this monsoon. The people of Pamunugama are those only of the district who practise this kind of fishing. They generally go in boats provided with a sufficient quantity of chools (torches) at 9 or 10 in the night, and having arrived at a favourable spot in the sea, they hold the blazing torch in one hand, and with the other employ a small attangua, to take in the prawns and other fish that come up to the surface, dazzled by the light. This fishing continues till the following morning, and prawns are taken in the largest number. The principal and peculiar net of the north-east monsoon is the *má-dela*, which is made of coir, except the bag, which is of hemp. One end of this net is left on shore in charge of one party, and the rest, laden in a boat, is rowed off, and while it proceeds, the net is gradually paid off ; a semicircle is described as the boat approaches the shore, and the enclosed space is dragged. All kinds of fish, from the largest to the smallest, as they enter the bag are caught in this net. " The *núl-dela* is made of three sections; the ends are of coir, the next section of hemp, and the middle section of cotton thread. The meshes are very small, and allow no escape, even to the smallest fish. This is also laid in the same manner as the preceding. The fish, principally small, are caught not only in the bag, but in the meshes also all throughout. The *pulunu-dela* is carried by a boat to which one of its ends is attached, and it is paid off as the boat moves on. It is kept straight stretched by floats and weights, and the boat being made stationary at the extreme end, a few hours' time is allowed for the fish to entangle themselves in the meshes, when it is hauled into the boat. *Kaṭṭa-del, suḍa-del, uḍu-del,* and *kaṭṭumarang-del* are also used during the north-east monsoon. The fishing operations above described may be divided into two classes, the legitimate and the illegitimate. The legitimate or proper kind of fishing nets to be used during the south-west monsoon are the *móra-del, kumkuṭu-del,* and *kaṭṭa-del;* and the *má-del*, the *baru-del* or *gahaneng-del*, during the north-east monsoon. The other description of nets already enumerated are all more or less calculated to damage the fishing by scaring away the fish that enter the harbour, or by disturbing and destroying the spawn. The pulunu-dela, though an old net, is now made broader, and it has completely changed its character. The kaṭṭumarang-dela is undoubtedly the most injurious to the fishing interests of the people. The *pavenné-dela* is also a most objectionable description of net. The third description of objectionable net is the *súda-dela*. The fourth description of objectionable net is the *udu-dela*. The fifth description of objectionable net is the *núl-dela*. *Fishing by torchlight* is also calculated to injure the interests, both of the legitimate net and boat-fishers. Another objectionable method of fishing is adopted by some of the fishermen at a particular season. This system of fishing is highly injurious to the interests of the common fishermen, as these boats start almost in the midnight, and sail about the sea. The fish mostly being found close to the surface at this time are frightened by the noise and rapid motion of the boats at this unusual hour."

115. The following is a synopsis of some of the evidence given Native opinions respecting by the *natives*, mostly fishermen. *Appu* deals the fisheries. in fish, and has some coast fishermen under him. They come here in gangs of five or six holding one net among them. They bring hardly any money. In Colombo, there are five or six capitalists who advance them money, and thus at once acquire a control over them, provide them dwellings, and requisites for fishing. When they bring fish on shore, the landlord makes his own selections, taking about one-third in kind during the sale, which he conducts. The sale over, he takes $\frac{1}{12}$ th of the receipts. At *Batticoloa* a disturbance arose amongst the fishermen, one party having infringed the old modes of fishing by drawing a seine net within the bar. The parties made charges and counter-charges for breaking customs, which all admitted were as follows:—" (1.) No seine or drag net to be used within the bar. (2.) No nets are ever to be cast in those parts of the lake set aparts for fishing with hook and line. (3.) Cast-nets (sanel vallei) may, with the foregoing exceptions, be used at all times during the day, but not at night, until the 22nd night of the moon's age, the use of this net being prohibited for the first 21 days of the moon's age reckoning from the new moon. (4.) Fish spears, made of Palmyra wood may not be used, the descriptions permitted being those the shafts of which are made of Samandale and Vinangu wood. The reason for this is, that Palmyra spears being very heavy, cause so much disturbance that the fish are frightened away by their use." Many of the fishermen consider that the use of small-meshed nets has largely increased during the last few years, and with great injury to the fisheries, by destroying the small ones or bait, consequent upon which the larger fish that prey upon them are either frightened away, or not finding sufficient food have migrated elsewhere. Natives of India appear to be largely concerned in this mode of capturing fish, and their captures are sold cheaper than the larger and more wholesome varieties. "The Malabars lay their nets at night and draw them up early in the morning, securing all the small fish which would otherwise have served as bait for the larger fish

which come towards the shore early, and it is these large fish which would have served for fishing with the Ma-del." A fish-dealer and net-holder thus explains why he rather deals with natives of India than with the Sinhalese: "We can't get so much out of the Sinhalese fishermen; they would not give us anything. We have only to make a sham of giving a Tamil a crack on the head, and he will give anything. The Sinhalese will only give the share for the Roman Catholic churches, because the rent is sold, and if the share is not paid, the rites of the church are refused." Many persons suggest prohibiting fishing from kattumarangs during August and September, so that the fish may be permitted to come in peace towards the shore and congregate; subsequently there does not appear to be so much objection raised to their use.

116. *Bennett*, writing upon the 'Capabilities of Ceylon (1843, p. 162),'

Bennett's remarks on the sea fisheries.

observes "that a method of curing fish upon an improved system to the primitive one that has obtained in the island from immemorial time would both ensure very ample profits to those concerned in it, and prove a blessing to the colony." In salting fish in the tropics air and shade are necessary, open platforms of split bamboo canes raised on tiers, so that if salt only is necessary it can be done, or fish smoked "by lighted wet rice straw being laid under the lower tier" (p. 167). Sand, he observes, is present in Indian cured fish, likewise incipient putrefaction; for this last they should be boiled with charcoal. "I therefore earnestly endeavour to impress upon the attention of individual capitalists that there is no speculation more certain of success amongst the many other very encouraging ones that present themselves at Ceylon than that of establishing factories for the curing of fish, at such places upon the coast as have the best fishing and little or no demand for fresh fish" (p. 186).

BENGAL.

117. The sea-board of the Bengal Presidency, unlike that of Madras, is intersected with rivers debouching into the sea. The most southern of these is the Mahanuddi, or great river of Orissa, and several others intervene between it and the Sunderbunds, where the various mouths of the Ganges emerge in the Bay of Bengal, through a country which has become depopulated and is now in most parts dangerous to reside in, due to endemic diseases and other causes. In such a district one could hardly expect the sea fisheries to be carried on very vigorously; but one might have imagined something could have ere this been discovered to remedy the almost total apathy which exists, the deadening effects of which are extended even into Chittagong, also under the Bengal Government. It is not until we step over the boundary into Burma that any activity in the fisheries is perceptible. Having mentioned some of the physical difficulties to carrying on the occupation of fishermen, we will next briefly consider whether the fisheries themselves are of any importance. Along the coast of Orissa and the Midnapore districts, which are well inhabited, fish are found in great abundance, but due to several causes, which will be pointed out, there does not appear to be a disposition on the part of the natives to avail themselves of the bounties of nature. However, before we blame the fishermen, it might be as well to take into consideration whether the rulers of the country are or are not responsible for much at least of this apparent apathy. If we investigate the Sunderbunds, we find every water-course and estuary filled with excellent varieties of fish, many growing to a very large size; but the crocodile and other vermin obtain far more sustenance from them than does man.

Bengal sea-board not adapted for fisheries, due to some local causes: its sea and estuary fisheries.

118. The fishermen appear to be of two classes—the inhabitants and Madrassees who were induced to settle at Pooree, and probably elsewhere, on certain conditions, which the local officials have since disregarded.

The fishermen.

119. The fisheries appear to be but little worked to what they might be, consequently the amount of fish obtained is nothing like what it ought to be. This is apparently due to the want of a local market; fresh fish finds a sale, but the salting of fish is a trade but little carried on. When I say fresh fish I must modify my remarks, for the putrid substances sold in Balasore and elsewhere I have been, are in such a condition that even an approach to them is most undesirable. The Collector of Balasore pithily remarks on the natives "not being averse to fish far advanced in decomposition." Salt being evidently too dear to be procured, they are stated to dry their fish "in the sun, and eat it when it is quite putrid. They like it in this way, and there is no reason why they should be interfered with." A predecessor of his enquiring as to the sanitary state of these fishermen from a local zemindar, was

The fisheries, how they are unworked.

informed "Cholera seems made for these people." Fish captured in their stake nets were cheap, but why? because they had no means of preserving them; if they are too large to sun-dry, they must be eaten or sold fresh or else spoil.

120. What is the price of Government salt is rather an important consideration, although the Collector of Balasore, who would feed those people on putrid fish, observes "the price of salt had nothing to do with their not salting it." As he gives no answer as to the price of salt, I conclude it is about what prevailed when I was there, from Rs. 4-8 to 5 a maund. Out of six European officials whose opinion are recorded, three consider the high price of salt has acted detrimentally on the curing of fish; 2 that people do not like salt fish; and the remaining one, as remarked on above, that the price of salt has nothing to do with it. Here I would observe that there are two varieties of salt in Orissa, the English imported and the sun-dried native-made, the latter being considered by Hindus on religious grounds as the correct article to be employed. The high price of salt appears to be one of the reasons that fish are merely sun-dried along the coast; for, as observed, it is not only the cost of the article that has to be considered, but the ruinous interest fish-curers would have to pay the usurers who lend the money whereby only they can obtain this antiseptic.

<small>The high price of salt and its relationship to fish-curing.</small>

121. But other causes have been adduced for the absence of a good salt-fish trade, *viz.*, a decrease of sea fish; this, of course, may be owing to a real diminution of the fish themselves, or that the fishermen are not reaping the harvest of the sea. In some places, as at Balasore, "the fishermen state that the supply of fish has much decreased of late; but I do not find that this statement is borne out by facts.[1] They probably make it in the interest of their trade." A local cause may have been at work at this place, for the chief mode of capturing fish is here by stake nets and in-shore fishing, as previously described and the fish been scared away. Or the decreased supply may be due to a less demand, owing to the inability to cure the surplus, and this having eventuated in less exertions being made to capture the fish. One reason brought forward by Mr. Geddes at Poorce is alluded to in para. 127, *viz.*, that a Hindu deity's temper has been upset by the killing of cows and other abominations, and that in anger the earth, &c., does not bring forth its wonted increase. This cause and my not advocating devil worship I have alluded to further on.

<small>Other causes adduced for diminished supply of fish.</small>

122. *What remedies have been proposed?* Dr. W. Hunter has suggested that something should be done to cheapen the salt to the fish-curers, but it does not appear that any one else has proposed such a course. The usual objections to smuggling have been raised, and the matter becomes dropped: for such only refers to the loss of good food to the teeming thousands who are of or below the caste of karans or writers.

<small>Remedies that have been proposed.</small>

---

[1] It would have been more to the point, if *the facts*, whatever they may be, had been stated, also the length of time and opportunities that have given the Collector such cause to deny the fishermen's statements.

cxvii

123. In answer to my enquiry whether it would be practicable or advisable to construct enclosures in favorable localities, where fish could be salted and salt disposed of inside at rates just remunerative for the *bonâ fide* salting of fish, the five answers have been as follow. First, that as there is no hope of decreasing the price of salt, it is useless to discuss the question further. The four others consider such as either visionary, not practical, not advisable, or inapplicable. The two chief reasons adduced are, difficulty of preventing smuggling, and that either the people do not know how to salt fish, or the supply is merely equal to the demand.

Respecting the sale of salt at a reduced rate within enclosures for the purpose of salting fish.

124. *The Commissioner of Orissa* observed (December 4th 1865) during the late famine, "whilst the condition of the residents of this place, where my camp is, which is called Nijhur Bulbudderpore or Kentesakee, is somewhat easier as living by their fisheries, they are not so affected by present circumstances." But on the other hand their condition in some places, I was informed, was "not better than that of their neighbours, as at Bhudruck, Balasore, and Pooree they suffered greatly, but this suffering seems most to have affected the sea fishermen, who perhaps found the sea too rough to fish in in their exhausted state." (Report on Fisheries of Orissa.)

Opinions of European officials in Orissa.

125. *The Collector of Cuttack* (October 28th 1872) stated that fish as food is generally preferred in its fresh state: near the estuaries of the different rivers it is salted and dried, but the process is very incomplete, and the fish when offered for sale has an offensive strong smell.

Opinion of the Collector of Cuttack.

126. *The Officiating Collector of Pooree* observed (July 27th 1858). "*Shamùdra machdiya.* The fishings in estuaries, like that facing Mr. Drares and Captain Saxton's bungalows here, are not salt-water fisheries, as I find the entry exists in the account of the estates previous to the settlement, that to the present time the Telinga Lulliyas are the only parties who fish in the sea, and it is well known that they were first brought here in Mr. Wilkinson's time, or subsequent to the time to which these accounts alluded to refer * * *Sárpat machdiya,* or fishing on flooded plains. I would here observe that, as a general rule, all fishings have been left unassessed. The zemindars now demand rent from these people on account of the sea fishery, but their land only went to the sea. Lulliyas cultivate no land, but live on the sands of the sea-shore, and are boatmen as well as fishermen : none but these men are capable of managing surf boats." The inducement to settle held out 'was the promise of certain employment in boating, salt for four months, and free leave to fish in the sea and collect shells for burning into lime for the remaining eight.' 'There is one other point which suggests itself while considering the matter, and that is by declaring such fishing free, you may enormously benefit the whole country; you allow of better nets and tackle being used, and improved methods introduced in curing the fish ; you not only increase production, but you lower prices and raise the standard of diet throughout the country.

Opinions of Officials in Poree.

Indeed, the matter appears to me of such moment, involving as it does the consideration of the best means of availing ourselves of a 'harvest' hitherto allowed to remain untouched, that I have often wondered that the Indian Government did not follow the example of the British and French, and make the advancement of sea-fisheries a special object of legislation and care." The *Revenue Board* wrote to the Commissioner of Cuttack (August 17th 1858), in reply, "that the Board quite agree with you, that no parties have an exclusive right to sea-fishing, or to shells and other productions of the sea between high and low water mark." In a letter from the *Collector of Pooree*, 1862, it is observed that the zemindars collect rents for stake nets in estuaries. Dr. W. W. Hunter, (Orissa, i, p. 45) observes how the fisheries are carried on "on the northern point of Párikud, where the channel opens towards the sea. I came upon a region of endless shallows and stake fisheries, in which even my light draught pinnace had to be pushed by main force through the mud. The stakes form close wattle fences, about five feet high, of which two-thirds are under water. They are arranged as three sides of an oblong, or as two of a triangle, sometimes a mile in length, with narrow-mouthed baskets opening from their sides, like the pockets along a billiard-table. The tide flowed in with a gentle ripple, bringing up the fish, who swam along the fences till they came to the mouths of the baskets, which they forthwith entered and were caught. The most lucrative enclosure or pocket was of course the one in the angle where the stake lines met. Fishers in box-shaped canoes punted from one basket to another, and bagged the prey." Respecting these fishermen, he remarks (p. 31) that the "fishing communities, and especially the hamlets of boatmen, are Telinga settlers from the Madras coast. They speak a different language, observe different customs, and worship different gods from those of the agricultural population; and here I may notice, as I shall not again have to recur to the subject, that the boating villages along the whole coast, up to the Deví river and even above it, consist of similar Telinga colonies from the south; they are indistinct remnants of the aboriginal races; and although their separate origin is forgotten, they are excluded from the Hindu commonwealth. They disregard a thousand customs and scruples, religiously observed by the Hindus. There is a little village of them on the sands of Puri, which although situated within holy precincts, is denied an entrance into the great national temple of Jagannáth." Respecting the effects of the salt duty in Orissa, he observes (ii, p. 160) of fish, that "the climate renders it impossible to keep them in a fresh state long enough to reach the cultivated parts of the province; and the high duty on salt renders it impossible to cure them. * * I have heard two objections urged to my views on this point. The first is, that any system of drawbacks, which would allow the curing of fish with cheap salt, would lead to smuggling, and give rise to a host of difficulties. One of these difficulties would unquestionably be, that, owing to the high duty on salt, it would pay people to cure fish with the duty free commodity, and then dissolve the salt out again. (1.) But this might be overcome by a differential rate, calculated so as to

---

(1.) If this were so, why is it not done with the salt-fish imported duty free into Bombay and elsewhere? Salt, in some of these foreign places, is sold at two annas a maund or less, yet such a course would not pay!

enable the fishermen to salt, and yet to render it unprofitable for the consumer to dissolve out the saline ingredients. I do not deny the grave practical difficulties of such a proposal; but Government has to consider whether it is not worth while to encounter and overcome these difficulties, rather than to continue to deprive the often famine-stricken population of the delta of a great staple article of its natural food. The second objection is, that even although the people could get salt fish, they would not eat it. Undoubtedly this is the fact as regards the Bráhmans and the high castes. But it is most certainly not the fact with regard to four-fifths of the population. All castes in Orissa, below the *Karans* or writers, would gladly use salted fish; and at this moment they consume great quantities of fish imperfectly cured in the sun, and more or less rotten. With the *chásas*, or peasant population, who form the great body of the people, this is a favourite article of food; indeed, almost the sole relish which they can afford to their monotonous rice diet. The husbandman stores up his supply of dry fish in reed baskets, and sparingly doles out the decomposing mass as a luxury to his frugal household throughout the year."

127. *Mr. J. C. Geddes, Officiating Collector of Pooree,* (August 29th 1872) observes that—" near the Chilka Lake the people subsist largely on fish all the year round. At a distance from the Chilka lake, fish is a rarity and a delicacy, except during the rainy season, at which period every chásá plies his net, basket or trap for small fry." " There is some trade in salted fish from the Chilka with the Gurgats. People have a liking for salted fish, both there and on the plains, but there is some complaint amongst the fishermen of being hampered by the greater stringency of late years in matters of salt tax. It is a pity that there is this difficulty; were it not for this, the people of the interior, who at present have little animal food of any kind during the dry season, might share in the ample supplies of the Chilka Lake. This is all the more to be regretted at a time, when, owing to various circumstances, which I need not recount, milk diet of any kind is becoming less and less accessible to the bulk of the population. With regard to the question whether the supply of fish is falling off, the universality and unamimity of popular opinion seems to me to demonstrate the fact. Among other reasons assigned for this, irrigation anicuts are mentioned, but this, of course, is stated to be only a local aggravation of a general hardship. The usual reason assigned is that the divinity is becoming less gracious to mankind in these days, and that the earth is no longer bringing forth its wonted increase; now that perjury is rampant in the courts, kine are slaughtered and other wickednesses abound. This sort of reasoning will of course be set down as theological superstition ([1]) by the Director General of Indian fisheries, but it seems to me to be relatively more scientific than Dr. Day's postulate of an innate viciousness, impelling the natives to a reckless destruction of fish.[2] The common Hindu notions about the

Mr. Geddes' opinions.

([1]) I must decline admitting this to be 'reasoning', or being responsible for setting such down to so fine a term as 'theological superstition'.

([2]) I must be permitted to remark that this idea imputed to me is no emanation of mine.

Kali Júg (fourth or degenerate age), with its blight over nature, may be silly, but in every relative and scientific view it is at least as reasonable as the similar English nonsense which has been written up in our day, about the golden age of the Indo-European antecedent caste, polygamy, and polytheism, when the Ayran herdsman was milking the Ayran cow high on the table land of Asia, and praising the true God in monotheistic Rig-Vedas. At a time when rubbish of this sort passes for learning, I see no reason for requiring a hypothesis more scientific than that current among the Urias, as to the decaying supply of fish. At all events, the official view of Dr. Day, that natives are more cruelly inclined than they used to be against finned creatures, is not worth setting up in the place of the anger of Parameshwara.[1] Until the facts have been properly co-ordinated, it is useless to propose or discuss a remedy. It may or may not be hopeless to propitiate Parameshwara,[2] which would be the natural remedy according to the Hindu explanation, but it is more certainly hopeless to police net and trap throughout the empire, which is what Dr. Day, I believe, has actually proposed. With respect to one conspicuous though local evil of recent introduction, viz., the undue exposure of fishes at irrigation weirs, the Hindus have often mooted a practical suggestion which the Inspector General of Indian Fisheries has adopted from them, viz., that fishing at these places should be placed under stringent restrictions."[3] Subsequently (November 26th 1872) he continued "that there are certain customs among the fisher people whereby a sort of close or breeding season is meant to be secured. Thus there are certain better kinds of fish which it is considered improper to net between the Kartick purnima and the Dol purnima, or say from November to February. During this period it is customary to seek for certain small fry, whose breeding season does not come on until after February. In the same way they have a close season for

---

([1]) The decrease of fresh-water fish being admitted, of course, different persons may consider such due to as many various causes. My opinion is, that in certain localities increased modes of capture and a decreased size of the meshes of nets has eventuated in a diminution of the fresh-water fish supply. This however with Hindus may not be worth setting up against the Divine anger of one of their idols.

([2]) As Mr. Geddes and I appear to have different opinions as to the cause, so we certainly have as to remedies. I think if the plans which have been adopted with success in other countries, in South Canara by Mr. H. S. Thomas, and in the Doon by Mr. Ross, were tried in Orissa, there would be more reason to hope for beneficial results than were we to turn to *Devil worship*, and attempt to propitiate Parameshwara. I may add that a great difficulty in Bengal appears to be to induce European officials to obtain replies from their native subordinates, who ought to be able to give native opinions. However, when such answers as the following are returned, it becomes a labour to unravel them. When asked as to what difficulties there are in regulating the size of the mesh of nets, and what minimum size is considered advisable? I received this reply:—"The only difficulty would be that of supervision. Economy forbids increase of expenditure, and justice demands avoidance of oppression: any prohibition would be sure to eventuate in thoroughly sumptuary hyperregulation."

([3]) Mr. Geddes here indulges in an incorrect assertion. He states "irrigation anicuts" as one cause given by the natives for a decrease of fish in Orissa, a Tamil term which Ooriahs would hardly employ. Next, he would have found in English law works that restrictions as to fishing near weirs in England was in existence in 1861, prior to the first English-constructed weir in Orissa. Lastly, had he gone into irrigated fields during the rainy or drying-up season, he would find miniature weirs of all size, and so far from Hindus sparing breeding fish and fry they trap them by every device. If, therefore, stringent regulations are necessary at British-constructed weirs, on what grounds are they unnecessary at Native ones?

some particular fish during the Ruth Jatra. The rule is expressed in theological form, thus, that the thakuranie will be wroth if such or such a fish be caught at this or that festival, but substantially it is correspondent to a close season. At the same time it is notorious that the rules are not well observed. The absence of sufficient arrangements for curing fish is not conclusive to economy or to forbearance in this matter. On the Chilka lake the salting of fish would be increased enormously if salt were in free use for the purpose. As there is no hope of securing this privilege, it is useless to discuss the matter further. The proposed immunity for salt curing is incompatible with the present stringency of salt administration. The present retail price of salt is 9 seers per rupee. A small amount of duty-paid salt is employed in curing fish, but the salting, such as it is, is done with saline earth. The Pooree fisher class come from southern districts, and are rather strangers here, so that their customs are not easy to get at in this district." The title of headman is hereditary, if the heir is competent, if not another is elected by the people or selected by the zemindar. Boats for fishing are imported from Ganjam. " The sea fishermen have diminished in numbers of late years. Their trade has been hampered by the greater stringency of the salt laws in recent years; also many fishermen were cut off by the famine of 1866."

128. The *Collector of Balasore* some years since remarked :—" the coast fisheries are exceedingly rich, and contribute in an important degree to the diet of the people of the town of Balasore." When I was at Balasore in 1868, I found that the hilsa forms a portion of the fish taken throughout the year, excepting during the time they are ascending the rivers to breed : while in the rivers north of Midnapore the the young hilsa were being taken in thousands. Consequently, if weirs are to be erected across the large rivers without gaps or fish passes, permitting neither the old fish to ascend to their breeding grounds nor the young to descend to the sea, the species must of a necessity be exterminated, as it only breeds in fresh water. At Balasore the Borabolung river which though fresh is tidal, and the mesh of nets employed here is above 6 inches in circumference. The amount of sea fish in it was asserted to be stationary. Besides investigating the Borabolung river at Balasore, I went three times to its mouth at Chanderpore, about nine miles from the former station. The spring tides were present, and quantities of fish were brought into the market, but it was always those captured the previous day which were offered for sale; the fishermen and coolies admitted that " the smell was a little high, but they were only for native consumption and not for the Europeans, so a slight amount of putridity was immaterial." Whether this has much effect in a sanitary point of view it would be out of place to enquire into here. A gentleman in the station, however, informed me that not long since he was near the coast, and made some remarks to a zemindar on the horrible smell from the drying fish, enquiring if it was not very unhealthy? " Yes," rejoined his companion, " cholera seems made for these people." The mode in which the Chanderpore fishing was carried on was as follows :—Stakes extending about ¾ths of a mile are so placed as to surround a semicircular portion of the sea, the base being formed by the shore.

Opinion of the Collector of Balasore.

To these stakes are attached the nets, and at points where the water is deepest is a second net placed parallel to the first, and also supported by stakes. Here I saw a good number of hilsa captured, and the fishermen assured me that they are always present, except during the southwest monsoon, when they ascend the large rivers to breed. I took a full sized one out of season, and another half-grown one that appeared very healthy, therefore it may fairly be concluded that were this species of fish destroyed the coast fishery must suffer. To show how cheap they were, I may mention that I purchased a roe-ball (*Polynemus tetradactylus*), weighing about 30lbs., for 8 annas. The *Collector of Balasore* (July 27th, 1872) observed that " in this district sea, river, and tank fish are procurable in considerable abundance. On inquiry the fishermen state that the supply of fish has much decreased of late, but I do not find that this statement is borne out by facts. They probably make it in the interest of their trade. Three-fourths of the people use fish; it 'is generally preserved by drying in the sun and mixing it with salt'; these dried fish are used as a relish. ' The fish that are caught are brought at once to the nearest market, and meet with a tolerably ready sale, though the people are not in a hurry to get rid of them, not being averse to fish far advanced in decomposition. Even in the town of Balasore, only six miles from the sea, the fish sold in the markets is so stale that no European would touch it, and much of it is putrid. Dried fish is always taken to the Tributary States during the winter for sale." He continued (September 16th 1872):—" No Government interference is required. The sea is regularly fished, and large quantities of fish are daily exposed for sale in the town and at the hâts in the rural parts of the district. The people of this district do not salt their fish; they dry it in the sun, and eat it when it is quite putrid. They like it in this way, and there is no reason why they should be interfered with. They do not like salt fish, and the price of salt has nothing to do with their not salting it. With regard to para. 8 of Dr. Day's report, I would therefore briefly reply as follows:— Question 1.—No. Question 2.—No. Question 3.—No." With this brief reply the answers from native officials, if collected, were not forwarded.

129. The *Commissioner of the Presidency Division* (February 10th 1873) observes—"no deep-sea fishery is carried on here, and only those sea-fish are caught which come up the Hooghly, Mutlah, Echamutty, and other tidal estuaries." The proposal for forming enclosures for salting fish is, therefore, "neither advisable nor practicable here. It does not seem that the capture is in excess of the local demand, and the only fish salted here is the 'hilsa,' especially at the time it finds its way up the streams for breeding purposes. It might be practicable, though not without difficulty, to carry out the proposal at places especially selected on the sea-coast, near a port to which salt is imported, and near which there are great takes of fish, but not in this division." *Mr. Verner, the Joint Magistrate of the 24-Pergunnahs*, remarks— "from the wording, it is difficult to be sure what that proposition exactly is. It is perhaps this, that the Government should make large enclosures [1] where fish could be salted, and should sell salt within

Opinion of European officials in the Presidency Division.

---

[1] The question was ".would the proposition in paragraph 7 be advisable or practicable in your district?" The proposition was—" if large enclosures were made near favour-

these enclosures at reduced prices. If this is what is meant, I think the proposition is neither advisable nor practicable. The enclosures and the necessary establishment would cost much. Reduced prices can only mean removal or reduction of duty, and in this case a considerable establishment would be required to prevent the salt being smuggled away for other purposes. The only form the proposition could take, not to be utterly visionary, would be a lease to a Company (1). The Company would have to meet the cost of the establishment appointed by Government for supervision, and would have to pay largely (2) for the concession of reduced duty, such a Company would pre-suppose consumers of salt fish." *Mr. Smith, the Collector of Jessore*, replies as to whether the fishermen have increased, decreased, or remained stationary? "This is a very difficult question to answer, as we have nothing like statistics on this point; there are two castes of fishermen, the *Hindu Jalais who are* mostly employed in catching fish."

130. From *Bhaugulpoor.*—The *Officiating Collector* of *Monghyr* observes (March 7th 1873)—"fish in this district is very cheap, if we may judge from the price, an average of one anna a pound throughout the year. I think there is no reason to suppose that the supply is falling off. The markets are invariably well supplied, excepting perhaps during the height of the rains, when, as every one knows, it is very difficult to catch fish of any kind. All classes in this district eat fish,—Christians, Mussalmans, and Hindoos, from the Brahmins to the lowest castes. From the abundance of fresh fish procurable there is little inducement to dry or salt it. The people in Pergunnah Farkya, however, where fish is very abundant, dry their superabundant stock, and export it in the countries lying to the east, and to Calcutta. The fish thus dried is of every description. Fish is not salted in this district, probably partly in consequence of the high price of salt, and partly because salt fish is not very palatable."

Bhaugulpoor Commissionership. Opinion of Collector of Monghyr.

131. The *Commissioner of Dacca* forwards report from the *Officiating Collector* of *Backergunge*, who observes (November 16th 1872)—"that he is not aware that breeding and immature fish are destroyed to any extent in the tidal estuaries of the district. " I do not think that the salting of sea-fish could be much developed in this district, as the amount of sea-fish which can be caught is not very great. I do not think that the proposition in paragraph seven (see *ante*, enclosures) would be practicable in this district. The selling price of salt is Rs. 5 a maund. Government salt is employed when salt is used at all, but very often the fish are simply dried. Sea water or salt earth is not employed. The practice of salting fish, I believe, is stationary. I never heard of fishermen having any priviliges. There are no headmen.

---

able localities (away from large towns), and where fish could be salted, would the following be impracticable? That salt might be sold inside them, at rates just remunerative for the *bonâ fide* salting of fish." This gentleman observes that—"from the wording it is difficult to be sure what that proposition exactly is."

[1] Without entering into the comparative expenses of a single or more capitalists and that of companies, it may be well to refer to where officials both European and Native understood the proposition made and considered that a trial might lead to beneficial result for the food of the poor of India.

[2] If they are to *pay largely* for the concession, and be put to all sorts of expenses, it would appear the result must be to entirely vitiate the proposition.

The fishing in the tidal estuaries is, I believe, perfectly free. The fishermen supply themselves with boats and nets; their numbers are stationary. In conclusion, I have the honor to state that there are no deep-sea fishings in the district, and that the fishings in estuary of the Megna and along the coast of the Sunderbunds do not appear to be of any great value. In the cold weather, a number of boats come from Jessore, and fish on the coast off Chapli, and other parts to the south of this district, and Mugh boats come from Chittagong, and catch and dry fish."

132. *Dr. Cantor* observes (Journal Royal Asiatic Society, V., p. 170, 1839) that—"with a view to ascertain how far the locality and climate would favour the process of salting and drying fishes on the coasts of Bengal, Captain Richard Lloyd, who as Marine Surveyor-General always has evinced a strong desire to enquire into the natural products and resources of those localities, which by his indefatigable zeal have been surveyed, caused a series of experiments to that effect to be tried on board his own vessel. The materials submitted to trial were either purchased from fishermen at the rate of Rs. 3 a hundred, or supplied by the nets belonging to the fishing boats attached to the navy. The experiment turned out so satisfactory that I feel convinced that the process of curing, salting, and drying fishes may be easily accomplished there during the north-east monsoon, that is, during the period from 15th October to the 15th April."

Opinion of the late Dr. Cantor.

133. From the Chittagong Division, the *Collector of Chittagong* replies (April 29th, 1873) that "at present—a small quantity of hilsa fish is salted and sent to Chittagong from Moiscal, Chuckeria, and Ramoo ; but it appears that the whole is consumed locally, and there is no export. The fishermen object on the score of the high price of salt : and even were this objection removed, the Collector doubts whether the fishermen would be inclined to come and prepare their fish for market* within the proposed enclosures. Moreover, these enclosures would have to be carefully guarded to prevent the low-priced salt being smuggled out. In short, the Collector does not consider that the action suggested in Dr. Day's para. 7 would be advisable. Both breeding and immature fish are supposed to be destroyed to a considerable extent in the district of Chittagong, no restriction being placed on the operations of the fishermen. The Collector thinks, that if the demand for salt fish were to increase, the supply would increase in the same proportion.† The latest quotations of salt in the district of Chittagong are as follows: Sudr station, Rs. 4-6 per maund ; at Koomercia, Rs. 5 per maund ; at Hathazaree, Rs. 4-6 per maund ; at Cox's bazar Rs. 5 a maund ;" it is

Chittagong Division. Opinions of European officials.

---

* It is not proposed that any one should be *obliged* to cure their fish inside these enclosures, but whether if those who did so were allowed previleges, open to every one, some would not avail themselves of such. The prepared fish would be more for export than local consumption, and if it answered it would increase the local demand and with such the fishermen's work, for had they a good market, doubtless it would be adequately supplied. Now Chittogong fishermen go to Akyab to carry on their trade, because salt is cheap there instead of working off their own coast.

† I would here suggest that this must greatly depend on the price of the salt. If the charge is so much that the curers cannot purchase it, how can they make use of it ? or if they can employ it at such a cost, who would be able to buy the salt fish ?

used for curing fish, and as far as our information goes it only is employed. The practice of curing fish has probably remained stationary. The system of headmen amongst fishermen is not known to prevail in Chittagong: they buy their boats, make their own nets, and it is believed their numbers have continued the same of late years. *The Collector of Noakhally* answers, that the suggestion of enclosures for salting fish in—" would be inapplicable in his district, both because the people are not skilled in salting fish and because the supply of fish barely meets the requirements of the local markets. The destruction of breeding and immature fish in the tidal estuaries of the district, is estimated to average 200 or 300 a day. Sea-fish is not salted. The present average selling price of salt is Rs. 5-2 per maund. Government salt is used for curing the hilsa for home consumption, but to a very limited extent, as this particular species is rarely procurable: neither salt earth nor sea water is employed. Fishermen had no privileges in the earliest times which they have not at present. But after the resumption, and before the release of the fisheries by Government, rents were more strictly exacted from the fishermen than either in the oldest times or at present. There are headmen of the fishing caste, termed 'Sirdars' who possess an hereditary right to the title. A vacancy occurring on the death of an heirless headman, is generally the cause of much dispute between those who consider themselves entitled to the right of possession, and is filled up on the decision and nomination of their zemindars. The duties of the headmen are to preside over marriages, religious ceremonies and feasts, and to decide all social disputes, for which they receive from one to four rupees, and at times, both money and cloth according their rank; the sea-fisheries are owned by Government and by zemindars. The fishermen purchase their boats from Chittagong and Tipperah, and manufacture their own nets from hemp grown in the district. The number of the fishermen is believed to have remained stationary.

134. From the *Presidency Division*, the following replies to the questions sent to *tehsildars, or native officials*, have been received. The selling price of salt is from Rs. 4-6 to Rs. 5 a maund; it is used for curing fish in Jessore, but not in the 24-Pergunnahs, but neither salt earth nor sea water are. As to whether the practice of salting fish has increased or not, it is observed that there are no available statistics from which a definite answer could be given to this question. It is not the practice now in the 24-Pergunnahs to salt fish, and there is nothing to show that it ever was the practice there. There are no headmen to the fishing castes. Nets are made by the fishermen, and they buy boats or wood to construct them of, or hire them at about Re. 1-8 a month each. As to whether the fishermen have increased, it is observed, " there are no regular sea fishermen here, though occasionally fishermen pass through the Sunderbunds and catch fish in the open water."

Presidency Division. Opinion of Native officials.

BURMA.

135. The sea coast of British Burma extends nearly 1,000 miles
Sea coast and fisheries of Burma, how they are now but little worked. along the eastern shore of the Bay of Bengal, from the Naf estuary, which is about 20° 50′ north latitude, forming the boundary between Arracan and Chittagong, to the Pakchan stream, that separates Tenasserim from Siam, a little south of the 10° of north latitude. In many respects the seas of this coast differ considerably from what we find elsewhere in India (if we except the portion off the Sunderbunds). The shore is very low, the waters very muddy, and abound in animal life; crustacea are in myriads, and marine fish which prey upon them, are in abundance. It is, I think, evident to everyone who has been in Burma that these deep-sea fisheries are but slightly worked; it may be that some stimulus is requisite to incite the Burmese to capture the fish, or even to a slight extent it may be due to taxes existing on their fishing nets. Although, doubtless, it is quite reasonable to tax fixed engines, the general belief throughout the world seems to be that marine fishing nets that are not fixed should be tax-free. One reason why the deep-sea fishing is not carried on, doubtless, is, that these people do not much care to go far out to sea or incur the expense of making deep-sea nets; but it seems strange that with such magnificent fisheries they are scarcely worked. One reason adduced for this is, that persons migrating from inland to the sea shore are liable to suffer severely from fever. The monsoon of course totally prevents this form of fishing from being carried on during some months of the year, and the fixing of weirs is an easier and less dangerous occupation than venturing out to sea in the fine months. The Chinese and people from Madras and Chittagong fish in some of these waters.

136. What the amount of cured fish may be there are no returns to show, but as the customs house between *Fishing increasing.* British and Upper Burma appear to demonstrate that the exports are annually exceedingly large. (See para 141) There is a complaint that fishermen are migrating from Tavoy to Mergui, but this is probably due to the latter town being on the sea shore, where fish abound, whereas the former is up a river.

137. Finally, it may be remarked that the Indian salt tax is not in existence in Burma and this condiment, it *Price of salt.* is stated, in Arracan costs from 10 annas to 12 annas a maund, 8 annas sometimes; in Rangoon, imported salt from 12 annas to Re. 1-4 a maund, or of the country-made article, from Re. 1-8 to Rs. 3-12 for the same quantity; in Tavoy, Re. 0-9-7 a maund.

138. *The Officiating Secretary to the Chief Commissioner of British Burma* (February 21st 1873) observes:—" It *Opinion of the Chief Commissioner.* would appear that as a result of the previous investigations made by Dr. Day, he was led to

believe that sea fisheries in those localities far removed from large towns required some stimulus, in order that the harvests of fish might be rendered more conducive to the general good. He thought this object would best be gained by the fish being cured and exported to other places where it would find a ready sale. He found it asserted by several officials that the monopoly price of salt is such as to interfere with the extensive manufacture of salt fish, and he asks whether that is the case in British Burma, or whether there be other causes preventing the development of this particular trade. The Chief Commissioner desires me to observe that there is already an exceedingly large production of prepared fish for consumption in British Burma and the neighbouring inland native States. The particular article in use among the Burmese is called 'nga-pee,' and its preparation is described by Dr. Day in the 'Digest' of his previous investigations. This 'nga-pee' is extensively employed as a condiment by the Indo-Chinese races. The quantity consumed in our own territory is not known accurately, but the average yearly export to Upper Burma during the past four years has been upwards of sixteen thousand tons, with a value of £170,000; and the home consumption in British Burma is certainly far greater than the quantity exported. The sea coast of the delta of the Irrawaddy is the locality where this nga-pee is principally prepared; and from the two districts of Rangoon and Bassein, which there furnish the sea board, there is derived an annual revenue from the rent of the sea-fisheries of £22,000. It does not appear to the Chief Commissioner that there is any special call for stimulus as regards the traffic in prepared fish in this Province and the neighbouring States, and there is no difficulty thrown in the way of the extension of the trade by reason of the high price of salt. There is no Government monopoly of salt in Burma, and an ample quantity can always be obtained by the fish curers at a very moderate cost. It is very certain, as remarked by Dr. Day, that new markets would produce increased manufacture of cured fish in places where at present they only prepare for home consumption, and this result will, in the opinion of the Chief Commissioner, gradually be arrived at from improved means of carriage. The district of Mergui at the southern extremity of the province has long been famous for its cured fish, which it has exported in considerable quantities, but the district is much isolated, and has a difficult and dangerous coast. Quite lately, however, a merchant steamer has commenced to run between Rangoon and Mergui *, and undoubtedly that will have the effect if the line be kept up of increasing the export of fish from that district. Mr. Eden has lately returned from a tour in the Mergui Archipelago, and was very much struck by the enormous supply of fish among these uninhabited islands; these fisheries are now untouched and unvisited by any fishermen. The sea is literally alive with fish, and it at

---

* It appears to have been overlooked that up to 1851 a monthly steamer ran between Mergui and Moulmein. At this period, due to the Burmese war, a sailing vessel was employed. After the war another monthly steamer plied, and only stopped for a few months during 1857. From Moulmein to Rangoon and Calcutta, &c, the mails have been carried regularly for years in Peninsular and Oriental or British India Steam Navigation Companies steamers consequently it may be open to a doubt whether a fresh line from Rangoon to Mergui will have very much effect in stimulating this trade. Steamer rates and contingent charges are high.

once occurred to the Chief Commissioner to suggest to the Government of India the possibility of establishing a colony of fishermen and fish-curers among these islands. Dr. Day will probably be able to say what prospect there would be of inducing a colony of 100 or 200 fishermen to come over from Madras and establish themselves. It might be necessary to make them advances for the purchase of nets and boats, and to guarantee them a certain minimum of earnings. * If a system of this sort could be established and carefully maintained, Mr. Eden believes that a supply of salt-fish would be obtained which would lead to a very large export to India; and it seems to him that a small portion of the proceeds of the fishery tax could not be employed in a more legitimate manner. With this exception, it seems to the Chief Commissioner that the expansion of trade in the article of salted or cured fish may be left to national development, and requires no exceptional measures on the part of Government. It will be observed that the Deputy Commissioner of the Rangoon district remarks that the tax on sea fisheries in his district is considered too high, or at least that the manufacture would be increased if the tax were reduced and altered in its mode of collection. The Chief Commissioner will cause further enquiry to be made into this question, and the effect it would have on the revenue derived at present from the fisheries, and will take into consideration the advisability or otherwise of lowering the present rates."

139. *Dr. Helfer*, in his fourth report on the Tenasserim district, published in the Journal of the Asiatic Society of Bengal (1840, p. 183), observes, respecting the sea fisheries of Mergui, "it seems that a fishery carried on by Europeans on a similar scale with that of Newfoundland would be much more remunerative, as a ready market here and in India will always be found, cured fish forming a universally relished condiment and ingredient of native dishes." *Crawford* in his 'Indian Archipelago' "asserts that fish are almost if not always dried without salt, but this

Modes of preserving fish.

---

* There are two countries from which fishermen could probably be introduced, *viz.*, Madras or China. Some from the east coast of the former part of India have already migrated in considerable numbers to Burma, and it might be advisable to enquire whether they might not be available for this purpose, provided they had certain inducements offered them, such as being permitted to manufacture salt in these islands free of duty, &c. If Government guarantee a minimum of earnings in a sea fishery, it does not appear improbable that the measure would be a failure, but if the captures in excess of those required by the fishermen were purchased at a certain rate, and fish salted as proposed, more satisfactory results might be obtained. Or the salted or cured article might be bought at a fixed price, for the worker must have an interest in his own success. Another important consideration is, that the siluroids or cat-fish, which are of a very inferior description as food, but are easily captured, may be passed off in selling the excess to Government: this however would not signify if the trade were for Ceylon. Rules would be necessary for the sorts of fish, as some of the very finest species exist in this Archipelago. If the fresh fish were bought, arrangements would have to be made for salting them on the spot, this occupation being usually carried on by the women and children of the families of the fishermen, but in Burma it is improbable that the Madras fishermen would be provided with these encumbrance, whilst, were men solely thus employed, it would raise the price of the article. It must not be forgotten that those islands at present are unsuited for living on during the months of the South-west monsoon, but could fishermen be induced to go there with their families, doubtless they might become populated in time. Fish obtained in the waters of these islands might be sold by auction or contract in the Moulmein or Mergui bazar, leaving it to the enterprise of the local merchants to develop the expansion of the trade.

is not quite correct,* for although the greater portion of the fish taken are sun-dried or smoked, some salt is employed. The fishermen proceed to some of the islands to the south of Mergui, and there capture and prepare their spoils, but their plans are rude and primitive, probably having remained unchanged for ages, consequently the article is very inferior." It will, from the foregoing, be observed that the salt-fish is very inferior, whilst, due to the moisture in the Burmese climate, it would require a fully salted and prepared article to keep well in such a country. The mode in which the Burmese salt their fish is peculiar, and termed nga-pee, a description of which I give, and it will be easily understood that such an article is not, or very little, affected by their climate.

140. '*Nga-pee*' is a Talaing word of the following component parts:—

Nga-pee, two kinds.

nga, *fish*, pee, *three*, this last syllable having reference to the three ingredients out of which it is formed,—fish, salt, and the action of the sun's rays, the last not being essentially necessary. It is divisible into two sorts, *nya-pee-goung* or *nga-pee*, 'whole;' and *nga-pee-toung*, or *nga-pee*, 'pounded.' The general mode pursued, without specifying details, in making the 'nga-pee-goung' is, after having removed the head, the fish, if large, is split in two, then washed, and dried in the sun for a few hours: good salt is now rubbed in, and the fish (along with others similarly prepared) is packed in a jar, from which the next day they are all removed, again dried in the sun, have more salt rubbed in, and are then re-packed as before. In some inland fisheries, where large quantities have to be thus prepared, they are laid in long bamboo baskets in alternate layers of salt and fish, and buried for some time underground. The best species out of which this compost is manufactured in British Burma are considered to be as follows: nga-myin (*Pseudeutropius taakree*, Sykes): nga-khoo (*Clarias magur*, H.B.): nga-kyee (*Saccobranchus fossilis*, Bloch): nga-bat (*Wallago attu*, Bloch): nga-yan-gyee (*Ophiocephalus striatus*, Lacépéde): nga-byays-ma (*Anabas scandens*, Dald.): nga-tha-louk (*Clupea palasah*, C. and V.): nga-oik (*Macrones menoda*, H.B.): nga-phe (*Notopterus kapirat*, Bonn.): kabaloo: kakooyan: nga-wek-ma (*Nandus marmoratus*, C. and V.): nga-gyein (*Cirrhina mrigala*, H. B.): nga-hoothan (*Labeo calbasu*, H. B.): kakadit: (*Lates calcarifer*, Bloch.): nga-thaing (*Catla Buchanani*, C. and V.): nga-chong-ma (*Barbus chola*, H. B.): nga-phyin-thaleb (*Trichogaster fasciatus*, Bl.) The second sort of 'nga-pee,' termed '*nga-pee-toung*' or 'pounded nga-pee,' is either made of fish or prawns; the latter being chiefly prepared at Tavoy and Mergui, and is known in Rangoon as '*Ba-lee-chong*'. When fish are used, they and the salt are placed together in a large trough, which has a tube or pipe opening from its floor. Men wearing wooden clogs now tread the fish and salt into one pulpy mass. The superabundant fluid runs off by the tube, is collected for sale, and employed as a condiment. The roughly trodden down or pounded 'nga-pee' is either used as it is or undergoes refinement. This last is accomplished by pounding it up with the bark of the on-don, which gives it a colour, and now more white salt and flour are mixed with it. When it is desired to prepare this nge-pee at once, the fish having

---

* *Report on the Sea-fisheries of Burma*, 1869, p. 9.

been cleaned, and their heads removed, are pounded with salt in a mortar. Different species of fish, the quality and amount of the salt employed, and other causes, give a varying value to the nga-pee. Some of the siluroid or scaleless fishes are most approved of for the manufacture of 'nga-pee.' In Bengal I have observed that some of the same class of fish (as *Saccobranchus fossilis* and *Clarias magur*, both having an accessory apparatus for breathing) are esteemed highly nourishing. In making nga-pee with siluroid fish no scales can become mixed with the preparation, which, if present, must occasion intestinal irritation. It is observed that the smaller the species of scaled fresh water fish that are employed the more inferior is the article. It is a very common, though according to my local investigations, a very erroneous statement, that the poorer classes prefer making nga-pee from very small fish, such being only due to their being permitted to slaughter the immature and smaller kind. Whilst for nga-pee-toung the usual length of the fish is from 8 to 10 inches. I observed that the term *balachung* was applied in Rangoon to that form of 'nga-pee-toung' which is made from prawns at Mergui, but at the latter station they reserve this term for the superior Malay article, which is of the consistence of chocolate cakes, instead of being a paste, although the two are otherwise much the same, being prepared from identical substances. At Mergui there are two kinds of nga-pee,—the first made of very small prawns or shrimps, which are there captured in enormous quantities, coming in shoals and being taken in boatloads at a time. When the manufacture commences in the morning, these crustacea are mixed with about one-third of their weight of white salt, and dried on mats in the sun for 5 or 6 hours; then they are pounded up in a wooden mortar; or if the process is carried on in the island, they are trodden down into a mass by persons wearing wooden clogs. This mass, if possible, should be again dried for 4 or 5 hours in the sun the same day, and the pounding is repeated in the evening. The next day the drying has to be again gone through, and it becomes the 'nga-pee toung.' The superfluous fluid is collected and termed 'nga-pya-gee', and is much esteemed as a condiment. *Dampier* describes this condiment and 'balachaun' in his Voyages and Travels, observing, "the mashed fish that remains behind is called balachaun, and the liquor poured off is called *nuke-mum*. The poor people eat the balachaun with their rice. 'Tis rank scented, yet the taste is not altogether unpleasant, but rather savoury, after one is a little used to it. The *nuke-mum* is of a pale brown colour, inclining to grey, and pretty clear. It is also very savoury, and used as a good sauce for fowls, not only by the natives, but also by many Europeans, who esteem it equal with soy." The second form of nga-pee made at Mergui is termed *damin*, and for its preparation small sea fish are employed. They are first well washed and cleaned, or the substance will be of a dark colour and inferior quality, due to the contained blood. These small fish are mixed with salt, dried in the sun, and pounded in a similar manner to the prawns as already described. *Smoking fish* is likewise carried on in the island, round Mergui, where firewood is obtainable without cost and an inexhaustible supply of fish close at hand. Open platforms of wood are raised, which are generally and should always be in the shade; the firewood is burnt beneath, and being green emits a cxxxi

dense smoke. *Drying fish in the sun* is also very common. When the flesh of large fish, as sharks, is thus prepared, it is cut into strips and dried upon the rocks, but I could only ascertain that this was done to the young ones from two to three feet in length. *Tamarind fish* prepared at Mergui is held in high estimation, and is chiefly made from the white pomfret (*Stromateus*). The fish I understood is first boiled, then the bones are removed, when having been highly spiced it is left to soak in vinegar, &c., for some days, and subsequently packed in small jars or chatties containing about one viss in each. The exact receipt could not be obtained. A small amount of *fish oil* is made along the coasts of Burma. The *fins of sharks* are dried and exported to China, as are also the air vessels or *sounds* of some fish, especially the siluroids and the *Bola pama*, H. B.

141. As regards the exports and imports of nga-pee, dried fish, and salt, I have only obtained those existing between Upper and Lower Burma, as the British territory is the highway to the Native State. My figures, obtained locally from the custom house authorities, widely differ from those of the Secretary to the Chief Commissioner, which I can only account for by supposing he merely refers to nga-pee, whilst the custom house authorities gave me 'nga-pee and dried fish.' Amongst the latter I found large quantities of the sepia *Octopus* imported, it being highly esteemed by the Chinese.

Exports of salt fish and salt from British into Upper Burma.

Imports from British to Upper Burma.

YEARS.	NGA-PEE AND DRIED FISH.		SALT.		Annual value.
	Maunds.	Value.	Maunds.	Value.	
		Rs.		Rs.	Rs.
1861-62	5,53,706	16,64,343	1,51,885	1,67,946	18,32,289
1863-64	4,62,102	14,34,102	2,28,747	2,78,967	17,13,069
1864-65	5,25,751	18,50,296	2,42,320	2,88,281	21,38,577
1865-66	3,96,035	14,15,585	2,05,618	3,43,379	17,58,964
1866-67	3,41,980	14,64,180	2,32,192	5,19,783	19,83,963
1867-68	4,21,394	20,01,774	2,74,691	5,35,364	25,37,408
1868-69	4,32,638	18,91,482	3,17,965	4,69,916	23,61,398

Thus, in the three last years shown in the above return, the imports of dried fish and nga-pee from British into Upper Burma give Rs. 53,57,436, or £535,743-12s., against £463,978-6s. in the previous three years, ostensibly showing an increasing trade. But if we examine the cxxxii

weight imported we find that the increased trade is due to a higher value having been placed on the articles. Thus in 3 years ending 1866, 13,83,888 maunds were valued at Rs. 46,99,983: whereas in 3 years ending 1868-69 the smaller amount of 11,96,062 maunds were valued at 53,57,436.

142. The Burmese, although excellent fresh-water sailors and boat-men, do not appear, as a rule, to venture far out to sea for fishing purposes. In fact, they cannot be termed good salt-water fishermen, but mostly take what they are able within an easy distance from the shore, sometimes by means of nets, more commonly by fixed engines, and not, so far as I could ascertain, by means of hooks and lines. The Chief Commissioner's observation that in the Mergui Archipelago the sea is literally alive with fish, is only what has again and again been brought to the notice of the authorities, but without inducing any steps being taken to utilize this harvest. The Burmese, however, so long as they can capture fish near the land, will not venture out to the best fishing grounds, and it now will be necessary to give a short description of the modes employed by them in the capture of sea fish. The following remarks are from the Administration Report in 1867-68. In Ramree, Arracan, it is observed that a revenue is levied in this district on the various descriptions of nets used in the sea and river fisheries, showing a decrease of 22 nets." The Arracan fisheries or net tax is insignificant and apparently stationary. In Rangoon, it is observed—"During the year under review I have had opportunities of visiting the circles where the greater portion of the sea-fishery is carried on, and where inland fisheries exist. To carry on fishing on the sea-board requires a capital of some Rs. 500 to purchase a seaworthy boat. The necessary fishing apparatus, termed 'hunyoon,' having either been worked from bamboos during the leisure moments of the fishermen during the rains, and the month previous to the actual time they proceed down to the sea-board station ; or if the person is a capitalist he purchases it from others. In these boats they proceed out to sea to their fishing grounds, and fix stakes to attach the hunyoon traps for the fish. These grounds are visited daily. * * On the whole this kind of fishing is attended with danger, and generally conducted by needy men with borrowed capital, and who have to pay high rates of interest. The average number of traps to each boat is 15, and each trap pays a revenue of Rs. 4. The 'dameng' fishing is carried on in the main river by boats, which have broad bamboo traps thrown out on each side, and let down at the ebb tide ; each boat is assessed at Rs. 4." Fish are taken in the Tavoy and Mergui districts in rather a different manner from other localities. The *tsanda* is in common use (see Fresh Water Fishery Report, p. 212, &c.). *Colonel Fytche* observed in 1853—" These stakes are sometimes worked all the year round, but not generally; for, being situated near the mouths of rivers, the freshes during the south-west monsoon drive away the salt-water fish." This must be accepted with a reservation that many migratory sea and littoral species remain even when the water is fresh. "There are generally eight or ten men employed in the working of a tsanda, and they work one tide in the twenty-four hours, sometimes bringing in as much as 60 rupees worth of fish in one trip ; but the stakes,-

Burmese not good sea fishermen.

owing to the attacks of acephalous testaceous molluscs, which abound in these waters, require to be renewed twice or thrice in a year, and it takes the party from two to three months (weeks?) to erect new ones. The fish are divided in kind, the headman, or he that finds the capital, taking two shares to the other's one, and each man generally makes his own bargain with the people who keep stalls in the bazar. A single share in a tsanda, when fairly worked, is said to be worth Rs. 180 a year. It is not worked during neap tides." *Hloot-gyee.*—" Some seven or eight men are employed on this trap, which is erected on a spot dry at low water. It is constructed similar to the tsanda, but without a stage. The fish go up the passage, which is closed shortly after high water, and the fish are taken as the tide recedes." *Hloot-ngay* " resembles the former, but much smaller." *Lamoo-gyee.*—" The stakes for this trap are erected in a somewhat similar manner to those of the tsanda, but the stakes are smaller and made of split bamboo and cane, instead of the wood of the mangrove, which is principally used for the tsanda, and are always erected close in-shore, whereas the tsanda is sometimes three or four miles out. It is worked in like manner to the tsanda, but does not require so many men. The fishing by the *Lamoo-gyee* does not last more than six months in the year; hardly any fish are caught during the neap tides, and it is comparatively more expensive to erect than a tsanda, whilst it is not quite so profitable." *Henyan-paik-gyee.*—" A large net consisting often of twenty separate pieces, a set being worth about Rs. 80. These nets are taken out during the spring tides in a boat by two men. They are floated by buoys and allowed to drift with the tide. These nets are not used during the south-west monsoon. Large hauls of fish are often made with these nets, but the yield is uncertain. One man occasionally owns two or three sets of the henyan-paik, and each set is said to be worth to him Rs. 200 a year. *Tshwai-paik-gyee.*—" A similar net to that used in the Sandoway and Henzada districts, and used exactly in the same manner as the seines used in salmon fisheries." *Lamoo-byay.*—" These are moveable screens made of bamboo and cane; these cost about two months' labour, and do not last more than four months; these screens are put up on sand and mud flats, which are left exposed during spring tides. The lamoo-byay are taken out to some eligible spot in the archipelago where a small fishery station is made, and the fish caught, dried, and smoked for conveyance to the nearest market. A large number of skate and sharks are caught in these nets. This mode of fishing is nearly as profitable as the tsanda." *Gawa.*—" This is a net in a frame, which is lowered over the side of a boat, and is only used for about three months in the north-east monsoon. It is used for catching prawns. It costs about Rs. 4; when worked by an industrious man would bring him in from Rs. 80 to 100 during the season." *Gwy-gawa.*—" A net very similar to the one last described. It is used for catching shrimps or small prawns, and in the same way as the shrimping net is used in England, the only difference being, that the man who uses it pushes a canoe along with him instead of carrying a basket. During the season a man can earn about eight annas a day with it, but he must then be at work whenever the tide suits for wading over the mud or sand, in water about two or three feet deep." At Akyab, Major Spilsbury furnished me with the following return of the taxed implements employed in the fisheries. *Khyan-lamoo.*—Temporary bamboo screens, erected on the shore and used throughout the year; they are worked by from six to fifteen men, and one or two boats. *Laig-wen-paik.*—Bamboo screens which stretch across a stream, and are used throughout the year. *Bashoo-paik.*—A floating net, employed both in the rivers and sea, used in one set from 10 to 14 pieces, each 32 by 8 cubits in extent; it is worked by five men and one boat. *Ran-paik* is a net stretched along the shore, suspended by posts, and captures fish which come in with the high tide; it is used throughout the year. *Tansoung*—A long floating net used by 10 men in a boat in the sea during the fine season of the year. *Hsway-paik-kyee*, or *paik-blook-kyee.*—A single net about 12 or 16,000 cubits long, and 12 deep; when employed in the sea its lower margin is weighted; it is kept open by means of props or sticks, fastened to both sides of the opening; two boats are employed, manned by 10 or 12 persons. *Kyan-myet-kweng* are permanent bamboo screens, from 8 to 1,600 cubits long, and four deep; they are fixed on the shore, and used throughout the year. A system of levying taxes on nets used in the sea and tidal creeks existed at Akyab until September 1866, after which it was levied on a set of nets. "Major Broadfoot, Commissioner of the Tennasserim Provinces, in 1843, published an order abolishing all the then existing fish taxes, both on nets and traps, and levying in lieu thereof a tax of so much per head; in short, the present capitation tax, there first levied, and which to the present day is called by the people of the country a 'fish tax,' and is so termed also both in the English and Burmese translations in the printed Revenue Rules for the Tennasserim Provinces" (Colonel Fytche's report, 1862). At this date it was decided to levy a general capitation tax throughout Burma, so the nets and traps employed by the fishermen were again subjected to the revenue laws. *Major Spilsbury* describes some other descriptions of engines used for taking fish in Tavoy, but which are also employed in the remainder of the districts. *Hmyone.*—"Cylindrical bamboo baskets from 3 to 6 feet in length, and from 1 to 2 feet in diameter, with a trap-door on both ends in which fish are caught, used in the streams (and tideways). *Khaya* is a long cone-shaped basket, held under water in a strong tide; the fish get in and are jammed at the sharp end, and when the tide is slackened the basket is taken up by means of a rope; distance between the rods or screens ⅜ths of an inch."

143. *The Commissioner of Arracan* forwards reports with remarks

Arracan. Opinions of the European officials.

(dated July 22nd 1872). *The Deputy Commissioner of Akyab* observes that shrimps are taken in large quantities from October to May, the greater portion of which are used in making nga-pee. The nets in which these shrimps are taken is roughly wove, like cloth, the holes being about $\frac{1}{10}$th of an inch in size. *The Deputy Commissioner of Sandoway* (December 24th 1872) replies—" I am of opinion that a reduction in the price of salt would neither increase the local sale nor induce larger exports of fish. Salt is already cheaper in this mart than in Bengal, and yet the people prefer agriculture to the pursuit of fishing. The fishing being merely carried on to provide the local consumption, it is probable that a good many breeding and immature fish are destroyed in the tidal estuaries, but to what extent it is impossible to say. I do not think the salting of sea-fish could be increased in this district. Salt is cheap, and scarcely ever exceed Rupees 8 (he probably means annas 8) per maund, and can often be purchased at the rate of Rs. 2-8 and Rs. 3 per 100 viss, which is the equivalent of 4 maunds.* *The Deputy Commissioner of Ramree* observes (1st February 1873), that breeding and immature fish are destroyed in the tidal estuaries, but not to any great extent. Does not think the salting of sea-fish could be increased in his district.

144. *In the Tenasserim Division, the Commissioner* replies (24th December 1872) that—"fish are caught at all seasons of the year, but chiefly on the coast during December, January, February, March, and April." Subsequently (3rd January 1873), he observes—"that the price of salt does not interfere with the curing of salt-fish in this division, and that no arrangement is required by which salt, free of duty, would be sold to fishermen. The trade in salted fish is generally very brisk, especially at Mergui, where some 500 boats are employed every season, the profits in each boat averaging from Rupees 6 to 800 a season. For each net a yearly tax of Rupees 20 is levied. Breeding and immature fish are, in large numbers, destroyed in the tidal estuaries of this division. The sea along the coast, in this divison, is 'swarming with fish. To increase the salting of fish, all that is required are men to catch them, and an increased demand for the increased supply when the fish are salted." *The Deputy Commissioner of Tavoy* (14th January 1873) replies, that—"breeding and immature fish are destroyed in the tidal estuaries to a small extent. The salting of sea-fish could not be increased, because the fish caught are not so abundant as to require salting; and besides fresh fish sells better than salted fish in this district, and therefore very little is salted. As salt is cheap—and can be procured without difficulty in any place, whilst the people do not go far from the town for fishing, what they get is brought to market, and they are not in the habit of salting fish. A great quantity of salt fish is imported from Mergui and other parts of Burma; little is here salted."

145. *The Deputy Commissioner of Rangoon* (24th December 1872) observes, that—" breeding and immature fish are not, it is believed, destroyed to any great extent in the tidal estuaries; a few might be caught for consumption, but not for sale. It is believed that the salting of sea-fish might be increased by a reduction in the rate of tax, and perhaps by changing the method of collecting it. The tax at present is Rupees 4 a trap, and this is considered high. Persons report to the Thoogye of the circle in which they reside the number of traps they are going to use, and they have to pay according to this number, whether they use them or not; some no doubt use more traps, but some again lose them in bad weather and have to pay all the same. The tax is collected some time after fishing operations have been carried on, by which time many have lost their gains in gambling and do not like to return. A fixed sum to be paid for a license for the season, at the time of taking out the license, without reference to the number of traps to be used, would, it is believed, be a better arrangement, and be likely to encourage

---

* The Burmese viss of 100 tikals (365lb. avoirdupois) is exactly equal to 140 tola or rupees weight of India.

persons to take to this occupation. Salt used for curing fish is not imported, but manufactured in the country."

146. I now propose detailing what I personally witnessed in 1869,
Personal observations. or information which I locally obtained mostly from the Myookes. At *Mergui*, there were about 700 fishermen in the township, amongst whom about 326 live entirely by this occupation. These fishermen are Burmese, Mussalmen, Chinese, Portuguese, and Malays. When I was there, the inhabitants were suffering so severely from small-pox that but little sea fishing was going on, except at the fixed stakes and weirs. There were two bazars in the town, which were said to be well supplied with fish during the spring tides, as at that period the fixed engines came into use; but during neap tides fish were more or less scarce. All the people, with the exception of some Hindoos, consume fish, which article had no fixed price, but varied from nine annas a viss to Re. 1-12-0, in accordance with its size and quality. In the bazars I specially noticed several species of pomfret, *stromateus*; horse-mackerel, *caranx*; the begti, *lates calcarifer*; a sea perch, *serranus suillus*; and the *bola pama*, all of which appeared to find a ready sale for food. The skates and rays did not seem to meet with much approval. I enquired respecting sharks, for the purpose of making fish-liver oil, but the larger ones, though numerous, were but seldom captured, the usual size being from 2 to 3 feet in length. The Myookee observed that—"more small fish were caught at the fishing stakes during the rains than during the dry season," this being the period much of the young are about and seeking their food in the shallows. With many contiguous islands, a quiet sea for above half the year, a very limited population and myriads of fishes, generations after generations must pass away before any regulations need be framed to protect those in the open sea; augmentation of capture is required, and a long period will probably elapse before Burmese fishermen will be found who will make deep-sea fishing an occupation. They appear satisfied with placing fixed engines in the shallows, dredging at the prawns, and netting such fish as come near the shore, and thus they kill the food that would bring the finer species inland and frighten them into the deep sea or off to the neighbouring islands, instead of capturing them for sale. Fish variously prepared were being exported to Tavoy, Rangoon, and Moulmein. When steaming along between the islands and the mainland, many seir fish and enormous numbers of siluroids were seen. Along the coast of Moulmein and as high as the north of the Rangoon river, the character of the fish was much the same, and in no place were large sharks taken to any extent; neither did the oil sardines appear in sufficient numbers to render their capture for the preparation of oil of any commercial importance. At Akyab, the Kolodyne river enters the sea, and fishing in-shore was carried on to a considerable extent, but the strict sea-fisheries were very little worked. *Captain Porter*, the Master Attendant, in opposition to the views of the fishermen, considered these fisheries had become impoverished during the 13 years he had resided at this place, and which he attributed to the small mesh of the nets and the minuteness of the distance between the split bamboos forming the fixed fishing screens which were placed across every small creek, opening, or available spot. The smallest fly could not escape; the meshes in those of

the large fixed nets I saw were $1\frac{3}{10}$ inches in circumference, and that of the bag $\frac{8}{10}$ of an inch. Fish were salted at Akyab, but only to a limited extent, and chiefly by persons who came for that purpose from Chittagong about November, at which time the sea is pretty smooth. Bombay ducks (*Harpodon nehereus*), mackerel (*Scomber kanagurta*), various species of horse mackerel (*Caranx*), seir fish (*Cybium*), also some fine *Trachynotus* and pomfrets (*Stromateus*) being the sorts which were mostly cured by salting, drying, or converting into nga-pee. The isinglass-producing fish were not rare, as *Otolithus*, *Bola*, and *Polynemus*, but their air-vessels were said not to be prepared, but they must be in reality, as they as well as sharks' fins are exported from this port. But whilst I was there the daily supply of fresh fish merely equalled the daily demand. It must not be omitted to be mentioned that along the sides of rivers, across all creeks, openings or available spots within the influence of the tides, as well as above such, weirs or screens of finely-woven split bamboos were permitted, and which allowed the fry of fish and those seeking food to pass up these places, but as the tide receded left them floundering in the drying up mud, from whence the owner of this fixed engine could take what he required.

147. *In Tenasserim the native officials of Tavoy* report that the average selling price per maund of salt is Re. 0-9-7, the tax being Re. 0-1-2; that the salt manufactured in the district and which has paid tax being employed in salting fish, whilst neither salt earth nor sea-water is used. Only fish which cannot be sold fresh is salted. There are no headmen to the fishermen. The sea-fishermen have gradually decreased during the past five years, because fishing in Mergui waters is more profitable, and fish there more plentiful.

<small>Tenasserim. Opinions of native officials.</small>

148. *From the native officials of Rangoon* the following reply has been forwarded. The price of imported salt varies from Rs. 3 to 5 per 100 viss, and of country manufactured salt from Rs. 6 to 15. Imported salt is not used for curing fish; it is sometimes used in the manufacture of nga-pee. The use of salt earth or sea-water is not known in the district. There is a very great increase in the practice of salting fish, compared with the Burmese time. There is said to be sufficient nga-pee manufactured to meet the demand, but not salt fish. Persons go out (to sea), and fix their stakes where they please, the bolder ones out of sight of land, and the more timid ones near land. Some get advances on the promise of supplying a certain quantity of fish at a low rate, and with this money purchase boats and other necessaries; others again hire boats, and some purchase them with their own money. A headman generally employs three or four labourers, to whom he pays Rs. 50 to 80 for the season. As to the number of fishermen, they vary every year; if there has been a bad season, or an outbreak of fever, as is said to occur every three or four years, the numbers drop off for a year or two, and then increase again. On the whole, the number of those whose regular occupation is that of sea fishing is supposed to be much the same, as those only who have been living for any time near the sea are able to stand the climate. Persons living inland are generally subject to fever when they first reside near the sea.

<small>Rangoon. Opinion of native officials.</small>

ANDAMAN ISLANDS.

149. These islands have been fully reported upon, so merely a few
remarks are necessary. In 1789, Lieutenant
Blair first established a convict settlement
here, which, on account of its unhealthiness, was abandoned in 1793.
Again, in 1857, it was re-occupied by Dr. Walker as a penal settlement
for mutineers and convicts. Port Blair is in 11°42′ N. Lat., and 93° E.
long; it has a harbour of moderate size in the deeper parts, but extending
a long distance in its shallower portions; across its mouth and running
nearly north and south is the island of Ross, protecting it from the full
force of the easterly gales, but leaving a passage for vessels on either side
between it and the mainland on the north-east and south-west. The
waters are beautifully clear, except during the stormy months of the year,
whilst where corals and rocks do not exist the bottom is sand. The
inhabitants consist of two great elements, the aborigines and the immigrants, the latter being principally of the convict class.

Situation, &c., of Andaman Island.

150. The *Chief Commissioner of the Andamans and Nicobars*
observes (October 31st, 1872), that the fishing
operations of this colony are at present
strictly confined to its local requirements,
owing to the demand for labour on account of the more important works
on hand. Independently of this, however, the difficulties which must
continue to exist under the present constitution of the settlement are so
great and irremediable that the methods now employed are capable of but
little improvement. The fisheries are of necessity restricted to the inner
harbour at Port Blair, because it is impossible to trust even term-convicts with really sea-worthy boats. A few of the oldest of these term
convicts are now allowed to ply their trade in the deep sea outside, but
we shall soon have none of this class available, and our operations must
then be rigidly confined to the harbour, unless police guard boats could
be employed as escorts to the fishing gangs when at sea. Again, the harbours and coasts of the Andamans are so hemmed in and fringed by rocks
and coral reefs, that seines and drag-nets of any size cannot conveniently
be used, and in those cases where such nets are employed they are constantly torn by sharks if they escape the rocks. Under these circumstances, it will be seen that it is neither expedient nor practicable, in the
present condition of the colony, to attempt more than is needed for the
wants of our population, and to this extent our operations are fairly
adequate."

Opinion of Chief Commissioner.

151. The fishes of the Andamans are exceedingly varied in their
character, and chiefly remarkable by the absence of siluroids, which revel in mud, and
also of carps, due to a deficiency of fresh water. The modes of fishing
are stake enclosures, drag-nets, cast-nets, and hooks and lines. *Stakefishing* is carried on by means of bamboo fences or screens, which are
fixed in the muddy bed of some appropriate locality; a large piece of
water is enclosed, and at its centre, opening inwards, are three chambers

Fisheries of the island.

communicating one with another, the innermost having a door which can be closed at pleasure; it is cleared out at low water. The upper edge has to be about one foot above high-water mark. Owing to the large number of barnacles and other destructive animals, they rarely last above a few months, whilst, unless kept clean, fish will not enter. They have to be shifted every ten days or a fortnight, as the fish become too knowing to enter after it has been standing a few days. Owing to the large amount of coral, it is not always possible to discover a place where bamboos can be driven in. The captures are greatest during the rainy season, when the water becomes turbid, or from June to September. *Drag-nets* can be employed in places, but are seriously interfered with by coral reefs. *Cast-nets* are at times exceedingly productive, especially when the oil sardine, *Clupea Neohowii*, the species with a black tipped tail, *C. melanurus* and an anchovy *Engraulis boelama*, arrive in droves, not only around the islands, but also extending their range within the limits of the harbour. These fish are used as baits, or for food, and their best season appear, to be from November to March. There appears to be no difficulty during this period in capturing, in appropriate places, quantities of these fish by means of cast-nets, but as food they are considered inferior to those fishes which are taken by hooks and lines, or the large species captured in the enclosures. Port Mouat seemed to be especially visited by these fishes; but as they rapidly putrify they cannot be carried to the larger station, and are, generally dried when caught. Thus, one day we sent out four convicts with cast-nets, and in four hours they returned with about 260℔. weight, their canoe not being able to hold more. They asserted that they could continue taking at this rate for months had they a market, and it may be worthy of consideration whether fish-oil could not be manufactured during the season these fish are in sufficient condition for the purpose, and nga-pee at other times for the Burmese jails. *Hooks and lines* are employed amongst the coral reefs and rocks, and out at sea. The description of fish thus obtained being excellent.

ISINGLASS.

Isinglass, what it is, and where procured.

152. Isinglass appears to have been discovered many ages since, for certainly it was known to the Romans, being mentioned by Pliny. It is obtained in several parts of the world from the air-vessels (termed 'sounds' or 'maws') of various species of sea, estuary, and fresh-water fishes, England procuring her best from Russia, where it is principally collected from the family *Accipenseridæ* or sturgeons, and the following species according to Brandt and Ratzeburg (1829). *Accipenser sturio*, the common sturgeon; *A. huso*, the great sturgeon: *A. Guldenstadtii*, the Osseter; *A. rathenus*, the sterlet; *A. stellatus*, the Sevruga or starred sturgeon, in which account are likewise included the *A. brevirostris*; *A. schypa*; *A. Ratzeburgii*: *A. Lichtensteinii*, also *A. maculosus*, and *A. oxyrhynchus* from North America. * The Russians are also said to obtain it from the Acanthopterygian *Lucioperca Wolgensis*, Pall.; from the siluroid, *silurus glanis*; and from certain Barbels, *Barbus*, amongst the carp family. Inferior descriptions are likewise received from the Brazils and North America.

Discovery of the export of this substance from India. Dr. Royle's pamphlet on the subject.

153. Dr. *McClelland* discovered, about the year 1839, that the Chinese had been importing isinglass from India in enormous quantities and from immemorial ages; and an investigation was commenced into this subject. He ascertained that from one village, six miles south-east of Calcutta, from eight to nine hundred maunds yearly, valued at from Rs. 25 to 40 the maund, were exported. Lord *Auckland*, when Governor-General, sent some specimens to England as a new export, and, according to Dr. Royle, gave " a general view of Indian fisheries, and the propriety of attending more extensively to the curing of fish". † Dr. Royle, in 1842, in a pamphlet " on the production of isinglass along the coasts of India," gives a *resume* of what had been previously accomplished, as well as some very interesting figures and experiments on the value of this article, as received in London, both in an economic and financial point of view.

Its uses, and the forms that are brought to market.

154. Isinglass, the purest known form of animal jelly, has, in a measure, had its consumption checked by its high price, and substitutes are employed such a gelatine (of which it is itself the purest form). It is of a highly nutritious and unirritating nature, admirably adapted for the sick room, and the preparation of some forms of confectionary and cookery, besides being employed both externally and internally in medicine; in the preparation of court plaister, in some arts and manufactures, but more extensively for

---

* In looking over the last published work detailing all the fishes of this family (British Mus. Catal. Vol VIII.) I see these eleven species are reduced to eight by considering *A Lichtensteinii* the young of *A. sturio*, and *A. oxyrhynchus*, a variety—by uniting *A. Ratze. Burgii* with *A. stellatus*, whilst *A. schypa*, Brandt and Ratz. is placed as a doubtful species.

† This despatch I have not been able to trace.

clarifying or fining wines and beer. The brewer employs it as follows:—some having been finely divided, is dissolved in sour beer, to the consistence of a thick mucilage, and a portion is added to the fluid which it is intended to clarify, and after a longer or shorter period suspended substances subside. Some suppose that all floating particles become entangled in the isinglass, and, uniting with it, form an insoluble compound which becomes precipitated; others, that when dissolved in a fluid it lessens its affinity for the suspended particles, which, being thus set free, subside. The finest description of isinglass is thin, tough but flexible, white, semitransparent, and destitute of both taste or smell; it almost entirely dissolves in boiling water, and provided " it contains as much as $\frac{1}{100}$ of its weight of gelatine, has the property of gelatinizing or assuming the form of a soft, tremulous solid as it cools" (*Solly*). The inferior sorts are thick, opaque, white, or yellow, having a fishy taste and smell, and only partially dissolve. The commonest kind, termed cake-isinglass, is of a brownish colour, having an unpleasant smell, and is only used in the arts, and for the preparation of glue. Isinglass, observes Dr. Royle "is brought to the market in different forms, sometimes in that of simple plates, at other times rolled up into different shapes, or cut into fine threads" (p. 10). The Brazilian is very inferior to the Russian, and is in the form of pipe, block, honey-comb, cake, and tongue isinglass. The North American is like long ribbons, produced from the air vessel of the *Otolithus regalis*. Bl. Schn.

155. The species of fishes from which this useful substance has been obtained in India do not appear in all instances to have been correctly defined, as it has been suggested as derived from some which are destitute of air vessels. There are two great and very different orders from which they are derived;—the best being collected from the (1.)—Order ACANTHOPTERYGII in its more percoid families, as *Percidæ* and *Polynemidæ*: and the inferior from the (2.)—Order PHYSOSTOMI and its family *Siluridæ* or cat-fishes. These fishes, suited for the production of isinglass, are not all found in the same localities, nor in the like proportions in each different district. Without entering too minutely into the subject, it may be stated that the first division are most numerous towards Sind, and as we come to Bombay they are somewhat equal with those in the second division. Along the western coast, and down Malabar, the siluroids are in the majority; but wherever large rivers debouch into the sea, there the *Polynemi* are captured. As we advance up to the eastern coast, at first the Acanthopterygians are in excess, whilst off Masulipatam, to the north again, the Polynemi become numerous, especially off the Sunderbunds. In Burma, due to the character of the water, the siluroids again obtain the predominance.

Indian fishes from which this substance is obtained.

156. Amongst the isinglass-producing Acanthopterygians, the *Polynemi* are most noted, but the species constituting this genus are peculiar, in having filamentous prolongations at the base of each pectoral fin; these are remarkably elongated in *Polynemus paradiseus*, known in Bengal as the *Tupsi mutchi* or mangoe-fish. By correctly

The Polynemi fish which is useful in this manufacture.

ascertaining the number of these prolongations in a specimen, a decision may be arrived at whether the species possesses or is destitute of an air vessel; or, in short, if isinglass can, or cannot be obtained from it. The following are the Indian species with the synonyms given by Dr Royle : (1.) *Polynemus paradiseus*, Linn., *P. risua*, H. B., VII pectoral appendages; it attains nine or ten inches in length, and is destitute of an air vessel. (2.) *P. heptadactylus*, Cuv. & Val., VII pectoral appendages, attains about six inches in length; no air vessel. (3.) *P. xanthonemus*, Cuv. & Val., VII pectoral appendages; no air vessel. (4.) *P. sextarius*, Bloch. VI pectoral appendages, attains about seven inches in length, has a small air vessel. (5.) *P. Indicus*, Shaw, *P. uronemus*, Cuv. & Val., *Maga booshee*, Russell, *P. sele* H. B.; *P. lineatus*, McCelland, V pectoral appendages, attains about 3 feet in length; has an air vessel, the form of which is oval, and its membranes thick and silvery; it occupies the whole length of the abdomen, terminating posteriorly in a very sharp point, which penetrates the muscles of the tail over the first inter-spinal of the anal: from either side of the ventral surface proceed from 28 to 35 appendages. (6) *P. plebejus* Gm. Linn., V pectoral appendages; it has a simple air vessel. (7) *P. tetradactylus*, Shaw, *P. teria*, H. B., *Maja jellee* Russell, IV pectoral appendages, attains six feet in length, has no air vessel. Thus out of seven species at present recognised in the seas or estuaries of the coasts of India, merely two are useful for this manufacture, and they are the only ones which have five pectoral appendages; consequently, unless a large *Polynemus* possesses five only of these filamentous appendages at the base of its pectoral fin, it will be useless looking for its air vessel; it has none, and isinglass cannot be prepared from it. When dried, the air vessel is tongue-shaped, as are also those of others of the Acanthopterygians whose air vessels are loose in the cavity of the abdomen, have no communicating duct leading to the pharynx and are not attached to the vertebræ.

157. Amongst other estuary or marine isinglass—producing fishes of the Acanthopterygian order, we have a considerable number. At Gwádur it is chiefly obtained from the *Sciæna axillaris*, and *S. diacanthus*; at Karachi, in Sind, the sea-perches, *Serrani*, the Begti, *Lates calcarifer*, the *Sciænas*, *Otolithus*, and *Polynemus* were the forms from which it was taken, never, so far as I could ascertain, from the siluroids. Of Bombay, the Malabar, and Coromandel coasts I have already alluded to. Of Bengal and Burma the *Sciænoides pama* ascends with the tides all the large rivers, and its air-vessel is of a good quality. The same may be said of the *Otolithus ruber* found along the coasts of India.

Other estuary or marine fishes from which isinglass is prepared.

158. But if we examine the air-vessels of the *siluroid* or scaleless cat-fishes which are used for isinglass we find them entirely different. They are like short rounded bags with an open mouth, this latter being where they have been torn away from their adhesions to the vertebræ. The fishes which furnish these descriptions of air-vessels are mostly found in muddy waters, estuaries, and the mouths of rivers, but do not thrive where the sea is clear. The *Rita ritoides*, C. & V., or *Pimelodus rita*, H. B., which attains a large size, and is found far up rivers is said

Isinglass from siluroid fishes.

to afford this substance in large quantities. The fish maws, sounds, or air-vessels that the India House received were 3 or 4 inches across in both diameters, something in the shape of short purses with spring clasps,* of a light colour and semi-transparent. Having seen the specimens, no doubt can be entertained but that they have been derived from siluroid fishes, the principal genera from which they are collected being *Arius, Osteogeniosus*, and perhaps *Macrones*.

159. *Mr. Remfrey* sent two specimens of Indian isinglass to the India House, one being simply the air-vessel removed from the fish, and sun-dried; the second being that organ laid open when fresh, its lining membrane removed, then washed with cold water, and subsequently beaten flat. *Dr. McClelland* despatched 46 seers taken from the *Polynemus Indicus*, and the *Sciænoides pama*; it sold at one shilling and seven pence a pound. It cost in India one shilling and a penny a pound, but charges in India and England were so heavy, that the receipts did not quite amount to one third of the outlay. Specimens were sent in entire pieces cut into shreds, whilst some chalk had been added to preserve it dry and free from insects. He remarked—" the sounds, when received fresh, are opened, and stripped of the vascular covering and internal membrane, washed, and at once made into any form the manufacturer finds most convenient for packing. * * When dry, before it reaches the manufacturer (which is commonly the case, the fish being caught at a distance towards the sea), the sound is to be opened, and as much of the lining membrane as possible removed by the hand. A large earthen vessel is then filled with sounds, and water poured into it, and the whole covered up for twelve hours, when the sounds will have been brought back to their original soft state, in which they may be as perfectly cleaned as if they had been obtained fresh." It seems more than probable that this will account for the fishy odour of this isinglass, as the sounds should be quite fresh when prepared. *Dr. McClelland* bleached his specimens in alum water, (one ounce to four or five gallons), soaking then a short time, and, when saturated, removing them to a linen or cotton cloth, likewise saturated with alum water. In this they were tightly rolled up and set aside for twelve hours, the process being repeated until they were white. Some were sprinkled or dusted with chalk, in case of exposure to damp in their homeward voyage ; it can be easily rubbed off. At Gwádur, as I have already observed, the air-vessels were soaked in brine before being dried ; but elsewhere I have seen them simply removed and dried in the sun.

Indian isinglass in the home market.

160. *Jackson*, reporting on isinglass and its uses in 1765, observed that its mode of preparation in Russia was then unknown to either Russian or British merchants at St. Petersburg, whilst any imprudent enquiries on the subject might render the investigator liable to "banishment, imprisonment, or a worse fate." But, in 1783, he was enabled to report fully on this subject in the 'Philosophical Transactions.' The sound, he states, was—" taken from the fish while sweet and fresh, slit open, washed from the slimy sordes, divested of every thin membrane

How it is manufactured in Russia.

---

* Mr. Yarrell thought they must have been obtained from some species "allied of the Gurnards," in which conclusion he was evidently mistaken.

which envelopes the sound, and then exposed to stiffen in the air." When the sounds of the cod or ling are prepared, the only difference is, that—"they are slit open, washed in lime water in order to absorb their oily particles, and then in clean water, when they are laid upon nets to dry." In the present Russian factories of the Caspian and Volga, the fresh sounds are first split open, well washed to separate the blood and impurities, "then spread out and exposed to the air to dry, with the inner, silvery-white membrane turned upwards. This, which is nearly pure gelatine, is carefully stripped off, laid in damp cloths (or left in the outer covering), and forcibly kneaded with the hands. It is then taken out of the cloths, dried in the form of leaf isinglass, or rolled up, and drawn out in a serpentine manner into the form of a heart, horse-shoe, or lyre (long and short staple) between three pegs, on a board covered with them; here they are fixed in their places by wooden skewers. When they are somewhat dried there they are hung on lines in the shade, till their moisture is entirely dissipated. The oblong pieces are sometimes folded in the form of book-isinglass. In order to obtain good isinglass, it is necessary to have well-arranged rooms to dry it in, as at Astrachan.* * According to *Pallas*, at the lower parts of the Volga, a fine gelatine is boiled out of the fresh swimming-bladders, and then poured into all kinds of forms. In Gurief, a fine boiled fish-glue is prepared, perfectly transparent, having the colour of amber, which is cast into slabs and plates. The Ostiaks also boil their fishglue in a kettle. The common cake-isinglass is formed of the fragments of the other sorts; these are put into a flat metallic pan, with a very little water, and heated just enough to make the parts cohere, like a pan-cake, when it is dried" (*Royle*). The lining membranes of the air vessel of the sturgeon, as already noticed, yields the best isinglass, but it has been rejected in the Indian forms, which accounts for its more fibrous nature, although not proving that this lining portion in India is of the best quality.

161. East Indian isinglass possesses all the characteristics of isinglass, for which reason it is valued by the Chinese and imported into their empire from the ports of India. Yet it has some positive defects, such as retaining a fishy smell, besides being partially insoluble, perhaps due to some portions of the albuminous membranes remaining. In fact, it requires more care in its preparation, which should be undertaken whilst it is quite fresh; and greater caution is necessary in the drying process. Should "it not be properly dried, it might possibly undergo a slight change or decomposition, and become partially converted into a more insoluble form of gelatine. A more important objection is the smell, which, however, may likewise, to some extent, be traced to the preparation"* (*Royle*). Care should be taken

Characteristics of East Indian isinglass.

---

* In 1868, I removed the air-vessel from an *Otolithus ruber*, and having simply dried it in the shade sent it to *Mr. Broughton* at Ootacamund. Having analysed it, he replied (May 2nd, 1868) "I have examined the fish bladder sent: it contains about 80 per cent. of gelatine (isinglass containing about 90 per cent). It will set to a jelly in about 26 times its weight in water. I do not know which part of the membrane is richest in gelatine, but it will require very different preparation to the specimen you sent, which possessed a most fishy odour, and that extended to the jelly, and would quite preclude its use for the ordinary applications of isinglass."

that it is not contaminated by the animal fluid of the fish, for subsequently it becomes very difficult to purify. Likewise, it is too thick, which may be obviated by beating or pressure, as is now done with some American and Brazilian kinds. "The extra labour that this would require," observes *Royle*, "might be profitably saved by not tearing it into fibres, in which state it is disapproved of in the market; but it might still be cut or rasped into a state fit for domestic use." The same authority likewise states that—"it is preferable, and will be cheaper to prepare the article and send it as sheet-isinglass, that is, in the form of the slit sounds themselves, or their purest membrane, washed, cleaned, and dried in the best manner.* * Isinglass cut into threads is unsuitable for the English market, because there is a great prejudice against purchasing wholesale, things in a cut and powdered state, in consequence of the innumerable methods adopted for falsifying and adulterating almost every drug.* * By these means, or by others which will no doubt suggest themselves, when the objections to the Indian isinglass are known, the manufacturers will be able to improve it to the degree requisite to enable it to occupy a permanent as well as high place among the isinglass imported into the principal markets of Europe. Though the first quantities sent from India brought only one shilling and seven pence, others have been sold for three shillings, and a few samples have been valued at four shillings per pound. Besides this opening to an extensive sale in the European markets, even in its present state, we know there is always a constant demand in China for the isinglass of Bengal. This will, no doubt, afford sufficient encouragement to persevere in the extension and improvement of this newly established and highly promising article of the export trade of India to Europe.

Exports.

162. As an export from Madras and Bombay to China it is thus recorded for the years 1837-38 :—

From	Weight.		Value.
	Cwts.	lbs.	Rs.
Bombay	5,088	39	2,55,145
Madras	1,082	... ...	34,407
Total ...	6,170	39	2,89,552

Unfortunately fish-fins and fish-sounds are mixed up together in the Sind returns, but I find the quantities given thus. For four years ending 1869-70, cwts. 4,896 of the value of Rs. 2,60,672 for 1870-71, ,, 1,514 ,, ,, ,, 73,024 cxlvi

FISH-OIL.

163. The following return* shows the annual amount of the exports of fish-oil from the undermentioned Presidencies and Provinces during the past official year. As no exports occur in Bengal, that Presidency has been omitted.

Fish-oil exported from India.

YEAR.	MADRAS.		BOMBAY.		SIND.		TOTAL.	
	Number of lbs.	Value in Rs.	Number of lbs.	Value in Rs.	Number of lbs.	Value in Rs.	Number of lbs.	Value in Rs.
1861-62 ...	1,431,328	80,469	272,586	30,932	...	...	1,703,914	111,401
1862-63 ...	81,414	5,452	138,336	14,229	...	...	219,750	19,681
1863-64 ...	1,387,188	163,568	285,109	32,220	...	...	1,672,297	195,788
1864-65 ...	3,194,672	203,829	96,156	69,235	...	...	3,290,828	273,064
1865-66 ...	20,801	2,074	1,004	135	...	...	21,985	2,209
1866-67 ...	1,133	225	13,031	1,755	19,328	2,365	33,492	4,345
1867-68 ...	820	120	...	...	25,773	3,549	26,593	3,669
1868-69 ...	10,383	1,676	38,523	7,064	4,898	758	53,804	9,498
1869-70 ...	6,781	815	9,617	1,364	58,586	9,039	74,984	11,218
1870-71 ...	1,425,531	191,220	Statement not received.					
TOTAL ...	7,560,141	649,448	854,452	156,934	108,585	15,711	7,097,647	630,873

164. As very considerable misapprehension appears to exist, not only in respect to the character of these oils, but how they are prepared, I propose to give a résumé of the information collected. These Indian fish-oils are mainly of two descriptions (1) medicinal fish-liver-oil, and (2) common fish-oil. Slight variations in the mode of preparing it occur in different places, but the main features will be here alluded to.

Different varieties of Indian fish-oils.

---

* From a report by *Dr. Balfour*, Inspector General of Hospitals, Madras, dated 8th September 1871. Burma is not shown. The average export value in lbs. is given thus — Madras Rs. 0-1-4; Bombay Rs. 0-2-11; Sind Rs. 0-2-3; average Rs. 0-1-5, due to the largest exports being from Madras, where the article is cheapest.

cxlvii

165. *Medicinal fish-liver oil* has for some years been prepared on the Western Coast of India mainly from the livers of the "Cartilaginous" or *Chondropterygious* sub-class of fishes* This manufacture was originally commenced in Cochin (about 1852) by the Civil Surgeon of that station, when the cost of the article was Rs. 0-2-2 per lb. Owing to the limited space in British Cochin, the making of this oil was removed to Calicut and placed under the supervision of the Civil Surgeon of Malabar. On October 24th, 1871, the Madras Government resolved that its manufacture should be discontinued, as it was ascertained that, whereas its cost about 20 years previously was Rs. 0-2-2 or $3\frac{1}{4}d.$ a lb., from 1863 to 1871 it had averaged Rs. 0-13-6 or $1s.$ $8\frac{1}{4}d.$ a lb.; whilst in 1869-70 a consignment of cod-liver-oil was delivered in Madras at Rs. 0-9-1 or $\frac{1}{8}$th of a penny over $1s.$ $1\frac{1}{2}d.$ per lb. Taking these figures as the average cost, the loss had become upward of $6\frac{1}{2}d.$ a lb. It was observed by Dr. Balfour that Indian fish-oil was being sold at Rs. 0-1-5 per lb.; "it is an article valued in the currying of leather; but whether or not sent for that purpose, or to be purified and whitened by passing it through boiling water with the aid of sulphuric acid, or of chlorine or other gas, and hot filtering through animal charcoal, can only be surmised. The Home Government might, however, be asked to trace its ultimate disposal, though I have no doubt that it ultimately finds its place amongst the cod-liver oils of Europe. I, however, bring to the notice of Government the extensive export from India of fish-oil, because its manufacturers would doubtless readily enter into contracts for the supply of any quantity of it, and Government would get it in that way both cheaper and of better quality than they now obtain it of their own manufacture." That the cost of the manufacture of Indian fish-liver oil has become too heavy there can be no question, but an equally good article could *not* be obtained from the native manufacturer at the price mentioned, for he only, as I shall presently show (para. 173), makes a most horribly fœtid common fish-oil.

166. About the year 1854, the manufacture of this oil was commenced at Calicut; the building, which was a mere shed, was soon blown down, and various temporary edifices were used until 1866, on the 27th October of which year, another building, costing Rs. 854-12-4, was constructed. It was made of stone pillars, thatched with leaves, and surrounded with mats. However, on June 5th 1868, it shared the fate of its predecessor, being blown down, and a third building was raised and completed in August 1869, at a cost of about Rs. 1,080. A wall was sanctioned in 1870 to enclose it for the purpose of preventing cattle breaking in and destroying the bottles and jars.

167. The cost of the raw livers, from which this oil is manufactured, has risen from 8 pie a pound to $1\frac{1}{2}$ annas; to $1\frac{3}{4}$ annas in 1869-70, whilst it fell to 1 anna in 1870-71. Now, as every pound of liver is computed to yield half a pound of oil, it can hardly be admitted that this must be the only cause of the increased cost of the Indian oil, from Rs. 0-2-2 a lb. to Rs. 0-13-6

---
* It has been erroneously asserted that the livers of the seir fish, *Cybium*, are employed for this purpose, and the error is still being re-copied and thus reproduced.

cxlviii

a lb., especially as the processes employed are much the same, for 4 pie a lb. difference in the cost of the raw livers would only be Rs. 0-1-4 a lb. rise in that of the oil, or if the article could be manufactured at Rs. 0-2-2 a lb., then, other things being equal, it would be now at Rs. 0-3-6, or less than one half the cost of cod-liver oil

168. I will now trace out how this augmented cost has come about.

The reason why the cost has augmented.
In the earlier days of the manufacture, the working months were from December to March, when the sharks and other appropriate species of fish are about, and for that period only was the working establishment kept up at an outlay of about Rs. 564 per annum, the oil costing from 4 to 5 annas a pound. But in 1863 an increased establishment was sanctioned, augmenting it to Rs. 1,176 a year, although the amount of the manufactured article continued the same (or about 4,500 lb. a year), thus raising its cost by more than 2 annas a lb., which being deducted from Rs. 0-11-1, its cost in 1871 reduces it to slightly below Rs. 0-9-1, the price of the cod-liver oil referred to. The following table shows the amount of oil manufactured, with the total cost, taken from a return furnished me at Calicut by Dr. Roberts :—

Year.	Number of lbs. manufactured.	Expenditure for			Cost of establishment.
		Material.	Packing.	Freight.	
		Rs. A. P.	Rs. A. P.	Rs. A. P.	Rs. A. P.
1863-64	3,820	1,145 9 3	133 1 1	...	2,174 3 4
1864-65	5,469	1,131 12 1	415 12 0	645 2 8	1,176 0 0
1865-66	5,000	2,502 2 2	413 10 8	548 6 3	1,193 0 0
1866-67	5,350	2,082 6 5	587 9 4	663 8 10	1,126 0 0
1867-68	2,026	1,412 4 7	166 8 0	293 2 4	1,176 0 0
1868-09	1,871	866 5 7	169 0 3	167 14 5	1,076 1 0
1869-70	2,390				

The cost per lb. has been as follows from the figures furnished me :—

Year.					Rs. A. P.
1863-64	...	...	...	...	0 15 0
1864-65	...	...	...	...	0 9 10¼
1865-66	...	...	...	...	0 14 10¾
1866-67	...	...	...	...	0 13 4¼
1867-68	...	...	...	...	1 2 5
1868-69	...	...	...	...	1 4 3½
1869-70	...	...	...	...	0 15 9¼*

* It was stated in 1870 "that the last fish-oil received from Calicut cost Re. 0-10-7 per lb.; cod-liver oil from England Re. 0-11-2," whilst Dr. Balfour gives the last at Re. 0-9-1.

cxlix

169. The consignments of this article on an average of four years from 1867-68 to 1870-71 have been as follows:—

The amount prepared.

Bombay	lbs.	1,530
Calcutta	,,	1,400
Madras	,,	404
TOTAL lbs.		3,334

170. The oil is manufactured from the livers of sharks and saw-fishes. They are mostly obtained during the months of October and November, when sardines are plentiful; some are likewise taken in July and August. The supply of these fishes evidently depends very much, if not entirely, upon the presence or absence of the oil sardines (*Clupea neohowii*, C. & V.), which are variable in their arrivals and departures, sometimes forsaking the coast for years, and then as suddenly returning in droves, consequently in some seasons sufficient sharks cannot be captured for the requirements of the factory. Shark-livers appear equally good at any season of the year, but the size has much to do with the comparative amount of oil which they yield; thus other things being equal, small livers give only one-third their weight of oil, but large ones nearly one-half. No livers under 40lbs. weight were purchased. Sometimes very large ones were received, one of 290lbs. weight was brought, and another of a female saw-fish, 14 feet long, weighed 185lbs. There are three recognized qualities of livers (1) the best is firm and pinkish'; (2) the medium is also firm, greyish externally, and reddish when cut into; (3) whilst the most inferior is flabby, whitish externally as well as internally, and is useless, or nearly so.

From what the oil is prepared.

171. Owing to the large size of the fish from which such livers are taken, the Malabar fishermen, unlike those off Sind, are unable to capture them with nets. Putrid beef, or porpoise flesh is employed, large pieces being buried for a day or two previous to their being used. The hook is attached by a chain to the line, whilst the fishing is carried on as described in para. 82.

How the fish are captured.

172. The process employed in the manufacture of this medicinal oil has undergone change. Formerly the oil was dark-coloured, had a bad, exclusive of a fishy odour, whilst it did not keep. It deposited a dark sediment, irrespective of a whitish one (squalin), whilst further changes occurred, which often rendered it unfit for internal administration. Some few years since this was remedied by the employment of a thicker species of thick twilled cotton cloth (termed satin-cloth in the bazar) in some parts of the straining. In 1854, Dr. Barker observed, that if this oil were kept for any length of time in casks, it re-acquired its fishy odour and became like train oil. Livers had to be received at the factory within six hours of the death of the fish, the gall-bladder was at once removed, and the gland thoroughly washed. The veins were slit up to admit the blood's draining off, and for this purpose the liver was left in a large, flat, copper vessel, and water frequently poured over it until it was no longer discoloured by blood. Then it was cut up into pieces

Process of manufacture.

of about 4℔s. weight each, two being placed in au earthen vessel, capable of containing from four to five gallons; over it was poured about a quart of water, or just sufficient to cover the liver with 1½ inches of fluid. This had now to be placed for 15 or 20 minutes over a slow fire, and when the temperature reached 130° it was stirred up; as froth began to ascend, the vessel had to be at once removed from the fire and placed on sand to cool. The oil was soon perceived floating, and was skimmed off into large glazed earthen-ware jars, by means of a wooden ladle formed by half a small cocoanut shell, attached to a bamboo handle. This oil, or *rough oil*, had now to be strained through flannel, what did not pass being rejected. The straining-stands resembled chatty-stands, having four posts five feet high, at each angle of a square, and three transverse connecting bars about one foot long, with the same vertical distance between each. This rough oil had now to stand three or four days, when it was again strained through four layers of long-cloth and thick satin-cloth, and four layers of flannel to remove the stearine, &c. After fifteen or twenty days the straining was repeated through two layers of long-cloth and thick satin-cloth in each bar of the strainer, and two layers of flannel in addition in the centre bar, and this had to be repeated four times at intervals of twenty or twenty-five days between each. During these intervals the oil had to remain quiet in the large glazed earthen-ware jars, and if any crystals or portions of stearine were seen, the straining had to be repeated. The sixth, and generally the last straining was through cloth, and filtering paper, in funnels direct into the bottles in which it was bottled off, and sent away. This oil should have no deposit, be of a light, clear, straw-colour, its odour much resembling pure cod-liver oil, although it occasionally is somewhat stronger.

173. *Secondly.—Common fish-oil.* This may be prepared from the sea or fresh-water fishes, but the latter seem mostly to be used for this purpose in Burma.

Common fish-oil, how prepared.

If we turn to the table of exports of Oil* from India (para. 163 *ante*), one cannot help being struck with the very great variations in different years. Now these variations are not caused by the oil-producing fishes migrating along the Indian coast, so that if they were absent one year in Malabar they might be expected to be present at another spot, as Sind for instance, but the years they are plentiful in one portion of the coast so they also are at another, and when rare at one spot an equal paucity is perceived everywhere. The oil sardine, however, is the basis of all the oil, for if not prepared from it, a great amount is from the sharks and fish who live upon them. "When Dussumier was in Malabar, probably about 1827, he observed that those not eaten were used for manuring the fields, as they were too fat to salt well; at the present time *(Fishes of Malabar,* 1865), mostly from this species, an average of upwards of seven thousand pounds worth of fish oil is annually exported from Malabar. But the oil-sardine is very capricious as to its arrival and departure. Thus in 1855-56, from the port of Cochin, only 45 cwts. of fish-oil were exported; the next year merely 181 cwts.; but in the year 1857-58, 68,499 cwts., which augmented in the succeeding

---

* The whole of this is common fish-oil; not a pound is the medicinal substance.

year to 102,924 cwt. and the year after to 133,143 cwts.* Again, it gradually fell off, until in 1862-63 as little as 115 cwts. were exported, but the shoals suddenly returned, and in 1863-64 the exports rose to the enormous quantity of 148,206 cwts." In 1864-65 still larger exports were made, whilst during the next five years comparatively nothing was done in this trade, but in 1870-71 the shoals re-appeared as abundantly as ever, and with these shoals returned the sharks whose absence had increased the cost of the manufacture of medicinal fish-oil so much at Calicut, by the expenses being the same, but the out-turn so much below the average. When the sardines arrive off the coast to breed it is most fortunate that they are deficient in fat, as they would be turned into oil; but it is not until they have done breeding that deposits of fat commence, and by October, sometimes before, and for a couple or three months they are fat, and well adapted for making oil from; subsequently they again become lean. This fish is believed, as a rule, only to come to the Western Coast of India to breed, to Ceylon and the Andaman Islands, but small portions show themselves along the shores of the Bay of Bengal. They are captured either by long float nets attached at either end to a boat, and by making a circuit a shoal is surrounded, or else several canoes put off together, and pull off to a shoal of these fish, which they take by cast nets.

174. A boat-load of sardines is computed to hold 14,000, and at
Common fish-oil continued. the seasons when they do not afford oil they are much cheaper than when it is present. Unless their livers are fat it is useless attempting to manufacture oil from them, and it is prepared either by removing the livers or by decomposition. The natives prepare fish-oil from the livers of sharks, skates, sawfishes, rays, cat-fishes, as *arius, &c.*, oil-sardines, and some other varieties. The cat-fishes' livers have the most oil about January, just before they are breeding. Should the livers of any of these fishes alone be employed, they are heated up to about 130° in water, having about 1½ inches in depth over them; after about 15 or 20 minutes, on being stirred, the froth rises, and it, with the oil, is skimmed off into large vessels, in which state it is sold as fish-oil. There is no washing of the livers, fresh or semi-putrid, bloody or clean they are put into the pot; the oil undergoes no straining or purification, and is exported to Europe, where I understand it is worth about £30 a ton. In Calicut this oil is worth about four and a half rupees a maund in the bazar, or less than one anna a ℔. If oil is to be obtained by decomposition the following process is adopted: a boat about three quarters full of fat sardines is placed in the sun, and about midday sufficient boiling water is poured over the fish to cause the oil to float; it is then skimmed off, and this is repeated on three successive days. West of Bombay† the fish-oil may all be said to come from fish of the sub-class Chondropterygii or sharks, saw-fishes and skates, the oil-sardines in that quarter being said to be unsuited for the purpose. When I was at Gwádur and

---

* These figures were taken direct from the Customs House returns in Cochin, and do not coincide with the Collector of Malabar's report. But I feel satisfied that the amounts given here were actually exported.

† In answer to some questions I sent to the Collector of Sea Customs, Kurrachee, Mr. *Cole* observes that all the oil is made from the livers of sharks and such-like fish. "The flesh of the largest is thrown away, the middle sized ones are salted and eaten, the smallest are eaten, fresh and salted,"

Karachi about January they had just arrived, and certainly had no fat ; probably those which had done breeding were migrating, if so, the shoals go to the west towards Sind and Beluchistan; perhaps up the Persian Gulf or off the south-east side of Arabia. Unfortunately our knowledge of the fishes of the Persian Gulf has yet to be acquired. Its presence is not noticed off the eastern coast of Africa in the fishes of Zanzibar, neither have Ruppell or others observed it in the Red Sea. If we enquire more to the south-east of Burma, neither Cantor nor Bleeker appear to have seen it amongst the Malayan fishes.*

175. The *Officiating Magistrate of Rangoon* (January 28th 1867), observed that the average quantity of fish-oil obtainable in the town of Rangoon is about 500 viss a month (or 77 tons and 20 lbs.), but from November to May much larger quantities are procurable, it being only made at these times. The price fluctuates from Rs. 40 to Rs. 50 per 100 viss (a viss is 3·65 lbs. avoirdupois). It is used for lamps, even for curries, and frying fish, never as medicine. It is obtained by boiling the intestines of some fish, the heads of others, and even whole fish, in an iron vessel with water in it, and the fatty substance, as it floats, is skimmed off again into another pan and boiled again until the oil floats. It is said to be chiefly extracted from the Nga-bya-ma, *Anabas scandens ;* and the Nga-khou-ma, *Barbus chola,* which species, and other small ones, are boiled entire. It is also extracted from the Nga-tha-louk, *Clupea palasah'* and the intestines of the Nga-yan, *Ophiocephalus striatus.* Some oil is likewise made from the livers of the saw-fish. This is a simple fish-oil of no better description than that described in the last paragraph.

Burmese fish-oil.

---

* If *Clupea lemuru,* Bleeker, is the oil-sardine, this fish would appear to be found in the Malay Archipelago.

SEA-FISHES OF INDIA AND BURMA.

176. The sea-fishes of India and Burma are divisible into* the strictly marine ones, and those which ascend within or even above tidal influence, either to deposit their ova in suitable localities, or else to obtain sustenance. In some of the strictly marine forms the fry are hatched along the coast, and subsequently pass into small estuaries, creeks, rivers and streams, in order to find security and also food suitable to their infantile condition. In the ensuing list, some fishes, recorded amongst those which are taken in the fresh-waters, must of necessity find a place; but to obviate filling up more space than is absolutely necessary, they will only be referred to in accordance with the number under which they are enumerated in the fresh-water fishery report. This list must be merely looked upon as a compilation showing those sea-fishes whose existence in the waters of India, Burma or Ceylon has been distinctly recorded, or else which I have personally satisfied myself of their presence. A very large proportion of my specimens having been sent to Europe, to my collection there, I have not, as yet, had an opportunity of comparing many which appear to me doubtful forms, with types of some insufficiently described species, consequently they will be omitted, but included in my 'Fishes of India', should I subsequently publish such a work. Many fish also have been introduced upon the statements of those who have examined the types which I have not yet seen, whilst some have been omitted, as they have either been asserted by others to be merely synonyms, or else believed to be so by myself.

The sea-fishes of India, Burma, and Ceylon. A compilation of those recorded.

Sub-class—TELEOSTEI.

177. Fishes having an osseous skeleton, completely separated vertebræ, and the posterior extremity of the vertebral column either bony or armed with bony plates. Bulb of aorta simple, with a pair of valves at its origin. Branchiæ free.

Fishes with bony skeletons, &c.

Order—ACANTHOPTERYGII.

A portion of the dorsal, anal and ventral fin-rays unarticulated, forming spines, air-vessel, when present, completely closed, not having a pneumatic duct.

Family—PERCIDÆ, Cuv.

Percoidei, pt. Cuvier : *Percidæ*, pt. et *Theraponidæ*, pt. Richardson. Branchiostegals from five to seven: pseudobranchiæ, as a rule, present. Form of body generally oblong. Eyes lateral. All, or some of the opercles (except in Apsilus) serrated or armed. Mouth in front of snout having a lateral cleft, occasionally situated on the lower side. A barbel on the lower jaw in Pogonoperca. Teeth villiform or conical in the jaws, canines occasionally present, the vomer and generally the palatines armed with teeth. Anterior portion of dorsal fin spinous: ventrals thoracic, each having one spine and five rays. Scales ctenoid. Lateral line, when present, continuous, except in some species of Ambassis. Air-vessel usually present, and when so, simple. Pyloric appendages in varying numbers.

---

* See appendix to report on the Fresh-water Fish and Fisheries of India and Burma, p. ccxlviii *et seq.*

Genus—*LATES*, Cuv. and Val.

Branchiostegals seven: pseudobranchiæ. Pre-orbital, and shoulder bone serrated: preopercle with strong spines at its angle, and denticulated along its horizontal limb: opercle spinate. Teeth villiform on jaws, vomer and palatine bones, tongue smooth. Two dorsal fins united at their bases, the first with seven or eight spines, the anal with three: caudal rounded. Scales of moderate size. Cæcal pylori few.

Lates calcarifer, Bloch. (See No. 1. F. W. F. Report.) *To-dah*, Andamanese:—*Koduwa*, Tam.

Genus—*SERRANUS*,* Cuv.

Branchiostegals seven: pseudobranchiæ. Eyes lateral, of moderate size. Preopercle with its vertical limb more or less serrated, its horizontal one entire, opercle with two or three flat spines. Teeth villiform in the jaws. Tongue smooth. Dorsal fin single, having from eight to twelve spines: anal with three: caudal deeply forked, emarginate, truncated or rounded, whilst its two central rays may be prolonged. Scales small. Pyloric appendages many, in moderate numbers, or few.†

Fishes of this genus may be divided in accordance with the shape of their caudal fins and the number of spines in the dorsal. But colouration cannot be accepted as a means for grouping species, because vertical bands, the same as in many of the *Carangidæ* or the *Osphromenus* are frequently merely a sign that the specimen is immature. The fins of these fishes sometimes alter with age, the spines not increasing in length so rapidly as the rays, consequently they become comparatively shorter in the adult; but even the rays in the adult are less in their proportionate height to the length of the fish, then they are in the young. The form of the preopercle is not invariably identical amongst specimens of the same species, or even in the opposite sides of the same specimen. Most of the *serrani* or "Sea perches" appear to attain to a large size and are esteemed as food, whilst the air-vessels of some are used as isinglass.

2. *Serranus altivelis*, Cuv. & Val. D. $\frac{11}{18-19}$, A. $\frac{3}{9-10}$, Upper profile of head concave. Preopercle finely and evenly serrated. Caudal rounded: dorsal and anal fins elongated. Yellowish, covered with widely separated black spots margined with white. China and Port Essington, a large specimen in the British Museum, is reputed to have come from the East Indies..

---
* Fishes of this genus are termed *Cullawah*, Tamil.

† In the Proc., Zool. Soc., March, 12th 1868, I described a new species of percoid fish from Madras as the type of a fresh genus. *Priacanthichtys* differing from *Serranus* in the existence of a preopercular spine, and a serrated ventral one, it was as follows:—*P. Madraspatensis*, Day. B. VII, D. $\frac{11}{11}$, A. $\frac{3}{7}$, L. l. 70. Violet, with two bluish bands, one from the upper edge of the orbit to the middle of the soft dorsal; one from its lower edge to the centre of the caudal fins, I observed; "appearance that of a *Serranus* except in the preopercular and ventral spines." It may be related to *S. grammicus* as the specimens hardly exceeded 1½ inches in length. I have, however, not as yet seen any young *Serrani* with the ventral spine serrated. Dr. *Günther*, "Ann. and Mag. Nat. Hist.," November 1871, says: "*Priacanthichthys* has proved to be the young of *Serranus*." This proof I have not yet seen. I suggested they were like; he asserted they were identical, and it is a subject worth investigating. I therefore place this genus as a doubtful one, but probably the young of *Serranus*.

3. *Serranus lineatus*, Cuv. and Val. D. $\frac{11}{17\text{-}18}$, A. $\frac{3}{8\text{-}9}$, Cæc. Pyl. above 50. Preopercle with two or three denticulations at the angle, rather well developed. Caudal rounded. Brown with four or five blue longitudinal bands. India and China, attaining at least four feet in length.

4. *Serranus hexagonatus*, Forst: *Naambu*, Bel. *Pulli cullawah*, "spotted perch," Tam. D. $\frac{11}{15\text{-}17}$, A. $\frac{3}{8}$, L r. 105, Cæc. Pyl. 32. Preopercle with strongest serrations at the angle. Caudal rounded. Brown covered with large hexagonal or rounded spots. Red Sea, East coast of Africa, Seas of India, Malay Archipelago to the Pacific.

5. *Serranus flavo-cœruleus*, Lacép. *Mungil cullawah*, Tam. D. $\frac{11}{16\text{-}17}$, A. $\frac{3}{8}$. Serrations on preopercle weak, strongest at its angle. Caudal slightly emarginate. Purplish blue, tail and fins gamboge yellow, ventral and anal with black tips. From the East coast of Africa throughout the seas of India.

6. *Serranus oceanicus*, Lacép. D. $\frac{11}{16}$, A. $\frac{3}{8}$. Preopercle rather strongly serrated, most so at its angle. Caudal rounded. Orange, with five cross bands : the dorsal with a black edge. One specimen 9 inches long exists in the Madras museum. Red Sea, Mauritius and Madras.

7. *Serranus argus*, Bl. Schn. D. $\frac{11}{15}$, A. $\frac{3}{8}$, L. l. 95, Cæc. pyl. 8. Preopercle very finely serrated ; three opercular spines well developed, the central one the longest. Caudal rounded. Reddish brown, usually with cross bands, head, body and all the fins covered with numerous small blue dark-edged spots. Seas of India, very common at the Andamans, amongst the coral reefs.

8. *Serranus Hoëvenii*, Bleeker. D. $\frac{11}{15}$, A. $\frac{3}{8}$, L. l. 80. Preopercle serrated. Caudal rounded. Greyish olive, darkest along the back. Body and head covered with irregularly-sized pearly-white spots, whilst a black line exists on the maxilla. Fins dark grey, externally nearly black; but the margins of the pectoral, ventral, soft dorsal, and caudal have a very narrow white border. The whole of the dorsal fin with white spots, as on the body. East coast of Africa, seas of India, and Burma, to the Malay Archipelago.

9. *Serranus sexfasciatus*, Cuv. and Val. *Damba*, Sind, and *Chaancha*, Beluch. D. $\frac{11\text{-}12}{15\text{-}16}$, A. $\frac{3}{8\text{-}9}$, L. l. 100, Cæc. Pyl. 11. Preopercle with strong teeth at its angle. Pinkish brown on the back, rose coloured on abdomen. Six vertical dark bands, the first on the head. Fins with dark margins. Found throughout the seas of India to Java. Is very common in Sind, and specimens reach 18 inches or more in length.

10. *Serranus marginalis*, Bloch. D. $\frac{11}{15\text{-}16}$, A. $\frac{3}{8}$, L. l. 80-90. Similar to *S. Oceanicus*, except in colouration. Bright red, with four or five darker cross bands. A dark red band from snout to base of dorsal spines; another to the angle of the preopercle. Spinous portions of dorsal and sometimes of the caudal dark edged : soft dorsal with a yellow margin. East coast of Africa through the seas of India to the Malay Archipelago.

11. *Serranus lanceolatus*, Bloch. *Gussir*, Scindee : *kurrupu*, Mal. *Commaaree*, if young,*wutla-cullawah*,or " Sore-headed perch," Tam. D. $\frac{11}{15\text{-}16}$, A. $\frac{3}{8}$, L. l. 95. Cæcal pylori numerous, but very short. When young it is gamboge yellow, with five blackish blue cross bands. Fins yellow

with black bands and spots. As it becomes adult the bands become broken up into irregular markings, and the yellow colour disappears, except from the fins, in which the black becomes also broken up into black spots (*S. horridus*. C. V). East coast of Africa, seas of India to the Malay Archipelago. Very numerous at Kurrachi: it attains a large size.

12. *Serranus erythrurus*, Cuv. & Val. D. $\frac{11}{16}$, A. $\frac{3}{8}$. Preopercular border rounded and finely serrated in its vertical portion. Fins rounded. Head and back greenish shot with red; under surface of the body silvery. Dorsal greenish; pectorals, ventrals, and anal yellowish; tail and free portion of caudal reddish. Specimen 8 inches in length, but said to attain 4 feet. Malabar.

13. *Serranus dermochirus*, Cuv. & Val. D. $\frac{11}{16}$, A. $\frac{3}{8}$. Said to be of a short and thick set form, and that the pectoral rays are invested in a thick skin, this fin being large and rounded. Dorsal spines short and very stout. Serrations at the angle of the preopercle very weak. Uniform brown without spots. A specimen 9 inches in length from the Coromandel coast.

14. *Serranus bontoo*, Cuv. and Val. *Madinawa bontoo*, Tel.: *Row-jedah*, And. D. $\frac{11-12}{16 \cdot 17 \cdot 18}$, A. $\frac{3}{8}$, L. r ca. 120, Cœc. pyl. 50-60. Vertical limb of preopercle finely serrated, having from four to seven coarse teeth at its angle Fins rounded. Brownish grey on the back becoming lighter towards the abdomen. Whilst living there are from five to six dark vertical bands, usually dividing on the abdomen to two each; they commonly fade after death. The whole of the head and body, covered with large round dark brown or black spots, the largest approaching to blotches being on the sub- and inter-opercles and the lower jaw. Fins dark grey, often spotted like the body; dorsal darkest in its upper half. Iris with a narrow golden edge. East coast of Africa, seas of India to the Malay Archipelago. In the Fishes of Zanzibar, p. 5, this species is given as a doubtful synonym of *S. suillus*, Cuv. and Val., *S. coiodes*, H. B.

15. *Serranus coiodes*, H. B. *Bontoo*, Tel. *Rab-na-dah*, or *O-ro-tamdah*, Andamanese: *Punni-Cullawah*, Tam. D. $\frac{11}{16}$, A. $\frac{3}{8}$, Cœc. pyl. 50-60. Vertical limb of preopercle serrated, strongest at the angle. Fins rounded. Brownish, with about eight cross bands, the first over the the head, the second over the nape. Head and body covered with large round yellow spots, that usually become brown in dead specimens; yellow spots also on the dorsal fin, which sometimes coalesce and form bands. East coast of Africa, Seas of India to the Phillipines. It attains a very large size.

16. *Serranus salmonoides*, Lacép. D. $\frac{11}{16 \cdot 16}$, A. $\frac{3}{8}$, L. l. 90, *Cul-la-wa*, Tam. Vertical limb of preopercle serrated with three or four coarse teeth at the angle. Fins rounded. Brownish yellow: body and fins entirely covered with black or yellow spots. From the Red Sea through the Seas of India to the Malay Archipelago.

17. *Serranus diacanthus*, Cuv. & Val. D. $\frac{11}{16 \cdot 16}$, A. $\frac{3}{8}$, L. l. 90. Vertical limb of preopercle strongly serrated and with two or three spinous teeth at its angle. Caudal rounded. Brownish, with five vertical cross bands: spotted with orange or gamboge yellow, which becomes brown in dried specimens. Seas of India to China, attaining a large size.

18. *Serranus semipunctatus*, Cuv. & Val. D. $\frac{11}{15}$, A. $\frac{3}{10}$. The serrations of the preopercle are fine. Caudal rounded. Body with six or seven broad cross bands; head and fins only are spotted. Pondicherry, to 1 foot in length.

19. *Serranus summana*, Forsk. D. $\frac{11}{15\text{-}16}$, A. $\frac{3}{8}$. Canine teeth small. Preopercle serrated, with a shallow notch above its angle. Second anal spine longest and strongest: caudal rounded. Brown, body and vertical fins covered with small, round, white dots. Scarcely any spots on the head: a black streak above the maxillary. Red Sea, East coast of Africa and Andaman Islands, where it is very common. This species requires examining with *serranus merra*, which is stated in Cuv. and Val. to be found at the Seychelles, &c.

20. *Serranus radiatus*, Day. D. $\frac{11}{15}$, A. $\frac{3}{8}$, L. l. above 120. Preopercle with three strong teeth at its angle. Caudal rounded. Greenish olive, becoming dull yellow on the abdomen; several irregular bluish-white bands radiate from the orbit or exist on the head, whilst others are seen on the body. Madras, from whence a single specimen, 4 inches long, was procured.

21. *Serranus glaucus*, Day. D. $\frac{11}{15}$, A. $\frac{3}{8}$, Cæc. pyl. 13-14. Three strong denticulations at angle of preopercle. Caudal lunate. Greyish, head and body covered with large closely-set yellow spots. Fins spotted, and all, except the pectoral, with black white edged margins. Andamans.

22. *Serranus dispar*, Playfair. D. $\frac{11}{14\text{-}15}$, A. $\frac{3}{8}$, L. l. ca. 80. Canine teeth feeble in the upper and not apparent in the lower jaw. Vertical limb of preopercle rather strongly serrated, but more coarsely at its angle: third anal spine longest, but not so strong as the second: caudal rounded. Greyish, with brown spots of a larger or smaller size irregularly disposed. East coast of Africa: Andaman Islands.

23. *Serranus nouleny*, Cuv. & Val. D. $\frac{11}{12}$, A. $\frac{3}{8}$. Preopercle, with fine serrations, coarsest at the angle: scapular strongly denticulated. Canine teeth in the upper jaw very strong. Back golden yellow: abdomen rosy: head and paired fins reddish: caudal yellowish. Coromandel coast, to 5 inches in length.

24. *Serranus grammicus*, Day. D. $\frac{11}{12}$, A. $\frac{3}{8}$, L. l. 90. Preopercle serrated, more coarsely at its angle. Caudal fin cut nearly square. Greyish, with a golden gloss about the head. Three narrow black bands; the superior passes from the upper edge of the orbit to the last dorsal spine; the second from the upper third of the orbit over the superior opercular spine to the base of the sixth dorsal ray; and the third from the lower edge of the orbit to below the middle opercular spine, and on to the upper third of the caudal fin, where it takes the form of rounded blotches. Dorsal fin with a row of black spots along its centre, and edged with black; anal and caudal edged with black, the latter with numerous black spots. Madras to at least 15 inches in length.

25. *Serranus lemniscatus*, Cuv. & Val. D. $\frac{11}{12}$, A. $\frac{3}{8}$. Height of body 1/3 of its length. A brown band from the eye to the caudal fin, and an obsolete one below. Ceylon, to 4 inches in length.

26. *Serranus pavoninus*, Cuv. & Val. D. $\frac{10}{12}$, A. $\frac{3}{8}$. Muzzle pointed. Preopercle serrated with long and strong spines at its angle. Caudal cut square. Reddish: under the first four soft rays of the dorsal is a black ocellus, surrounded by a brilliant silvery ring: caudal yellow, with a

small vertical black line at its base. This species was named from a single specimen one inch in length. Bombay.

27. *Serranus formosus*, Shaw. *Verri-cullawah*, Tam. D. $\frac{9}{16\text{-}16}$ A. $\frac{3}{8\text{-}9}$, L. r. about 90. Preopercle most coarsely serrated at its angle. Caudal rounded. Yellowish brown: snout pale blue: lips and throat spotted with a darker blue: and about five fillets of the same colour diverge from the orbit and cross the opercles. Tortuous blue lines along the body. Seas of India to the Malay Archipelago and China.

28. *Serranus cyanostigmatoides*, Bleeker. D. $\frac{9}{16\text{-}16}$, A. $\frac{3}{9\text{-}10}$, L. r. 150, Cæc. pyl. 12 (Madras) to 16 (Andamans). Sub- and inter-opercles serrated, as is also the vertical limb of the preopercle: opercle, with three spines, the upper the shortest. Caudal rounded. Scarlet: body, cheeks, dorsal, caudal, and anal fins covered with large blue spots. Two dark streaks from the orbit along the snout: fins darkest at their outer edges. Two rows of large blue spots along the hard dorsal, and six or eight over the soft and the anal. Madras, Andamans, to the Malay Archipelago.

29. *Serranus guttatus*, Bloch. D. $\frac{9}{15\text{-}16}$, A. $\frac{3}{8\text{-}9}$. Preopercle not emarginate: edge very slightly if at all serrated: brownish black, head body and all the fins with round blue black-edged, spots, caudal, anal, and the posterior half of the dorsal with a white edge. Red Sea, seas of India to the Malay Archipelago, China, and Australia.

30. *Serranus Sonnerati*, Cuv. & Val. *Siggapu cullawah*, Tam. D. $\frac{9}{15}$, A. $\frac{3}{8}$, L. r. 120, Cæc. pyl. 11 or 12. Vertical limb of preopercle finely serrated: caudal rounded. A dull lake colour, with the head and jaws covered with reticulated bright blue lines, enclosing spaces equal to about one-sixth of the diameter of the orbit. Some very indistinct spots over the whole of the body. Fins lake colour, darkest at the edges. Soft dorsal, anal, and caudal with some lightish blue badly-marked spots. East coast of Africa, seas of India to Sumatra, and the Louisiade Archipelago.

31. *Serranus urodelus*, Forst. D. $\frac{9}{13}$, A. $\frac{3}{8}$, L. l. 85. Vertical limb of preopercle finely serrated: caudal rounded. Of a sanguineous colour, with a purplish tinge: caudal and anal with pale red and blue spots. Two oblique lines with white borders on the caudal, converging posteriorly. A specimen in the British Museum is reputed as having come from India. Malay Archipelago.

32. *Serranus louti*, Forsk. D. $\frac{9}{15\text{-}16}$ A. $\frac{3}{8}$, L. l. 70. Vertical limb of preopercle slightly emarginate and finely serrated. Caudal rounded. Madder brown, with seven or eight darker cross bands, and a dark spot between the two upper opercular spines. Fins brown, the vertical ones with black edges and white margins. East coast of Africa, Ceylon.

33. *Serranus Homfrayi*, Day. D. $\frac{9}{13}$, A. $\frac{3}{8}$. Caudal rounded. Whitish, with roseate spots. A dark band over the free portion of the tail: the longest specimen (7½ inches) having the body with nine bands. Fins spotted with red and edged with white: upper half of caudal with a dark margin. Andamans, longest of two specimens 7½ inches.

Genus—G*RAMMISTES* (*Artedi*.), *Cuv*.

Branchiostegals seven: pseudobranchiæ. Body oblong and compressed. Opercle and preopercle unserrated but spinate. Eyes lateral. Teeth villiform in the jaws and palatines: no canines. Tongue smooth. Two dorsals, the first with seven spines, the anal spineless. Scales minute, adherent, and enveloped in the epidermis. Pyloric appendages few.

35. *Grammistes Orientalis*, Bl. Schn. D. $7/\frac{1}{15}$, A. 9. Three spines on preopercle. Caudal rounded. Deep brown, with seven narrow longitudinal white bands, which anteriorly are continued on to the head. Seas of India, Philipines, and Australia ; it attains a few inches in length.

Genus—*DIPLOPRION (Kuhl. and V. Hass.), Cuv.*

Branchiostegals seven: pseudobranchiæ. Body oblong, compressed. Eyes lateral. Opercle spinate ; preopercle with a double denticulated limb. Teeth villiform in jaws and palatines : no canines. Tongue smooth. Two dorsals, the first with eight spines, anal with two. Scales small, adherent. Pyloric appendages few.

36. *Diploprion bifasciatum* (Kuhl. & V. Hass.) Cuv. & Val. D. 8/15, A. $\frac{2}{12}$, Vert. 12/13, Cæc. Pyl. 3. Yellowish with two broad black cross-bands. Seas of India, Malay Archipelago and China; attaining six inches and more in length.

Genus—*GENYOROGE, Cantor.*

Diacope, sp. Cuv.: *Mesoprion*, sp. Bleeker.

Branchiostegals seven. Snout somewhat elongated, the preorbital rather high. Preopercle serrated, having a notch above its angle, as deep as broad, receiving a spinate knob of the interopercle. Opercle with two or three flat spines. Villiform teeth on both jaws and palatines : with canines in either jaw. Tongue smooth. Dorsal fin single, with from ten to eleven, but rarely more, spines : anal with three spines. Scales of moderate size. Cæcal pylori, few or absent.

37. *Genyorogæ Sebæ*, Cuv. & Val. *Viri-cut-ta-lay*, or *nai-kerruchi*, "smelling like a dog," Tamil. D. $\frac{11}{15\text{-}16}$, A. $\frac{3}{9\text{-}11}$, L. l. 40, Cæc. pyl. 4-5, Vert. 10/14. Caudal fin emarginate. Reddish, a black band passes from before the dorsal fin through the eye to the snout: a second from the summit of the second to the sixth dorsal spines to the ventral fin : a third from the soft dorsal, curving downwards to the lower half of the caudal fin : a black band along the upper half of the caudal : ventral and lower half of anal black. Seas of India to the Malay Archipelago, &c.

38. *Genyoroge Amboinensis*, Bleeker. D. $\frac{11}{13}$, A. $\frac{3}{8}$, L. l. ca. 70, Cœc. pyl. o. Caudal slightly emarginate. Rose coloured, five or six yellow longitudinal stripes along the sides : fins yellow : a black blotch immediately above the lateral line, not apparent until after death in Andamanese specimens. Andamans to the Malay Archipelago.

39. *Genyoroge Bengalensis*, Bloch. *Viri-keechan*, Tam. D. $\frac{11}{14}$, A. $\frac{3}{8}$, L. l. 48, L. tr. 8/18. Caudal slightly emarginate, yellowish brown superiorly, becoming yellowish white inferiorly. A light finger mark, black in the immature, exists on the lateral line, under the first portion of the soft dorsal fin. Five bright blue bands pass from the orbit across the opercles : the superior to the ninth dorsal spine : the second to the fourth dorsal ray: the third to the last dorsal ray : the fourth to the centre of the base of the caudal: and the fifth across the base of the

pectoral to the posterior extremity of the anal: fins yellowish. Seas of India, Malay Archipelago, China, &c.

40. *Genyoroge rivulata*, Cuv. & Val. *Kalee maee*, Tel.: *Cuttupirriun*, Tam. D. $\frac{10}{1\frac{2}{3}}$, A. $\frac{3}{8\text{-}9}$, L. l. 45-50, L. tr. 9/19, Cœc. pyl. 5. Caudal slightly emarginate. Colours (in the *immature*), back olive, with a slate coloured spot in the centre of each scale, thus forming lines passing upwards and backwards: abdomen greyish with horizontal golden lines crossing the centre of each scale, and vertical dark ones along their bases. Several bright blue lines pass downwards and backwards over the preopercle and opercle, and two larger ones along the snout. A large white blotch on the lateral line opposite the third to the fifth soft ray, having a wide black edge anteriorly and posteriorly in its upper third. This white mark covers four scales transversely, is one below and three above the lateral line. Dorsal, slate coloured, superiorly reddish with a narrow white edge. Pectorals reddish: ventrals slaty with a dark edge. Caudal bluish, tipped with red. (In the *adult*) as about 15 inches in length, the mark on the lateral line has wholly or entirely gone, the golden shade is wanting, but the blue spots remain. The white edge to the fins is also usually absent. Seas of India to the Malay Archipelago.

41. *Genyoroge notata*, Günther. D. $\frac{10}{12}$, A. $\frac{3}{8}$. Second anal spine longer and stronger than the third. Back reddish brown, a black blotch above the lateral line, and bright blue lines along the body. East coast of Africa and seas of India according to Günther and Playfair. But it seems very doubtful if some error has not occurred and the *Mesoprion Russellii* got mixed up with this Zanzibar species which is not Rüppell's fish. (See No. 57.)

42. *Genyoroge melanura*, Rüpp. D. $\frac{10}{14}$, A. $\frac{3}{8\text{-}9}$, L. l. 55-60. Cœc. pyl. 4. Preopercle serrated, strongest at the angle: the notch well developed. Uniform crimson: dorsal, caudal and anal fins with a black margin having an external white edge: pectoral and ventral yellow: basal portion of the caudal black. Red Sea: Andamans.

43. *Genyoroge marginata*, Cuv. & Val. *Cul-meen,* Sungarah, and Vekkerday*, Tamil. D. $\frac{10}{12}$, A. $\frac{3}{8}$, L. l. 50-55, L. tr. 6/14. Pectoral reaching to above the first anal spine: second anal spine strongest and rather longer than the third: caudal emarginate. Purplish yellow, caudal fin deep purple with a white edge: no lateral blotch. Seas of India to the Malay Archipelago, is not a large species.

44. *Genyoroge grammica*, Day. D. $\frac{10}{12}$, A. $\frac{3}{7}$, L. l. 45, Cœc. pyl. 5. Second anal spine longest and strongest: caudal lunated. Yellow with five blue lines on the body, the upper three going to the dorsal fin, the fourth to the middle of the caudal, and the fifth to the end of the base of the anal. Four blue lines on the head, two from the eye join the second and third body lines: two from the snout become the fourth and fifth on the body. A black finger-mark exists on and above the lateral line, opposite the commencement of the soft dorsal fin. Andamans, to about 6 inches in length.

---

* *Cul-meen* or 'stone fish,' Tamil, is a very common term, applied to some species the fishermen know, and many they have not previously observed.

Genus—*MESOPRION,* Cuv.

Diacope, sp. Cuv. & Val.: Rüppell, &c.
Branchiostegals seven. Snout somewhat elongated, the preorbital rather high. Preopercle serrated, without or with a very open notch. Opercle with three or two, or more rarely one indistinct point. Villiform teeth in both jaws with canines, teeth on the palate: tongue smooth. Dorsal fin single, with from nine to twelve spines: anal with three. Scales of moderate size. Cæcal pylori few or absent.

45. *Mesoprion dodecacanthoides,* Bleeker. D. $\frac{12}{13}$, A. $\frac{3}{8}$, L. l. 46. Spinous portion of dorsal fin higher than the soft. Caudal very slightly emarginate. Rose coloured, with seven oblique brownish streaks: a brown spot at the base of the tail. Fins yellow. Madras and Malay Archipelago.

46. *Mesoprion chirtah,* Cuv. & Val. *Soosta,* Ooriah: *Rettum-pirriun,* Tam: *An-na-kah-ro-dah,* And. D. $\frac{11}{13\text{-}14}$, A. $\frac{3}{8}$, L. l. 54, L. tr. $1\frac{4}{23}$, Vert. 10/14, Cæc. pyl. 5-6. Third anal spine longest. Caudal truncated. Colours (in the *immature M. annularis*) crimson with orange reflections, a dark mark along the base of the dorsal fin commencing at the opercles: a black band across the back over the free portion of the tail behind which it is of a whitish colour. Fins, except the pectoral, with a fine black edge. In specimens about five inches in length there are many longitudinal black lines. (In the *adult*) they become of a uniform rose colour, somewhat orange on the fin, the edges of which are darkest. . When alive, even in some large specimens, the remains of the mark on the tail is perceptible.

47. *Mesoprion Malabaricus,* Bl. Schn. D. $\frac{11}{13\text{-}14}$, A. $\frac{3}{8}$, L. l. 52. Second anal spine longest and strongest: caudal truncated. Rose coloured with oblique yellow streaks above the lateral line and longitudinal ones beneath. A bluish violet band along the back at the base of the dorsal fin, and a reddish violet spot on the back of the tail: dorsal and caudal fins with a black edge. Seas of India. This, it is suggested, may be a variety of the *M. annularis.*

48. *Mesoprion Mitchelli.* Günther. D. $\frac{11}{14}$, A. $\frac{3}{9}$, L. l. 50, L. tr, 9/18. Body elevated and compressed: lower jaw considerably the longer. Caudal lunated: anal spines weak, second and third of equal length. Yellow, red along the back, becoming rosy below the lateral line: dark olivaceous stripes along the rows of scales above the lateral line, but more yellow below it. Fins olive edged with black. Madras up to 10 inches in length. This fish appears to closely resemble genus *Odontonectes,* Günther.

49. *Mesoprion rubellus,* Cuv. & Val. *Jahngarah,* Tel. D. $\frac{10\text{-}11}{14\text{-}13}$, A. $\frac{3}{8\text{-}9}$, L. l. 46, L. tr. $\frac{7}{13}$. Second anal spine strongest but shorter than the third: caudal emarginate. Back greyish brown, chest orange, abdomen and sides of a light violet, each scale with a white edge. Spinous portion of the dorsal greyish, but the soft with a more yellow tinge: pectoral reddish: dorsal spines greyish, and caudal tinged with brown. Seas of India and Red Sea, attaining four feet in length.

50. *Mesoprion Johnii,* Bloch. *Doondiawah,* Tel.: *Chembolay,* Mal D. $\frac{10}{14}$, A. $\frac{3}{8}$, L. l. 48, L. tr. 7/10. Second anal spine longest and strongest:

caudal slightly emarginate. Yellowish, lightest on the abdomen, with a large black finger-mark on the lateral line between the 22nd and 30th scales. A dark line in some localities is observed along the centre of each row of scales, and then the finger-mark is usually badly developed. Fins yellowish dashed with red. Seas of India, Malay Archipelago, China, Australia, and the Pacific. The very young has a long spine at the angle of the preopercle, as is observed in *M. pomacanthus*, Bleeker.

51. *Mesoprion sillaao*, Cuv. and Val. *Sillaao*, Tel. D. $\frac{10}{14\text{-}15}$, A. $\frac{3}{9}$, L. l. 50, L. tr. 7/12. A slight knob at the upper angle of the interopercle. Second anal spine strongest, the third the longest: caudal slightly emarginate. Back brownish red, the base of each scale darkest, below the lateral line of a bright lake colour.' A tinge of orange along the lower surface of the body. A blue zig-zag line over the sub-and preorbitals. Spinous portion of dorsal and anal greyish, the soft orange-scarlet. Pectoral scarlet. Seas of India, attaining a large size.

52. *Mesoprion flavipinnis*, Cuv. and Val. D. $\frac{10}{4}$, A. $\frac{3}{9}$. Greyish on the back, becoming whitish on the abdomen, with a general silvery tint. All the fins are yellowish. Pondicherry, attaining five feet in length. This may be *M. sillaao*.

53. *Mesoprion rangus*, Cuv. and Val. *Rangoo*, Tel.: *To-go-re-dah*, And. D. $\frac{10}{13\text{-}15}$, A. $\frac{3}{8}$, L. l. 46-50, L. tr. $\frac{7}{16\text{-}16}$, Cæc. pyl. 4. A slight tuberosity on the interopercle. Second anal spine longest and strongest: caudal slightly emarginate: first ventral ray prolonged. Dark reddish brown on the back, becoming dull cherry red below the lateral line. Fins reddish. A blue line along the sub-orbital ring of bones. The young have eight or nine irregular and very narrow white bands: an orange streak along the edge of the spinous dorsal: and a white streak on the external side of the ventrals. Seas of India and Malay Archipelago. Attains two feet and upwards in length.

54. *Mesoprion vitta*, Quoy and Gaim. D. $\frac{10\text{-}11}{12\text{-}14}$, A. $\frac{3}{9}$, L. l. 55-62. A tuberosity on the interopercle: second anal spine stronger than the third, but the two of nearly equal length: caudal slightly emarginate. Yellowish, with oblique streaks above the lateral line, and horizontal ones below it. A black band from the eye to the base of the caudal fin: dorsal and caudal blackish. Andamans, Malay Archipelago to China and Japan, &c.

55. *Mesoprion gembra*, Bloch. D. $\frac{10}{13\text{-}14}$, A. $\frac{3}{8}$, L. l. 46. Second anal spine slightly longer and stronger than the third: caudal truncated. Olive, scales with dark bases: dorsal, anal, and ventral with darker margins. Immature with cross bands. Seas of India to the Moluccas.

56. *Mesoprion aurolineatus*, Cuv. and Val. D. $\frac{10}{13\text{-}14}$, A. $\frac{3}{8}$, L. l. 46. Second anal spine longest and strongest: caudal slightly emarginate. Olivaceous: abdomen with several brilliantly golden horizontal lines, divided by a darker one, and running along each row of scales. A large black blotch on the lateral line from the 21st to the 31st scales, two-thirds of it being below the line. Fins yellowish. Seas of India.

57. *Mesoprion fulviflamma*, Forsk. *Antika doondiawah*, Tel. D. $\frac{10}{13\text{-}14}$, A. $\frac{3}{8}$, L. l. 48—54, L. tr. 7/16, Cæc. pyl. 4—6. Second and third anal species of about equal length and strength: caudal slightly

clxiii

emarginate. Olivaceous yellow or rosy along the back, with four narrow, and brilliant golden bands passing obliquely upwards and backwards from the lateral line: four similar golden bands exist below the lateral line, the first proceeding from the posterior edge of the orbit to the lateral finger mark; the second from the middle of the opercle to opposite the end of the soft dorsal, where it is lost on the lateral line; the third from below the orbit to the base of the caudal; and the fourth from below the base of the pectoral to the base of the anal. A large black finger-mark exists on the lateral line opposite the commencement of the soft dorsal fin from the 23rd to the 26th scales. Prof. Kner. (Novara fische, p. 35) appears to have correctly identified this species with Russell's pl. 98, which is the same as *Diacope notata*, Cuv. and Val. Seas of India, East coast of Africa, Malay Archipelago and China.

58. *Mesoprion carui*, Cuv. and Val. *Karooi*, Tel. D. $\frac{10}{13}$, A. $\frac{3}{8}$, L. l. 50, L. tr. 6/12. Lower jaw slightly the longer. Second anal spine strongest, but not quite so long as the third: caudal lunated. Yellowish red along the back, becoming rosy below the lateral line: olive stripes passing obliquely upwards and backwards above the lateral line, and brilliant yellow ones below it. Fins orange. Coromandel coast. This species requires comparing with *M. flavipinnis*, Cuv. and Val.

59. *Mesoprion decussatus*, Cuv. and Val. *Jeu-win-dah*, Andamanese. D. $\frac{10}{13\text{-}14}$, A. $\frac{3}{8}$, L. l. 54, L. tr. 6/17, Cæc. pyl. 3. Third anal spine slightly the longest and strongest: caudal forked. Whitish, with six black horizontal bands along the body, and six badly-marked short vertical ones in its upper third from the dorsal fin, the crossing of the two sets of bands forming large white spots. A deep black mark at the base of the caudal fin. A white band across the occiput, which is continued on to the preopercle. Fins greyish. Seas of India, Malay Archipelago to the Philipines.

60. *Mesoprion multidens*, Day. D. $\frac{10}{11}$, A. $\frac{3}{8}$, L. l 52, L. tr. 7/17, Cæc. pyl. 5. Six canines in the lower jaw: some smaller ones in the upper. Last dorsal and anal rays elongated; second anal spine strongest, third the longest: caudal deeply forked. Rosy, with about six longitudinal yellow bands along the body, and one golden one from the inferior angle of the eye to the snout, and another across the forehead. Andamans obtaining a large size.

Genus—*AMBASSIS** (Comm), *Cuv. and Val.*

Chanda, Hamilton Buchanan; *Bogoda*, pt. Bleeker.

Branchiostegals six. Body compressed, more or less diaphanous. Lower limb of the preopercle with a double serrated edge: opercle without a prominent spine. Villiform teeth on the jaws and palate: generally no canines. Two dorsal fins, the first with seven spines, the anal with three: a recumbent spine directed forwards in front of the base of the dorsal fin. Scales of moderate or small size, frequently deciduous. Lateral line complete, interrupted, incomplete, or absent.

A. *Lateral line continuous.*

61. *Ambassis Commersonii*, Cuv. and Val. D. $7/\frac{1}{9\text{-}11}$, A. $\frac{3}{9}$, L. l. 30-33, Vert. 9/15. Preorbital rather strongly serrated: serratures

---
* *Gu-nas-si*, Mugh.

passing downwards and slightly backwards. Vertical limb of preopercle finely serrated: its inferior limb with its double edge also serrated: two or three coarser teeth on the angle. Inferior edge of interopercle finely serrated. Two or three small and very blunt denticulations at the posterior superior angle of the orbit. Second dorsal spine the longest: second anal spine the strongest, and about as long as the third. Silvery, with purplish reflections: a bright silvery line from the eye to the caudal fin. Red Sea, East coast of Africa, through the seas of India to those of North Australia: ascends estuaries, attains 7 inches in length.

62. *Ambassis urotænia*, Bleeker. D. 7/$\frac{1}{9\text{-}10}$, A. $\frac{3}{9\text{-}10}$, L. l. 27, L. tr. 4/10. Second spine of first dorsal fin rather above half the height of the body: third anal spine the longest in the fin but not so long as the second of the dorsal. Silvery, with a burnished lateral band. The interspinous membrane between the second and third spines black: a black longitudinal band on each lobe of the caudal. Andamans and Malay Archipelago.

B. Lateral line interrupted.

63. *Ambassis Dussumieri*, Cuv. & Val. D. 7/ $\frac{1}{9\text{-}10}$, A. $\frac{3}{9\text{-}10}$, L. l. 27, L. tr. 3/6. Preorbital with six strong denticulations directed downwards and backwards on its anterior edge, and a few serratures on its posterior: anterior serratures on the double serrated edge of the preopercle the largest. Two or three strong spines directed backwards at the posterior superior angle of the orbit, sub-and inter-opercle entire. Third anal spine longer, but not quite so strong as the second. Lateral line interrupted after about the eleventh scale. Silvery, with a bright lateral band: blackish between second and third dorsal spines: a blackish edge to the caudal. East coast of Africa, seas of India to the Malay Archipelago and China.

64. *Ambassis Buruensis*, Bleeker. D. 7/ $\frac{1}{8\text{-}9}$ A. $\frac{3}{9\text{-}10}$ L. l. 28, L. tr. $\frac{3\text{-}4}{7\text{-}8}$. Above the eye two small spines directed backwards: interopercle with a small spine at its angle. Second spine of first dorsal ¼ of the total length: third anal spine considerably longer than the second. Lateral line interrupted below the commencement of the soft dorsal fin. Two broad bands of scales on suborbitals. Silvery, with a lateral band: fins yellowish, blackish between second and third dorsal spines. Andamans and Malay Archipelago.

C. Lateral line incomplete.

65. *Ambassis macracanthus*, Bleeker. D. 7/$\frac{1}{9\text{-}10}$, A. $\frac{3}{9\text{-}11}$, L. l. 27-29. Anterior margin of orbit serrated, two spines at its posterior superior angle: preorbital serrated. A double serrature on vertical limb of preopercle. Sub-opercle with four denticulations at its angle. Second dorsal spine one-half the length of the body: third anal spine the longest in the fin. Lateral line incomplete. Silvery, without any lateral stripe. Second dorsal spine bright orange: the interspace between it and the third black. Estuaries in the Andamans, Malay Archipelago.

Genus—*APOGON*, Lacép.

Branchiostegals seven: pseudobranchiæ absent. Opercle spinate: preopercle with a double serrated ridge. Teeth villiform in jaws vomer and palatines, no canines: tongue smooth. Two separate dorsal fins, the first with six or seven spines, the anal with two. Lateral line commences opposite the upper edge of the opercle. Scales large, deciduous.

66. *Apogon nigripinnis*, Cuv. & Val. D. 7/$\frac{1}{9}$, A. $\frac{2}{8}$. Caudal rounded. Dorsal, ventral, and anal fins black. Seas of India, Malay Archipelago to China; attaining at least 2½ inches in length.

67. *Apogon chrysotænia*, Bleeker. D. 7/$\frac{1}{9}$, A. $\frac{2}{8}$, L. l. 26, L. tr. $\frac{2}{8}$. Caudal notched. Brilliant golden with black head. A silvery white median band on the head, which divides, one branch proceeding to the upper half of the tail on either side: a second band goes from the orbit to the middle of the tail: a third to its lower half: and a fourth from the angle of the mouth to the base of the pectoral. Fins orange. Andamans and Nicobars amongst the coral reefs; also in Malay Archipelago.

68. *Apogon maculosus*, Cuv. & Val. D. 7/$\frac{1}{8}$, A. $\frac{2}{7}$. Brown with four rows of brown spots on either side: fins also brown spotted with darker, none on the head. This fish is said generally to have come from the seas of India, and to attain three inches in length.

69. *Apogon Novæ Guineæ*, Valen. D. 7/$\frac{1}{8}$, A. $\frac{2}{8}$, L. l. 24-25, L. tr. 8-9. Fourth dorsal spine the longest: caudal slightly notched. Head with brown spots: body and fins yellowish. Seas of India, Malay Archipelago, &c.

70. *Apogon annularis*, Rüpp. D. 7/$\frac{1}{9}$, A. $\frac{2}{8}$, Cæc. pyl. 4. Body pinkish shot with gold, a broad black or dark band round the free portion of the tail. Fins reddish. The first dorsal tipped with black: ventral spine black, and a narrow black edge to second dorsal, anal and caudal. Some fine black or brown spots on the snout and head: a black horizontal band through the eye. Red Sea, seas of India to the Malay Archipelago. Attains at least 4½ inches in length.

71. *Apogon quadrifasciatus*, Cuv. & Val. D. 7/$\frac{1}{9}$, A. $\frac{2}{8}$. Caudal notched. Silvery red, on either side two longitudinal bands, the lowest going from the mouth and eye to the middle of the base of the caudal fin: the highest from above the orbit to the upper portion of the base of the caudal. Fins yellowish, the upper half of first dorsal black. From the East coast of Africa through the seas of India to the Malay Archipelago and China.

72. *Apogon fasciatus*, White. D. 7/$\frac{1}{9}$, A. $\frac{2}{8}$, L. l. 26, L. tr. 8-9. Caudal notched. Olive, a black stripe from the upper edge of the orbit to the end of the second dorsal: a second below it to the tail above the lateral line: a third parallel to it below the lateral line: a fourth from the upper part of the lower jaw to the root of the pectoral. Dorsal yellowish tinged with darker. Ventral, anal, and caudal reddish. East coast of Africa through the seas of India to the Malay Archipelago and Australia.

73. *Apogon Bleekeri*, Günther. D. 6/$\frac{1}{5}$, A. $\frac{2}{5}$ (14-17), L. r. 20. Caudal slightly forked. Whitish, having a pink tinge on its fins: and a rather large black mark on the lateral line at the root of the caudal: opercles silvery. Seas of Madras and the Malay Archipelago.

74. *Apogon Ceylonensis*, Cuv. and Val. D. 6/$\frac{1}{9}$, A. $\frac{2}{1\frac{1}{4}}$. Caudal slightly forked. Reddish shot with gold, but neither spots nor transverse bands. Ceylon, to 2 inches in length.

75. *Apogon hyalosoma*, Bleeker. D. 6/$\frac{1}{9}$, A. $\frac{2}{8}$, L. l. 24, L. tr. 2$\frac{1}{2}$/8$\frac{1}{2}$. Caudal slightly forked. Of an olive colour, with a darkish blotch at the root of the caudal fin: between second and third dorsal spines black. Seas of India and Malay Archipelago, to at least 6 inches in length.

76. *Apogon thermalis*, Cuv. and Val. D. 6/$\frac{1}{9}$ A. $\frac{2}{8}$. Uniform yellowish, with a black blotch at the side of the tail: between second and third dorsal spines black. This species, due to having been taken in warm springs at Cania in Ceylon, is considered to be a different sort from the preceding: a comparison is needed

77. *Apogon arbicularis* (Kuhl. and v. Hass.), Cuv. and Val. D. 6/$\frac{1}{8}$, A. $\frac{2}{8}$, L. l. 25, L. tr. $\frac{3}{7}$. Olive: a dark zone round the body in front of the first dorsal fin: head spotted with black: a cloudy band below the second dorsal: free portion of the tail spotted: first dorsal with black spots: base of second dorsal cloudy: ventrals nearly black. Andamans and Malay Archipelago.

Genus—APOGONICHTHYS, Bleeker.

Branchiostegals seven: pseudobranchiæ absent. Opercle spinate: preopercle with a double ridge but destitute of any serrature. Teeth villiform on the jaws and palatines: no canines: tongue smooth. Two separate dorsal fins, the first with six or seven spines, the anal with two. Scales large, deciduous.

78. *Apogonichthys auritus*, Cuv. and Val. D. 7/$\frac{1}{5}$, A. $\frac{2}{7}$, L. l. 23. Caudal rounded. Lateral line ceases under the middle of the soft dorsal fin. Body and head spotted and marbled all over with brown, a round black spot on the opercles having a white lower edging. Andamans and Nicobars, also Red Sea and the Mauritius.

Genus—CHEILODIPTERUS, Lacép.

Branchiostegals seven. Opercle without spines: preopercle having an interior ridge and generally a double serrature. Villiform teeth in both jaws, and usually canines as well: teeth on the palatine bones. Two dorsal fins separated by an interval, the first with six spines, the anal with two. Scales large, deciduous.

79. *Cheilodipterus lineatus*, Forsk. D. 6/$\frac{1}{9}$, A. $\frac{2}{8}$, L. l. 28. Caudal notched. Lower preopercular edge serrated. Silvery, with from 7 to 16 horizontal black bands, apparently dependent upon age; a black mark on the side of the base of the tail: a black blotch between the second and third dorsal spine: fins red. Red Sea, East coast of Africa, seas of India, to Malay Archipelago.

80. *Cheilodipterus quinquelineatus*, Cuv. and Val. D. 6/$\frac{1}{9}$, A. $\frac{2}{8}$, L. l. 25. Five black bands along the sides: a black spot at the root of the caudal, with a bright yellow ocellus round it. Nicobars, Malay Archipelago.

Genus—DULES, Cuv. and Val.

Branchiostegals six. Eyes of moderate size. Opercles spinate, without a membranous lobe: preopercle serrated: chin moderately prominent. Villi-

clxvii

form teeth on jaws and palatines: no canines. One dorsal fin with ten spines having a deep notch between the ninth and tenth: anal with three spines. Scales ctenoid and of moderate size.

81. *Dules rupestris*, Lacép. D. $\frac{10}{10\text{-}11}$, A. $\frac{3}{10\text{-}11}$, L. l. 40-42, L. tr. 5/8. Preorbital and inferior limb of preopercle finely serrated: opercle with two spines. Fourth and fifth dorsal spines longest: the third of the anal longer than the second: caudal slightly emarginate. Greyish, becoming silvery beneath the lateral line: all the fins edged with white, and having a dark grey band along their bases: sometimes every scale with a black central spot: the vertical fins brown spotted, edged with black. Andamans, Mauritius, Malay Archipelago, and Feejee Islands.

82. *Dules tæniurus*, Cuv. and Val. D. $\frac{10}{9}$, A. $\frac{3}{7}$, L. l. 50-55, L. tr. 6/12. Fourth and fifth dorsal spines the longest: caudal deeply notched: third anal spine longer but not so strong as the second. Bluish, becoming silvery on the abdomen: soft dorsal with a brown anterior and superior edge: caudal brown, with an oblique white band on either lobe, they converge posteriorly. Andamans, Malay Archipelago, and China.

83. *Dules Bennetti* (Bleeker), Peters. D. $\frac{10}{9}$, A. $\frac{3}{10}$, L. l. 50, L. tr. 6-12. Similar, if not identical, with the last, but with a black streak along the centre of the caudal fin, and two oblique bands on either lobe. East coast of Africa and seas of India.

Family—PRISTIPOMATIDÆ.

Percoidei, pt., *Sciænoidei*, pt., *Sparoidei*, pt., et *Mænides*, pt. Cuv.: *Theraponidæ*, pt., *Hæmulonidæ*, pt. Richardson.

Branchiostegals from five to seven: pseudobranchiæ generally well developed. Body oblong, compressed. Eyes of medium size, lateral. Mouth moderately or very protractile, placed in front of the snout, and having a lateral cleft. Muciferous system of the head rudimentary, or slightly developed. Preopercle entire or serrated. Barbels absent. Teeth in villiform bands, with conical canines in some genera, but neither molars nor cutting ones in the jaws: palate usually edentulous. A single dorsal fin: the length of the bases of the spinous and soft portions being of about equal extent, the first containing strong spines, or being continuous with the soft: anal mostly with three spines, its soft portion similar to that of the dorsal: lower pectoral rays branched: ventrals thoracic, with one spine and five rays. Scales finely ctenoid or cycloid, extending over the body and head: cheeks not cuirassed. Lateral line continuous. Air-vessel present, more or less simple, being divided by a constriction in some species into an anterior and posterior portion. Stomach cæcal. Pyloric appendages few or in moderate numbers.

Genus—THERAPON,* *Cuv.*

Datnia, sp. Cuv. and Val.: *Pelates*, sp. Cuv.

Branchiostegals six: pseudobranchiæ. Eyes of moderate size. Opercle with spines: preopercle serrated. Teeth villiform in both jaws, the outer row being sometimes the larger: deciduous ones on the vomer and palatines. Dorsal fin single, but more or less notched, having eleven to thirteen spines: anal with three. Scales of moderate size. Air-vessel divided by a constriction. Pyloric appendages in moderate numbers.

84. *Therapon puta*, Cuv. and Val. Keelputa, Tel.: Keetchan, Tam. and Mal. D. $\frac{11\text{-}12}{10}$, A. $\frac{3}{8\text{-}9}$, L. l. 90—100, L. tr. $\frac{15}{24}$, Cæc. pyl. 7, Vert. $\frac{10}{13}$. Preopercle with five strong denticulations on its vertical limb, the

---

* *Sabah-za*, Mugh.

middle usually the longest. Opercle with two spines, the inferior longest
and strongest. No teeth on the palate, except in the very young.
Greyish, with three or four longitudinal straight blackish-brown bands;
spinous portion of dorsal in its upper three quarters blackish between
the third and seventh spines. Two oblique bands pass across the upper
caudal lobe, and one, sometimes two, across its lower one. Seas of India
to the Malay Archipelago. This species is the *T. trivittatus*, of Cantor
and Günther: Hamilton Buchanan's original drawing of his *Coius trivittatus*
having been copied by Hardwicke is considered another species
T. servus, a conclusion open to discussion.

85. *Therapon servus*, Bloch. D. $\frac{11\text{-}12}{10}$, A. $\frac{3}{8\text{-}9}$, L. l. 80—86. Vert.
$\frac{10}{15}$ (see 'F. W. Fishery Report,' No. 8, p. ccl.)

86. *Therapon quadrilineatus*, Bloch. D. $\frac{11\text{-}12}{10}$, A. $\frac{3}{10}$, L. l. 70, L. tr.
$\frac{13}{20}$, Cæc. pyl. 18. Preopercle with rather stronger serratures at its angle:
lower opercular spine the strongest. Five horizontal black bands along
the body: the first passes backwards to the anterior portion of the soft
dorsal, the second to the posterior end of its base, the third to the upper
third of the base of the caudal, the fourth to its lower third, the fifth
to the end of the base of the anal. A large black blotch exists on the
shoulder. Dorsal with a black mark between its third and sixth spines:
a black tip to the soft dorsal, and a badly-marked band along its centre:
anal and caudal with black edges. Seas of India, Malay Archipelago,
and China, attaining 6 inches in length.

87. *Therapon argenteus*, Cuv. & Val. D. $\frac{12}{10}$, A. $\frac{3}{9}$, L. l. 56, L. tr.
25, Cæc. pyl. 11. No palatine teeth. As Cuvier observes, this fish has a
more elevated body than the typical *Therapons*, a concave profile, a
pointed snout, the dorsal spines being stronger and their base occupying
a comparatively greater length of the back than the rays, and there being
a very slight notch between the last two. In fact, it is his first species
of *Datnia*. Silvery, darkest on the back: a narrow black edge to the
spinous dorsal, and a blackish blotch on the soft anal. Seas of India
and Malay Archipelago.

88. *Therapon theraps*, Cuv. & Val. D. $\frac{12}{10}$, A. $\frac{3}{8}$, L. l. 50-55, L. tr.
$\frac{11}{17}$. Serrations at angle of preopercle rather coarser than along the rest
of its vertical margin, but of an even size. Lower opercular spine longest,
but not equalling that of *T. servus*. Three horizontal blackish brown
lines, the first from the second to the last dorsal spine: the second from
the nape to the second and third dorsal ray, and continued on to the fins;
the last to the upper third of the tail. Dorsal with a black mark
between the third and sixth spine, a dark band along the upper portion
of the soft rays: a brown band along the centre of the caudal, and two
oblique ones across either lobe. East coast of Africa, seas of India
through the Malay Archipelago to China. It attains 6 inches in length.

89. *Therapon squalidus*, Cuv. & Val. D. $\frac{12}{10}$, A. $\frac{3}{9}$, Cæc. pyl. 13.
Preopercle almost rectangular. Vomerine and palatine teeth (in a specimen
3½ inches long). Of a pale colour, with two broad silvery bands: two black
bands on each lobe of the caudal, anal with two large black spots.
Indian Ocean; locality not given.

90. *Therapon transversus*, Cuv. & Val. D. $\frac{12}{10}$, A. $\frac{3}{8}$, Cæc. pyl. 11.
No palatine teeth (specimen 3 inches long). Yellowish brown, with pale

bands, and five or six indistinct cross bars : a black spot on the spinous dorsal, another on the soft rays: five bands on the caudal, and a spot on the anal. Malabar and the Indian Ocean.

91. *Therapon virgatus*, Cuv. & Val. D. $\frac{12}{12}$, A. $\frac{3}{8}$. Preopercle strongly denticulated: preorbital smooth. The length of the base of the spinous portion of the dorsal fin is double that of the soft. No teeth on the palate. Brown, with small bluish spots : three longitudinal yellow bands along the body, the central one being prolonged on to the caudal fin, two other oblique bands on each of its lobes. Spinous dorsal with a long blackish spot : soft dorsal and anal, each with two brown spots. Bay of Bengal, to 3 inches in length.

92. *Therapon cinereus*, Cuv. & Val. D. $\frac{12}{10}$, A. $\frac{3}{8\cdot10}$. Preopercle rounded, with moderate serratures, coarsest at the angle. Uniform colour, with a black blotch between the third and sixth dorsal spines. India.

Genus.—PRISTIPOMA, *Cuv.*

Branchiostegals seven : pseudobranchiæ. Body oblong, compressed. Eyes of moderate size. Cleft of mouth horizontal: gape not very wide: intermaxillaries moderately protrusible: jaws of nearly equal length. A central longitudinal groove below the chin. Preopercle serrated: opercle with indistinct points. Teeth villiform without canines : palate edentulous. Dorsal with eleven to fourteen spines, and sometimes with a very deep notch between the last two : anal with three spines. Vertical fins scaleless or only scaly along their bases. Scales of moderate size, ctenoid. Air-vessel without any constriction, simple. Pyloric appendages in small numbers.

93. *Pristipoma paikeeli*, Cuv. & Val. *Paikeeli*, Tel. D. $\frac{12}{13}$, A. $\frac{3}{8}$, L. l. 55-60. Silvery, with six light brown parallel bands edged with black along the sides: fins with black dots and darker edges. Seas of India; Malay Archipelago.

94. *Pristipoma stridens*, Forsk. D. $11/\frac{1}{13\cdot15}$, A. $\frac{3}{7\cdot8}$, L. r. 63, L. tr. $\frac{10}{22}$, Cæc. pyl. 5-6. Its colours much resemble those of a *Therapon :* purplish on the back, becoming dirty white on the abdomen : a golden band from the eye to the centre of the caudal fin, and two more lighter ones above it : a dark blotch at the upper third of the opercle : the lower half of the anal stained darkish. Red Sea, along the Meckran Coast and very common at Kurrachee.

95. *Pristipoma operculare?* Playfair. D. $11/\frac{1}{14}$, A. $\frac{3}{8}$, L. l. 57, Cæc pyl. 5, very long. Height of body equals the length of the head and 2/7 of the total. Length of snout equals 1¼ diameters of the orbit: otherwise its form agrees with *P. operculare*, Playfair, Fishes of Zanzibar. Silvery, a black blotch at the upper and posterior corner of the opercle : upper, half of body with numerous black spots, which in some do, in others do not, form undulating bands : a black spot at the base of each dorsal spine and ray: dorsal and caudal dark edged. East Coast of Africa? Meckran Coast, and Kurrachee.

96. *Pristipoma Dussumieri*, Cuv. and Val. D. $\frac{12}{14}$, A. $\frac{3}{7}$, L. l. 46-50. Fourth dorsal spine the longest: caudal slightly lunated. *P. Neilli* is identical with this species. Its colours are greyish, becoming white along the abdomen. A brilliant yellow band passes from the eye to the caudal fin : a similar one parallel to and above it. Dorsal and caudal

greyish: soft dorsal with a light edge. Pectorals, ventrals, and anal yellowish. Coast of India.

97. *Pristipoma guoraka*, Cuv. and Val. *Guoraka*, Tel. D. $\frac{12}{13\text{-}14}$, A. $\frac{3}{7\text{-}8}$, L. l. 44, L. tr. 6/11. Diameter of eye 1/3 of length of head: 1 diameter from end of snout. Fourth dorsal spine the longest: second anal spine thick, a little longer than the fifth of the dorsal, striated in grooves along its anterior edge, and one-fourth longer than the third spine of fin. Greyish above: abdomen silvery white: minute brown points on the membrane of the dorsal fin: its edges rather dark. Seas of India and Malay Archipelago, attaining 2 feet in length.

98. *Pristipoma hasta*, Cuv. and Val. *Coroua*, and *Corake*, Tam.: *Coompoo*, Bel. D. 11/$\frac{1}{7\text{-}8}$, A. $\frac{3}{-8}$, L. l. 47, L. tr. 7/10, Cæc. pyl. 6-7. Diameter of eye 1/5 of length of head, 1⅓ diameter from end of snout. Fourth dorsal spine the longest: second anal spine longest and strongest. Four or five lines of dark grey along the sides, and three or four above the lateral line: sometimes they coalesce and form bands. Two or three rows of spots along the dorsal fin, and in the adult a single row of full blotches at the base of the fin. Red Sea, East coast of Africa, Seas of India, Malay Archipelago to North Australia, and attains 1½ feet or more in length. This fish appears to have been described under several different names. *P. kaakan, Commersonii,* and *hasta*, Cuv. and Val., seem to be identical. *Coius gudgutia*, Ham. Buch. pp. 94,370: *Mesoprion gutgutia*,C. V., or *Polotus nitidus,* = *P. gutgutia*, Blyth, and is evidently this species.

99. *Pristipoma maculatum*, Bloch. *Çaripe*, Tel.: *Erruttum corah*, Mal., *Curutche*, Tam. D. 11/$\frac{1}{11\text{-}14}$, A. $\frac{3}{7}$, L. l. 52-56, L. tr. 8/12, Cæc. pyl. 6. Posterior preopercular limb emarginate and serrated. Fourth dorsal spine the longest: second anal spine longer and stronger than the third: caudal emarginate. Greyish, becoming white beneath, and having a purplish tinge about the head: a blackish band over the snout: a second from the occiput touches the posterior edge of the orbit, and descends over the opercles. A vertical black band, about eight scales wide, passes over the nape, and terminates about three scales below the lateral line: posterior to this are six black blotches, three or four above, and two or three below the lateral line, not forming bands, but placed like squares on a chess board. First dorsal with a large black mark in its centre between its fourth and seventh spines: upper edge of both dorsals stained with black, as is also the caudal. From the Red Sea and Eastern Coast of Africa through the Seas of India to the Malay Archipelago and New Guinea: it attains at least 16 inches in length.

100. *Pristipoma argyreum*, Cuv. and Val. D. $\frac{12}{13}$, A. $\frac{3}{7}$, L. l. 45. Eyes half a diameter from end of snout. Angle of preopercle slightly produced, serrated. Third dorsal spine longest, being 2/3 the height of the body: second anal spine strong, as long as the first ray: caudal cut square. Silvery: a dark blotch on the opercle: membrane of first dorsal blackish. Seas of India to the Malay Archipelago.

Genus—DIAGRAMMA, *Cuv.*

Plectorhynchus, (Lacép.) Cantor.

Branchiostegals six or seven: pseudobranchiæ. Body oblong, compressed, with the upper profile of the head parabolic. Eyes of moderate size. *Mouth small, slightly protractile, the lips thick, and folded back. Preopercle serrated: suborbitals entire. Four or six pores on the under surface of the lower jaw, but no median groove. Teeth in jaws villiform without canines. One dorsal fin with from nine to fourteen spines, anal with three: caudal not forked. Scales ctenoid, usually small, but in some species of a moderate size. Air-vessel without any constriction, simple. Pyloric appendages few.*

101. *Diagramma altum*, Day. B. vi, D. $\frac{14}{15\text{-}16}$, A. $\frac{3}{7}$, L. l. 56-59, L. tr. 10/24. Eyes one diameter from end of snout. Third and fourth dorsal spines the longest: second anal spine longer and stronger than the third: caudal rounded. Purplish: tips of dorsal spines, outer third of the dorsal, anal and caudal rays, pure white. Coasts of India, Andamans, and Burma.

102. *Diagramma nigrum*, Cuv. & Val. D. $\frac{14}{17}$, A. $\frac{3}{7}$, L. l. 45-47, L. tr. 10/17, Cæc. pyl. 6. Eyes $1\frac{1}{2}$ diameters from end of snout. Fourth dorsal spine the longest: second anal spine stronger and one-third longer than the last. Greyish, or slate colour, with a brassy tinge on the body, and a violet one on the head. A few irregular coppery spots on the body, and a tinge of the same colour over the hard dorsal: the other fins of a violet slate colour, lightest along their centres. Seas of India, Malay Archipelago to North-West Australia. This fish appears identical with *Pristipoma nigrum*, a term given it by Cuv. & Val. from a Russian drawing by Mertens, brought from Manilla. Cantor thought he recognised the species in the present (his specimen is still in existence in the British Museum), but overlooking it being a *Diagramma*, termed it in 1850, as Cuv. & Val. had in 1830, *Pristipoma nigrum*. In 1859, in the Catalogue of the fishes of the British Museum, it was re-named *Diagramma affine*, Günther. In 1865 in the Proceedings of the Zoological Society, page 14, I observed, on the *D. nigrum*, C. & V., being identical with *D. affine*, Günther: in the Fishes of Zanzibar, by Dr. Günther and Playfair, this name is, however, retained, although it is admitted that Cantor's specimen is a *diagramma*, and identical with this species, but it is still doubted whether it is Cuvier's fish. Anyhow, if not *D. nigrum* (Mertens) C. & V., it certainly should be *D. nigrum*, Cantor, and not *D. affine*, Günther, unless a still older designation is brought to light.

103. *Diagramma Orientale*, Bloch. D. $\frac{}{17\text{-}18}$, A. $\frac{3}{8}$, L. r. 65. A white mark across the snout, a second over the nape, a third near the last dorsal spine, and a fourth round the free portion of the tail. Dorsal fin black with white spots: caudal white with a black edge, and a longitudinal band of the same colour. Ceylon and Coromandel Coast, Malay Archipelago.

104. *Diagramma griseum*, Cuv. & Val. D. $\frac{12}{19\text{-}21}$, A. $\frac{3}{7}$, L. l. 74, Cæc. pyl. 9. Eyes $1\frac{1}{2}$ diameters from end of snout. Third dorsal spine the longest: second anal spine very strong and longer than the third. Uniform grey, with the fins nearly black. East Coast of Africa, Meckran Coast, and throughout the seas of India, attaining at least 18 inches in length.

105. *Diagramma lineatum*, Linn. Gm. B. vi, D. $\frac{12}{20}$, A. $\frac{3}{7}$, L. r. 75. Fourth dorsal spine the longest: caudal rounded. Brownish, with six

longitudinal yellow bands, the first on occiput, the second from the eye to the soft dorsal, the next three to the caudal fin, the sixth from the mouth to the anal. Fins yellow: anal and dorsal with black margins and bases. Caudal with a median and a lateral band on either side. Pectoral with three black blotches: ventral with one. Red Sea, seas of India, to the Malay Archipelago.

106. *Diagramma Blochii*, Cuv. & Val. B. vi, D. $\frac{10}{25}$, A. $\frac{3}{7}$. Second to the fourth dorsal spines equal to $\frac{1}{4}$ the height of the body: the last two anal spines also equal. Orange yellow: a brown band along the base of the dorsal fin: a second from the orbit divides into two, uniting again above the pectoral and continued to the posterior part of the soft dorsal: the third and broadest from the eye to the root of the caudal: the fourth below and parallel: the next two from the muzzle to the end of the abdomen: the seventh from the gill opening to the end of the base of the anal fin. Dorsal edged with black, a black spot between its third and fourth spines: an oblique band along the fins uniting with the first of the body: caudal with black spots. Anal and ventral greyish, edged with black. Pectoral orange, with black bands. Seas of India and Malay Archipelago.

107. *Diagramma punctatum* (Ehrenb.), Cuv. & Val. B. vii, D. $\frac{10}{22\text{-}23}$, A. $\frac{3}{7}$, L. r. 90-105, L. tr. 15/20. Second and third dorsal spines the longest: third anal spine somewhat the longest, the second the strongest. Body and vertical fins with numerous round brown spots: ventrals stained in their outer half. Red Sea, East Coast of Africa, seas of India, Malay Archipelago to China; attaining at least 20 inches in length.

108. *Diagramma pictum*, Thunb. D. $\frac{9\text{-}10}{23}$, A. $\frac{3}{8}$, L. r. 90-100, Cæc. pyl. 5. Second anal spine longest and slightly strongest. Back and sides brown, with about four white longitudinal bands. Spinous dorsal black, but white between the first three spines: rays with a white band: caudal with three black bands: anterior half of anal black, the remainder white: ventrals black in their external half. East Coast of Africa, seas of India, Malay Archipelago, and China.

109. *Diagramma pœcilopterum*, Cuv. & Val. D. $\frac{9}{25}$, A. $\frac{3}{7}$. White, with six or seven longitudinal bands on either side, alternately complete and interrupted. The dorsal and caudal with round or irregularly shaped black spots: ventrals and anal nearly black. Found in Pondicherry during the north-east monsoon. Seas of India, Moluccas, and Japan: attains at least 8 inches in length.

Genus—LOBOTES, Cuv.

Branchiostegals six: pseudobranchiæ. Body and fins somewhat elevated: upper profile of head concave. Eyes rather small. Snout rounded: mouth moderately protractile, its cleft oblique, lower jaw the longer. Opercle with obtuse points: preopercle serrated. Villiform teeth in the jaws, without canines. One dorsal with twelve spines, anal with three. Scales ctenoid, of moderate size. Air-vessel without any constriction, simple.

110. *Lobotes Surinamensis*, Bloch. *Parrandee*, Mal. *Musalli*, Tam. D. $\frac{12}{16\text{-}16}$, A. $\frac{3}{11\text{-}12}$, L. l. 48, L. tr. 8/11, Cæc. pyl. 4. (3), Vert. 13/11. Brassy brown, blotched with darker and having the extremity of the

caudal dirty white. East Coast of Africa, seas of India to the Malay Archipelago : it attains at least 2½ feet in length. This fish has several synonyms : in its immature state it appears to be *L. Farkharii*, Cuv. & Val: in its adult stage, *L. incurvus*, Richardson : also *L. auctorum*, Günther.

Genus—DATNOIDES, Bleeker.

Branchiostegals six : pseudobranchiæ. Body elevated. Eyes of moderate size. Intermaxillaries very protractile. Preopercle serrated: opercle with short spines. Villiform teeth in the jaws, without canines. One deeply notched dorsal fin having twelve stout spines, anal with three : caudal rounded. Scales ctenoid, rather small. Air-vessel simple. Pyloric appendages few.

111. *Datnoides polota*, Ham. Buch. D. $\frac{12}{13}$, A. $\frac{3}{6-9}$, L. l. 70, L. tr. 12/25, Cæc. pyl. 5. Fifth and sixth dorsal spines the longest : second anal longest and strongest. Brownish glossed with copper, having six or seven narrow blackish brown vertical bands on the body, as well as some similar ones radiating from the orbit. From the estuaries of the Ganges to the Malay Archipelago.

Genus—GERRES, Cuv.

Catochœnum, Cantor: *Diapterus* and *Synistius*, Gill.

Branchiostegals six : pseudobranchiæ. Body elevated or oblong, compressed. Mouth very protractile and descending when protruded. Preopercle mostly entire. Eyes rather large. Villiform teeth in the jaws, without canines. Inferior pharyngeal bones firmly united by a suture.([1]) *Length of the base of the spinous and soft portions of the dorsal fin of nearly equal extent, having a scaly sheath into which it may be almost entirely received. The spines nine or ten, the rays ten or eleven: anal with three spines : caudal forked. Scales of moderate size, when ctenoid very slightly so. Air-vessel simple. Pyloric appendages few.*

112. *Gerres setifer*, Ham. Buch. D. $\frac{9-10}{10}$, A. $\frac{3}{7}$, L. l. 38, L. tr. 5/10. Eyes not quite one diameter from the end of snout. Intermaxillary groove scaleless, and extending behind the front edge of the orbit. The third dorsal spine the highest, its height being four-ninths of that of the body. Free portion of the tail higher than long. Silvery, a narrow dark edge to the dorsal interspinous membrane. This appears to be *G. altispinis*, Günther.

113. *Gerres punctatus*, Cuv. & Val. D. $\frac{9}{10}$, A. $\frac{3}{7}$, L. l. 48. Second dorsal spine prolonged to sometimes as much as three-fourths of the height of the body. Silvery, dorsal with a narrow black edge and a brown spot on the middle of each ray. The young with transverse brown bands. Seas of India: attaining at least nine inches in length.

114. *Gerres filamentosus*, Cuv. & Val. *Oodan*, Tam. D. $\frac{9}{10}$, A. $\frac{3}{7}$, L. l. 45-47, L. tr. 6/14. The second dorsal spine prolonged, sometimes extending as far as the caudal fin: second anal spine stronger, but shorter than the third, or half the length of the head. Seas of India, Malay Archipelago to Australia.

([1]) Due to this circumstance the genus (Family *Gerridæ*, Günther) would be amongst the *Acanthopterygii pharyngognathi* when such a sub-order is recognised, a division of which the advisability appears more than doubtful.

115. *Gerres acinaces*, Bleeker. D. $\frac{9}{10}$, A. $\frac{3}{7}$, L. l. 45, L. tr. 6/11. Eyes not quite one diameter from end of snout. Intermaxillary groove scaleless, extending beyond the front edge of the orbit. Second dorsal spine longest, 2/7 of height of body: third anal spine slightly the longest. Silvery, with indistinct and interrupted longitudinal bluish brown bands along the sides. East Coast of Africa, seas of India, Malay Archipelago; attaining at least 9 inches in length.

116. *Gerres limbatus*, Cuv. & Val. D. $\frac{9}{10}$, A. $\frac{3}{7}$, L. l. 37, L. tr. 5/10. Eyes rather more than one diameter from end of snout. Intermaxillary groove scaleless, extending beyond the front edge of the orbit. Second dorsal spine longest, being above half the height of the body. Second anal spine stronger and a little longer than the third. Silvery, caudal with a dark edge: spinous dorsal with a narrow black margin. Seas of India and Malay Archipelago.

117. *Gerres oyena*, Forsk. D. $\frac{9}{10}$, A. $\frac{3}{7}$, L. l. 35-38, Cæc. pyl. 3. Eyes rather more than one diameter from end of snout. The second dorsal spine is half the height of the body and twice as long as the second anal spine which is not longer than the third. Silvery, with interrupted and very indistinct longitudinal spots: caudal posteriorly edged with black. Red Sea, East Coast of Africa, seas of India and Malay Archipelago.

118. *Gerres poeti*, Cuv. & Val. D. $\frac{9}{10}$, A. $\frac{3}{7}$, L. l. 35, L. tr. 5/10. Eyes not quite one diameter from end of snout. Dorsal spines strong and broad, the third the longest, and 4/9 the height of the body: second anal spine very strong, nearly as long as the third. Silvery. Seas of India.

119. *Gerres abbreviatus*, Bleeker. D. $\frac{9}{10}$, A $\frac{3}{7}$, L. l. 33, L. tr. 5/10. Second dorsal spine highest, nearly as long as the head. Second and third anal spines half the height of the body. Silvery, dorsal with a black edge. Andamans and Malay Archipelago.

120. *Gerres lucidus*, Cuv. & Val. D. $\frac{9}{10}$, A. $\frac{3}{7}$. Said to be very similar to *G. limbatus*, but with a shorter body: the dorsal spines less curved. Silvery, with the back a little reddish: caudal without any black border: ventrals yellow. Pondicherry.

121. *Gerres oblongus*, Cuv. & Val. D. $\frac{9}{10}$, A. $\frac{3}{7}$. Height of body less than 1/4 of the total length. Spines weak. Silvery, with five or six indistinct brown bands. Ceylon: to five inches in length.

Genus—SCOLOPSIS, *Cuv. & Val.*

Branchiostegals five: pseudobranchiæ. Body oblong. Eyes of medium size or large. Mouth moderately protrusible, jaws of equal length anteriorly: cleft of mouth horizontal. Infraorbital arch with a spine directed backwards: preopercle as a rule serrated: opercle with a weak spine. A single dorsal fin with ten spines, anal with three: caudal forked. Scales ctenoid. Air-vessel without any constriction, simple. Pyloric appendages few.

122. *Scolopsis bimaculatus*, Rüpp. D. $\frac{10}{9}$, A. $\frac{3}{7}$, L. l. 46, L. tr. 6/14. Slaty brown becoming dull white on the abdomen. A broad white opercular band. Branchiostegal membranes blood red. A brownish band over the snout and two blotches on the lateral line, the first is large being from the 11th to the 22nd scale, the other smaller and behind the posterior extremity of the dorsal. Fins orange. Seas of India and China.

123. *Scolopsis bilineatus*, Bloch. D. $\frac{10}{9}$, A. $\frac{3}{7}$, L. l. 46, L. tr. 4/14, Cæc. pyl. 5. A yellow brown-edged band passes from the mouth to the commencement of the soft dorsal and two or three more exist on the head. A large yellow blotch below the last half of the soft dorsal, which latter fin is edged with black: anal black in its anterior, white in its posterior half. Andamans and Malay Archipelago.

124. *Scolopsis phæops (?)* Bennett. D. $\frac{10}{9}$, A. $\frac{3}{7}$, L. l. 45, L. tr. 5/16. Greenish olive, lightest below. A light band along the base of the dorsal fin. A wide bright blue band from the eye to the angle of the mouth, another to the axil, where it ends in a blue spot. Fins reddish. Sind and the Mauritius.

125. *Scolopsis monogramma* (Kuhl & v. Hass.) Cuv. & Val. D. $\frac{10}{9}$, A. $\frac{3}{7}$, L. l. 44, L. tr. 5/14. Olive with a deep black band, one scale in width, passing through the eye to above the base of the caudal fin, until opposite the end of the dorsal fin it is below the lateral line. Fins immaculate. Andamans, Malay Archipelago.

126. *Scolopsis Japonicus*, Bloch. *Cundul*, Tam. *Kurite*, Tel. D. $\frac{10}{9}$, A. $\frac{3}{7}$, L. l. 44, L. tr. 5/14. Back reddish yellow, the rest of the body slaty. A light buff band over the nape opercles and on to the branchiostegal rays. Opercular spine with a blood red mark having a dark margin: inside of mouth red. Fins dusky yellow. Red Sea, East Coast of Africa, seas of India to the Malay Archipelago: attaining 8 inches or more in length.

127. *Scolopsis cancellatus*, Cuv. & Val. D. $\frac{10}{9}$, A. $\frac{3}{7}$, L. l. 42, L. tr. 3/12, Cæc. pyl. 6, Vert. 10/14. Back marbled with greyish brown: two longitudinal yellowish bands from the upper half of the orbit: a black spot between the first three dorsal spines. Andamans, Malay Archipelago, &c.

128. *Scolopsis auratus*, Mungo Park. *Kundul*, Tam. D. $\frac{10}{9}$, A. $\frac{3}{7}$, L. l. 42, L. tr. 5/12. Of a pale flesh colour, becoming rosy on the fins and head. A whitish streak extends from the upper edge of the opercle to opposite the middle of the soft dorsal: in the anterior half of the body each scale has a dark mark along its centre. Seas of India, and the Malay Archipelago.

129. *Scolopsis ciliatus*, Lacép. D. $\frac{10}{9}$, A. $\frac{3}{7}$, L. l. 40, L. tr. 4/15, Cæc. pyl. 5, Vert. 10/14. Greenish olive above, becoming lighter on the abdomen: a silvery white line extends between the lateral line and the back from near the head to opposite the commencement of the soft dorsal: the scales below the lateral line have each a golden spot. Fins reddish. Andamans, Malay Archipelago, &c.

Genus—DENTEX, *Cuv.*

Synagris, pt., Günther.

Branchiostegals six: pseudobranchiæ. Body oblong, rather elongated and a little elevated. Mouth moderately protractile, its cleft more or less horizontal: jaws of nearly equal length. Preopercle entire, or very indistinctly serrated: distance between the eye and the angle of the mouth considerable. Generally strong canines in both jaws, almost invariably present in the upper. One scaleless dorsal, having from ten to thirteen spines, anal with three: caudal forked. Scales ctenoid, of moderate size, more than three rows on preopercle (Dentex), or only three (Synagris).

clxxvi

Air-vessel not constricted, but posteriorly notched. Pyloric appendages few.

A. *More than three rows of scales on preopercle (Dentex).*

130. *Dentex hasta*, Cuv. & Val. D. $\frac{11}{11}$, A. $\frac{3}{8}$. Six large canine teeth in either jaw. Dorsal and anal spines strong. Greenish brown on the back, becoming silvery on the abdomen: ten to twelve violet lines on the sides: fins blue shaded with violet. Malabar, to 8 inches in length.

B. *Three rows of scales on preopercle (Synagris).*

131. *Dentex furcosus*, Cuv. & Val. D. $\frac{10}{9}$, A. $\frac{3}{7}$, L. l. 48-50, L. tr. 4/14. Six canines in either jaw. Fourth to sixth dorsal spines longest, and about 3/7 of the length of the head: posterior dorsal and anal rays slightly elongated. Rosy with golden reflections: caudal bright red, its lower lobe margined with orange: dorsal rosy and orange. Ceylon, Malay Archipelago to Australia.

132. *Dentex notatus*, Day. D. $\frac{10}{9}$; A. $\frac{3}{7}$, L. l. 48, L. tr. 3/10. Four canines in the upper, six in the lower jaw. Fifth and sixth dorsal spines the longest, nearly 1/3 as long as the head. Rosy, with a brilliant spot on the first five scales below the lateral line, the upper half red, the lower yellow. Five or six longitudinal yellow bands below the lateral line, and three silvery white ones. A broad purplish band below the eye leading to the shoulder mark. A yellow band along the base of the dorsal and anal fins. Andamans.

133. *Dentex grammicus*, Day. *Chungarah*, Tam. D. $\frac{10}{9\text{-}10}$, A. $\frac{3}{7}$, L. l. 48, L. tr. 4/10. Preopercle finely serrated in its lower half. Six canines in the upper jaw. Dorsal spines from the third, continue about the same length, or 1/3 of that of the head. Yellowish red, with longitudinal red lines along each row of scales. Dorsal fins yellowish, with a grey base and pinkish margin. Seas of India.

134. *Dentex luteus*, Bl. Schn. D. $\frac{10}{9}$, A. $\frac{3}{7}$, L. l. 40. Canines, 6 or 8 in the upper jaw. Body red, abdomen silvery. Third ray of the caudal prolonged into a filament. Coromandel Coast, to 7 inches in length.

135. *Dentex filamentosus*, Rüpp. D. $\frac{10}{9}$, A. $\frac{3}{8}$. The posterior dorsal spines the longest, being 1/3 of the height of the body: upper caudal lobe sometimes prolonged. Reddish with longitudinal yellow streaks. Red Sea, Coromandel Coast.

136. *Dentex striatus*, Bloch. D. $\frac{10}{9}$, A. $\frac{3}{8}$. Scales less than 40. Preopercle not serrated. Reddish. Tranquebar.

Genus—SMARIS, *Cuv.*

Branchiostegals six: pseudobranchiæ. Body oblong or cylindrical. Eyes of medium or large size, mouth very protractile. Preopercle entire. Teeth in the jaws, none on the vomer. A single, sometimes deeply notched scaleless dorsal fin, with from nine to fifteen feeble spines, anal with three. Scales ctenoid, rather small. Air-vessel not constricted, but generally forked posteriorly. Pyloric appendages few.

137. *Smaris balteatus*, Cuv. & Val. D. $\frac{12}{10}$, A. $\frac{3}{10}$. Dorsal fin deeply notched: caudal forked. Along the back reddish brown, with small but brilliant silvery spots. A silvery band from the eye to the caudal. Ceylon, to 4 inches.

clxxvii

Genus—CÆSIO (*Comm.*), *Cuv.*

Branchiostegals six or seven: pseudobranchiæ. Body oblong. Mouth moderately protractile, its cleft oblique, lower jaw sometimes slightly the longer. Preopercle entire or minutely serrated. Teeth in the jaws, none on the palate. A single dorsal, more or less scaled, with the anterior portion the higher, and having from nine to thirteen feeble spines: anal with three. Scales ctenoid, of moderate or small size. Air-vessel not constricted, simple. Pyloric appendages few.

138. *Cæsio curing*, Bloch. B. vi., D. $\frac{10}{13}$, A. $\frac{3}{11}$. Of a yellowish colour, stated to have come from the Indies.

139. *Cæsio aurolineatus*, Day. B. vi, D. $\frac{10}{14}$, A. $\frac{3}{9}$, L. l. 72, L. tr. 8/16. Height of body 2/7 of the total length. Third and fourth dorsal spines the longest. Above the lateral line of a light lake colour: from the eye to the base of the caudal a shining golden band: three scales deep anteriorly, decreasing to one posteriorly: below this band pinkish white: caudal fin tipped with black. Madras.

140. *Cæsio cærulaureus*, Lacép. D. $\frac{10}{13}$, A. $\frac{3}{13}$. Height of body 1/5 of total length. Bluish above, with a golden longitudinal band along the lateral line: a black band along the middle of either caudal lobe. A black spot at the axil. Red Sea, seas of India.

Family—SQUAMIPINNES, *Cuv.*

Chætodontidæ, pt. Richardson.[1]

Branchiostegals six or seven: pseudobranchiæ well developed. Body elevated and compressed. Eyes lateral and of moderate size. Mouth generally small, with a lateral cleft, and situated in front of snout. Teeth villiform or setiform: neither incisors nor canines: in most of the genera the palate is edentulous. Soft portion of the dorsal fin of greater extent than the spinous, sometimes considerably more, rarely slightly so: anal with three or four spines, its soft portion similar to that of the dorsal: lower pectoral rays branched: ventrals thoracic, with one spine and five rays. Scales cycloid, or very finely ctenoid, extending to a greater or less extent over the vertical fins, but occasionally absent from the spinous portion. Air-vessel present, generally simple. Intestines usually much convoluted: stomach cæcal. Pyloric appendages in moderate numbers.

Genus—CHÆTODON, *Cuv.*

Branchiostegals six. Body elevated and strongly compressed. Snout of moderate length, or short. Preopercle entire, or slightly serrated and destitute of any spine at the angle. Palate edentulous. Spinous and soft portions of the dorsal fin not separated by a notch, the rays of slightly or considerably larger numbers than the spines, none of the latter elongated: anal with three or four spines. Scales of large, moderate or small size. Lateral line continuous, sometimes incomplete. Air-vessel may be constricted or with horns. Intestines much convoluted.

A. With four anal spines.

141. *Chætodon plebejus*, Gmelin. D. $\frac{14}{17}$, A. $\frac{4}{15}$, L. l. 50. Eye nearly one diameter from end of snout. Preopercle finely serrated. The

---

[1] Genus *Tholichthys*, Günther, is the young state of some fishes of this family which have large suprascapular, humeral, and preopercular plates. They exist in the seas of India, some are immature *Chætodons*, others probably *Holacanthi* or *Heniochi*, whilst in the fry of the *Scatophagus argus*, a bony ridge terminating in a spine passes from the eye above the opercle to the shoulder.

dorsal and anal fins rounded posteriorly. Yellow, the ocular band black, having a white edge. An ocellus at the base of the caudal fin having a white margin. Andaman Islands and the South Seas.

B. *With three anal spines.*

142. *Chætodon vagabundus*, Linn. *Pah-noo-dah*, Andam. D. $\frac{13}{23}$, A. $\frac{3}{20}$, L. l. 30. Eyes, diameter slightly shorter than the snout. Preopercle finely serrated. Soft dorsal and anal fins rounded. A black ocular band descends to the angle of the interopercle. Numerous dark bands pass downwards and forwards to the centre of the depth of the body, when they pass backwards. Dorsal and anal fins margined with black. Two black vertical bands on the caudal, the anterior of which is concave. From the Red Sea through the seas of India to Polynesia.

143. *Chætodon auriga*, Forsk. D. $\frac{12}{23}$, A. $\frac{3}{21}$, L. l. 33. Eyes, 1½ diameters from end of snout. Preopercle finely serrated. Third to fifth dorsal rays elongated. A brown ocular band having a white anterior edge passes through the orbit to over the interopercle. Body with darkish lines, passing upwards and backwards in the anterior third of the body, and downwards and backwards posteriorly. A darkish band proceeds from the base of the soft dorsal across the free portion of the tail and to the lower half of the anal. A dark band goes through the middle of the anal fin. Dorsal anal and caudal edged with white. Red Sea to seas of India.

144. *Chætodon setifer*, Renard. D. $\frac{13}{23-24}$, A. $\frac{3}{20-23}$, L. l. 33. Snout equals 1¼ diameters of the orbit in length. Preopercle finely serrated. Fifth dorsal ray elongated. The ocular band widens after it has passed the orbit and extends to the interopercle. Body with darkish bands passing upwards and backwards in the anterior third of the body, and downwards and backwards posteriorly. A dark ocellus on the centre of the middle dorsal rays: fins dark edged. From the Red Sea through the seas of India to Polynesia.

145. *Chætodon guttatissimus*, Bennett. D. $\frac{13}{22}$, A. $\frac{3}{10}$, L. l. 32. Eyes rather above one diameter from end of snout. Preopercle serrated. Dorsal and anal fins rounded posteriorly. A brown ocular band narrower than the orbit, superiorly edged with yellow, passes from the nape to the lower edge of the interopercle. Body yellow: each scale with a black spot: dorsal and anal purplish brown, edged with yellow, and dotted with black: caudal with a black crescentic mark across its centre. Ventrals yellowish. Ceylon.

146. *Chætodon vittatus*, Bl. Schn. D. $\frac{13}{21-22}$, A. $\frac{3}{20}$, L. l. 36. Eyes 3/4 of a diameter from end of snout. Preopercle finely serrated. Dorsal and anal fins rounded. A dark line passes over the snout, separated by a thin white band from the ocular one, which is half as wide as the orbit, and passes to the chest. Another dark line exists parallel to it, with an intermediate fine light one. Body with about fifteen fine lines passing backwards. A dark band along the base of the dorsal fin, becoming wider over the free portion of the tail: another band along the centre of the soft dorsal. A light yellow band along the base of the anal, with a dark one above it. Fins margined with dark and edged with yellow. A dark vertical band on the posterior third of the caudal. Ventrals white. From the Red Sea to Polynesia.

147. *Chætodon prætextatus*, Cantor. D. $\frac{12}{26\text{-}28}$, A. $\frac{3}{20\text{-}22}$, L. l. 20. Length of head 2/7, height of body 1/2 of the total length. Eyes, 1/2 a diameter from the end of snout. A few fine serrations at the angle of the preopercle. Brownish olive, each scale light citron colour in its centre. A bluish-white band passes from in front of the dorsal fin over the opercles and on to the throat, where it expands: a second across the preorbital and over the cheek to the throat: opposite the orbit it gives off another branch which passes to the angle of the mouth and the throat. Another similar line exists on the forehead, and is lost opposite the anterior edge of the orbit. Dorsal and anal fins tinged with reddish violet, the upper fourth of the soft portion being margined with six coloured bands in the following order from without: white, black, scarlet, black, pearl white, and black: anal tipped with three rows, white, black, and scarlet. Posterior half of caudal pearly white, divided by a black band from a scarlet base. Ventrals black. Cochin and Malay Archipelago.

148. *Chætodon xanthocephalus*, Bennett. D. $\frac{12}{23}$, A. $\frac{3}{17}$, L. l. 38, L. tr. 8/15. Snout longer than the diameter of the eye. Preopercle indistinctly serrated. Dorsal and anal fins rounded. A small dark blotch above the orbit. Body brownish yellow, with five indistinct dark vertical streaks. Dorsal and anal fins blackish, having white edges: caudal and ventrals yellowish. Ceylon.

149. *Chætodon pictus*, Forskal. D. $\frac{12}{23}$, A. $\frac{3}{20\text{-}22}$. Length of head above 2/7, height of body 2/3 of the total length. Eyes, diameter 2/5 of length of head, one diameter from end of snout. Soft dorsal and anal fins angular. Snout with a black band: a dark ocular one descends through the eye over the interopercle to the chest: a third from the two anterior dorsal spines passes downwards at first forwards, then bends back to the base of the ventral fin. Numerous fine dark lines descend downwards and forwards to the middle of the body, where they change their direction to downwards and backwards. Dorsal and anal fins dark, having a black margin and a light external edge: the dark line is continued over the posterior third of the body: there is another dark line over the free portion of the tail, and a dark semilunar mark on the caudal fin. Ventral darkest in its centre. Red Sea, through the seas of India to the Malay Archipelago.

150. *Chætodon biocellatus*, Cuv. & Val. D. $\frac{12}{23\text{-}25}$, A. $\frac{3}{18\text{-}20}$, L. l. 42. Eyes, 3/4 of a diameter from end of snout. Preopercle indistinctly serrated. Dorsal and anal fins rounded posteriorly. A dark ocular band scarcely so wide as the orbit, crosses the occiput, and going through the eye is lost on the cheek, posterior to it is a wider white one. Along the base of the dorsal spines and first half of the rays is a broad brown band, which passes down to the abdomen: another similar band exists at the base of the caudal, with a white one anterior and posterior to it: and another is present posterior to it. Soft portions of dorsal and anal have black tips. A large black ocellus with a white edge is present in the centre of the soft dorsal fin: the other fins white. Andaman Islands to the Malay Archipelago.

151. *Chætodon lunula*, Lacép. D. $\frac{12}{25\text{-}26}$, A. $\frac{3}{19}$, L. l. 33, L. tr. 8/14. Length of head 2/7, height of body 1/2 of the total length.

Eyes, 1¼ diameters from end of snout. Preopercle finely serrated. Ocular band of a deep chestnut colour, rather wider than the orbit: a second brown band passes from the five first dorsal spines, and unites with the occipital one: a third band from the fifth and sixth dorsal spines gradually widens and goes as low as the base of pectoral fin: a band along the base of the soft dorsal passes over the free portion of the tail. Caudal with a dark band in its posterior third: dorsal and anal with a dark edge and white margin. Andamans, seas of India to the Malay Archipelago.

152. *Chætodon oligacanthus*, Bleeker. D. $\frac{6\text{-}7}{29\text{-}30}$, A. $\frac{3}{19\text{-}22}$, L. l. 46-48. Length of head 1/4, height of body 2/3 of the total length. Eyes, diameter 2/5 of length of head, 2/3 of a diameter from end of snout. Lower limb of preopercle minutely serrated, occasionally some serratures on its vertical one. Lateral line ceases opposite the posterior fourth of the dorsal fin. Yellowish white, with five vertical bands: the ocular one is brown and narrower than the orbit: three more similar bands pass from the back to the abdomen: at the upper part of the back is a dark ocellus at the base of the 8th to 12th dorsal rays inclusive: the fifth band is over the free portion of the tail. The posterior half of the ventrals is sometimes black. Ceylon and the Malay Archipelago to the Philipine Islands.

Genus—CHELMO, Cuv.

Branchiostegals six or seven: pseudobranchiæ. Body elevated and compressed. Muzzle produced as a long round tube, the gape of the mouth anteriorly being small. Preopercle without any spine. Teeth on the jaws: none on the palate. One dorsal with from nine to thirteen spines, none being elongated: anal with three. Scales of moderate or small size.

153. *Chelmo longirostris*, Broussonat. D. $\frac{12}{23}$, A. $\frac{3}{8}$, L. l. 75, L. tr. $\frac{12}{30}$. Length of snout much more than ½ that of the head. Yellow: a black triangular patch extends from the base of the first dorsal spine to the snout, its lower margin going through the eye to the opercle: a small round black spot at the posterior angle of the anal fin: caudal grey. East Coast of Africa through the seas of India, to the Malay Archipelago.

154. *Chelmo rostratus*, Linn. D. $\frac{9}{30}$, A. $\frac{3}{27}$, L. l. 47-50, L. tr. 9/20. The anterior margin of the orbit is in the middle of the entire length of the head. Head and body with five orange cross bands edged with brown, and the last two with white: a round black white-edged spot in the middle of the soft dorsal and within the fourth cross band: a dark band round the free portion of the tail. Soft dorsal caudal and anal with blue and white edges. East Coast of Africa, through the seas of India, Malay Archipelago, and China.

Genus—HŒNIOCHUS, Cuv & Val.

Taurichthys, Cuv. and Val.

Branchiostegals five: pseudobranchiæ. Body elevated and strongly compressed: mouth short, or of moderate length. Preopercle finely serrated or entire. Teeth villiform: none on the palate. A single dorsal fin, with from eleven to thirteen spines, the fourth of which is elongated and filiform:

anal with three. Scales ctenoid or cycloid, of moderate size, and more or less covering the vertical fins. Lateral line continuous. Air-vessel present. Pyloric appendages few.

155. *Hæniochus macrolepidotus*, Artedi. *Chuddukun*, Tam.: *Pah-nodak*, Andam. D. $\frac{11-12}{24-25}$, A. $\frac{3}{17-18}$, L. l. 60, L. tr. 9/22, Vert. 10/14, Cæc. pyl. 6. No knob over the orbit. Pearly white, with a dark purplish band over the summit of the snout, another over the eyes: a third broad one extending from the three first dorsal spines and posterior two-thirds of the opercle, passing downwards includes the whole of the ventral fin and extends backwards to the anal: the last commences at the summit of the fifth dorsal spine, passes downwards to the base of the seventh, is as wide as to the first ray, and ends on the posterior third of the anal fin. Pectoral, soft portions of dorsal, anal, and caudal fins bright yellow. East Coast of Africa through the Indian Ocean and Malay Archipelago to Australia: is said to attain 18 inches in length.

Genus—HOLACANTHUS, Lacép.

Branchiostegals six: pseudobranchiæ. Body compressed and as a rule much elevated. Preopercle with a strong spine at its angle. No palatine teeth. A single dorsal fin with from twelve to fifteen spines: anal with three or sometimes four. Scales of moderate or small size, more or less covering the vertical fins. Air-vessel with two horns posteriorly. Pyloric appendages many.

156. *Holacanthus imperator*, Bloch: *Chippee*, Tam. D. $\frac{14}{21-22}$, A. $\frac{3}{20-21}$. Preopercular spine reaches to the base of the pectoral. Body blue, having a greenish tinge along the back: about nineteen canary-coloured lines pass upwards to the dorsal, horizontally to the caudal, or downwards towards the anal fin. A large black descending band with a blue anterior edge, on the shoulder: chest chestnut. A light blue stripe across the snout, round the cheeks and to the preopercular spine which is nearly black. A brown band superiorly edged with blue crosses the eye and passes on to the preopercle. Opercle yellow edged with blue and the branchiostegals black. A narrow black edge to the caudal: ventral dark with orange-coloured rays: pectoral blackish. Scales very small. From the East Coast of Africa through the seas of India, to the Malay Archipelago.

157. *Holacanthus semi-circulatus*, Cuv. and Val. D. $\frac{13-14}{21}$, A. $\frac{3}{20-21}$. Preopercular spine smooth, extending to opposite the posterior edge of the opercle. Scales minute. Reddish brown, with semi-circular light lines alternately white and bluish, the white being the broader. These lines are continued on the head, body, dorsal, anal, and caudal fins. East coast of Africa, seas of India to the Malay Archipelago, &c.

158. *Holacanthus xanthurus*, Bennett. D. $\frac{14}{18}$, A. $\frac{3}{18-19}$, L. l. 46, L. tr. 7/22. Preopercular spine smooth, reaching the base of the pectoral fin. Brownish: caudal yellow. Ceylon and Madras.

159. *Holacanthus diacanthus*, Boddært. D. $\frac{14}{18}$, A. $\frac{3}{17}$, L. l. 50. Preopercular spine reaches the base of the pectoral fin. Yellow, with about eight blue, broadly brown edged vertical body bands continued on to the dorsal fin: anal with bluish streaks running parallel to its base: caudal yellow. Seas of India and Malay Archipelago.

160. *Holacanthus annularis*, Bloch. *Sahni-tschapi*, Tel. D. $\frac{13}{20-21}$, A. $\frac{3}{19-21}$, L. l. 66. Preopercular spine rather flat, and reaching to below the posterior edge of the opercle. Of a brownish green colour, with a blue ring on the shoulder. Two blue streaks from the opercle, the one through, the other below the eye. Body with six or seven arched blue bands radiating from the pectoral: caudal yellow. Seas of India, &c.

161. *Holacanthus striatus*, Ruppell. D. $\frac{12}{23-24}$, A. $\frac{3}{21}$. Preopercular spine smooth (but both in my specimen are bifurcated at their extremities), reaching to below the posterior edge of the opercle. Deep blue, with curved concentric, alternately light and darker bluish transverse lines, the lighter ones being broader: caudal white with a narrow black edge. Red Sea, Bombay, where I procured a single specimen three inches long.

Genus—SCATOPHAGUS, Cuv. & Val.

Cacodoxus, Cantor.

Branchiostegals six: pseudobranchiæ. Body much compressed and elevated: snout of moderate length. Preopercle spineless. Palate edentulous. Two dorsals, united at their base, the first having ten or eleven spines, and anteriorly a recumbent one directed forwards: the soft dorsal covered with scales: anal with four spines. Scales very small. Air-vessel simple. Pyloric appendages rather numerous.

162. *Scatophagus argus*, Linn, *Qu-ee*, Sind: *Chitsilloo* and *Eesputti*, Tel.: *Nutchar char*, Mal.: *Nga-pa-thoom*, Mugh: *Po-ra-dah*, Andam. D. 10/$\frac{1}{16-17}$, A. $\frac{4}{14-15}$, Cæc. pyl. 18 (20). Purplish, becoming white on the abdomen: large round blackish or greenish spots on the body, most numerous along the back. First dorsal brownish-blue, having a few minute spots: second dorsal yellowish, with slight brown markings between the rays. In the very young, a bony ridge, ending in a spine, passes from the eye to above the opercle on to the shoulder. Indian Ocean to China and Australia, attaining a foot in length: it enters backwaters and rivers, but is a foul feeder.

Genus—EPHIPPUS, Cuv.

Ilarches, Cantor.

Branchiostegals six: pseudobranchiæ. Body much compressed and elevated. Snout short, the upper profile parabolic. Preopercle without a spine. No teeth on the palate. Dorsal with eight or nine flexible and elongated spines, receivable into a groove at their base: interspinous membrane deeply cleft, and a deep notch between the spinous and soft portions of the fin: three anal spines: pectoral short. Scales of moderate or small size, some over soft dorsal, anal, and caudal fins. Air-vessel bifurcated anteriorly, and with two long horns posteriorly. Pyloric appendages few.

163. *Ephippus orbis*, Bloch. *Nulla torriti*, Tam.: *Kol-lid-dah*, Andam. D. 8-9/$\frac{1}{19}$, A. $\frac{3}{17}$, L. r. 35, L. tr. 7/15, Cæc. pyl. 4. Back and head greyish green: sides and abdomen silvery shot with pink: fin membranes diaphanous, finely dotted with black, more especially in their marginal halves: rays bluish white. Seas of India and Malay Archipelago, attaining at least 6 inches in length.

Genus—DREPANE, Cuv. & Val.

Harpochirus, Cantor.

Branchiostegals six: pseudobranchiæ. Body elevated and much compressed. Snout short. Preopercle spineless. Palate edentulous. Dorsal having anteriorly a concealed spine directed forwards, and eight or nine spinous rays which, as well as those of the anal, are receivable into a groove at their bases: interspinous membrane deeply notched: pectoral long and falciform. Scales of moderate size. Air-vessel posteriorly prolonged into two horns. Pyloric appendages few.

164. *Drepane punctata*, Linn. *Pun-nur*, Sind.: *Shuk*, Beluch.: *Pündthee*, Mal.: *Pulli* or *Torriti*, Tam.: *Thetti*, Tel. D. $\frac{8-9}{20-21}$, A. $\frac{3}{17-18}$, L. l. 50, L. tr. 14/33, Cæc. pyl. 2 (3). Silvery, with a gloss of gold: fins yellow, their edges being slightly stained with grey. About six dark greyish bands pass downwards from the back, and in them are several round black spots, none of which are below the middle of the body. The older the specimens the less apparent the bands, but usually the more decided are the spots. Sometimes, however, in adults spots are absent. Red Sea, East coast of Africa, seas of India, to the north-west coast of Australia. It attains at least 15 inches in length, and in most places is esteemed as food.

Family—NANDIDÆ, *Günther*.

Pseudochromides pt., et *Mænoidei*, pt. Müll. & Trosch.

Branchiostegals from five to six: pseudobranchiæ present in marine genera, but sometimes concealed or absent in those of the fresh-water. Body oblong and compressed. Teeth feeble, but dentition more or less complete. Dorsal fin single: the length of the base of the spinous portion of greater or equal extent to that of the soft portion: anal with three spines, its rays similar to those of the dorsal: ventrals thoracic, with one spine and four or five rays. Scales ctenoid, covering the body. Lateral line interrupted or absent. No super-branchial organ. Air-vessel present. Pyloric appendages few or absent.

Genus—PLESIOPS, *Cuv.*

Pharopteryx, Rüpp.

Branchiostegals six: pseudobranchiæ. Body oblong and compressed. Mouth moderately protractile. None of the bones of the head serrated. Small teeth on the jaws, vomer, and palatines. Eleven to twelve spines in the dorsal fin: three in the anal: ventral with one spine and four rays, the first being elongated and bifid. Scales cycloid, of moderate size. Lateral line interrupted. Air-vessel present. Pyloric appendages absent.

165. *Plesiops corallicola* (K. & v. H.), Bleeker. D. $\frac{12}{6-7}$, A. $\frac{2}{8}$, L. l. 26-29, L. tr. 8/10. Brownish, each scale with a blue centre. Opercles with a large black ocellus, generally lost after death. Short vertical or oval bluish spots sometimes exist on the caudal and vertical fins. Andamans and Malay Archipelago. This is considered as probably a variety of *P. nigricans*, Rüpp.

Family—MULLIDÆ, *Bleeker*.

Branchiostegals four: pseudobranchiæ. Body rather elongate. Profile of head more or less parabolic. Eyes of moderate size, lateral. Mouth in front of snout rather small, and with a lateral cleft. Two stiff barbels below the chin belonging to the hyal apparatus. Teeth feeble and variously inserted. Two dorsal fins situated at some distance asunder: the anal similar to the second dorsal: ventral with one spine and five rays. Scales large, very feebly ctenoid. Air-vessel, when present, simple. Pyloric appendages few or in moderate numbers.

Genus—*UPENEOIDES*, *Bleeker*.

Upeneus, sp. Cuv.

Definition as in the family, except :—teeth fine in the jaws, vomer, and palatine bones.

166. *Upeneoides vittatus*, Forsk. D. 8/$\frac{1}{8}$, A. 7, L. l. 39, L. tr. 3/6. No air-vessel. Two or three bright yellow longitudinal bands along the sides: first dorsal fin black-edged and with two blackish bands: upper caudal lobe with six yellowish-brown bars with dark edges across it, and a black tip, whilst on the lower lobe there are three oblique dark bars and a white tip. Red Sea, seas of India, and Malay Archipelago.

167. *Upeneoides bivittatus*, Cuv., *Chah-ti-ing-ud-dah*, Andam. D. 8/$\frac{1}{8}$, A. 7, L. l. 36-38, L. tr. 3/5, Cæc. pyl. 11. An air-vessel present. Darkish chestnut along the back, two or three yellow longitudinal bands along the sides: upper caudal lobe with four dark bands, the lower with three: the internal one of each lobe horizontal, the others oblique. Seas of India and Malay Archipelago.

168. *Upeneoides tragula*, Richardson. D. 8/$\frac{1}{8}$, A. 7, L. l. 30, L. tr. 2/5, Cæc. pyl. 6. Vomerine teeth in two lateral stripes. Dorsal fin with dark bands, each caudal lobe with five or six oblique black bands: the other fins, head and body irregularly spotted with brownish. A brown longitudinal band from the snout through the eye to the base of the caudal fin. East Coast of Africa, Andamans, Malay Archipelago.

169. *Upeneoides guttatus*, Day. D. 7/$\frac{1}{8}$, A. $\frac{2}{7}$, L. l. 32, L. tr. 3/5. Vomerine teeth in two lateral stripes. Air-vessel absent. Chestnut along the back, becoming golden on the abdomen: head reddish: a silvery stripe from the eye to the centre of the caudal fin, with a row of red spots above it and another below it. Dorsals tipped with black, and having two reddish bars across them: caudal with four reddish oblique bars on the upper lobe, the lower being destitute of any. Pectorals, ventrals, and anal yellow. Madras. This may be identical with the last species.

170. *Upeneoides cæruleus*, Day. D. 7/$\frac{1}{8}$, A. 7, L. l. 32, L. tr. 3/5. Teeth in a single row on vomer and palate. Air-vessel present. Leaden colour superiorly, becoming dirty white below. First dorsal with a black tip, a whitish band along its centre and a badly marked one at its base. Second dorsal dark with a light band along its centre. Extremities of caudal stained with black and a band across the upper lobe: pectoral, ventral and anal yellowish. Madras, to 4 inches in length.

171. *Upeneoides Bensasi*, Tem. & Schleg. D. 7/$\frac{1}{8}$, A. 7, L. l. 30. Vomerine teeth in an uninterrupted angular band. Rose coloured: dorsal fins with two or three deep red longitudinal bands: upper caudal lobe with four oblique ones. Seas of India and Malay Archipelago.

172. *Upeneoides fasciolatus*, Day. D. 7/$\frac{1}{7}$, A. 2/5, L. l. 36, L. tr. 2$\frac{1}{8}$/6. Teeth on vomer in an uninterrupted semilunar band. No air-vessel. Of a reddish chestnut colour on the back, becoming silvery along the abdomen. A brilliant golden stripe, two-thirds as wide as a scale, passes from the orbit to the upper third of the tail. First dorsal milk white, edged with black, having two horizontal yellow lines finely

dotted with black. Second dorsal with only one band. Caudal reddish, with a black white-edged tip. Madras, to 3 inches in length.

173. *Upeneoides tæniopterus*, Cuv. & Val. D. 7/$\frac{1}{3}$, A. 7, Cæc. pyl. 2. Teeth on the vomer, and in two oval groups on the palatines. Air-vessel large. Back reddish, becoming white on the abdomen. A large triangular reddish spot on the base of the free portion of the tail. First dorsal with three brownish bands, the second dorsal also banded: caudal with six longitudinal and parallel streaks on either lobe, and black at the tip. Ceylon, to 9 inches in length.

Genus—*MULLOIDES*, Bleeker.

Upeneus, sp. Cuv. & Val.

Definition as in the family, except that the teeth in the jaws are in several rows: palate edentulous.

174. *Mulloides flavolineatus*, Lacép. D. 7/$\frac{1}{8}$, A. 8, L. l. 34-36, L. tr. 3/6, Cæc. pyl. 18. No air-vessel. Upper surface of head and back of a reddish chestnut, becoming whitish along the sides, and tinged with yellow on the abdomen. A brilliant golden longitudinal band passes from the posterior margin of the eye to the centre of the caudal fin, it is rather above one scale in width. Fins flesh coloured, the outer edge of the lower caudal lobe grey. Red Sea, seas of India to China.

175. *Mulloides Zelonicus*, Cuv. & Val. D. 7/$\frac{1}{8}$, A. 7. Red: the first dorsal yellowish blended with brown, the second pure yellow with a black line at its base: on both sides of the base of the caudal is a triangular patch of a carmine colour. Ceylon, where it is said to be very common.

Genus—*UPENEUS* (*Cuv. & Val.* sp.), *Bleeker*.

Definition as in the family, except that there is only a single row of teeth in either jaw, and the palate is edentulous.

176. *Upeneus barberinus*, Lacép. D. 8/9, A. 7, L. l. 29-31, L. tr, 2/7, Vert. 10/14. A dark band passes from the eye along the lateral line, ceasing below the end of the soft dorsal: a dark mark at the root of the caudal fin: a black band along the base of the soft dorsal: caudal with a black edging. Red Sea, seas of India to the Malay Archipelago.

177. *Upeneus Indicus*, Shaw. *Rahtée goolivinda*, Tel.: *Mussara*, Tam. D. 8/9, A. $\frac{1}{7}$, L. l. 30, L. tr. 3/7. Red, with a shining golden spot on the lateral line disappearing after death, and a dark mark on either side of the free portion of the tail between the end of the dorsal and the base of the caudal fin. Seas of India and China.

178. *Upeneus Malabaricus*, D. 8/9, A. 7, L. l. 30, L. tr. 3/7. A large whitish oval spot on the lateral line, above the end of the pectoral fin: a dark spot on the side of the root of the tail. Malabar and Philipines.

179. *Upeneus trifasciatus*, Lacép. D. 8/9, A. 7, L. l. 30, L. tr. 3/7, Vert. 10/14. A black vertical band from the anterior portion of the second dorsal fin, a second over the free portion of the tail, the intermediate parts yellow or white. Sometimes another vertical black band passes from the base of the first dorsal. An oblong black spot behind the eye: a black band along the lower half of the second dorsal: the anal with dark bands. Seas of India to Polynesia.

180. *Upeneus displurus*, Playfair. D. 8/9, A. ⅓, L. l. 29, L. tr. 2/7. Red, with a black band over the free portion of the tail, from the base of the dorsal to this there is a light blotch. East Coast of Africa and Kurrachi, attaining at least 11 inches in length.

181. *Upeneus pleurotænia*, Playfair. D. 8/9, A. 7, L. l. 29, L tr. 2½/7 Reddish, scales with dark edges. Two shining light longitudinal bands, the first from the orbit to the middle of the soft dorsal, and the second, which is broader, from the upper lip to the middle of the back. A light mark over the free portion of the tail. First dorsal marbled with brown, the second with four, and the anal with three reddish bands. East Coast of Africa and Kurrachi, attaining at least 8 inches in length.

182. *Upeneus spilurus*, Bleeker. D. 8/⅓, A. 8, L. l. 28. Rose coloured, with a black blotch above the lateral line on the free portion of the tail. Andamans and Malay Archipelago.

183. *Upeneus cinnabarinus*, Cuv. & Val. Of a red vermilion colour: darkest on the back, paler on the abdomen: dorsal and anal rays yellow, the membrane reddish: upper caudal lobe orange, the lower red. A large purple spot covers the opercle and descends on to the subopercle. Barbels rosy. Ceylon, where it is said to be very common.

Family—SPARIDÆ, *Cuv.*

Squamipinnes pt. Cuv.: *Chætodontidæ*, pt. Richards.

Branchiostegals from five to seven: pseudobranchiæ well developed. Body oblong and compressed. Eyes of moderate size, lateral. Mouth in front of snout, having a lateral cleft. Bones of the head with a rudimentary muciferous system. No teeth on the palate (except in Genus *Pimelepterus*): more or less broad and cutting or conical teeth in front of the jaws, or a lateral series of molars, or both conjoined, A single dorsal fin formed by a spinous and soft portion, their bases being of nearly equal extent: anal with three spines: lower pectoral rays generally branched, but not so in some genera: ventrals thoracic, with one spine and five rays. Lateral line continuous, not extending on to the caudal fin. Scales cycloid or minutely ctenoid.

Genus—CRENIDENS, *Cuv. & Val.*

Branchiostegals five: pseudobranchiæ. One or two rows of broad teeth in both jaws, with the cutting edge crenulated: a band of granular teeth posteriorly but no pointed lateral ones: neither molars nor vomerine teeth. A single dorsal with eleven spines which can be received into a groove: three anal spines: lower pectoral rays branched. Scales of moderate size, covering cheeks and opercles, but not the vertical fins. Airvessel simple. Pyloric appendages in small numbers.

184. *Crenidens Indicus*, Day. *Keen-see*, Beluch. D. $\frac{10-11}{10-11}$, A. $\frac{3}{11}$, L. l. 53, L. tr. 7/15, Cæc. pyl. 5. Height of body 2/5 of the total length. Two rows of scales on the cheeks. Silvery, outer edge of scales darkest, also the margin of the vertical fins: pectorals yellow. The *C. Forskalii* is said to have A. ⅔, Cæc. pyl. 3., &c. Kurrachi and Madras, to 12 inches in length.

Genus—SARGUS, (*Klein*) *Cuv.*

Branchiostegals five or six: pseudobranchiæ. Opercles not armed. A single row of cutting teeth in the front part of the jaws, and several lateral rows of rounded molars. A single dorsal with from ten to thirteen spines receivable into a groove along its base: anal with three. Scales of moderate size covering the cheeks. Air-vessel sometimes notched anteriorly and posteriorly. Pyloric appendages few.

185. *Sargus noct* (Ehren.), Cuv. and Val. *Keen-see*, Beluch. D. $\frac{12}{13-14}$, A. $\frac{5}{13}$, Cæc. pyl. 8 (5). On either side of the free portion of the tail a black band. Red Sea to Sind, attaining 12 inches in length.

Genus—*LETHRINUS*, Cuv.

Branchiostegals six: pseudobranchiæ. Cardiform teeth in front of the jaws, as well as canines: lateral teeth in a single row conical or molar-form, sometimes very obtuse. Dorsal with ten spines: anal with three. Scales of moderate size: none on the cheeks. Air-vessel generally notched posteriorly, and with short lateral appendages. Pyloric appendages few.

186. *Lethrinus rostratus* (Kuhl and v. Hass.), Cuv. and Val. D. $\frac{10}{9}$, A. $\frac{3}{8}$, L. l. 50, L. tr. 6/16., Cæc. pyl. 3. Height of body 1/4 of the total length. Eyes 2½ diameters from end of snout. Canines of moderate size. Lateral teeth conical and pointed, only the most posterior ones molariform. Olive, with a black blotch between the pectoral and the lateral line: rays of vertical fins with dark streaks: ventral rays white, membrane blackish. Seas of India and Malay Archipelago.

187. *Lethrinus karwa*, Cuv. and Val. *Karwa*, Tel. D. $\frac{10}{9}$, A. $\frac{3}{8}$, L. l. 46, L. tr. 5/16. Height of body 1/3 of the total length. Eyes 2¼ diameters from end of snout. Teeth as in the last species. Olivaceous brown, becoming lighter on the abdomen: the centre of each scale having a cœrulean blue spot forming lines in the direction of the rows of scales. Pectoral flesh coloured: base of second ray a bright blue. Dorsal and anal slate coloured, margined with orange. Inside of mouth bright orange. Coromandel coast, to 15 inches in length, at least.

188. *Lethrinus harak*, Forsk. *Po-tang-dah*, Andam. D. $\frac{10}{9}$, A. $\frac{3}{8}$, L. l. 47, L. tr. 6/14, Cæc. pyl. 3. Height of body 3/10 of the total length. Eyes 1¾ diameters from end of snout. Canines anteriorly, with distinct molars laterally and posteriorly. Greenish olive, with an oblong lateral blotch of a blackish colour below the lateral line and opposite the middle of the dorsal fin. Red Sea, through the seas of India.

189. *Lethrinus ramak*, Forsk. D. $\frac{10}{9}$, A. $\frac{3}{8}$, L. l. 48, L. tr. 6/15. Height of body 4/17 of the total length. Eyes 2 diameters from the end of snout. Teeth rather small: canines anteriorly, with distinct molars laterally and posteriorly. Olive, with an indistinct longitudinal streak of a shining yellow colour. Red Sea, seas of India.

190. *Lethrinus xanthotænia*. Bleeker. D. $\frac{10}{9}$, A. $\frac{3}{8}$, L. l. 48-50, L. tr. 6/16. Height of body 4/13 of the total length. Eyes 1 diameter from end of snout. Canines small: posterior teeth tubercular. Greenish olive, with five or six yellow longitudinal bands: opercular membrane, dorsal, and caudal fins red, the others yellow. Andamans and Malay Archipelago.

191. *Lethrinus fasciatus*, Cuv. and Val. D. $\frac{10}{9}$, A. $\frac{3}{8}$. Eye small. Cheeks blue, the sides with 7 or 8 longitudinal bands of yellow and blue alternately. Abdomen rosy with five or six blackish stripes. Dorsal and caudal violet: pectorals orange, with a violet spot in the axil: ven-

trals blackish: the interior of the mouth orange. Trincomalee, to 7 inches in length.

192. *Lethrinus frenatus*, Cuv. and Val. D. $\frac{10}{9}$, A. $\frac{3}{8}$. Teeth small. Back greenish, becoming white on the abdomen. Eighteen to twenty yellowish-olive longitudinal bands: above the lateral line are bluish spots. Head olive: before the eyes on the preorbital are three blue or violet oblique lines, another more anteriorly, whilst a fifth is on the suborbitals. Inside of mouth orange. Dorsal fin violet, mixed with orange. Ceylon, to 9 inches.

193. *Lethrinus korely*, Cuv. and Val. Said to be very similar to the last. One of the teeth is a little larger and more rounded. It has only two streaks between the end of mouth and the eye. When fresh it is said to be of a vinous brown colour on the back, with a white abdomen, and the dorsal spotted with red. Pondicherry, to 2 feet in length.

194. *Lethrinus maculatus*, Cuv. and Val. Body more oval than the last: snout a little shorter, and the teeth smaller and more pointed. Superiorly the head and the back are reddish or of a vinous brown colour: below the eye two rows of brown points: on the sides below the lateral line is a black spot, and five or six badly-marked cloudy brownish bands. Pondicherry, to 7 inches.

195. *Lethrinus cinereus*, Cuv. and Val. Is said to have a more elevated body than the four preceding ones, with a shorter snout, and more rounded teeth. Reddish on the back and greyish on the sides and beneath. Fins grey: the caudal with some traces of blackish vertical bands: pectoral yellowish. Ceylon, to 6 inches in length.

196. *Lethrinus geniguttatus*, Cuv. and Val. D. $\frac{10}{9}$, A. $\frac{3}{8}$, L. l. 45, L. tr. 21. Height of body about 3/10 of the total length. Eyes, 1 diameter from the end of the snout. Canine teeth anteriorly, conical pointed ones laterally, with three molarform ones posteriorly. Reddish on the back, becoming silvery on the sides and abdomen: some pearly spots along the back: some white spots on the cheeks, and red spots on the dorsal. Seas of India, to 7 inches.

Genus—SPHÆRODON, *Rüpp.*

Chrysophrys sp., Cuv. and Val.: *Pagrus*, sp. Bleeker.

Branchiostegals six: pseudobranchiæ. Jaws with conical canines anteriorly and a single row of molars laterally. A single dorsal fin, with ten spines receivable into a groove at their base: three anal spines. Scales of moderate size extending on to the cheeks. Pyloric appendages few.

197. *Sphærodon heterodon*, Bleeker. D. $\frac{10}{10}$, A. $\frac{3}{8}$, L. l. 48, L. tr. 5/14. Dorsal spines not elongated. Rose coloured: scales with darker edges: base of pectoral violet. Ceylon and Malay Archipelago.

Genus—PAGRUS, *Cuv.*

Branchiostegals six: pseudobranchiæ. Jaws with an anterior row of conical canines, and laterally two rows of rounded molars. A single dorsal fin with from eleven to twelve, sometimes elongated, spines, receivable into a groove at their base: anal with three. Scales of moderate size, extending on to the cheeks. Air-vessel simple. Pyloric appendages, when present, in small numbers.

clxxxix

198. *Pagrus spinifer*, Forsk. *Soh-ru*, Beluch: *Kooroota*, Tel. *Pununtalai*, Tam. D. $\frac{12}{10}$, A. $\frac{3}{8-9}$, L. l. 53, L. tr. 7/15, Cæc. pyl. 5. Dorsal spines from the third usually flexible and elongated. Whitish, with pinkish bands passing along the centre of every scale becoming rather indistinct below the middle of the height of the body. In specimens up to 4 inches in length there are five vertical bands on the body. Red Sea, East coast of Africa, Seas of India, to the Malay Archipelago.

Genus—CHRYSOPHRYS, *Cuv.*

Branchiostegals six : pseudobranchiæ. Body oblong, compressed. Four to six conical or blunt teeth anteriorly, and three or more rows of rounded molars laterally in either jaw. A single dorsal fin, with from eleven or twelve spines, receivable into a groove at their base: anal with three spines. Scales of moderate size, extending over the cheeks. Air-vessel sometimes notched or with very short appendages. Pyloric appendages few.

199. *Chrysophrys hasta*, Bl., Schn. *Mutti-wyan*, Tam. D. $\frac{11\text{-}11\text{-}12\,(13)}{12\text{-}11\text{-}10\,(9)}$, A. $\frac{3}{8\text{-}9}$, L. l. 42-48 L. tr. 5/9. Eyes, two diameters in the adult from the end of snout. The molar teeth are numerous, being in four or five rows in the upper, and three or four in the lower jaw. Second anal spine longest, very strong, and 2/3 the length of the head. Silvery, scales with dark bases: dorsal fin black tipped, the other vertical fins dark. Coasts of India to China.

200. *Chrysophrys sarba*, Forsk. *Tin-til*, Beluch.: *Chitchillée*, Tel. D. $\frac{11}{12\text{-}13}$, A. $\frac{3}{11}$, L. r. 55, L. tr. 7/14. Eyes 1¾ diameters from end of snout. The molars in four rows in either jaw, with a large oval posterior one. Second and third anal spines of nearly equal length and 4/11 of the length of the head. Silvery, with dark streaks along the rows of scales. From the Red Sea through the seas of India.

201. *Chrysophrys haffara*, Forsk. D. $\frac{11}{11\text{-}(13)}$, A. $\frac{3}{9\text{-}10\,(11)}$, L. l. 48, L. tr. 6/16. Cæc. pyl. 3. Eyes 2 diameters from end of snout. Molars in three rows in the upper and two in the lower jaw, and having an outer row of conical ones, with a large ovate one posteriorly. Second anal spine much the strongest slightly longer than the third, and 2/5 of the length of the head. Greyish, a black edge to the dorsal fin and a dark mark on shoulder. Sind, to 15¼ inches long.

202. *Chrysophrys calamara*, Cuv. & Val. *Dun-de-a*, Sind : *Calamara*, Tel.: *Aree*, Mal.: *Coorrie*, Tam.: *Kala mudwan*, Hind.: *Nga-wah*, Mugh: *Moo-roo-kee-dah*, Andam. D. $\frac{11}{11}$, A. $\frac{3}{8\text{-}9}$, L. r. 35. Eyes 1 to 1½ diameters from end of snout. Four rows of molars in the lower, five in the upper jaw, the outer ones the largest. Second anal spine very strong, about 2/5 of the length of the head. Greyish, scales with dark edges: fins edged with blackish. Indian Ocean to the Malay Archipelago: attaining 12 inches or more in length.

203. *Chrysophrys bifasciata*, Forsk. *Bah-mear*, Beluch. D. $\frac{11}{13}$, A. $\frac{3}{10\text{-}11}$, L. l. 48-50, L. tr. 7/15, Cæc. pyl. 2. Eyes two diameters from end of snout. Molars more numerous in the anterior part of the lateral band, but larger posteriorly. Second anal spine longer and stronger than the third, being 4/11 of the length of the head. Silvery, with two black cross bands, the first through the eye, the second over the posterior edge of

the opercle. Red Sea and seas of India, attaining at least 15 inches in length.

Genus—PIMELEPTERUS (Lacép) Cuv.

Kyphosus (Lacép.), Cuv.

Branchiostegals seven: pseudobranchiæ. Preopercle as a rule serrated. Villiform teeth in the jaws, with an outer row of cutting ones: fine teeth on the vomer, palatines, and tongue. A single dorsal with eleven spines, anal with three. Scales of moderate size, fine ones over the soft portions of the vertical fins. Air-vessel divided posteriorly into two long processes, sometimes notched anteriorly. Pyloric appendages few or very numerous.

204. *Pimelepterus tahmel*, Forsk. Thendala, Tam. D. $\frac{11}{12}$, A. $\frac{3}{1}$, L. l. 60, L. tr. 10/18, Cæc. pyl. short and numerous. Eyes 2/3 of a diameter from end of snout. Spines of dorsal not quite half so high as the rays. Silvery grey, with a dark band between each row of scales. Fins slate colour, nearly black. Air-vessel as described by Cuv. & Val. in *P. Dussumieri*, which appears to be the same species. Red Sea, seas of India.

Family—CIRRHITIDÆ, Gray.

Branchiostegals three, five, or six. Body oblong and compressed. Mouth in front of snout having a lateral cleft. Eyes of moderate size: cheeks not cuirassed. Teeth in the jaws villiform or pointed, sometimes canines as well: vomerine and palatine teeth present or absent. A single dorsal fin composed of spines and rays of nearly equal extent: anal with three spines. Lower pectoral rays simple, and generally thickened: ventrals thoracic, at some distance from the insertion of the pectorals, and having one spine and five rays. Scales cycloid: lateral line continuous. Air-vessel absent, or with many appendages. Pyloric appendages few.

Genus—CIRRHITES (Comm.), Cuv.

Branchiostegals six. Preopercle denticulated: opercle unarmed. Villiform teeth in both jaws: canines also: teeth on the vomer: none on the palatines. A single dorsal fin with ten spines: the lower five to seven pectorals rays unbranched. Scales of moderate size. Air-vessel absent. Pyloric appendages few.

205. *Cirrhites Forsteri*, Bl. Schn. D. $\frac{10}{11}$, A. $\frac{3}{6}$, L. l. 50, L. tr. 5/12, Cæc. pyl. 4, Vert. 10/16. Seven unbranched pectoral rays. Head and chest with black spots: a broad black band from the middle of the body to the upper half of the caudal fin: a wide yellow band from above the pectoral to the lower half of the caudal fin. East Africa, seas of India, &c.

206. *Cirrhites fasciatus*, Cuv. & Val. D. $\frac{10}{9}$, A. $\frac{3}{5}$. Dorsal interspinous membrane very deeply notched: five simple pectoral rays. Greyish, becoming white below. Head, back, and dorsal interspinous membrane with small white spots. Pondicherry.

Genus—CIRRHITICHTHYS, Bleeker.

Branchiostegals six. Preopercle denticulated: opercle spinate. Villiform teeth and canines in the jaws: teeth also on vomer and palatines. A single dorsal fin with ten spines: anal with three: five to seven of the lower pectoral rays unbranched. Scales of moderate size. Air-vessel absent. Pyloric appendages few.

207. *Cirrhitichthys Bleekeri*, Day. *Shun-gun*, Tam. D. $\frac{10}{12}$, P. 8 + VII, A. $\frac{3}{7}$, L. l. 45-46, L. tr. 5/10. Eyes 3/4 of a diameter from end of snout: fifth and sixth dorsal spines the longest. Rosy, with a large badly-defined dark blotch below the soft dorsal, extending half away down the side: a small black dot behind the upper edge of the preopercle: dorsal, caudal, and anal fins more or less banded: soft dorsal darker than the spinous portion. Madras, up to 4 inches.

208. *Cirrhitichthys maculatus*, Lacép. D. $\frac{10}{11}$, P. 7 + VII, A. $\frac{3}{6}$, L. l. 40-42, L. tr. 4/9, Vert. 10/16. Palatine bones with a very small batch of teeth anteriorly. Body and vertical fins spotted with brown: a row of dark spots along the base of the dorsal. Red Sea, seas of India, to the Pacific.

209. *Cirrhitichthys oxycephalus*, Bleeker. D. $\frac{10}{12}$, A. $\frac{3}{6}$, P. 9 + VI, L. l. 40, L. tr. 5/10. Eyes not quite one diameter from end of snout. Fourth and fifth dorsal spines the longest. Rosy, head with two transverse bands. Body with five oblique cross bands and with spots intermediate. Madras and Malay Archipelago.

Family—SCORPÆNIDÆ.

Sclerogenidæ, pt. Owen.

Branchiostegals five to seven: pseudobranchiæ. Body oblong, compressed or subcylindrical. Eyes lateral. Cleft of mouth lateral. Some of the bones of the head armed: suborbital ring articulated with the preopercle. Teeth in villiform bands, occasionally small canines. Two separate dorsal fins or two distinct portions in the fin: the anal usually similar to the soft dorsal: ventrals thoracic. Body scaled or scaleless, sometimes with plate-like scales. Air-vessel not always present. Pyloric appendages when present, few or in moderate numbers.

Genus—SCORPÆNA, Artedi.

Scorpænopsis, Heck.: *Scorpænichthys*, Bleeker.

Branchiostegals seven: pseudobranchiæ. *Head large with a scaleless groove on the occiput, armed with spines and usually with skinny flaps. Villiform teeth on the jaws, vomer, and usually on the palatines. A single dorsal fin deeply notched, dividing the two portions, having twelve spines, and three in the anal which latter fin is not elongated: pectoral large, without free rays. Air-vessel absent. Pyloric appendages few.*

210. *Scorpæna polyprion*, Bleeker. D. $11/\frac{1}{10}$, A. $\frac{3}{5}$, L. l. 47. Interorbital space deeply concave and with ridges that do not terminate in spines: vertex with a shallow groove, broader than long, surrounded with spines: preorbital with a skinny flap: no orbital tentacle. Palatine teeth present. Fourth dorsal spine about 1/3 of length of head and shorter than the second anal. Brown, marbled with darker: axil with or without lighter spots. Ceylon, Malay Archipelago.

211. *Scorpæna rosea*, Day. D. $11/\frac{1}{10}$, A. $\frac{3}{5}$, L. r. 43, L. tr. 7/18. Interorbital space deeply concave, with a ridge on either side that do not terminate in spines. Vertex surrounded with spines: preorbital with a spine and small skinny flap: an orbital tentacle. No palatine teeth. Fourth dorsal spine equals half the length of the head, but shorter than the second anal. Rosy, marbled with greyish: one or two irregular vertical grey bands on the caudal fin: dorsal, anal, and ventral also banded: pectoral with numerous dark spots. Madras.

212. *Scorpæna venosa*, Cuv. & Val., may be either of the foregoing, but the description is too brief: no orbital tentacle was observed however, and no remarks are made as to the dentition.

Genus—SEBASTES, Cuv. & Val.

Branchiostegals seven: pseudobranchiæ. Head and body somewhat compressed. No groove on the occiput, usually a few small spines: preopercle armed. Villiform teeth on the jaws, vomer and usually on the palatines. Fins not elongated: a single dorsal, but the spinous portion separated from the soft by a notch, spines twelve to thirteen: anal not elongated, with three spines: no free rays to the pectoral fin. Scales present and of moderate or small size, extending as far forwards as the orbit or even beyond: no skinny appendages. Air-vessel, as a rule, present. Pyloric appendages few or in moderate numbers.

213. *Sebastes polylepis*, Bleeker. D. 12/$\frac{1}{9}$, A. $\frac{3}{7}$, L. l. 35, diameter of eye from 1/3 to 2/7 of the length of head: no groove beneath it. Preorbital spines obtuse, but prominent ones on the crown of the head. No palatine teeth. Brown, marbled with darker. Malay Archipelago, and said to have been brought from the East Indies.

214. *Sebastes strongensis*, Cuv. & Val. D. 11—12/$\frac{1}{9}$, A. $\frac{3}{7}$, L. l. 45, L. tr. $\frac{5}{17}$. Eyes 3/4 of a diameter from end of the snout: no groove below the orbit. Three strong spines on the preorbital, and one on the shoulder: supraorbital ridge spinate and others exist on the crown of the head. No palatine teeth. Brownish, banded with darker, the first passing downwards through the eye: a large brown spot on opercle: fins irregularly banded in dotted lines. Ceylon, Andamans, Malay Archipelago.

Genus—PTEROIS, Cuv.

Branchiostegals seven: pseudobranchiæ. Head rather large, armed with spines and having skinny flaps: no occipital groove. Villiform teeth in jaws and on vomer: none on the palate. A single dorsal fin, deeply notched, having from twelve to thirteen spines: anal with two or three spines and few rays: rays and sometimes spines elongated: no pectoral appendages. Air-vessel large. Pyloric appendages few.

215. *Pterois kodipungi*, Bleeker. D. 12/$\frac{1}{11-13}$, A. $\frac{3}{7}$, L. l. 70. Eyes, diameter 2/7 of length of head, and one diameter apart. An orbital tentacle of moderate length. Pectorals reach to or beyond the caudal. Reddish, with eleven or twelve dark vertical bands, and intermediate narrow ones in the posterior portion of the body. Fins reddish unspotted: ventrals almost black. Madras, Malay Archipelago.

216. *Pterois muricata*, Cuv. & Val. D. 12/$\frac{1}{11-12}$, A. $\frac{3}{7}$. Eyes, diameter 1/4 of length of head, 1/$\frac{1}{2}$ from end of snout, and 1 apart: a supraorbital filament. Many spines on the head well developed. The ventrals reach beyond the pectorals, which do not extend to the caudal. Scarlet, with many bands across the cheeks and on the body. Pectorals black with white spots: rays of vertical fin spotted with black. Red Sea, Seas of India, to the Malay Archipelago.

217. *Pterois miles*, Bennett. D. 12/$\frac{1}{10}$, A $\frac{3}{7}$. Eyes, diameter nearly 1/4 of length of head, 1/$\frac{1}{4}$ from end of snout and nearly one apart. No supraorbital filament. Many well developed spines on the head, which

along the cheek have numerous spines upon them. Pectoral reaches to the middle of the anal rays. Of a much darker colour than *P. muricata*: the bands more numerous, darker, and broader: lines on the lower jaw curved: two of an S shape over the chest. Dorsal spines annulated with black: rays black spotted, as are also those of the caudal and soft anal: pectoral stained with black: ventral nearly black, with some white spots: a few also over anal spines. Both coasts of Madras and Ceylon.

218. *Pterois volitans*, Linn. *Purrooah*, Mal.: *Cheeb-ta-ta-dah*, Andam. D. $12/\frac{1}{10\text{-}11}$, A. $\frac{2\text{-}3}{7\text{-}6}$, L. r. 90, Cæc. pyl. 3, Vert. 10/14. Eyes 1/¾ diameters from end of snout, and one apart. Pectoral reaches beyond the base of the caudal fin. Red, with vertical bands of reddish brown: spines and rays of the vertical fins spotted: a large white spot in the axil. East coast of Africa, through the seas of India to Australia.

219. *Pterois cincta*, Rüpp D. $11/\frac{1}{11\text{-}12}$, A. $\frac{3}{6}$, L. l. 45, L. tr. 7/25. Eyes one diameter from end of snout, 1/3 of a diameter apart. Orbital tentacle distinct. Pectoral reaches as far as the end of the caudal. Snout uncoloured: a deep brown band edged with white extends from the eye to the angle of the interopercle: the second encircles the neck, and there are six more on the body. A dark band at the base of the pectoral, which is also stained in its outer half. Ventral greyish, its spine white: caudal spotted. Red Sea, Andamans.

Genus—*APISTUS*, Cuv.

Polemius & *Cocotropus*, Kaup.: *Prosopodasys*, Cantor: *Pentaroge* & *Tetraroge*, Günther.

Head large, it and the body compressed. Several bones of the head armed, more especially the preorbital and preopercle. Villiform teeth in the jaws, vomer, and sometimes on the palatine bones. One or two dorsal fins, with from twelve to seventeen spines: anal with three: pectoral elongated, and without or with one filamentous appendage. Scales ctenoid, when present, of moderate size, small, or rudimentary. Air-vessel present. Pyloric appendages few.

This *genus* in its East Indian species has been subdivided as follows:—

a. A cleft behind the fourth gill: one pectoral appendage: body scaled ... ... *Apistus*.
b. A cleft behind the fourth gill: one pectoral appendage: body scaleless ... ... *Minous*.
c. No cleft behind the fourth gill: no pectoral appendage: one dorsal fin ... ... *Tetraroge*.
d. No cleft behind the fourth gill: no pectoral appendage: two dorsal fins ... ... *Prosopodasys*.

A. *A cleft behind the fourth gill: one pectoral appendage: body scaled. (Apistus).*

220. *Apistus alatus*, Cuv. & Val. *Woorrah-minoo*, Tel. D. $\frac{5}{8}$, A. $\frac{3}{7}$, L. r. 70. A long barbel under the symphysis of the lower jaw, and another on either limb of mandible. A notch between the two portions of the dorsal fin. A single pectoral appendage. Air-vessel thick, constricted in the centre. Body greyish along the back, becoming rosy on the abdomen: pectorals deep black: appendage milk-white: dorsal diaphanous, tinged with grey and edged with black, a deep black blotch

from the 8th to the 14th spines: three oblique brownish streaks on the soft dorsal, which also has a brown edging: caudal with four vertical black bands: anal greyish, with a yellow horizontal band. Madras, Andamans, to the Malay Archipelago. *Apistus carinatus*, Bl., appears to be this species, although it is said to have only 13 dorsal spines.

B. *A cleft behind the fourth gill: one pectoral appendage: body scaleless. (Minous.)*

221. *Apistus (minous) monodactylus*, Bl. Schn. Cul-plaachee, Tam. B. vii., D. $\frac{9\text{-}11}{12\text{-}0}$, A. 9-11. No scales. Head greyish-brown above, sides and abdomen lighter or flesh coloured, with dark blotches and marks: dorsal fin light brown margined with black: anal buff, with its outer half black: pectoral black, its appendage white: caudal buff, with three vertical brown bars. Seas of India to China: attaining 4 or 5 inches in length.

C. *No cleft behind the fourth gill: one dorsal fin, no pectoral appendage. (Tetraroge.)*

222. *Apistus (tetraroge) tænianotus*, Lacép. B. v., D. $\frac{17}{8}$, A. $\frac{3}{7}$, Cæc. pyl. 4. Dorsal commences between the eyes, high anteriorly, no notch, and slightly continuous with the caudal. Scales rudimentary. A brown mark between the fifth and sixth dorsal spines. Seas of India and Japan.

223. *Apistus (tetraroge) macracanthus*, Bleeker. B. vii., D. $\frac{15}{9}$, A. $\frac{3}{6}$. Eyes 3/4 of a diameter from end of snout, one diameter apart. Palatine teeth present. Dorsal commences before the eye, is high anteriorly, the first three spines being at some distance from the others, it is slightly joined to the caudal. Scales rudimentary. Brown. Andamans and Malay Archipelago.

224. *Apistus (tetraroge) echinata*, Cantor. B. v., D. $\frac{17}{11}$, A. $\frac{3(2?)}{6(8?)}$. C. 12. Eyes 1½ diameters from end of snout, and one apart. No palatine teeth. The single dorsal commences over the anterior half of the orbit, the first spine being the longest, and no division between the two portions. Buff colour; upper edge of dorsal purple. Some large brownish blotches in the upper half of the body: all the fins more or less dotted with brown: five brown lines radiate from the eyes. Andamans and Pinang.

225. *Apistus (tetraroge?) Belengerii*, Cuv. & Val. D. $\frac{12}{9}$, A. $\frac{3}{7}$. Height of body 1/3 of its length. Dorsal commences opposite the posterior border of the eye, this fin is not united to the caudal. Scales very small. Body grey, finely dotted with brown, except on the abdomen, which is white. A black spot on the dorsal fin, from the fifth to the eighth spine. Malabar, to 2¼ inches in length.

D. *No cleft behind the fourth gill: no pectoral appendage: two dorsal fins. (Prosopodasys.)*

226. *Apistus (prosopodasys) niger*, Cuv. & Val. Pom-tho-cho-roguedah, Andam. D. 3/$\frac{9\text{-}10}{8}$, A $\frac{3}{7}$. Fine teeth on vomer and palate. Eyes one diameter from end of snout. Scaleless, except a few rugosities here and there in the skin. Brownish black: caudal yellowish white striated with brown, having a dark band in its last fourth and a white external edge. Coromandel coast, Andamans, Malay Archipelago.

227. *Apistus (prosopodasys) dracæna*, Cuv. & Val. B. vii., D. 3/$\frac{8}{8}$, A. $\frac{3}{6}$. Fine teeth on vomer and palate. Eyes one diameter from end of

snout. Scales minute. Greyish brown: dorsal and anal fins black: pectoral black in its last half. Malabar, common in South Canara, to 3 inches in length.

Genus—*AMPHIPRIONICHTHYS*, Bleeker.

Branchiostegals four (six). Body compressed. Preorbital, preopercle and opercle armed, the two last small and not covering the gill membranes. Villiform teeth in the jaws, the vomer and palatines edentulous. A single dorsal fin, with a moderately deep cleft, having seven or eight spines, anal with two. Body scaleless.

228. *Amphiprionichthys Zelonicus*, Day. B. iv., D. 7/14, P. 13, A. $\frac{2}{17}$, C. 15. Eyes $1\frac{1}{4}$ diameters from end of snout. A strong preorbital spine: and two on the opercle. Bluish along the upper half of the body, becoming dirty brown on the abdomen. An irregular series of about eight yellow blotches along the back, increasing in number towards the abdomen. Fins light coloured. Dredged off Point de Galle.

Family—TEUTHIDIDÆ, Cuv.

Teuthyes, pt. Cuv.

Branchiostegals five: pseudobranchiæ well developed. Body oval and strongly compressed. Eyes of moderate size, lateral. Mouth slightly cleft, and but little protractile. A single row of cutting incisors in either jaw: palate edentulous. One dorsal fin with the spinous portion more developed than the soft: anal with seven spines. Ventrals thoracic, with two spines and three intermediate soft rays. Scales minute. A complete lateral line, but no armature on the side of the free portion of the tail. Air-vessel present. Pyloric appendages few.

Genus—*TEUTHIS* (1), *Linn*.

Siganus, Forsk.: *Centrogaster*, Houtt.: *Amphacanthus*, Bl. Schn.: *Buro* (Comm.) Lacép.

Branchiostegals five: pseudobranchiæ. Body oval, strongly compressed. Teeth small, denticulated. A single dorsal fin with thirteen spines and a horizontal one anteriorly: anal with three: each ventral with two, an outer and an inner one. Scales minute, cycloid. Air-vessel large, forked both anteriorly and posteriorly. Pyloric appendages five or six.

229. *Teuthis java*, Linn. *Thar-oar-dah*, Andam.: *Worahwah*, Tel. D. $\frac{13}{10}$, A. $\frac{7}{9}$. Height of body 2/5 of the total length. Head, back, and sides of a dark brownish neutral tint, becoming lighter on the abdomen. On the head and back many pale grey rounded spots, becoming more elongated on the sides and abdomen. The upper spots are not so wide as the ground colour. No streaks on the head: fins immaculate. Seas of India, Malay Archipelago, &c.

230. *Teuthis concatenata*, Cuv. & Val. *Thar-oar-dah*, And. D. $\frac{13}{10}$, A. $\frac{7}{9}$, Cæc. pyl. 5-6, Vert. 10/13. Height of body 2/5 of the total length. Dark greyish brown, covered all over with light orange spots, which along the back are larger than the interspaces, but decrease in size towards the abdomen. A blue band extends from below the orbit to the angle of the mouth, and another passes along the preopercle. Andaman Islands and Malay Archipelago.

---

(1). *Worah*, Tam. *Nga-pron-ka*, Mugh.

231. *Teuthis vermiculata* (Kuhl & v. Hass.), Cuv. & Val. *Kut-e-rah*, Mal.: *Chow-lud-dah*, Andam. D. $\frac{5}{10}$, A. $\frac{7}{5}$. Height of body from 1/2 to 4/9 of the total length. Light brown, running into bluish green on the back, and nearly white on the abdomen. The whole of the body, head, and lips lineated with undulating bluish lines of about one-fourth the width of the ground colour, being broadest nearest the abdomen: caudal fin with brown lines. Seas of India, Malay Archipelago, &c. attaining at least 11 inches in length.

232. *Teuthis sutor*, Cuv. & Val. D. $\frac{5}{10}$, A. $\frac{7}{5}$. Height of body not quite 1/3 of its length. Greenish grey, spotted with dull blue. Seychelles and coast of Malabar.

233. *Teuthis albopunctata*, Tem. & Schleg. D. $\frac{5}{10}$, A. $\frac{7}{5}$. Height of body 2/7 of the total length. Brownish olive, with small scattered white (? blue) spots on the back and sides, much smaller than the interspaces. A brown shoulder mark. Andamans, Malay Archipelago to China.

234. *Teuthis stellata*, Forsk. D. $\frac{5}{10}$, A. $\frac{7}{5}$. Height of body nearly or quite 1/3 of the total length. Caudal fin forked. Greyish, covered all over with small angular spots of a purplish brown colour (a specimen in the Calcutta Museum, apparently of this species, from the Red Sea, has a fine white upper edge and dark lower one to each spot). Dorsal and anal also spotted with brown: the border of the soft rays and caudal yellowish: a greenish yellow spot in front of the dorsal fin. Red Sea, seas of India.

235. *Teuthis marmorata*, Quoy. & Gaim. D. $\frac{13}{10}$, A. $\frac{7}{5}$. Height of body 1/3 of its length. Body bluish superiorly, becoming white below: almost entirely covered by violet lines which are wider than the interspaces: those on the back form marblings and reticulations, whilst on the sides they are more longitudinal. Seas of India and Malay Archipelago.

236. *Teuthis Russellii*, Cuv. & Val. *Worahwah*, Tel. D. $\frac{13}{10}$, A. $\frac{7}{5}$. Scales stated not to be visible. White, darkish on the chest and abdomen: blackish, marbled with black on the rest of the body. Coromandel Coast.

237. *Teuthis virgata*, Cuv. & Val. *Tah-meer-dah*, Andam. D. $\frac{13}{10}$, A. $\frac{7}{5}$, Cæc. pyl. 4. Height of body 2/5 to 3/7 of the total length. Upper two-thirds of body coppery yellow, covered with round blue spots, and having blue lines on the head. A brown band, as wide as the orbit, extends from before the dorsal fin through the eye to below the jaws: a second from the sixth and seventh dorsal spines to the base of the pectoral, both these bands are edged with blue. Fins yellowish. Andamans, Malay Archipelago, China, &c.

Family—BERYCIDÆ, *Lowe*.

Branchiostegals from four to eight. Form of body oblong, or rather elevated and compressed. Opercles more or less armed. Head with large muciferous cavities. Eyes large, lateral. Cleft of mouth more or less oblique, extending to the sides of the muzzle. Teeth villiform in both jaws, and usually so on the palate. Dorsal fin, when single, having the spinous portion of less extent than the soft, or with isolated spines in front of the fin: or there may be two dorsals, the first being spinous. Ventrals thoracic, each with either less or more than five soft rays. Scales ctenoid, seldom bony or absent: none on the head. Pyloric appendages numerous or in moderate numbers.

cxcvii

Genus—*MYRIPRISTIS, Cuv.*

Branchiostegals eight, rarely seven. Eyes large, mostly lateral. Muzzle short, lower jaw prominent. Cleft of mouth oblique, in one species horizontal. Opercular pieces serrated: opercle generally with one spine: none on the preopercle. Teeth villiform on jaws, vomer, and palatines. Two dorsal fins, scarcely united: ventral with one spine and seven rays: anal with four spines: caudal forked. Scales large, ctenoid. Air-vessel transversely contracted near its centre. Pyloric appendages in moderate numbers.

238. *Myripristis kuntee*, Cuv. & Val. Sullanaroo kuntee, Tel. B. VIII, D. $10/\frac{1}{16}$, A. $\frac{4}{13}$, L. l. 28, L. tr. $3\frac{1}{2}/7\frac{1}{2}$. Rose coloured with scarlet fins: gill-opening deep blackish brown. Dorsal, caudal and anal fins with white margins, sometimes a deep black mark exists upon the anterior five dorsal rays extending to one-third of their height: another on the anal from the second to the fifth: occasionally the caudal lobes are tipped with black. Coasts of India and the Andamans: attaining at least 11 inches in length.

239. *Myripristis murdjan*, Forsk, Botche, Tel. B. VIII, D. $10/\frac{1}{13-14}$, A. $\frac{4}{13}$, L. l. 30, L. tr. 3/5, Vert. 10/15. Head scarlet, with a black blotch, having the appearance of congealed blood, behind the opercle. Body pink, each scale having the edges deeper coloured. Fins scarlet, with the outer ventral, anal and caudal ray milk-white. Tips of soft dorsal and anal black. Red Sea, seas of India to the Malay Archipelago.

Genus—*HOLOCENTRUM, Artedi.*

Corniger, Agassiz.

Branchiostegals eight. Eyes large lateral. Jaws of equal length, or the lower slightly the longer: snout of moderate length. Opercles and suborbitals serrated: opercle with two spines: generally a large spine at the angle of the preopercle. Villiform teeth on the jaws, vomer and palatines. Two dorsal fins scarcely united: ventral with one spine and seven rays: anal with four spines, the third being long and strong: caudal forked. Scales ctenoid, of moderate size. Air-vessel oval and simple. Pyloric appendages numerous.

240. *Holocentrum caudimaculatum*, Forsk. B. VIII, D. 11/14, A. $\frac{4}{10}$. Height of body 5/10 of total length. The lower opercular spine is the smaller: sub-opercle entire. Third anal spine strongest and longest, equalling 2/9 of the total length. Uniform red, with a white spot on the back of the tail. Red Sea and Ceylon.

241. *Holocentrum diadema*, Lacép. D. $11/\frac{1}{13}$, A. $\frac{4}{8}$, L. l. 48, L. tr. 3/7, Vert. 11/16. Height of body 5/18 of the total length. Opercle with two spines, the upper nearly equal to that of the preopercle. Third anal spine very strong and equalling about 1/5 of the total length. Red, with from eight to eleven longitudinal silvery bands: spinous dorsal black with a white longitudinal band: the other fins rosy. Red Sea, seas of India to China.

242. *Holocentrum Andamanense*, Day. D. 11/14, A. $\frac{4}{8}$, L. l. 42, L. tr. $3/7\frac{1}{2}$. Height of body 4/13 of the total length. Opercle with two flat spines and its lower edge serrated. Third anal spine strong and

equal to 1/6 of the total length. Uniform rosy scarlet. Andamans 8 inches in length.

243. *Holocentrum rubrum*, Forsk. *Cul-kah-catchee*, Tam. D. 11/12-13, A. ⅜, L. l. 35-37, L. tr. 3/6, Cæc. pyl. 20, Vert. 11/16. Height of body 1/3 of the total length. Third anal spine from 2/9 to 1/5 of the total length. Silvery white, with longitudinal dull rosy bands from the opercles, the second and third coalesce, as do also the fourth and fifth, opposite the end of the dorsal fin. Head more or less rosy, as are also the fins: anal spines white. Having examined a male it had no black blotch at the bases of the soft dorsal, anal, and the axil of the pectoral. Red Sea, seas of India to the Malay Archipelago.

244. *Holocentrum sammara*, Forsk. D. $10/\frac{1}{11-12}$, A. $\frac{4}{7-8}$, L. l. 40, L. tr. 4/7, Vert. 11/16. Height of body 1/4 of the total length. Third anal spine 3/14 of the total length. Spinous dorsal with a black mark between the first and fourth spines, and two rows of white spots, one along the base, the other along its upper edge. Red Sea, seas of India, Malay Archipelago.

Genus—R*HYCHICHTHYS* (1), *Cuv. & Val.*

Branchiostegals eight : snout more or less elongate, pointed, with the upper jaw prominent. Opercular bones serrated: opercle and angle of preopercle with distinct spines. Villiform teeth on the jaws, vomer, and palate. Two dorsal fins scarcely united: ventral with seven rays: anal with four spines, the third the longest: caudal forked. Scales ctenoid and of moderate size.

245. *Rhynchichthys ornatus*, Day. D. 12/⅛? A. ⅘, L l., 36, L. tr. 3½/6. Bluish silvery along the back and sides, rosy on the abdomen : fins yellowish : dorsal orange, with black spines: interspinous membrane between the first three spines and also between the sixth and last of a deep black colour.

Family—K*URTIDÆ*.

Branchiostegals seven : pseudobranchiæ absent. Body oblong and compressed. Eyes large. The infraorbital bones do not articulate with the preopercle. Cleft of mouth oblique : lower jaw prominent. Villiform teeth on jaws, vomer and palatines. A single dorsal fin, the spinous portion being of less extent than the soft : some spines may even be rudimentary : anal elongated, with two or three spines : ventrals thoracic with one spine and five rays. Scales of moderate or small size. Air-vessel present. Pyloric appendages few.

Genus—K*URTUS, Bloch.*

Branchiostegals seven: pseudobranchiæ absent. Body oblong and strongly compressed: back elevated. Cleft of mouth oblique and deep, the lower jaw prominent. Preopercle denticulated. Villiform teeth in the jaws, vomer, and palatines. A single dorsal fin of much shorter extent than the anal, its spines being rudimentary : between the ventrals is a horizontal backwards directed spine. Scales very small. Air-vessel present, enclosed

---

(1) Dr. Güuther observes in " Annals and Mag. of Nat. Hist," November 1871, that "I think before long Rhynchichthys will be shown to be the young of Holocentrum." This supposition is most probably correct. The elongated snout reminds one of the identical form in some of the immature Indian siluroids.

(2) My specimens being in Europe, I am unable to re-examine the number of the dorsal rays.

in a conical cavity made by the ribs, which are dilated, convex, and forming rings in contact with each other.

246. *Kurtus Indicus*, Bloch. *Kakasi*, Tel. : *Oordah, and Valliaul cutchul*, Tam. D. $\frac{6-7}{13-14}$, A. $\frac{2}{31-35}$, Vert. 8/15. Height of body 1/3 of the total length. Preopercle with three strong backward directed spines on its lower edge. Males have a cartilaginous arched process directed forward and somewhat downwards. Silvery, shot with steel blue or lilac reflections: back with fine black dots which behind the occiput form a rounded black spot. Seas of India, Malay Archipelago, and China.

Genus—PEMPHERIS, *Cuv. & Val.*

Branchiostegals seven. Body oblong, compressed: head obtuse. Eyes large. Cleft of mouth oblique, with the lower jaw prominent. Opercle with a small spine. Villiform teeth on the jaws, vomer, and palatine bones. A single short dorsal fin with six spines and nine rays: anal with three spines and many rays. Scales small, extended over the anal fin. Air-vessel divided into an anterior and posterior portion. Pyloric appendages few.

247. *Pempheris Otaitensis*, Cuv. & Val. D. $\frac{6}{9}$, A. $\frac{3}{40\,(42)}$, L. l. 55-65, Cæc. pyl. 6—7, Vert. 10/14. Height of body 2/5, of caudal 1/5, length of head 1/4 of the total length. Eyes, diameter rather above half the length of head. Preopercle entire. Rows of scales along the lateral line larger than the others. Silvery grey, upper third of dorsal rays black: pectoral orange, with a dark base: anal rather dark anteriorly and inferiorly: caudal with its posterior margin stained grey. Beluchistan, Sind, Malay Archipelago.

248. *Pempheris Molucca*, Cuv. & Val. D. $\frac{6}{9}$, A. $\frac{3}{43-46}$, L. l. 60. Height of body 1/3, length of head 3/13 of the total length. Eyes, diameter rather less than 1/2 the length of the head. Violet brown superiorly, becoming silvery below: fins rose coloured, with fine dots: axil and base of pectoral white. Seas of India and Japan.

249. *Pempheris mangula*, Cuv. & Val. *Mangula kutti*, Tel. D. $\frac{6}{9}$, A. $\frac{3}{38-40}$, L. l. 45. Height of body 1/3 of the total length. Brown with minute dark points : dorsal black anteriorly : anal and caudal with black edges : pectoral without any dark mark at its base. Coromandel coast, Malay Archipelago, &c.

Family—POLYNEMIDÆ, *Richards.*

Branchiostegals seven : pseudobranchiæ. Body oblong, somewhat compressed. Eyes large, lateral, more or less covered by an adipose membrane. Mouth on the lower side of a prominent snout, and having a lateral cleft. Muciferous system on the head well developed. Villiform teeth on the jaws and palatines, present or absent on the vomer. Two dorsal fins : several free and articulated appendages below the pectoral fin : ventrals thoracic, with one spine and five rays. Scales finely ctenoid or cycloid, and more or less covering the vertical fins. Lateral line continuous, continued on to the caudal fin. Air-vessel, when present, varying in form and structure. Pyloric appendages of varying numbers.

Genus—*POLYNEMUS*, (¹) *Linn.*

Preopercle serrated. Teeth villiform on the jaws, palatines, vomer and pterygoid bones. First dorsal fin with seven or eight weak spines: soft dorsal and anal of nearly equal extent. Scales rather small, extended on to the vertical fins. Air-vessel, when present, varying in form, size and structure. Pyloric appendages few, in moderate numbers, or many.

250. *Polynemus paradiseus*, Linn. D. 7/ $\frac{1}{13}$, P. 15+VII, A. $\frac{2}{12}$, L. l. 70, L. tr. 5/14. Cæc. pyl. 5 (10), Vert. 10/15. The seven free rays below the pectoral fin are longer than the fish: the three upper ones being frequently twice as long. No air-vessel. Colours golden. Seas of India, entering rivers as far as the tide reaches.

251. *Polynemus heptadactylus*, Cuv. & Val. D. 7/ $\frac{1}{13}$, P. 15+VII, A. $\frac{3}{14}$, L. l. 50, L. tr. 5/11. Cæc. pyl. 4. Of the seven pectoral appendages, the two highest and the lowest are of equal length and shorter than the remaining four, which reach to the third or fourth anal ray. No air-vessel. Silvery, with a yellowish tinge: pectorals deep black. Seas of India and Malay Archipelago, to at least 6 inches in length.

252. *Polynemus xanthonemus*, Cuv. & Val. D. 8 / $\frac{1}{11}$, P. 15+VI, A. $\frac{3}{11-12}$. Cæc. pyl. 12. The six pectoral appendages just reach beyond the end of the ventrals. No air-vessel. Silvery, pectoral black: other fins yellow, edged with black: end of free rays also black tipped. Seas of India: attaining 6 inches at least in length.

253. *Polynemus sextarius*, Bloch. D. 8/ $\frac{1}{12-13}$, P. 15+VI, A. $\frac{3}{12-13}$, L. l. 48. Cæc. pyl. long and rather numerous. The six pectoral appendages reach slightly beyond the end of the ventrals. Air-vessel of moderate size, simple. Sometimes a black spot on the shoulder: pectorals black. Seas of India and Malay Archipelago: to 6 inches in length.

254. *Polynemus Indicus*, Shaw. *Maga-boshe*, Tel.: *Tahlun kalah*, Tam.: *Yeta*, Mal. D. 8 / $\frac{1}{13-14}$, P. 20+V, A. $\frac{2-3}{11-12}$, L. l. 62-70, L. tr. 7/13, Vert. 5/19. Cæc. pylori numerous. The upper of the five pectoral filaments is the longest, reaching as far as the commencement of the anal fin, whilst the inferior or shortest one extends as far as the end of the pectoral. Air-vessel present, having many appendages. Back greyish green: abdomen silvery white: dorsal and caudal edged with fine black points. Seas of India, Malay Archipelago to Australia: attaining 3 feet or more in length.

255. *Polynemus plebejus*, Gm. D. 8/ $\frac{1}{13-14}$, P. 18+V, A. $\frac{2}{12-13}$, L. l. 50, L. tr. 5/10, Cæc. pyl. numerous. Of the five pectoral filaments some extend beyond the end of the fin. Air-vessel simple. Silvery in the young: a dark blotch near the commencement of the lateral line. Sind, through the seas of India, Malay Archipelago, &c.

256. *Polynemus tetradactylus*, Shaw. *Maga-jellee*, Tel.: *Polunkalah*, Tam.: *To-bro-dah*, Andam. D. 8 / $\frac{1}{13-15}$, P. 16+IV, A. $\frac{2}{15-17}$, L. l. 75-85, L. tr. 10/14, Cæc. pyl. numerous. No air-vessel. Silvery, a dark mark on the opercle. Seas of India, Malay Archipelago, &c.: attaining 6 feet or more in length.

(¹) *Kalah*, Tam. *Nut-tiah*, Mugh.

Family—SCIÆNIDÆ, *Cuv.*

Branchiostegals seven : pseudobranchiæ sometimes concealed, or even absent. Body somewhat compressed and rather elongate. Eyes lateral, of moderate or small size. Mouth in front of or below the snout. Cheeks unarmed: opercles sometimes with weak spines. Barbels present in a few genera. Muciferous system on the head well developed. Teeth in villiform bands: canines present in some genera but neither cutting nor molar ones in the jaws: palate edentulous. Two dorsal fins, the second much more developed than the first or than the anal: spines of first dorsal usually feeble : anal with two spines : pectoral rays branched : ventrals thoracic with one spine and five rays. Scales ctenoid. Lateral line complete, often continued on to the caudal fin. Stomach cæcal. Air-vessel, when present, as a rule with branching or elongated appendages. Pyloric appendages generally few.

Genus—UMBRINA, *Cuv.*

Branchiostegals seven: pseudobranchiæ. Body oblong. The upper jaw overlapping the lower. Preopercle, as a rule, serrated. A short barbel under the chin. Villiform teeth in the jaws. Two dorsal fins, the first with from nine to ten flexible spines: the anal with one or two. Scales of moderate size. Air-vessel with appendages.* Pyloric appendages in small numbers.

257. *Umbrina macroptera*, Bleeker. D. $10/\frac{1}{29\text{-}30}$, A. $\frac{2}{7}$, L. l. 49, L. tr. 6/15, Cæc. pyl. 11. Length of head 1/4 of the total and equal to the height of the body. Diameter of eye 1/5 of length of head, nearly 2 diameters from end of snout. Preopercle serrated. Barbels not quite 1/2 as long as the orbit. Pectoral and ventral fins of equal length : caudal rhomboidal. Greyish, becoming silvery below : fins yellowish : minute black points everywhere. Madras, Malay Archipelago.

258. *Umbrina Russellii*, Cuv. and Val. *Qualar-katchelée*, Tel. D. $10/\frac{1\text{-}2}{26\text{-}27}$, A. $\frac{2}{7\text{-}8}$, L. l. 52, L. tr. 6/13, Cæc. pyl. 7. Length of head 1/4, height of body 2/7 of the total length. Eyes rather more than 1 diameter from end of snout. Preopercle crenulated. Barbel slightly above 1/2 a diameter of the orbit in length. Caudal rhomboidal. Dark nearly blackish brown and shot with gold. Fine dark points all over the head and body : fins yellowish, also stained with dark spots especially the first dorsal and base of the pectoral. Seas of India and Malay Archipelago, attaining at least 10 inches in length.

259. *Umbrina Dussumieri*, Cuv. and Val. D. $10/\frac{1}{24\text{-}25}$, A. $\frac{2}{7}$, L. l. 51, L. tr. 7/19. Length of head 4/15, height of body 5/19 of the total length. Eyes much less than 1 diameter from end of snout. Preopercle entire. Barbels short. Back fawn colour with greenish-golden reflections : abdomen silvery : fins of a reddish tinge, except the ventral which is yellow. Coromandel Coast : attaining at least 6 inches in length.

Genus—*Sciæna (Artedi)*, *Cuv.*

Johnius, Bloch : *Corvina*, pt. Cuv : *Leoiostomus*, pt. Cuv. and Val : *Homoprion*, Holb.

Branchiostegals seven : pseudobranchiæ. Body compressed and rather elongate. Eyes of moderate size, with the interorbital space rather broad and

---

* Said to be absent in American species.

slightly convex. Snout rounded: cleft of mouth horizontal or slightly oblique: the upper jaw generally longer than the lower, or both may be of equal length. Barbels absent. The outer row of teeth usually the largest: no canines. Air-vessel present. Pyloric appendages few, or in moderate numbers.

This genus has been artificially sub-divided, in accordance with the length of the second anal spine, thus :—

    a. Weak, or about half the length of the first ray   *Sciæna.*
    b. Moderately strong, nearly two-thirds   ,,   ,,   *Johnius.*
    c. Very strong, about equal to   ...   ,,   ,,   *Corvina.*

(*a.*) *Second anal spine moderately strong nor much above 2/3 the length of the first ray,*—(*Johnius.*)

260. *Sciæna Dussumieri,* Cuv. and Val. D. 9-10,/$\frac{1}{28\text{-}29}$, A. $\frac{2}{7}$, L. l. 52, L. tr. 6/15, Cæc. pyl. 9, Vert. 10/14. Eyes diameter 1/4 of length of head, 1 diameter from end of snout. Upper jaw the longer. Preopercle finely serrated. Second anal spine moderately strong, 3/5 of the length of the first ray.. Greyish purple, with a golden gloss, and a dark mark on the opercle: vertical fins with dark edges. Seas of India to the Malay Archipelago.

261. *Sciæna sina,* Cuv. and Val. D. 10/$\frac{1}{27\text{-}28}$, A. $\frac{2}{7\text{-}8}$, L. l. 46. Eyes not quite 1 diameter from end of snout, rather more apart. Jaws of about equal length. Preopercle serrated, especially at its angle. Second anal spine 2/5 of the length of the head. Greyish, becoming white on the abdomen: a dark blotch behind the opercle, extending to the axil. First dorsal fin very dark. Seas of India to the Malay Archipelago: said to attain a foot in length.

262. *Sciæna maculata,* Bl. Schn. *Taan-tah,* Bel.: *Sari kullah,* Tel.: *Cooroowa,* and *Vari katchelee,* Tam. D. 10/$\frac{1}{23\text{-}24}$, A. $\frac{2}{7\text{-}8}$, L. l. 45-48, L. tr. 11/16, Cæc. pyl. 8. Eyes 1¼ diameters from end of snout. Upper jaw the longer, overlapping the lower. Preopercle with about six strong denticulations near its angle. Second anal spine of moderate strength, 2/3 as long as the first ray. Air-vessel with about 15 lateral processes on either side, each having two or three roots. Silvery, with five broad black bands passing downwards from the back, these are sometimes interrupted. Seas of India, attaining at least 10 inches in length.

263. *Sciæna diacanthus,* Lacép. *Nella* or *Cora katchelee,* Tel. D. 10/$\frac{1}{22\text{-}24}$, A. $\frac{2}{7}$, L. l. 52, L. tr. 7/18, Cæc. pyl. 8. Eyes 1¼ diameters from end of snout. Upper jaw the longer. A few badly marked serrations at the angle of the preopercle. Second anal spine of moderate strength, and at least 1/2 as long as the first ray. Brownish-grey in the adult, but slaty-grey in the young, with black blotches on its sides and fins: the immature are usually banded. Seas of India to China, attaining 5 feet in length.

(*b.*) *Second anal spine very strong, and about equal to the length of the first ray,*—(*Corvina.*)

264. *Sciæna semiluctuosa,* Cuv. and Val. *Suk-kun,* Beluch. D. 10/$\frac{1}{28(31)}$, A. $\frac{2}{7}$, L. r. 45, Cæc. pyl. 8. Eyes, 1¼ diameters from end

of snout. Upper jaw the longer. Margin of preopercle scarcely crenulated. Second anal spine nearly 1/4 shorter than the first ray. Deep grey, with a blackish band running along the centre of each row of scales: head tinged with purple: fins deep black, becoming lighter after death. Coasts of Beluchistan and India, attaining at least 8 inches in length.

265. *Sciæna lobata*, Cuv. and Val. D. 9-10/$\frac{1}{28\text{-}31}$, A. $\frac{2}{7}$ Cæc. pyl. 8. Eyes about 1 diameter from end of snout. Upper jaw overlapping the lower. Preopercle not serrated. Second anal spine moderately strong, but nearly 1/4 shorter than the first ray. Greyish, with five wedge-shaped or triangular crossbands descending along the upper half of the body: fins with dark margins. Malabar.

266. *Sciæna miles*, Cuv. and Val. *Tella katchelee*, Tel. D. 9-10/$\frac{1}{29\text{-}30}$, A. $\frac{2}{7}$. Eyes 1 diameter from end of snout. Upper jaw overlapping the lower. Preopercle feebly serrated. Second anal spine very strong, as long as the first ray. Greyish, dashed with green along the back, becoming white on the sides and abdomen. A small brownish spot in front of each dorsal ray: the external margins of all the fins nearly black, except the ventral, which is whitish. Seas of India to the Malay Archipelago, attaining at least 2 feet in length.

267. *Sciæna Belengerii*, Cuv. and Val. D. 9-10/$\frac{1}{28\text{-}30}$, A. $\frac{2}{7}$, L. l. 48, Cæc. pyl. 5. Eyes not 1 diameter from the end of snout. Upper jaw overlapping the lower. Second anal spine strong, 2/3 to 3/5 of the length of the first ray. Greenish brown along the back, becoming lighter on the abdomen: minute brown on black dots over the whole body and fins: margins of caudal, anal, and ventral fins blackish. Seas of India and Malay Archipelago.

268. *Sciæna cuja*, Ham. Buch. D. 10/$\frac{1}{27\text{-}29}$, A. $\frac{2}{7}$, L. l. 50-56, L. tr. 7/15. Eyes 1 diameter from end of snout. Jaws of nearly equal length. Second anal spine very strong, and nearly or quite as long as the first ray. Oblique streaks above the lateral line: horizontal ones below it: dorsals with two or three longitudinal rows of black spots. Ganges and Japan, attaining several feet in length.

269. *Sciæna carutta*, Bl. D. 10/$\frac{1}{27\text{-}28}$, A. $\frac{2}{7}$, L. l. 50. Eyes 1 diameter from end of snout. Upper jaw overlapping the lower. Preopercle indistinctly crenulated. Second anal spine very strong, 2/3 the length of the first ray: head and back of a brownish colour: upper half of first dorsal black: four large black blotches on the second dorsal: outer half of ventral, anal, and caudal black. Seas of India and Malay Archipelago.

270. *Sciæna axillaris*, Cuv. and Val. *Goal-mutchee*, Bel. D. 10/$\frac{1}{27\text{-}28}$, A. $\frac{2}{7}$, L. l. 43-50, L. tr. 8/14, Cæc pyl. 9. Eyes, 1 diameter from end of snout. Jaws of nearly equal length. Preopercle serrated, with two rather strong teeth at its angle. Second anal spine of moderate length, much shorter than the first ray, and 1/3 of the length of the head. Air-vessel large and anteriorly bulging out on either side like a hammer. Silvery, tinged with brownish along the back: a black spot above the axil: first dorsal blackish, more especially in its outer half: the other fins grey. Western coast of India and along the Mekran coast.

271. *Sciæna coitor*, Ham. Buch. *Bolahl* and *Pultheri-ki*, Ooriah: *Nga-pok-thin*, Burmese: *Sohlee*, Beluch. D. 10/$\frac{1}{d^{\prime}7}$, A. $\frac{2}{7}$, Cæc. pyl. 7-9. Eyes 1½ diameters from the end of snout. Upper jaw overlapping the lower. Preopercle slightly serrated. Second anal spine strong, 4/5 as long as the first ray. Silvery green above, becoming white below: fins darkest externally. Seas of India, also the larger rivers: attaining 12 inches in length.

272. *Sciæna albida*, Cuv. & Val. D. 9-10/$\frac{1}{24\text{-}25}$, A. $\frac{2}{7}$, L. l. 53. L. tr. 24, Cæc. pyl. 5. Eyes not quite one diameter from the end of snout. Jaws equal in front. Preopercle indistinctly serrated. Second anal spine strong, 4/5 as long as the first ray. Silvery, with a light streak along the centre of each scale: first dorsal with a dark edge: ventral, anal, and caudal yellowish. This is identical with *Corvina Neilli*, Day. Seas of India and China.

Genus—OTOLITHUS, Cuv.

Branchiostegals seven: pseudobranchiæ. Body oblong. Eyes of moderate size. Snout obtuse or a little pointed, the lower jaw being the longer. Preopercle crenulated, serrated or denticulated. No barbels. Conical canine teeth usually well developed in both jaws or merely in the upper. Two dorsal fins, the first with nine or ten weak spines: anal with one or two small ones. Scales of moderate or small size. Air-vessel present, some having on either side an anteriorly directed process: others with lateral appendages. Pyloric appendages few.

273. *Otolithus maculatus*, Cuv. & Val. *Birralli*, Ooryah. D. 10/$\frac{1}{30\text{-}31\,(34)}$ A. $\frac{2}{1\text{-}1}$. Eyes, 1½ diameter from end of snout. Preopercle finely denticulated. A pair of canines in either jaw. Air-vessel with about 54 lateral appendages on either side. Greyish in the upper part of the body, golden below: five or six rows of black spots on the body and caudal fin: the other fins stained at their edges. Bay of Bengal and Malay Archipelago: attaining at least 16 inches in length.

274. *Otolithus ruber*, Bl. Schn. D. 10/$\frac{1}{30\text{-}31}$, A. $\frac{1}{7}$, L. l. 54, Cæc. pyl. 4-5. Eyes one diameter from end of snout. Preopercle finely denticulated. On either side of the symphysis of the upper jaw a pair of large canines, the inner of which is the longer: on either side of the symphysis of the lower jaw is one canine. Air-vessel with 34 lateral processes on either side. Brownish red shot with silvery, becoming quite white below the lateral line: some fine black points on the fins. Seas of India and Malay Archipelago: attaining 2½ feet or more in length.

275. *Otolithus argenteus*, Cuv. & Val. *Bu-ru*, Sind. D. 10/$\frac{1}{25\text{-}28}$, A. $\frac{2}{7}$, L. l. 70, Cæc. pyl. 6. Eyes, 1½ diameter from end of snout. Preopercle finely denticulated. On either side of the symphysis of the upper jaw two or three canines, the posterior of which (or the centre one if three) is the longest, but all are shorter and wider apart than in *O. ruber*: on either side of the symphysis of the lower jaw is one very small canine. Air-vessel with 25 lateral processes on either side. Greyish superiorly with violet reflections, becoming golden on the abdomen: some fine black points on the fins. Seas of India, Malay Archipelago to China: attaining upwards of 2½ feet in length.

276. *Otolithus aneus*, Bloch. *Chal-burn-dah*, Andam. D. $10/\frac{1}{23\cdot24}$, A. $\frac{2}{7}$, L. l. 52, L. tr. 9/19, Cæc. pyl. 10. Bloch gives D 8, &c., but the plate shows D. 9, the first small spine having been overlooked as in *Bola coibor*, H. B., which seems to be the same species. Eyes one diameter from the end of snout. Preopercle crenulated. Canines not well developed. Silvery grey, becoming dirty white along the abdomen: first dorsal black tipped. Air-vessel with 30 lateral processes. Seas of India.

277. *Otolithus versicolor*, Cuv. & Val. *Pottee-kanasah*, Tel. D. $10/\frac{1}{2\cdot7}$, A. $\frac{1}{8}$ (²). Teeth as in the *O. maculatus*. Silvery. Coromandel coast.

278. *Otolithus brunneus*, Day. D. $9/\frac{1}{28}$, A. $\frac{2}{7}$, L. r. 102. L. tr. 21/34. Eyes, $1\frac{1}{2}$ diameter from end of snout. Preopercle scarcely crenulated. Canines small. Brownish, golden below; fins dark edged. Bombay, to 8 inches in length.

Genus—SCIÆNOIDES, Blyth.

Bola, pt, Ham. Buch. *Collichthys*, Günther, *Sciæna*, sp. Cuv. & Val. *Otolithus*, sp. Cantor.

Branchiostegals seven: pseudobranchiæ absent. Eyes small. Head broad with its upper surface very convex. Cleft of mouth oblique and deep. No barbels. Two dorsal fins, the second with many rays: two weak anal spines; caudal wedge-shaped. Air-vessel generally having a horn-like process on either side directed forwards, and with many lateral appendages. Pyloric appendages few or in moderate numbers.

Sciænoides pama, Ham. Buch. *Coii bola*, Bengali: *Ven begti* and *Botul*, Ooriah: *Nga-pouss-was*, Mugh. D. $10/\frac{1}{40\text{-}43}$, A. $\frac{9}{7}$, L. r. 70—80, L. tr. 9/25, Cæc. pyl. 9. Eyes two diameters from end of snout. Two skinny flaps to the opercles. Air-vessel dividing anteriorly into two short processes, whilst springing from near its posterior extremity are two more long processes which extend anteriorly as far as the auditory apparatus. Of a whitish colour. It is termed *whiting* in Calcutta. Bay of Bengal, entering estuaries and rivers: it attains at least 5 feet in length.

280. *Sciænoides lucida*, Richardson: *Otolithus bispinosus*, Blyth. D. $9/\frac{1}{35}$, A. $\frac{2}{8}$, L. l. 52, (L. r. 75), Cæc. pyl. 4. Eyes one diameter from end of snout. Air-vessel without any projections anteriorly, but with about 14 lateral appendages. Silvery. Seas of India to China.

281. *Sciænoides biaurita*, Cantor. D. $9/\frac{1}{27\text{-}32}$, A. $\frac{2}{7}$, Cæc. pyl. 13. Two skinny flaps to the opercles. Air-vessel with 25 lateral appendages. Silvery. Seas of India, Malay Archipelago, and China.

Family—XIPHIIDÆ, Agass.

Sword fishes.

Branchiostegals seven: pseudobranchiæ. Eyes lateral. Body compressed, the upper jaws (comprising ethmoid, vomer, and intermaxillaries) produced into a long, sword-shaped process: cleft of mouth deep. Teeth absent or rudimentary. One or two dorsal fins, without any distinct spinous portion: ventrals, when present, thoracic and rudimentary. Scales absent or in the form of rudimentary dermal productions. Air-vessel present. Pyloric appendages numerous.

Genus—HISTIOPHORUS, Lacép.

Notistium, Herm: *Tetrapturus*, Cuv. & Val.
Branchiostegals seven: pseudobranchiæ. Upper jaw much prolonged. Minute teeth on the jaws and palatines: vomer edentulous. Two dorsal and two anal fins, the anterior of each of which is the longer and composed of spines and ray: ventrals in the form of one or two spines. Scales absent, rudimentary dermal productions may be present. Air-vessel present. Pyloric appendages numerous.

282. *Histiophorus immaculatus*, Rüpp. *Yemungolah*, Tam. D. 47/7, A. 10-11/7, V 3. Length of head about 1/3, height of body 1/10 of the total length. Dorsal fin much higher than the body, the eleventh ray about the longest. Dermal productions lanceolate. Colouration uniform: dorsal fin blackish. Red Sea, seas of India: a specimen in the Madras Museum is nearly 6 feet in length.

283. *Histiophorus gladius*, Brouss. *Myl meen*, 'Peacock fish', Tamil. D. 40—50/7, A. 10/7, V. 2, Vert. 14/10. Length of head 1/4, height of body 1/7 to 1/8 of the total length. Dorsal fin much higher than the body. Dorsal fin of a bright Prussian blue, with darker spots: body bluish, becoming white beneath. Seas of India: I procured one 9 feet long at Madras.

284. *Histiophorus brevirostris*, Playfair. D. 35/7, A. 11/7, V. 2. Length of head 3/10 of the total length. Height of dorsal fin nearly equal to that of the body. Dermal productions lanceolate. Colour uniform. East coast of Africa, Madras: attaining upwards of 10 feet in length.

Family—TRICHIURIDÆ, *Günther*.

Branchiostegals seven to eight: pseudobranchiæ. Body elongated and compressed. Gill openings wide. Eyes lateral. Cleft of mouth deep. Teeth in jaws or palate, several being strong and conical. Dorsal fin, both in its spinous and soft portion, and the anal, many rayed: there may be finlets behind the dorsal or anal fins: ventrals, when present, thoracic, but sometimes they are rudimentary: caudal absent or present. Scales when present rudimentary. No prominent pupilla behind the vent. Air-vessel present. Pyloric appendages few or many.

Genus—TRICHIURUS (¹), *Linn.*

Branchiostegals seven. Body very elongate, strongly compressed, ribbon-shaped, tapering to a point at the tail. Cleft of mouth deep. Teeth in jaw and palatines, those in the jaw strong. A single long dorsal fin extending the whole length of the back: ventrals, when present, in the form of a pair of scales: anal spines minute, sometimes concealed beneath the skin: caudal absent. Pyloric appendages numerous.

285. *Trichiurus haumela*, Forsk. *Savala*, Tel.: *Sona-ka-wahlay*, Tam.: *Pa-pa-dah*, Andam. D. 127-133. Length of head 1/7, height of body 1/15 to 1/17 of the total length. Eyes, diameter from $1\frac{1}{2}$ to $2\frac{1}{3}$ in the length of the smout. Silvery, the upper half of the dorsal fin dark coloured. Seas of India and Malay Archipelago.

(¹) *Nus-sah-rue*, Mugh.

286. *Trichiurus savala*, Cuv. *Karte-ka-wahlah*, Tam. D. 112-130. Length of head 2/13 to 4/29, height of body 1/13 to 1/16 of the total length. Eyes, diameter 2/7 of the length of the snout. Silvery white. Seas of India and Malay Archipelago.

Family—ACANTHURIDÆ, pt. *Richards.*

Acronuridæ, Günther.

Branchiostegals from four to seven: pseudobranchiæ. Body oblong or elevated and compressed. Eyes of moderate size and lateral. Cleft of mouth very slight. Teeth in both jaws in a single compressed row, often lobate or serrated, and tapering incisors may be present: palate edentulous. A single dorsal fin with less spines than rays: anal with two or three spines: ventrals thoracic. Scales minute. Lateral line complete and continuous. The side of the free portion of the tail usually armed with one or more bony plates or spines, these are small or absent in the immature, developing with age. Air-vessel present, forked posteriorly. Pyloric appendages few.

Genus—ACANTHURUS [1], *Bl.*

Acronurus, Cuv. (?) young: *Keris*, pt., Cuv. (?) young.

Branchiostegals five: pseudobranchiæ well developed. Body elevated and strongly compressed. Teeth in a single row with lobate or crenulated margins. A single dorsal fin with less spines than rays: anal with three spines: ventral usually with one spine and five rays. Scales minute, ctenoid, sometimes spinate. Air-vessel large, posteriorly forked. Pyloric appendages five to seven.

287. *Acanthurus leucosternon*, Bennett. D. $\frac{9}{29\text{-}30}$, A. $\frac{3}{27\text{-}28}$. Height of body 9/5 of the total length, excluding the caudal fin. Five truncated and lobate teeth on either side of the upper jaw. Blue, head black: chest white, as is also a ring on the lower part of the mouth. A crescentic black band across the base of the caudal, a second along the margins of the fins. Ceylon, Mauritius, and Malay Archipelago.

288. *Acanthurus lineatus*, Linn. D. $\frac{8\text{-}9}{28\text{-}31}$, A. $\frac{3}{27}$. Height of body 1/2 the total length. Six lobate incisors on either side of the upper jaw. Ground colour canary yellow, with nine oblique blue bands passing from the head to the back or to the base of the caudal fin: two more curved blue bands pass from the eye to the base of the pectoral fin. Blue lines likewise on the caudal fin. Seas of India to the Malay Archipelago: common at the Andamans up to 10 inches in length.

289. *Acanthurus matoides*, Cuv. & Val. D. $\frac{9}{26\text{-}28}$, A. $\frac{3}{25\text{-}28}$. Height of body 4/9 of the total length. Seven or eight lobate incisors on either side of the upper jaws. Brown, with badly developed undulating lines which fade soon after death. Vertical fins yellowish with bluish lines or blotches. Dr. Günther considers *A. annularis*, Cuv. & Val., with a white caudal ring, "as immature specimens", but there are certainly large ones thus marked. Seas of India, Malay Archipelago, &c.

290. *Acanthurus rasi*, Cuv. & Val. *Mata*, Tel. D. $\frac{9}{25\text{-}27}$, A. $\frac{3}{23\text{-}25}$. Height of body 6/13 of the total length. The incisors of both jaws are rounded and with from 13 to 15 serratures on their sides and summits.

---

[1] *Kohli-meen*, Tam.

Colour brownish with a light ring at the base of the caudal fin. Coromandel coast, to 9 inches in length. There is a good specimen 7 inches long in the Madras Museum.

291. *Acanthurus nigrofuscus*, Forsk. D. $\frac{9}{24\cdot 27}$, A. $\frac{3}{23\cdot 25}$. Height of body 1/2 the total length, excluding the caudal fin. Eight or nine lobate incisors on either side of the upper jaw. Blackish brown. Ceylon, Malay Archipelago.

292. *Acanthurus melas*, Cuv. & Val., and *A. melanurus*, Cuv. & Val. D. $\frac{9}{27}$, A. $\frac{3}{25}$, &c., probably are the young of some of the foregoing or following species.

293. *Acanthurus Tennentii*, Günther. D. $\frac{9}{23}$, A. $\frac{3}{22}$. Height of body 1/2 the total length, excluding the caudal fin. Six lobate incisors on either side of the upper jaw, and eight on each side of the lower. Brown, with a roundish dark ring on the shoulder; caudal with a broad white edge. Ceylon.

294. *Acanthurus triostegus*, Linn. *Mootah*, Tel. *Kara hamoowah*, Cing. D. $\frac{9}{23\cdot 25}$, A. $\frac{3}{20\cdot 21}$. Height of body nearly 1/2 the total length. Seven lobate incisors on either side of the upper jaw, eight on each side of the lower. Greenish, with a brownish tinge along the back: one dark band along the snout; a second through the orbit: four along the body, and one over the base of the caudal fin, the last of which is in the form of two round spots, one above the other: fins stained darkish. Seas of India to the Malay Archipelago and beyond.

295. *Acanthurus ctenodon*, Cuv. & Val. D. $\frac{8}{(24)\ 28\cdot 30}$. A. $\frac{3}{(12)\ 26\cdot 27}$. Height of body nearly 1/2 the total length. Teeth setiform, dilated at their external 1/3, where they are laterally serrated. Lineated all over with blue and yellow lines, the latter being somewhat the wider. Numerous red spots about the head, more especially around the eyes. Dorsal and anal fins also lineated. East coast of Africa, seas of India to the Malay Archipelago and beyond.

296. *Acanthurus xanthurus*, Blyth. D. $\frac{5}{24\cdot 26}$, A. $\frac{3}{20\cdot 21}$. Height of body 2/5 of the total length. Ten lobate incisors or either side of the upper jaw. Blackish: caudal light yellow. East coast of Africa and Ceylon, to at least 8 inches in length.

297. *Acanthurus velifer*, Bloch. D. $\frac{4}{28\cdot 29}$, A. $\frac{3}{23\cdot 24}$. Height of body nearly 1/2 of the total length. Six lobate incisors on either side of the upper jaw. Greyish, head and body with ten vertical cross bands going from the back to the abdomen. Dorsal with four curved blue bands, six on the anal and four on the caudal. Andamans and Malay Archipelago.

Genus—NASEUS, *Commerson*.

Monoceros, Bl. Schn.: *Aspisurus*, Rüpp.: *Axinurus*, *Priodon* and (?) *Keris* pt. young, Cuv. & Val.

Branchiostegals four or five: pseudobranchiæ. Head sometimes with a bony prominence. A single row of compressed incisors which sometimes have serrated edges: palate edentulous. A single dorsal fin with from four to six spines, and many rays: anal with two spines, its rays similar to those of the dorsal: ventral with one spine and three rays. Scales minute, forming a rough covering. Side of the tail with from one to three bony plates, most developed in the adult. Air-vessel large and forked posteriorly. Pyloric appendages rather few.

298. *Naseus unicornis*, Forsk. D. $\frac{(5)\ 6}{(31)\ 28}$, A. $\frac{2}{20\text{-}27\ (29)}$, Cæc. pyl. 7-8. A horizontal horn-like protuberance from the forehead. Teeth not serrated. Two spinous plates on the side of the tail. Brownish grey: dorsal and anal fins with longitudinal blue stripes. From the Red Sea, through the seas of India, Malay Archipelago, and beyond.

299. *Naseus brevirostris*, Cuv. & Val. D. $\frac{5\text{-}6}{25\text{-}29}$, A. $\frac{2}{25\text{-}30}$. A horizontal horn-like protuberance from the forehead. Two horn-like plates on the side of the tail. Teeth small, serrated. Greyish brown posteriorly, with many short blue transverse lines: caudal with a light margin. East coast of Africa, through the seas of India to the Malay Archipelago and beyond.

300. *Naseus tuberosus*, Lacép. D. $\frac{5\text{-}6}{27\text{-}30}$, A. $\frac{2}{27\text{-}28}$. Snout with a convex and compressed hump above it. Teeth small, about twenty on either side of the upper jaw. Brown, with blue dots or short vertical lines: fins blackish, the vertical ones with light edges. East coast of Africa, seas of India, to the Malay Archipelago and beyond.

Family—CARANGIDÆ.

Scomberoidei, pt., et *Squamipinnes*, pt., Cuv.: *Scombrisidæ*, pt., Richardson.

Branchiostegals usually seven, occasionally less: pseudobranchiæ as a rule present, but absent in *Lichia* and *Trachynotus*. Body oblong, elevated, or sub-cylindrical and compressed. Gill-openings wide. Eyes lateral. Infraorbital bones do not articulate with the preopercle. Dentition varied. The length of the base of the spinous portion of the dorsal fin is of less extent than that of the soft, and is sometimes formed by isolated spines: the spinous may be continuous with, or distinct from the soft portion: the posterior portion both of the dorsal and anal sometimes consists of detached finlets: the soft dorsal and the anal of nearly equal extent. Anal spines, when present, may or may not be continuous with the soft portion. Ventrals, when present, thoracic, sometimes rudimentary. Scales usually small, unless absent. Lateral line may be wholly, partially, or not at all armed with shield-like plates. Air-vessel present. Pyloric appendages usually in large numbers. Vertebræ 10/14 (Naucrates 10/16).

Genus—CARANX ([1]), *Cuv. & Val.*

Olistus, Scyris, Blepharis, Gallichthys, et *Hynnis*, Cuv. & Val.: *Megalaspis, Decapterus, Selar, Carangoides, Leioglossus, Uraspis, Selaroides,* et *Gnathanodon*, Bleeker : *Carangus*, Girard.

Branchiostegals seven: pseudobranchiæ. Body oblong, sub-cylindrical, and more or less compressed. Eyes lateral. Dentition feeble. Two dorsal fins: the first continuous, with about eight weak spines, and sometimes rudimentary, having anteriorly a spine directed forwards: the second dorsal longer and similar to the anal: sometimes the last rays entirely or semi-detached: two pre-anal spines (which may be rudimentary) separated by a space from the rays. Scales minute. Lateral line with an anterior curved portion, and the posterior straight, having large plate-like scales, which are usually keeled and sometimes spinate. Air-vessel bifurcated posteriorly. Pyloric appendages in large numbers.

([1]) *Parah*, Tam.

1. *The second dorsal and anal with the posterior ray or rays detached.*

   a. *With eight or ten detached rays. (Megalaspis).*

   301. **Caranx Rottleri**, Bloch. *Woragoo* and *Sora parah*, Tel. D. $6-8/\frac{1}{9\text{-}11}$ + VIII-X, A. $2/\frac{1}{8\text{-}9}$ VIII, L. l. 55. (^{1}) Height of body 2/9 of the total length. Villiform teeth in jaws, vomer, and palatines. Lateral line makes a short and abrupt curve above the base of the pectoral fin, and opposite its first third proceeds direct to the base of the caudal. Back glossy green : abdomen silvery, tinged with yellow : a large black spot on the upper and posterior portion of the opercle. Fins yellow : dorsal and anal tipped with black. From the Red Sea and east coast of Africa, through the seas of India to the Malay Archipelago and beyond : is said by native fishermen to attain 5 feet in length.

   b. *A single ray of dorsal and anal fins detached. (Decapterus).*

   302. **Caranx kurra**, Cuv. & Val. *Kurra-wodagahwah*, Tel. D. $8/\frac{1}{30}$+I, A. $2/\frac{1}{27}$+I, L. l. 33. Height of body 1/6 of the total length. Teeth in a single row in both jaws; also on vomer and palate; three rows of sharp teeth along the middle of the tongue. Lateral line nearly straight, until opposite the end of the first dorsal, it then slopes downwards until opposite about the fifteenth ray, when it passes direct to the centre of the caudal. Bluish above, silvery beneath : a deep black spot on the upper margin of the opercle : upper surface of the head minutely dotted with black : fins yellow, darkest at their edges. Coasts of India : is a small species arriving at Madras about October. The *Caranx kiliche*, Cuv. & Val., from Pondicherry, may be this species: the teeth are not referred to : the fin rays are given thus, D. $8/\frac{1}{27}$+I, A. $2/\frac{1}{23}$+I, L. l. about 30.

2. *No dorsal or anal rays detached.*

   a. *None of the fin rays elongated.*

   303. **Caranx gymnostethoides**, Bleeker. *Tanga-parrah*, Tam. D. $8/\frac{1}{30\text{-}31}$, A. $2/\frac{1}{25\text{-}27}$, L. l. 25. Height of body 1/4 of the total length. Eyes two diameters from end of snout. Teeth in the jaws small, in a single row, also present on the vomer, palatines, and tongue. Lateral line very slightly bent, becoming straight below the middle of the soft dorsal fin. Chest scaleless. An indistinct opercular spot. Madras and Malay Archipelago.

   304. **Caranx macrurus**, Bleeker. D. $8/\frac{1}{24\text{-}27}$, A. $2/\frac{1}{20\text{-}22}$, L. l. 60-69. Height of body nearly 1/4 of the total length. Teeth on the jaws in a single row; also on the vomer, palatines, and tongue. Lateral line with a strong bend anteriorly, becoming straight below the commencement of the second dorsal : its plates are well developed. Chest scaled. An indistinct opercular spot. Madras and Malay Archipelago.

   305. **Caranx Djeddaba**, Forsk. D. $7-8/\frac{1}{24\text{-}26}$, A. $2/\frac{1}{20\text{-}22}$, L. l. 48-50. Height of body 3/11 of the total length. Lower margin of opercle

---

(^{1}) L. l. in the Genus *Caranx* denotes the number of plate-like keeled scales along the straight portion of the lateral line.

very concave. A single row of teeth on the jaws: teeth on the palate. Pectoral fin 1/4 of the total length. Lateral line with a strong bend anteriorly, becoming straight from below the first dorsal: the plates are well developed. Chest scaled. No opercular spot: a black edge to the second dorsal and anal. Red Sea, east coast of Africa, seas of India.

306. *Caranx vari*, Cuv. & Val. *Vari-parrah*, Tam. D. $8/\frac{1}{24\text{-}25}$, A. $2/\frac{1}{20\text{-}23}$. L. l. 55-58. Height of body 1/3 of the total length. Lower margin of opercle slightly concave. A single? row of teeth in the jaws and palate. Pectoral fin 2/7 of the total length. Lateral line with a strong bend anteriorly, becoming straight under the commencement of the second dorsal. None of the plates are more than 1/9 or 1/10 of the height of the body. A black opercular spot: and the first dorsal blackish. Coromandel coast and Malay Archipelago.

307. *Caranx Bidii*, Day. *Ramah-parrah*, Tam. D. $8/\frac{1}{24}$ A. $2/\frac{1}{22}$, L. l. 24. Height of body 1/4 of the total length. A band of fine teeth in the jaws: none on the vomer or palate. Lateral line is nearly straight until below the first third of the second dorsal, where it curves very gently downwards, and passes straight to the caudal. Plates small. Chest scaled. Silvery, with a broad golden stripe from above the eye to the upper edge of the tail: lower two-thirds of dorsal yellow, upper third dark: outer third of anal white, the rest yellow. A large black spot on the shoulder. Madras.

308. *Caranx calla*, Cuv. & Val. D. $8/\frac{1}{23\text{-}24}$, A. $2/\frac{1}{19\text{-}20}$, L. l. 42. Height of body 2/7 of the total length. Teeth in a single row in the jaws, also present on the vomer, palatines, and tongue. Lateral line with a strong bend anteriorly, becoming straight under the commencement of the second dorsal: the plates well developed. Chest scaled. A distinct black spot on the opercle. Red Sea, seas of India to China.

309. *Caranx xanthurus* (Kuhl. and v. Hass.), Cuv. & Val. *Matah-parrah*, Tam. D. $8/\frac{1}{24\text{-}25}$, A. $2/\frac{1}{19}$, L. l. 36-40. Height of body 3/11 of the total length. Teeth in a single row in the jaws, also present on the vomer, palatines, and the tongue. Lateral line with a slight curve anteriorly, becoming straight below about the sixth dorsal ray: plates well developed. Chest scaled. A distinct black spot on the opercle. From the east coast of Africa and seas of India to the Malay Archipelago and beyond.

310. *Caranx malam*, Bleeker., D. $8\text{-}9/\frac{1}{23\text{-}24}$, A. $2/\frac{1}{20}$, L. l. 50-56. Height of body about 3/10 of the total length. Teeth in a single row in the jaws, also present on the vomer, palatines, and the tongue. Lateral line with a strong bend anteriorly, becoming straight below the commencement of the second dorsal. Chest scaled. A distinct black spot on the opercle: first dorsal black. Madras, Malay Archipelago.

311. *Caranx melampygus*, Cuv. & Val. *Kuguroo-parah*, Tel. D. $8/\frac{1}{23\text{-}24}$, A. $2/\frac{1}{19}$, L. l. 34-36. Height of body 2/7 of the total length. Teeth in the jaws villiform, with an outer stronger row, also present on the vomer, palatines, and the tongue. Lateral line moderately bent anteriorly, becoming straight below about the fifth dorsal ray: plates well developed. Chest scaled. A small black spot on the opercle:

the dorsal and anal fins are black anteriorly. Mauritius, Coromandel coast, Malay Archipelago and beyond.

312. *Caranx hippos*, Linn. *Wotim-parah*, Tel. D. $7\text{-}8/\frac{1}{19\text{-}22}$, A. $2/\frac{1}{16\text{-}17}$, L. l. 30-36. Height of body nearly 1/3 of the total length. Teeth in the upper jaw in a villiform band, and an outer stronger row, those in the lower jaw in a single row but of irregular sizes. Lateral line with a strong bend anteriorly, becoming straight below the third dorsal ray: plates well developed. Chest scaled. An indistinct spot sometimes apparent on the opercle: top of the second dorsal blackish. Seas of India to the Malay Archipelago and beyond: it attains a large size.

313. *Caranx ire*, Cuv. & Val. *Ire-parrah*, Tam. D. $7\text{-}8/\frac{1}{23\text{-}25}$, A. $2/\frac{1}{20}$, L. l. 30-34. Height of body 3/10 of its total length. Teeth in the jaws in a narrow villiform band: teeth on the palate. Lateral line with a slight bend anteriorly, becoming straight below the first fourth of the dorsal: plates moderately developed. A large black spot at the summit of the anterior portion of the second dorsal fin: none on the opercle. Coromandel coast.

314. *Caranx melanostethos*, Day. D. $8/\frac{1}{23}$, A. $2/\frac{1}{19}$, L. l. 28. Height of body 1/4 of the total length. Teeth in the jaws villiform, also on vomer. Lateral line with a very gentle curve anteriorly, becoming straight below the middle of the second dorsal: plates not well developed. Chest scaled. Head and chest of a brownish black colour, no opercular spot: fins yellowish, with black points, the upper portion of the anterior part of the second dorsal black with a white summit. Seas of India.

315. *Caranx jarra*, Cuv. & Val. *Jarra-dandrée-parah*, Tel. D. $8/\frac{1}{22}$, A. $2/\frac{1}{18}$, L. l. 36. Height of body 2/7 of the total length. Teeth on the upper jaw in a band with an outer enlarged row: in a single row in the lower jaw, also on vomer, palatines, and tongue. Lateral line with a bend anteriorly and undulating in its course, it becomes straight below the fifth dorsal ray: the plates pretty well developed. No opercular spot: the tip of the upper lobe of the caudal black. Coromandel coast and Malay Archipelago.

316. *Caranx carangus*, Bloch. D. $7\text{-}8/\frac{1}{19\text{-}22}$, A. $2/\frac{1}{16\text{-}18}$, L. l. 30-33. Height of body 1/3 of the total length. Teeth on the upper jaw in a band, with an outer enlarged row, in a single row in the lower jaw. Lateral line with a bend anteriorly ending abruptly below the sixth dorsal ray: plates well developed. Chest scaleless. A dark opercular spot. Madras, Malay Archipelago, and the Atlantic provided the species is identical.

317. *Caranx ekala*, Cuv. & Val. *Ekala-parah*, Tel. D. $7\text{-}8/\frac{1}{19}$, A. $2/\frac{1}{18}$, L. l. 26. Teeth strong. Lateral line becomes straight under the first third of the second dorsal: plates well developed. Chest scaled. Upper angle of the second dorsal blackish: no opercular spot. Seas of India.

318. *Caranx talamparoides*, Bleeker. D. $7\text{-}8/\frac{1}{22}$, A. $2/\frac{1}{18}$, L. l. 25-30. Height of body $2\frac{3}{4}$—$2\frac{5}{8}$ in the total length. Profile before the eyes rather concave. Teeth in the jaws villiform, also on vomer,

palatines, and tongue. Lateral line moderately bent, becoming straight behind the middle of the second dorsal: plates slightly developed. Chest scaleless. A brown opercular spot. Ceylon and Coromandel coast.

319. *Caranx Malabaricus*, Bl. Schn. *Talem-parah*, Tel. D. $8/\frac{1}{22}$, A. $2/\frac{1}{17}$, L. l. 25-30. Height of body $2\frac{1}{4}$—$2\frac{3}{4}$ in the total length. Teeth in the jaws villiform, also on vomer, palatines, and tongue. Lateral line moderately bent, becoming straight behind the middle of the second dorsal: plates slightly developed. Chest scaleless. A black opercular spot: dorsals stained dark at their margins. Seas of India to the Malay Archipelago and beyond

320. *Caranx compressus*, Day. D. $8/\frac{1}{22}$, A. $2/\frac{1}{19}$, L. l. 13. Height of body 2/7 of the total length. Villiform teeth in the upper jaw, and in a narrow band in the lower, also on vomer, palatines, and tongue. Lateral line gently curved anteriorly, becoming straight below the tenth dorsal ray: plates well developed. Chest scaled. A small black opercular spot: a black edging to the vertical edge of the preopercle. Andamans.

321. *Caranx sansum*, Forsk. *Gundi-parah*, Tel. D. $8/\frac{1}{19\text{-}20}$, A. $2/\frac{1}{16\text{-}17}$, L. l. 28-30. Height of body from 1/3 to 2/7 of the total length. Teeth in the upper jaw in a villiform band with an outer larger row, those in the lower jaw in a single row. Lateral line with a moderate bend anteriorly, becoming straight below the seventh dorsal ray: plates well developed. Chest scaled. No black spot on the opercle. Red Sea, Mauritius, and Coromandel coast.

322. *Caranx nigrescens*, Day. D. $7/\frac{1}{19}$, A. $2/\frac{1}{17}$, L. l. 25. Height of body 2/7 of the total length. Teeth in villiform bands in both jaws, also on vomer and palate. Lateral line with a moderate curve to opposite the thirteenth dorsal ray, when it becomes straight: plates badly developed. Chest scaleless. A well marked black opercular spot: fins nearly black. Madras.

323. *Caranx speciosus*, Forsk. D. $7\text{-}8/\frac{1}{16\text{-}20}$, A. $2/\frac{1}{15\text{-}16}$, L. l. 34 (15 large ones). Height of body 1/3 the total length. Teeth in the adult almost or quite imperceptible. Lateral line with a gentle curve to below the sixth dorsal ray, from whence it proceeds straight: plates badly developed, the last 15 being most conspicuous. Chest scaled. Golden, with a greenish tinge along the back, with eight vertical greenish black bands: caudal lobes tipped with black. Red Sea, through seas of India to the Malay Archipelago and beyond.

324. *Caranx parah*, Cuv. & Val. D. $7/\frac{1}{23}$, A. $2/\frac{1}{21}$, L. l. 36-40. Height of body 3/13 of the total length. Diameter of eye more than 1/3 in the length of the head. Teeth almost imperceptible: present on the palate. Lateral line moderately curved, and under the first fifth of the second dorsal it becomes straight. Seas of India to the Malay Archipelago.

b.—One or more of the dorsal or anal rays elongated.

325. *Caranx atropus*, Bloch. *Mais-parah*, Tel.: *Canni-parah*, Tam. D. $8/\frac{1}{22}$, A. $2/\frac{1}{18}$, L. l. 32-35, Vert. $\frac{10}{4}$. Height of the body 2/5 of the total length. Teeth villiform in both jaws, also on vomer and palate. Some of the dorsal rays generally produced. Lateral line with a strong

bend anteriorly, becoming straight below the first fourth of the second dorsal: plates moderately developed. Chest scaleless. A black opercular spot: ventrals deep black: the young are vertically banded. Seas of India to the Malay Archipelago and China.

323. *Caranx oblongus*, Cuv. & Val. *Ro-thul-dah*, Andamanese. D. $8/\frac{1}{21}$, A. $2/\frac{1}{18\text{-}19}$ L. l. 36-40. Height of body about 1/3 of the total length. Villiform teeth in the jaws, also on vomer, palate and tongue. The first dorsal ray elongated. Lateral line moderately bent anteriorly, becoming straight below the second fifth of the soft dorsal: plates well developed. Chest scaleless. No opercular spot. Andamans, to Malay Archipelago and beyond.

327. *Caranx armatus*, Forsk. *Tohawil parah*, Tel. D. $6\text{-}8/\frac{1}{20\text{-}21}$, A. $\frac{1}{16\text{-}17}$, L. l. 20. Height of body about 1/3 of the total length. Teeth villiform in the jaws, also on vomer, palate, and tongue. The first dorsal ray and often some of the others elongated. Lateral line moderately bent, becoming straight under the middle of the second dorsal fin: the plates but little developed. Chest scaleless. Opercular spot moderately distinct: first dorsal blackish. In the immature, six vertical dark bands. Red Sea, east coast of Africa, seas of India to the Malay Archipelago and beyond.

328. *Caranx chrysophrysoides*, Bleeker. *Tanga parrah*, Tam. D. $8/\frac{1}{19}$, A. $2/\frac{1}{16}$, L. l. 24-28. Height of the body 1/3 of the total length. Eyes 2 diameters from end of snout. Teeth villiform in the jaws, on the vomer, palatine bones, and the tongue. Lateral line moderately bent, becoming straight below the last half of the second dorsal fin: plates not well developed. Chest scaleless. A black opercular spot. Specimens in the Madras Museum from Aden and Madras: it appears to resemble *C. chrysophrys*, Cuv. & Val. Seas of India to Malay Archipelago.

329. *Caranx ciliaris*, Bloch. D. $6/\frac{1}{19}$, A. $2/\frac{1}{16}$, L. l. 15. Height of body from 1/2 to 2/3 of the total length. Teeth villiform in both jaws, also on vomer, palatine bones, and the tongue. Spines of first dorsal fin rudimentary: rays of second dorsal and anal villiform and elongated. Lateral line with a bend anteriorly and becoming straight below the middle of the soft dorsal: plates but little developed. Scales absent, except on the lateral line. A black opercular spot: the very young banded. Red Sea, through the seas of India to the Malay Archipelago and beyond.

330. *Caranx gallus*, Linn. D. $\frac{1}{19}$, A. $\frac{1}{16}$, L. l. 8, Vert. 10/14. Height of body from 3/5 to 3/7 of the total length. Teeth villiform in both jaws, also on vomer, palatines, and the tongue. Anterior dorsal, anal, and ventral rays elongated, filiform. Upper profile very concave to the origin of the dorsal fin. Lateral line with a strong bend anteriorly, becoming straight below the middle of the dorsal fin: plates but little developed. Scales absent, except on the lateral line. Silvery: young with vertical bands. Red Sea, through the seas of India to the Malay Archipelago and beyond.

Genus—SERIOLA, Cuv.

Branchiostegals seven. Body oblong, and moderately compressed: abdomen rounded. Cleft of mouth sometimes deep. Preopercle entire:

Villiform teeth in the jaws, vomer, and palatine bones. First dorsal fin continuous, the spines not strong: the second dorsal and anal with many more rays: as a rule a pair of pre-anal spines, remote from the rest of the fin. Scales small or rudimentary. Lateral line unarmed. Air-vessel simple. Pyloric appendages many.

331. *Seriola nigro-fasciata*, Rüpp. D., 5-6 (7)/$\frac{1}{31\text{-}33}$, A. 0-2/$\frac{1}{15\text{-}17}$, Cæc. pyl. 7. Height of body nearly 1/4 of the total length. Eyes, diameter 2/7 of length of head, one diameter from end of snout. Back crossed by six brown cross bands. Red Sea, east coast of Africa, seas of India to the Malay Archipelago and beyond.

332. *Seriola Dussumieri*, Cuv. & Val. D. 5/$\frac{1}{27}$, A. 2/$\frac{1}{18}$. Body with seven vertical brown bands. Gulf of Bengal: specimens not above 2 inches in length.

Genus—SERIOLICHTHYS, Bleeker.

Branchiostegals seven. Body oblong, compressed: abdomen rounded. Cleft of mouth of moderate depth. Preopercle slightly crenulated or entire. Villiform teeth on the jaws, vomer, and palatines. First dorsal fin continuous, the second and the anal with many more rays, with one or two detached finlets posteriorly: a pair of pre-anal spines remote from the rest of the fin may be present or absent. Scales cycloid, small. Lateral line unarmed.

333. *Seriolichthys bipinnulatus*, Quoy and Gaim. *Kulul*, Tam. D. 6/$\frac{1}{24\text{-}25}$+I., A. 0-2/$\frac{1\text{-}2}{16}$, L. l. 96, L. tr. 16/28. Height of body 1/5 of the total length. Eyes 1¼ diameters from end of snout. Two longitudinal blue bands, one from the eye to the end of the dorsal fin, the second to the middle of the caudal fin. From the east coast of Africa, seas of India to the Malay Archipelago and beyond.

Genus—NAUCRATES, Cuv.

Nauclerus, Cuv. & Val. (young).

Branchiostegals seven. Body oblong, sub-cylindrical. Cleft of mouth moderate. In the very young there is a spine at the angle of the preopercle, which is absorbed as age advances. Villiform teeth in the jaws, vomer and palatine bones. The first dorsal fin, which is continuous in the young, becomes reduced to a few spines in the adult: the second dorsal and anal with many rays; no spurious fins: in the young two pre-anal spines remote from the soft fin, and which become lost with age. Scales small: lateral line unarmed: a keel on either side of the tail. Air-vessel present. Pyloric appendages in moderate numbers.

334. *Naucrates ductor*, Linn. D. 3-6/$\frac{1}{26\text{-}28}$, A. 0-2/$\frac{2}{16\text{-}17}$, Cæc. pyl. 12-15, Vert. 10/16. Height of body 1/4 of the total length. Bluish, with five or six dark vertical bands. Most of the seas of both tropical and temperate regions.

Genus—CHORINEMUS, Cuv. & Val.

Scomberoides, Lacép.

Branchiostegals seven or eight: pseudobranchiæ. Body oblong and compressed. Eyes lateral. Cleft of mouth moderate or deep. Teeth villiform on jaws, also present on vomer and palatines. Two dorsal fins, the first (preceded by an immoveable, recumbent spine, directed forwards), is formed by a few free spines, but in a less number than the rays of the second dorsal or anal, the posterior rays of both of which last are either detached or semi-detached, a pair of pre-anal spines separated by an interspace from the remainder of the fin. Dermal scales lanceolate. Lateral line continuous, not keeled. Air-vessel bifurcated posteriorly. Pyloric appendages numerous.

335. *Chorinemus sancti-Petri*, Cuv. & Val. *Pallagay*, "a plank," Mal.: *Toal parrah*, "leather or skinned horse mackerel," or *Tarelul*, Tam. D. $7/\frac{1}{20\text{-}21}$, A. $2/\frac{1}{18\text{-}19}$. Height of the body 2/11 of the total length. The maxilla reaches to rather behind the vertical from the middle of the eye. Eyes two diameters from end of snout. A bluish grey spot on the opercle and a row of from six to eight above the lateral line, sometimes another row below. Summit of second dorsal black. East coast of Africa, seas of India to the Malay Archipelago and beyond: it attains at least 20 inches in length.

336. *Chorinemus tol*, Cuv. & Val. *Tol parah*, Tel. D. $7/\frac{1}{20}$, A. $2/\frac{1}{19}$. Height of body 2/11 of the total length. The maxilla reaches to beneath the anterior edge of the orbit. Four or five vertically oval spots on the lateral line, the last opposite the commencement of the second dorsal: top of dorsal fin black. Seas of India and Malay Archipelago: attaining at least 15 inches in length.

337. *Chorinemus lysan*, Forsk. *Toal parrah*, Tam. D. $7/\frac{1}{19\text{-}20}$, A. $2/\frac{1}{17\text{-}18}$. Height of body 1/4 of the total length. Eyes 3/4 of a diameter from end of snout. The maxilla reaches behind the posterior end of the orbit to a distance equal to 1/3 of its diameter, it is widened and rounded posteriorly. Six or seven large greyish oval spots above the lateral line. Red Sea, seas of India to Malay Archipelago and beyond.

338. *Chorinemus tala*, Cuv. & Val. *Tala parah*, Tel. D. $7/\frac{1}{19\text{-}20}$, A. $2/\frac{1}{17\text{-}18}$. Height of body 2/7 of the total length. Eyes rather above 1/2 a diameter from the end of snout. The maxilla reaches to below the posterior edge of the orbit. Six to eight indistinct spots along the sides. Seas of India: attaining 2 feet in length.

339. *Chorinemus tooloo*, Cuv. & Val. *Tooloo parrah*, Tel. D. $7/\frac{1}{19\text{-}20}$, A. $2/\frac{1}{18\text{-}19}$. Height of body 1/4 of the total length. Eyes one diameter from the end of snout. The maxilla does not extend posteriorly so far as to below the posterior margin of the orbit. About six spots exist along the sides: the tips of the dorsal and anal, also the lobes of the caudal, blackish. Red Sea and seas of India: attaining at least 18 inches in length.

Genus—TRACHYNOTUS* (Lacép), Cuv. & Val.

Acanthinion & Cæsiomorus, Lacép.

Branchiostegals seven: pseudobranchiæ absent. Body compressed, more or less elevated, with the frontal region prominent. Eyes lateral.

---

* *Cutalli*, Tam.

Snout obtuse and swollen: cleft of mouth small. Teeth on the jaws, vomer, and palatines, usually lost with age. Two dorsal fins, the first composed of spines (having a small connecting membrane), and a horizontal one directed forwards, anterior to the fin: second dorsal and anal each with more rays than spines in the first dorsal: the anterior portion of the second dorsal and anal pointed and elongated: two pre-anal spines separated by an interspace from the remainder of the fin: no finlets. Scales small. Lateral line unarmed. An air-vessel, bifurcated posteriorly. Pyloric appendages numerous, or in moderate numbers.

340. *Trachynotus Baillonii*, Lacép. *Botla-para*, Tel. D. $6/\frac{1}{22\text{-}23}$, A. $2/\frac{1}{23\text{-}23}$. Cæc. pyl. 12. Height of body from 1/3 to 3/10 of the total length. The anterior dorsal rays when laid backwards reach to about the middle of the fin: the anal to the end of its base: length of caudal lobes 2/7 of the total length. Lobes of dorsal, anal and caudal black, some white likewise on those of the tail: a row of three or five blotches along the sides. Red Sea, east coast of Africa, seas of India to the Malay Archipelago: it attains at least 19 inches in length.

341. *Trachynotus oblongus*, Cuv. & Val. D. $6/\frac{1}{22\text{-}23}$, A. $2/\frac{1}{20}$. Height of body 2/7, length of the caudal lobes 2/9 of the total length. The anterior dorsal and anal rays when laid flat reach to about the middle of their respective fins. Lateral line nearly straight. Three or four, sometimes five, indistinct grey blotches above the lateral line: a blackish anterior border to the dorsal and anal and the inferior border of the caudal fin. Pondicherry and Malay Archipelago.

342. *Trachynotus ovatus*, Linn. *Mookalee parah*, Tel. D. $6/\frac{1}{18\text{-}21}$, A. $2/\frac{1}{16\text{-}19}$, Cæc. pyl. 12. Height of body 1/2 to 3/7 of the total length. The length of the dorsal and anal lobes varies with age. More or less golden colour: the upper half of the first five dorsal rays tipped with black, and the fin generally with minute black points: sometimes the anal and caudal are similarly marked. Red Sea, east coast of Africa, through the seas of India to the Malay Archipelago and beyond: it attains at least 20 inches in length.

Genus—PSETTUS (Comm.), Cuv. & Val.

Monodactylus, Lacép.

Branchiostegals six: pseudobranchiæ. Body much compressed and elevated. Eyes lateral. Cleft of mouth small, snout short. Teeth villiform on jaws, vomer, palatine bones and tongue. A single dorsal fin with seven or eight spines: anal with three, continuous with the rest of the fin: ventrals rudimentary. Scales small, covering the vertical fins. Lateral line unarmed. Air-vessel present, bifurcated posteriorly. Cæcal appendages numerous.

343. *Psettus falciformis*, Lacép. D. $\frac{8}{27\text{-}29}$, A. $\frac{3}{29}$, L. r. upwards of 120. Extreme height 3/4 of the total length. Eyes, diameter 1/2 of length of head, 1/5 of a diameter from end of snout. Silvery, with blackish marks on the dorsal and anal lobes. Red Sea, seas of India to China: attaining at least 9 inches in length.

344. *Psettus argenteus*, Linn. *Kauki-sandawa*, Tel.: *Nga-pussoond*, Mugh: *Oo-chra-dah*, Andam. D. $\frac{8}{28\text{-}30}$, V. $\frac{1}{3\text{-}5}$, A. $\frac{3}{29\text{-}30}$, L. l. 75.

Eyes, diameter about 1/3 of length of head, 1/2 a diameter from end of snout. Silvery, with a wide black band from the nape to the centre of the eye, and a second from the three first dorsal spines to the opercle : dorsal and anal lobes stained with black. Red Sea, east coast of Africa, seas of India to the Malay Archipelago and beyond: it attains at least 7 inches in length.

Genus—*PLATAX*, Cuv. & Val.

Branchiostegals six: pseudobranchiæ. Body compressed and much elevated. Eyes lateral. Cleft of mouth small: snout short. Teeth setiform, with an external larger row, trilobed at their summits, teeth also present on the vomer. A single dorsal fin with from three to seven spines, which are nearly hidden: anal with three, continuous with the rest of the fin: ventrals well developed. Scales of moderate size or small, extended on to the vertical fins. Lateral line unarmed. Air-vessel simple. Pyloric appendages few.

345. *Platax vespertilio*, Bloch. Sadyan, Tam. D. $\frac{5}{35\text{-}37}$, A. $\frac{3}{26\text{-}28}$, L. r. 64. The height of the body is more than its length. Snout obtuse. Dorsal, anal and ventral fins elongated. Brownish, sometimes spotted : a brown vertical band through the eye. Red Sea, east coast of Africa, through the seas of India to the Malay Archipelago and beyond.

346. *Platax orbicularis*, Forsk. D. $\frac{3\text{-}5}{36\text{-}37}$, A. $\frac{3}{26\text{-}27}$, L. l. 55-60. The height of the body is less than its length. Snout prominent. Dorsal, anal and ventral fins moderately elongated. Brownish, with a rather indistinct vertical band through the eye. Red Sea, seas of India to the Malay Archipelago and beyond : attains at least 18 inches in length.

347. *Platax Raynaldi*, Cuv. & Val. D. $\frac{5}{32\text{-}33}$, A. $\frac{3}{23\text{-}24}$. Height of the body 4/5 of the total length. Snout obtuse. Dorsal, anal and ventrals elongated. Silvery, with a vertical ocular band and another indistinct one passing to the pectoral fin : dorsal, anal and ventrals tinged with black. Ceylon and Pondicherry: attaining at least 5 inches in length.

348. *Platax Leschenaldi*, Cuv. & Val. *Kahi-sandawa*, Tel. D. $\frac{5}{31\text{-}32}$, A. $\frac{3}{23\text{-}25}$. Height of the body equals its length. Dorsal, anal and ventral fins moderately elongated. A brown ocular band, and a blackish border to the caudal. Coromandel coast.

Genus—*PSENES* (¹), Cuv. & Val.

Branchiostegals six. Body compressed and elevated, with the frontal region swollen. Eyes lateral. Cleft of mouth shallow, with a short snout. A row of fine teeth in the jaws, none on the palate. Two dorsal fins, the first continuous, the second with more rays and similar to the anal, which last has two or three spines joined to the soft portion of the fin: no finlets. Lateral line unarmed. Air-vessel bifurcated posteriorly.

(¹) This *Genus* is amongst family *Carangidæ*, Günther (1860), in which the vertebræ are given at 10/14, but *Naucleres* with Vert. 10/16 is included, probably because no skeleton existed. Unfortunately my specimens of *Psenes* are not at present available for examination, but I doubt the genus having Vert. 10/14, and consequently my species *Cubiceps Indicus* will require re-examination. If, however, we go by the Vertebræ it belongs to (*Nomeidæ*). But the question is, where should *Psenes* be located ?

349. *Psenes auratus*, Cuv. & Val. D. $10/\frac{1}{2\frac{1}{5}}$, A. $\frac{3}{2\frac{1}{4}}$, L. l. 55, L. tr. 5/18. Height of body 1/3, length of head 1/4 of the total length. Eyes, diameter 1/3 of length of head, 1/2 a diameter from end of snout. Maxilla extends to below the anterior third of the orbit. Lower edge of preopercle indistinctly crenulated. Leaden colour, fins black: pectoral yellow: caudal with a yellow tinge. This is probably *P. Javanicus*, Cuv. & Val. Bay of Bengal, very common in Madras about October.

Genus—EQUULA ([1]), *Cuv. & Val.*

Branchiostegals from five to six: pseudobranchiæ. Body oblong or elevated and strongly compressed. Eyes lateral. Mouth very protractile. Lower edge of preopercle serrated. Minute teeth of equal size in the jaws: palate edentulous. A single dorsal fin, having less spines (8-10) than rays (15-17): anal with three spines continuous with the soft portion, which has less rays (13-14) than the soft dorsal: no finlets: ventrals thoracic. Scales small, cycloid, and generally deciduous. Lateral line unarmed, usually complete, but in some species ceasing beneath the end of the dorsal fin. Air-vessel terminating anteriorly in two horns. Pyloric appendages few.

A. Without a black blotch on the dorsal fin.

350. *Equula caballa*, Cuv. & Val. *Tottah-karah*, Tel.: *kulli-karah*, Tam. D. $\frac{8}{16}$, A. $\frac{3}{14\cdot 15}$. Height of body 1/2, length of head 3/11 of its total length. Mandible very concave. Lower edge of preopercle finely serrated: a pair of small spines above the anterior edge of the orbit. The second dorsal spine as long as the head. Scales well marked. Silvery: no black on dorsal fin: axil blackish. Seas of India to the Malay Archipelago.

351. *Equula edentula*, Bloch. *Koma-karah*, Tel. D. $\frac{8}{15\cdot 16}$, A. $\frac{3}{14}$, L. l. 60. Height of body 2/5, length of head 2/7 of the total length. A pair of small spines directed backwards at the anterior-superior angle of the orbit. Mandible very concave. Lower edge of preopercle scarcely serrated. Second dorsal spine 1/6 of the total length, compressed, arched, and with a serrated process in its lower anterior third: second anal spine similar to the second of the dorsal except in being equal to only 1/7 of the total length. Silvery, greyish along the lateral line: the soft dorsal stained with grey on its upper edge. Red Sea, seas of India to the Malay Archipelago and beyond: attaining 10 inches and more in length.

352. *Equula Dussumieri*, Cuv. & Val. D. $\frac{8}{16}$, A. $\frac{3}{14}$. Height of body 2/5, length of head 1/4 in the total length. A pair of small spines above the anterior-superior angle of the orbit. Mandible very slightly concave. Lower edge of preopercle minutely serrated. Second dorsal spine strong (serrated?), about 1/7 of the total length: and the second anal spine shorter. Vertical lines from the back to a little below the lateral line: the fins colourless, except the anal which is orange. Coromandel coast.

353. *Equula fasciata*, Cuv. & Val. *Karah*, Tel. D. $\frac{8}{16}$, A. $\frac{3}{14}$. Height of body 3/7, length of head 1/4 of the total length. A pair of small spines above the anterior-superior angle of the orbit. Mandible

---

(1) Fishes of this genus are termed *Caraputty* in Madras, and are eaten salted by natives who are suffering from malarious fevers. At Akyab they are called *Nga-dan-gah*.

concave. Lower edge of preopercle indistinctly serrated. Second dorsal and anal spines long, sometimes considerably so. Scales distinct. Silvery, with vertical lines from the back. Red Sea, seas of India to the Malay Archipelago and beyond.

354. *Equula lineolata*, Cuv. & Val. D. $\frac{8}{17}$, A. $\frac{3}{14}$. Height of body 1/3, length of head 2/9 of the total length. A pair of small spines above the anterior-superior angle of the orbit. Mandible slightly concave. Lower edge of preopercle serrated. Second dorsal spine equals about 1/6 of the total length. Scales minute or absent. Some irregular dark spots on the back, but no black mark on the dorsal fin. Coromandel coast to the Malay Archipelago.

b. With a black blotch or mark on the dorsal fin.

355. *Equula insidiatrix*, Cuv. & Val. *Paarl coorchee*, Mal. D. $\frac{8\cdot9}{16}$, A. $\frac{3}{14}$. Height of body 2/5, length of head 1/5 of the total length. A pair of minute supraorbital spines: the middle third of the supraorbital region finely serrated. Mandible not concave. Lower edge of preopercle finely serrated. Scales minute in the pectoral region. The lateral line does not extend to the tail. Three or four horizontal lines of dark spots with bronze reflections, form eight or ten vertical bands along the upper half of the body. Dorsal tipped with black. Seas of India and Malay Archipelago.

356. *Equula bindus*, Cuv. & Val. *Bindoo-karah*, Tel. D. $\frac{8}{16}$, A. $\frac{3}{14}$. Length of head 3/10, height of body 1/2 of the total length. A pair of well marked supraorbital spines: supraorbital edge finely serrated. Second dorsal spine and second anal spine both 1/7 of the total length: ventrals equal one diameter of the eye in length. A dark line along the base of the dorsal, the upper edge of which is margined with black. Coromandel coast and the Malay Archipelago.

357. *Equula splendens*, Cuv. *Goomorah karah*, Tel. D. $\frac{8}{15\cdot16}$. A. $\frac{3}{14}$. Length of head about 1/4, height of body 3/8 of the total length. A pair of small supraorbital spines. Mandible slightly concave. Lower preopercular margin finely serrated. Second dorsal spine 1/5 of the total length, and the second anal 1/3 shorter. Scales small. A dark line along the base of the second dorsal fin: upper half of the first dorsal edged with black. Red Sea, seas of India to the Malay Archipelago.

358. *Equula brevirostris*, Cuv. & Val. D. $\frac{8}{17}$, A. $\frac{3}{14}$. Its proportions as *E. splendens*, but its spines shorter and less strong. A black spot on the dorsal fin. Malabar.

359. *Equula daura*, Cuv. *Dacer karah*, Tel. D. $\frac{8}{15\cdot16}$ A. $\frac{3}{14\cdot15}$. Height of body nearly 1/3, length of head 4/15 of the total length. A pair of small supraorbital spines. Mandible slightly concave. Lower preopercular margin serrated. Second dorsal spine 3/14 in the total length, and second anal 3/4 of it. A dark line along the base of the dorsal: a darkish triangular spot between the occiput and dorsal fin: upper half of first dorsal blackish. Ceylon and Coromandel coast to the Malay Archipelago.

360. *Equula Blochii*, Cuv. & Val. D. $\frac{8}{16}$, A. $\frac{3}{14}$. Body more oblong than *E. daura*, head higher and shorter: its mandible inferiorly concave and its dorsal and anal spines nearly as strong as in *E. edentula*.

A longitudinal silvery band separates the greyish back from the silvery abdomen: first dorsal with a black blotch. Malabar.

361. *Equula nuchalis*, Tem. & Schleg. D. $\frac{8}{16}$, A. $\frac{3}{14}$. Height of body 3/8, length of head 1/4 of the total length. A pair of small supra- or bital spines. Mandible slightly concave. Lower preopercular margin finely serrated. Second dorsal spine 1/6 of the total length. Scales minute. An oval black blotch on the upper half of the first dorsal fin, and a transverse blotch on the nape of the neck. Andamans to China.

Genus—*GAZZA*, Rüpp.

Equula, sp. Cuv. & Val.

Branchiostegals five: pseudobranchiæ. Body oblong, elevated and compressed. Mouth very protractile. Lower preopercular margin serrated. Teeth in the jaws of moderate size, with a pair of canines in the upper. A single dorsal fin, having less spines (8) than rays (16): anal with three spines continuous with the soft portion which has less rays (14) than the soft dorsal: no finlets. Lateral line unarmed.

362. *Gazza minuta*, Bloch. D. $\frac{8}{16}$, A. $\frac{3}{14}$. Canines of moderate size. Second dorsal spine 3/8 of the height of the body, the second anal longer and stronger. Silvery, axil black: spinous dorsal brownish anteriorly. Ceylon to the Malay Archipelago.

363. *Gazza equulæformis*, Rüpp. D. $\frac{8}{16}$, A. $\frac{3}{14}$. Canine teeth large. Second dorsal spine 1/5 of total length: the second anal is stronger and not so long. Silvery, axil of pectoral br°wn. Red Sea, through the seas of India to the Malay Archipelago.

Genus—*LACTARIUS*, Cuv. & Val.

Branchiostegals seven: pseudobranchiæ. Body oblong, compressed. Eyes lateral. Cleft of mouth deep, with the lower jaw prominent. Preopercular margins entire. Teeth in jaws small, with one or two pairs of strong canines. Two dorsal fins, the first with seven or eight feeble spines, second and the anal with many rays, and no finlets: anal with three spines joined to the remainder of the fin. Scales cycloid, of moderate size, some over the second dorsal and anal rays. Lateral line continuous, unarmed. Air-vessel bifurcated both anteriorly and posteriorly. Pyloric appendages few.

364. *Lactarius delicatulus*, Bl. Schn. *Purruwah*, Mal.: *Chundawah* and *Sudumu*, Tel.: *Sudumbu*, Tam. D. 7-8/$\frac{1}{21\text{-}22}$ A. $\frac{3}{25\text{-}28}$ L. l. 74-80, Cæc. pyl. 6. A black spot at the upper and posterior part of the opercle. Seas of India to the Malay Archipelago and beyond.

Family—STROMATEIDÆ.

Branchiostegals from five to seven: pseudobranchiæ. Body oblong or slightly elongated and compressed. Gill-openings wide. Eyes lateral. The infraorbital bones do not articulate with the preopercle. Small teeth in the jaws, palate edentulous: barbed teeth extend into the œsophagus. One long dorsal fin without any distinct spinous division, or with rudimentary spines anteriorly: ventrals, when present, thoracic. No prominent papilla near the vent. Air-vessel, when present, small. Pyloric appendages few, in moderate numbers, or numerous. Vertebræ exceed 10-14.

Genus—*STROMATEUS*, Artedi.

Peprilus Cuv.: *Apolectus*, Cuv. & Val. (young having ventral fins); *Rhombus* (ventrals reduced to a spine) (Lacép.) Cuv. and Val.: *Seserinus* (with minute ventrals,) Cuv. & Val.

Branchiostegals from five to seven: pseudobranchiæ. Body compressed, more or less elevated. Cleft of mouth narrow or of moderate depth. Teeth small, in a single row in the jaws: palate and tongue edentulous: œsophagus armed with numerous barbed teeth. A single long dorsal and anal fin, having rudimentary spines anteriorly: ventral fins not present in the adult stage. Scales small, covering the vertical fins. Lateral line, as a rule, smooth (keeled in S. niger). Air-vessel absent. Pyloric appendages numerous.

365. *Stromateus cinereus*, Bloch. *Sudi-sandawah*, Tel.: *Grey pomfret* of Europeans. B. vii, D. 7-8/$\frac{1}{38-41}$. A. 5/$\frac{1}{32-39}$. Height of body 1/2, length of head 1/5 of the total length. Dorsal spines appear above the skin in a truncated form, as do also those of the anal: both dorsal and anal pointed anteriorly, the latter being most produced: pectoral 3/8 of the total length and similar to the dorsal. Lower caudal lobe twice as long as the upper, and equalling the length of the body without the caudal fin. Lateral line unarmed. Of a grey colour on the back, becoming lighter below: dorsal and anal grey, with dark points: pectoral buff: caudal orange, with dark points. Seas of India to the Malay Archipelago.

366. *Stromateus argenteus*, Bloch. *Vella voval*, Tam.: *Tella sandawa*, Tel. *White* or *silver pomfret*. B. vii, D. 5-9/$\frac{1}{39-43}$, A. 6/$\frac{1}{34-41}$. Length of head 2/9, height of body 1/2 of the total length. Dorsal and anal spines appear above the skin in a truncated form: the caudal lobes are equal to 1/3 of the total length. Lateral line not armed. Of a greyish colour along the back, becoming white on the abdomen, but everywhere covered by minute dark points: dorsal and anal grey: caudal and pectoral yellowish white, all having minute dark points on them. Seas of India to the Malay Archipelago and China, &c.

367. *Stromateus Sinensis*, Euphrasén. *Atookoia*, Tel.: *Vella arwoolee*, Mal.: *Mogang voval*, Tam. D. $\frac{1}{43-50}$, A. $\frac{1}{39-42}$. Length of the head 1/5, height of body 2/3 of the total length. The height of the dorsal and anal lobes equals 1/4 of the total length, there are some concealed spines anterior to both these fins: caudal lobes equal 1/5 of the total length. Lateral line unarmed. Greyish, with minute brown points. Seas of India, Malay Archipelago to China.

368. *Stromateus niger*, Bloch. *Nala-sandawah*, Tel.: *Baal*, Ooriah: *Curroopoo-voval*, Tam.: *Kar-arwoolee*, Mal.: *Ko-lig-dah*, And. D. $\frac{5}{42-44}$, A. $\frac{3}{55-57}$. Height of body from 2/5 to 1/3, length of head 2/9 of the total length. Lateral line in its straight portion is keeled, as in *Genus Caranx*. Deep brown, fins edged with black. Seas of India to the Malay Archipelago and China.

Family—CORYPHÆNIDÆ.

Branchiostegals from five to seven: pseudobranchiæ present or absent. Body oblong or elevated and compressed. Gill-openings wide. Eyes lateral. The infraorbital bones do not articulate with preopercle. Teeth in the jaws, present or absent on the palate, none in the œsophagus. One long dorsal fin, without distinct spinous division: ventrals thoracic, except in *Pteraclis*, when they are jugular. No prominent papilla near the vent. Air-vessel present or absent. Pyloric appendages few or many. Vertebræ exceed 10/14.

Genus—CORYPHÆNA (*Artedi*), *Cuv. & Val.*
Lampugus (immature), *Cuv. & Val.*

Branchiostegals seven: pseudobranchiæ absent. Body rather elongated and compressed. Cardiform teeth in the jaws: villiform ones on vomer, palatines, and a patch on the tongue. A single long dorsal fin extending from the occiput nearly to the caudal, but without distinct spines, neither are they apparent in the anal: ventrals well developed. Scales small. Air-vessel absent. Pyloric appendages numerous.

369. *Coryphæna hippurus*, Linn. *Badahlan*, Tam. B. vii, D. 58-63, A. 25-27. Length of head 1/6, height of body 1/7 of the total length. The anal fin commences in the centre of the distance between the anterior edge of the orbit and the base of the caudal fin. Back greyish, shot with gold: abdomen golden, covered with blue spots, which become black after death: dorsal fin light blue at the base, becoming black towards the summit. These dolphins are not uncommon in Madras, and are eaten by the natives. Seas of India, Malay Archipelago, &c.

Genus—MENE, Lacép.

Branchiostegals seven. Body oval, strongly compressed, with a prominent and sharp-edged abdomen. Mouth very protractile. Villiform teeth in the jaws, palate edentulous. A single long dorsal fin, without any distinct spinous portion: anal spineless, the rays enveloped in skin and having very broad free extremities: ventral with one spine and five rays, the first of which is very elongate. Air-vessel large and bifurcated posterioriy. Pyloric appendages numerous.

370. *Mene maculata*, Bl. Schn. *Ambutan-parah*, Tam. D. $\frac{3-4}{40-43}$, A. 30-33, Cæc. pyl. 25-30. Anterior to the dorsal fin are three spines concealed in the skin. Deep blue along the back, becoming silvery white on the sides and abdomen: from two to three rows of large spots along the superior half of the body above the level of the pectoral fin. Seas of India to the Malay Archipelago and beyond.

Family—NOMEIDÆ.

Branchiostegals five or six. Body oblong, more or less compressed. Gill-openings wide. Eyes lateral. Cleft of mouth shallow or deep. The infraorbital bones do not articulate with the preopercle. Teeth in the jaws, present or absent on the palate. Dorsal fin in two distinct portions, the spinous either continuous with or separated from the soft part, finlets sometimes present: anal spines generally indistinct: ventrals thoracic, with one spine and five rays. Scales cycloid, of moderate or small size. No prominent papilla near the vent. Air-vessel present, and pyloric appendages numerous, at least in genus *Nomeus*. Vertebræ exceed 10/14.

Genus—CUBICEPS, Lowe.

Antimostoma, Smith: *Navarchus*, Fillipi.

Branchiostegals five or six. Body oblong and compressed. Cleft of mouth shallow: snout rather obtuse: margin of preopercle entire (or crenulated). No teeth on the palate in mature specimens. The first dorsal continuous, with about ten spines: the second and the anal with a larger number of rays: no finlets. Scales cycloid, of moderate size. Lateral line unarmed. Vertebræ exceeding 10/14.

371. *Cubiceps Indicus*,[*] Day. D. $10/\frac{1}{14}$, A. $\frac{3}{13}$, L. l. 33. Height of body 3/8, length of head 3/10 of the total length. Preopercle crenulated. Silvery. Madras, to 3 inches in length.

---

[*] See remarks under genus *Psenes*.

Family—SCOMBRIDÆ, *Cuv.*

Branchiostegals seven or eight: pseudobranchiæ. Body oblong or slightly elongated and compressed. Gill-openings wide. Eyes lateral. The infraorbital bones do not articulate with the preopercle. Teeth present in the jaws, absent or present on the palate. Two dorsal fins, the first being distinct from the soft, which has more rays than the first has spines: finlets present or absent: ventrals thoracic (jugular in *Hypsiptera*). No prominent papilla near the vent. Side of tail sometimes keeled. Scales, if present, small. Air-vessel present or absent. Pyloric appendages moderate, numerous, or dendritical. Vertebræ exceed 10/14.

Genus—SCOMBER, *Artedi.*

Branchiostegals seven: pseudobranchiæ. Body rather elongated and compressed. Cleft of mouth deep. Small teeth on the jaws, vomer, and palatine bones. Two dorsal fins, the first spinous and separated by an interspace from the second, behind which and also posterior to the anal are five or six finlets: less spines in the first dorsal than there are rays in the second dorsal or in the anal: ventrals thoracic. Two slight keels on either side of the root of the caudal fin. Scales small. Air-vessel, when present, simple. Pyloric appendages numerous.

372. *Scomber microlepidotus*, Rüpp. D. $10/12 + V$, A. $\frac{1}{11} + V$. Length of head 1/4, height of body 2/9 of the total length. Two rows of distinct black spots during life, and a black mark concealed by the pectoral fin. Red Sea, Kurrachee, and China.

373. *Scomber kanagurta*, Cuv. & Val. *Ila*, Mal.: *Karnakita*, or *Karnang-kullutan*, Tam.: *Kanagurta*, Tel.: *Nga-congree*, "Large head," Mugh.: *Look-wa-dah*, Andam. D. $8\text{-}9/\frac{1}{11} + V$, A. $\frac{1}{11} + V\text{-}VI$. Length of head 3/10, height of body 1/4 of the total length. Air-vessel present. Greenish, shot with purple. A row of about 16 spots along the back, sometimes badly marked yellow longitudinal bands. Vertical fins with dark margins. Red Sea to the Malay Archipelago, rarely exceeding 10 inches in length.

374. *Scomber Reani*, Day. D. $6/\frac{1}{12} + V$, A. $\frac{1}{11} + V$. Length of head 1/4, height of body 1/4 of the total length. Air-vessel present. Back bluish green, becoming silvery-white on the abdomen: a dark longitudinal band along the lateral line, three above it, and two yellow ones below it. Andamans, common up to 12 inches in length.

Genus—THYNNUS, *Cuv. & Val.*

Branchiostegals seven. Body oblong, somewhat compressed. Cleft of mouth deep. Small teeth on the jaws, vomer and palatine bones. Two separate dorsal fins, the spines weak and finlets behind the soft dorsal and anal. Scales small: those in the pectoral region form a kind of corselet. Lateral line unarmed, a longitudinal keel on either side of the tail. Air-vessel, when present, simple. Pyloric appendages numerous.

375. *Thynnus affinis*, Cantor. *Suraly*, Tam. D. $15/\frac{5}{12} + VIII$, A. $\frac{3}{11} + VII$. Length of head nearly 1/4, height of body 1/4 of the total length. Bluish along the back, with a number of undulating oblique bands, below silvery: first dorsal marked with brown. Madras and Malay Archipelago: one specimen of this Bonito in Madras was 23 inches in length.

Genus—CYBIUM, *Cuv.*

Branchiostegals seven: pseudobranchiæ. Body somewhat elongated. Cleft of mouth deep. Teeth large and strong in the jaws, villiform on vomer, palatines, and tongue. Two dorsal fins, the first with feeble spines, extending to the commencement of the second; more rays in the second dorsal than spines in the first: seven or more finlets behind the second dorsal or anal. Scales, when present, rudimentary. A slight keel on either side of the caudal lobes. Air-vessel present.

376. *Cybium guttatum*, Bl. Schn. *Buck-ku*, Beluch.: *Arrakeeah*, Mal.: *Wungeran* and *Velrah meen*, Tam.: *Wingeram*, Tel.: *Nun-ni-o*: *Mugh*. D. 16-18/19-21 + VIII-X, A. 20-22 + VII-IX. Height of body 1/5, length of head 3/16 of the total length. Teeth lancet-shaped and compressed. Lateral line becomes undulating opposite the commencement of second dorsal. Bluish above, silvery beneath: back and sides with numerous round and oval spots that are most apparent after death: the membrane between the first and eighth spines black, the rest pure white, edged with black. Seas of India to the Malay Archipelago and beyond: this *Seir fish* attains at least $3\frac{1}{2}$ feet in length.

377. *Cybium Commersonii*, Lacép. *Chumbum*, Mal.: *Konam, Mahwu-laachi*, or *Ah-ku-lah*, Tam. D. $16\text{-}17/\frac{2\text{-}4}{15\text{-}13} + $ IX-X, A. $\frac{2\text{-}4}{15\text{-}13} + $ IX-X. Length of head 1/4 to 1/5, height of body 2/11 to 1/7 of the total length. Teeth lancet-shaped. Lateral line forms an angle beneath the twelfth or thirteenth dorsal ray. Bluish above, silvery below: first dorsal minutely dotted with brown, its upper edge black: basal half of pectoral black. After death numerous vertical undulating lines and spots appear on the sides. Red Sea, east coast of Africa, seas of India to the Malay Archipelago and beyond: it attains at least 4 feet in length.

378. *Cybium interruptum*, Cuv. & Val. D. $16/\frac{1}{17} + $ IX, A. $\frac{1}{17} +_{a}$IX. Length of head 1/6 of the total length. Teeth rather conical. Lateral line undulating on the posterior third of the tail. Oblong longitudinal spots form three bands on either side. Pondicherry.

379. *Cybium lineolatum*, Cuv. & Val. D. !16/16 + IX, A. $\frac{2}{14}$ +X. Length of head 1/5, height of body 2/13 of the total length. Blue above, becomes silvery below, with numerous short black longitudinal streaks. Seas of India and Malay Archipelago.

Genus—ELACATE, Cuv.

Branchiostegals seven: pseudobranchiæ. Body fusiform: head depressed. Cleft of mouth of moderate depth. Villiform teeth on the jaws, vomer, and palatine bones. The first dorsal reduced to a few free spines: the second with many rays and somewhat similar to the anal: no finlets. Scales very small. No keel on the side of the tail. Air-vessel absent. Pyloric appendages dendritical.

380. *Elacate niger*, Bloch. *Cuddul-verarl*, Tam.: *Peddah mottah*, Tel. D. 8/28-35, A. 25-29, Vert. 12/13. Height of body about 1/8 of the total length. Back and upper half of sides deep brown, below a lighter color, giving the fish the appearance of being longitudinally banded. Red Sea, east coast of Africa, seas of India to the Malay Archipelago and beyond: attaining at least 3 feet in length.

Genus—ECHENEIS, Artedi.

Branchiostegals seven or eight: pseudobranchiæ. Body elongated, fusiform, head depressed and superiorly furnished with an adhesive organ.

Eyes lateral or directed downwards and outwards. Cleft of mouth deep. Villiform teeth on the jaws, vomer, palatine bones, and generally on the tongue. The first dorsal fin modified, on the summit of the head and occiput, into an adhesive disk: a long second dorsal and anal: no finlets: ventrals thoracic. Scales very small. No keel on the side of the tail. Air-vessel absent. Pyloric appendages in moderate numbers.

381. *Echeneis naucrates*, Linn. Putthu-muday, 'catching mouth' Mal.: *Ubbay*, Tam.: *Ala mottah*, Tel. D. 22-25/33-41, A. 32-38. Length of disk 1/5 of the total length, generally with 22 transverse laminæ. Brownish or with indistinct longitudinal bands. Red Sea, seas of India, Malay Archipelago and beyond: attains at least 3 feet in length.

382. *Echeneis remora*, Linn. D. 17-18/22-24, A. 25, Cæc. pyl. 6. Length of disk 2/7 in the total. Brown. Same localities as *E. naucrates*. The most usual place where these *sucking-fishes* are obtained is from the back or sides of sharks.

Family—T R A C H I N I D Æ, *Günther.*

Percoidei, pt., *Sciænoidei*, pt., *Gobioidei*, pt., Cuv.: *Uranoscopidæ* pt., Richardson.

Branchiostegals from five to seven: pseudobranchiæ present. Body low and more or less elongated. Gill-openings more or less wide. Eyes lateral or superior. The infraorbital ring of bones does not articulate with the preopercle. Villiform teeth in the jaws, canines present or absent: no molars or cutting teeth: palate with or without teeth. One or two dorsal fins, the spines being less in number than the rays, anal similar to the soft dorsal: ventrals with one spine and five or six rays: lower pectoral rays simple or branched. No prominent papilla near the vent. Lateral line continuous, incomplete, or interrupted. Air-vessel present or absent. Pyloric appendages, when present, few or in moderate numbers. Vertebræ exceeding 10/14.

Genus—URANOSCOPUS, *Cuv.*

Agnus, pt. Günther.

Branchiostegals six: pseudobranchiæ. Body somewhat cylindrical. Head large, broad, and partly covered with bony plates. The opercles usually armed. Besides the inferior gill-opening there is generally a second above the opercle. Eyes on the upper surface of the head. Cleft of mouth vertical, generally with a filament below or before the tongue. Villiform teeth on the jaws, vomer, and palatine bones; no canines. Two dorsals, the first with from three to five spines: ventrals jugular: lower pectoral rays branched. Scales, when present, small. Lateral line continuous. Air-vessel absent. Pyloric appendages in moderate numbers.

383. *Uranoscopus marmoratus*, Cuv. & Val. D. $4/\frac{1}{12}$, A. 13. Length of head nearly 2/7 of the total. Five spines on the preopercle: a strong spine at the shoulder. Chestnut brown, becoming bluish white on the abdomen: the whole of the upper part of the head and body covered with bluish spots: fins dark grey, edged with white: upper three-fourths of first dorsal black. Coromandel coast, to 6 inches in length.

384. *Uranoscopus guttatus*, Cuv. & Val. D. 4-5/12-13, A. 13, Cæc. pyl. 8. Length of head 1/3 of the total. Five or six spines on the preopercle, and two on the shoulder. Slaty brown, with two or three rows of bluish white spots along the back and half way down the sides. First dorsal pure white, with its upper two-thirds black: second dorsal black along its upper two-thirds: caudal with a wide vertical black band covering its middle third: anal white with a black base: pectoral black, its lower margin edged with white. Seas of India.

Genus—ANEMA, Günther.

Uranoscopus, sp., Cuv. & Val.
Branchiostegals six: pseudobranchiæ. Body somewhat cylindrical. Head large, broad, and partly covered with bony plates. No superior gill-opening. Eyes on the upper surface of the head. Cleft of mouth vertical. Some of the bones of the head armed. No filament below or before the tongue. Villiform teeth on jaws, vomer and palatines. One continuous dorsal fin with less spines than rays, the latter similar to the anal: ventrals jugular: lower pectoral rays branched. Scales small. Air-vessel absent. Pyloric appendages in moderate numbers.

385. *Anema inerme*, Cuv. & Val. *Nillum Koranjan*, "a diver into the mud," Tam. D. $\frac{4}{16\text{-}18}$, A. 16-19, Cæc. pyl, 9 (8). Length of head 2/7 of the total length. Of a yellowish buff colour, becoming white beneath: the upper half of the body with large round or oval white spots: the dorsal yellowish reticulated with brown: a dark bar across the pectoral and caudal fins. Seas of India to Japan.

Genus—PERCIS, Bl. Schn.

Branchiostegals six: pseudobranchiæ. Body rather elongated and sub-cylindrical. Eyes lateral, directed somewhat upwards. Cleft of mouth slightly oblique. Opercle with two small spines: preopercle sometimes slightly serrated. Villiform teeth and canines in the jaws, teeth also on the vomer, but not on the palatines. The first dorsal fin with four or five spines, more or less continuous with the second dorsal, which has a moderate number of rays similar to the anal: ventrals slightly anterior to the pectorals, the lower rays of which are branched. Scales ctenoid, rather small. Air-vessel absent. Pyloric appendages few.

386. *Percis punctulata*, Cuv. & Val. D. 5/21, A. 17, L. l. 60. Preopercle entire. Fourth dorsal spine the longest. Brownish grey, body banded: some white spots over the snout: neck and sides spotted with brown: spinous dorsal black with a white edge: three rows of brown spots along the second dorsal: caudal with three or four rows of black spots between the rays: anal with two rows of dark spots. Andamans and Mauritius.

387. *Percis hexophthalma*, Cuv. & Val. D. 5/19—21, A. 17-18, L. l. 62, L. tr. 8/21. Along either side of the abdomen three or more white spots having a black centre: a black spot at the base of the dorsal, three rows along the soft dorsal, and one along the anal: a large dark blotch on the caudal. From the Red Sea through the seas of India: attaining at least 8 inches in length.

388. *Percis millepunctata,* Günther. D. 4/21, A. 18, L. 1. 62, L. tr. 4/8. "The height of the body $7\frac{2}{3}$ in the total length: the length of the head four times. The width of the interorbital space is one-fourth of the length of the snout. The second and third dorsal spines are the longest, much shorter than the intermaxillary bone; the ventrals reach the vent. Body with blackish dots not banded: snout and lips blackish. Coast of Ceylon, 80 lines long: not good state." (*Günther*).

Whether the following Madras species* is identical is very doubtful without comparing the specimens. D. 5/22, A. 18, L. 1. 58, L. tr. 4/12. Length of head 2/9, height of body 1/7, of the total length. Eyes $1\frac{1}{2}$ diameters from end of snout, 1/2 a diameter apart. Preopercle entire. Fourth dorsal spine the longest, rather longer than the snout: caudal cut square, its upper ray rather prolonged: ventrals do not quite reach the vent. Reddish, with six crossbands of a darker colour: a light pinkish band along the side, becoming yellow along the centre of the caudal fin, which is dark reddish in its lower half: some fine vertical bars in its upper half. Head spotted with dark purplish-black marks superiorly: three well-marked vertical blue bands cross the sub and inter-opercles. Dorsal fins yellow: the fins black in their lower half: the soft with a single row of spots along its upper half: anal yellowish in its lower two-thirds, with five rows of round canary-coloured spots: its outer third reddish: ventrals reddish, stained at their edges. Several specimens from Madras, up to 5 inches in length. I see I have marked it *P. Grayi.* Spec. nov. in my notes.

Genus—SILLAGO, Cuv.

Branchiostegals six: pseudobranchiæ. Body elongated, somewhat cylindrical. Head rather pointed, with its muciferous system well developed. Gill-openings wide. Eyes lateral, or directed slightly upwards. Cleft of mouth slight: the upper jaw the longer. Preopercle serrated or crenulated: opercle ending in a point. Villiform teeth in the jaws and vomer: none on the palatines. Two dorsal fins, the first with nine to twelve spines, which are less than the rays of the soft dorsal, which last is similar to the anal: ventrals thoracic, with one spine and five rays: lower pectoral rays branched. Scales ctenoid. Air-vessel simple. Pyloric appendages few.

a. *With an elongated dorsal spine.*

389. *Sillago domina,* Cuv. & Val. *Yarra-soring,* Tel.: *Tool-danti,* Ooriah. D. $9/\frac{1}{25-(27)}$, A. $\frac{1}{26-27}$, L. 1. 90, L. tr. 6/13, Cæc. pyl. 4. Length of head 1/4, height of body 1/7, of the total length. An external row of enlarged teeth in either jaw: second dorsal spine elongated, usually reaching the base of the caudal. Greenish yellow, shot with purple. Bay of Bengal to the Malay Archipelago.

b. *Without an elongated dorsal spine.*

390. *Sillago sihama,* Forsk. *Soring,* or *Tella-coring,* or *Arriti-ki,* Tel.: *Cullingah,* Tam.: *Cudeerah,* Mal.: *Thol-o-dah,* And.: *Nga-rui,* Mugh. D. $10\text{-}11/\frac{1}{20-23}$, A. $\frac{1}{22-23}$, L. l. 70-74, L. tr. 4/11, † Cæc. pyl. 3-4.

---

* I have another species from Madras with 3 spines at the angle of the preopercle and otherwise hardly corresponding to described specimens.

† Either this species, as at present described, consists of several distinct ones, or else the number of scales between the lateral line and the dorsal fin varies between 4 and 7.

Length of head from 3/10 to 1/4, height of body 1/6 of the total length. Olive green along the back, becoming light on the abdomen, the whole having a brilliant purple reflection: a silvery longitudinal band: minute black points on the dorsal and anal fins. Red sea, seas of India to the Malay Archipelago and beyond. It is known as *Whiting* at Madras.

391. *Sillago maculata*, Quoy and Gaim. D. $11/\frac{1}{20}$, A. $\frac{1}{2}$,L. l. 70. Length of head 1/4, height of body 1/5, of the total length. A longitudinal band along the side, and seven or eight dark blotches: the second dorsal dotted with brown. Andamans to Australia.

Genus—OPISTHOGNATHUS, Cuv.

Branchiostegals six. Body rather elongated, and posteriorly somewhat compressed: the upper profile of the snout parabolic. Head not cuirassed. Gill-membranes conjoined inferiorly. Eyes large and lateral. Cleft of mouth deep, the maxilla being considerably produced backwards. Opercles unarmed. Teeth in the jaws cardiform and fine: palate edentulous. A single dorsal fin, with weak and flexible spines: ventrals jugular: lower pectoral rays branched. Scales small, cycloid. Lateral line incomplete. Air-vessel small. Pyloric appendages absent.

392. *Opisthognathus nigromarginatus.*—Rüpp. D. 24, A. 18. Length of head about 1/4, height of body 1/6, of the total length. Greenish yellow: a black lower edge to the maxilla. Base of dorsal and anal spotted: a large oval blue mark between the fourth and eighth dorsal rays. Red Sea, east coast of Africa, Pondicherry.

Genus—PSEUDOPLESIOPS, Bleeker.

Branchiostegals six: pseudobranchiæ. Oblong, and somewhat compressed: eyes lateral. Preopercle entire. Villiform teeth and canines in the jaws: teeth also on the vomer. A single spineless dorsal fin: ventrals spineless thoracic. Scales cycloid, of moderate size. Lateral line interrupted.

293. *Pseudoplesiops typus.*—D. 25, A 16, V. 5, L. l. 36. Height of body 3/10 of the total length. Two outer ventral rays elongated. Greenish yellow. Ceylon and Malay Archipelago.

Family—BATRACHIDÆ Günther.

Pediculati, pt. Cuv.

Branchiostegals six: pseudobranchiæ absent. Body low and more or less elongated: head large: the muciferous system well developed. Gills three. The gill-openings narrow, in the form of a slit before the pectoral fin. Opercle armed. Teeth conical, and of moderate size or small. First dorsal with few spines: the second and the anal with many rays: ventrals with two rays, jugular: pectorals not pediculated. Scales, when present, small. Air-vessel present.

Genus—BATRACHUS, Bl. Schn.

Branchiostegals six. Body anteriorly somewhat cylindrical, and posteriorly compressed: head broad, depressed. Gill-openings narrow. Eyes lateral. Gape of mouth wide. Gill covers with several spines. No canine teeth. First dorsal with three strong spines. Scales, if present, very minute. Air-vessel divided with two lateral parts. Pyloric appendages absent. Vertebræ $\frac{12}{17-27}$.

394. *Batrachus Gangene*, Ham. Buch. D. 3/20, A. 16-18, V. 1/2. Gill covers with four spines directed backwards: no foramen in the axil. No scales. Brown, marbled with darker.

395. *Batrachus grunniens*, Bloch. D. 3/21, A. 15-16. Gill covers with four spines directed backwards. A foramen in the axil. No scales. Yellowish, marbled with darker. This is the common species all along the western coast of India.

396. *Batrachus trispinosus*, Gunther. D. 3/22, A. 18, V. 1/5. Gill covers with three spines directed backwards: a foramen in the axil. No scales. Brownish olive marbled with darker. West coast of India to the Malay Archipelago.

397. *Batrachus Dussumieri*, Cuv. & Val. D. 3/20, A. 16, V. 1/2. Gill covers with three spines directed backwards. Foramen? No scales. Brown, with three indistinct cross bands.

Family—PEDICULATI, *Cuv.*

Branchiostegals five or six: pseudobranchiæ absent. Head and anterior portion of the body large, the former depressed or compressed. Gills two and a half or three and a half: gill-opening reduced to a small foramen, situated in or near the axil. Eyes superior or lateral. Teeth villiform or cardiform. The spinous dorsal, when present, composed of a few isolated spines: the carpal bones prolonged, forming a sort of arm for the pectoral fin: ventrals, when present, jugular, with four or five soft rays. Skin smooth, or covered with small spines or tubercles. Air-vessel present or absent. Pyloric appendages few or absent.

Genus—ANTENNARIUS, *(Comm.) Cuv.*

Chironectes, pt. Cuv.

Head large, elevated, and compressed. No cleft behind the fourth gill: only one half of the anterior branchial arch is provided with lamellæ. Cleft of mouth nearly vertical. Eyes lateral. Cardiform teeth in the jaws and palate. The spinous dorsal in the form of three isolated spines, the anterior of which, situated upon the snout, is modified into a tentacle: the second dorsal of moderate length: anal short. Body smooth, granulated, or covered with minute spines, and sometimes with cutaneous tentacles. Air-vessel present. Pyloric appendages absent.

398. *Antennarius marmoratus*, Schn. D. 3/12, A. 7, P. 10, Vert. 9/9. A small knob on the rostral tentacle, when it is present. Colours various. Red Sea, East coast of Africa, through the seas of India to the Malay Archipelago, and beyond.

399. *Antennarius hispidus*, Bl. Schn. D. 3/12, A. 7. Anterior dorsal spine of moderate length, terminating in a lanceolate, ciliated flap: a cutaneous fold from the third dorsal spine to the commencement of the second dorsal fin. Skin rough, spinate: a few short fringes on body and head. Yellow, with brown spots and streaks, some radiating from the eye, other small ones on the dorsal fin. Seas of India to the Malay Archipelago.

400. *Antennarius nummifer*, Cuv. D. 3/12, A. 7, P. 10. Anterior dorsal spine short, with a cuneiform extremity. Skin with fine spines and a few cutaneous fringes. Reddish, with darker cloudings and scattered violet colored spots, the largest of which is beneath the posterior part of the second dorsal fin: bases of dorsal and anal fins marbled with brown. Red Sea through the seas of India.

Family—COTTIDÆ.

Branchiostegals from five to seven: pseudobranchiæ. Body more or less elongated, posteriorly compressed: head often large. Eyes lateral, or directed upwards. Cleft of mouth almost horizontal, lateral, or even nearly vertical. Some of the bones of the head usually armed: the suborbital ring of bones articulates with the preopercle. Teeth villiform in the jaws, present or absent on the vomer and palatines. Dorsal fins sometimes in two distinct portions, but more or less connected: the spines may be in excess of the rays, but as a rule there are more of the latter: anal generally similar to the soft dorsal: ventrals thoracic: pectorals with or without appendages. Body scaleless, scaled, or with a single series of plate-like scales. Air-vessel present or absent. Pyloric appendages few.

Genus—*SYNANCIDIUM*, Müll.

Branchiostegals seven. Head monstrous and irregularly shaped, but without sharp spines. Villiform teeth on the jaws and vomer but not on the palatines. The soft dorsal continuous with the spinous, less rays (6-9) than spines (13): anal with few (5) rays: no pectoral appendages. Scales absent: body and sometimes the head with skinny flaps. Air-vessel small. Pyloric appendages few.

401. *Synancidium horridum*, Linn. D.13/6, A $\frac{3}{5}$, Cæc. pyl. 3, Vert. $\frac{10}{14}$. Crown of the head irregularly saddle-shaped : a deep groove on the cheek. Brownish fawn colour above, becoming lighter below. Seas of India to the Malay Archipelago and beyond.

Genus—*SYNANCEIA*, Bl. Schn.

Branchiostegals seven. Head monstrous, irregularly shaped but spineless. Villiform teeth on the jaws; vomer and palatine bones edentulous. The soft dorsal continuous with the spinous, less rays (5-6) than spines (13-16): anal with few (5-7) rays: no pectoral appendages. Air-vessel present. Pyloric appendages few.

402. *Synanceia verrucosa*, Bl. Schn. D. $\frac{13}{6}$, A. $\frac{3}{5}$, Cæc. pyl. 4, Vert. $\frac{10}{14}$. Interorbital space very concave: a slight groove on the cheek. Greyish, the pectoral and ventrals with white edges: two whitish vertical bands on the caudal. Red Sea, east coast of Africa, seas of India to the Malay Archipelago and beyond: attaining at least 10 inches in length.

Genus—*MINOUS*.

Branchiostegals seven. Head and body somewhat compressed. Several of the bones of the head, and especially the preorbital, are armed. Eyes lateral. Villiform teeth in the jaws and vomer but none on the palate. A single dorsal fin with a moderate number (9 to 11) of spines which may be more or less than the rays (9-12): anal with from one to three spines, its rays somewhat less than those of the dorsal: pectoral rather elongate with a single free ray inferiorly. Scales absent. Air-vessel present. Pyloric appendages few.

403. *Minous monodactylus*, Bl. Schn. *Toomba* and *Cul plaachi*, Tam.: *Worrah minoo*, Tel. D. $\frac{9 \cdot 11}{12 \cdot 9}$, A. 9/11, Cæc. pyl. 4. Length of head 1/3, height of body 1/4, of the total length. A barbel on either mandible, sometimes a few more shorter ones. Greyish-brown superiorly, sides and abdomen flesh coloured with dark blotches and marks : dorsal fin light brown edged with black: anal buff, with its outer half black : pectoral black, its appendage white : caudal buff with three vertical brown bands.

Seas of India to the Malay Archipelago and beyond: attaining 4 or 5 inches in length.

Genus—*PELOR*, Cuv. & Val.

Branchiostegals seven. Head irregularly formed. Villiform teeth in the jaws, vomer and palatine bones. The three first dorsal spines connected by a membrane and a little distant from the others (12—14), which are somewhat isolated one from another, due to the interspinous membrane being deeply cleft: two free rays at the base of the pectoral fin, having a connecting membrane: ventrals thoracic. Scales absent. Head, body and fins with skinny appendages. Air-vessel small. Pyloric appendages few.

404. *Pelor didactylum*, Pall. D. $3/\frac{13\text{-}14}{9}$, P. 10+II, A. 12-13. Length of head 2/7, height of body 2/7, of the total length. Interorbital depression divided by a transverse ridge. Caudal yellow, with a black vertical band across its middle, and another at its base which crosses the last dorsal rays. Andaman Islands and Malay Archipelago: attaining at least 5½ inches in length.

Genus—*CHORISMODACTYLUS*, Richards.

Branchiostegals six. Head and body compressed. Bones of the head with osseous ridges, the preorbital, preopercle, and opercle with spines: a groove on the occiput. Villiform teeth on the jaws: palate edentulous. A single dorsal fin with more spines (13) than rays (9): anal with two spines: pectoral fin with three free rays: ventrals with one spine and five rays. Scales absent: some skinny appendages on the body. Air-vessel absent. Pyloric appendages few.

405. *Chorismodactylus multibarbis*, Richards. D. $\frac{13}{9}$, P. 15+III, A. $\frac{2}{5}$. Length of head 1/4, height of body 1/3, of the total length. Two rather large barbels on the mandible. Brownish, with a yellow shoulder mark, and two or three vertical orange bands: base of ventral and anal with fine white spots. Fins blackish brown, with a light band between the fourth and sixth dorsal spines: margin of pectorals orange: caudal with a black band at its base: outer third of anal blackish: free rays black, with white in their middle. Madras and China, to about 4 inches in length.

Genus—*POLYCAULIS*, Günther.

Synanceia, sp. Bloch.

Branchiostegals seven. Body anteriorly sub-cylindrical, posteriorly compressed: head broad, rather depressed. Eyes directed upwards. Preopercle armed. Gill-opening with an opening above the opercle as well as posteriorly. Villiform teeth in the jaws. A single dorsal fin with flexible spines and rays: anal somewhat elongated: no pectoral appendages: ventrals thoracic. Scales absent.

406. *Polycaulis elongatus*, Cuv. & Val. D. $\frac{0}{14+15}$, A. 15, Vert. 10/17. Height of body 1/5 of its total length. Upper surface of head uneven: no groove on the cheeks. Preopercle with four blunt spines. No vomerine or palatine teeth. Brownish with white dots: pectoral and anal with a dark edge. Seas of India and Malay Archipelago.

407. *Polycaulis adhesipinnis.* D. $\frac{1}{1}\frac{0}{4}$, A. 11, P. 13. Length of head 2/9, height of body rather above 2/9 of total length. Upper surface of head uneven. Five blunt denticulations along the margin of the preopercle. Gill-openings with a small rounded superior orifice in addition to their posterior one. Villiform teeth in jaws, vomer, and palatine bones. Lower four pectoral rays unbranched. Brownish, lighter below: pectoral and caudal stained at their edges. Calcutta.

Genus—*PLATYCEPHALUS,* * Bl. Schn.

Branchiostegals seven: pseudobranchiæ. Head broad, depressed and armed with spines. Eyes lateral or superior. Villiform teeth in the jaws, vomer and palatines, sometimes intermixed with larger ones. Two dorsal fins, the first having a small isolated spine anterior to it: the soft portion similar to the anal: ventrals thoracic: no pectoral appendage. Scales present, ctenoid, small or rather so. Lateral line complete, in some species armed with spines. Air-vessel absent. Pyloric appendages in moderate numbers.

a. *Lateral line unarmed.*

408. *Platycephalus Malabaricus,* Cuv. & Val. D. 1/8/12, A. 12, L. l. 75. Length of head 4/13 of the total length. Two preopercular spines, the upper long, the lower short. Superciliary ridge serrated. Greyish brown, spotted with darker. Malabar and Malay Archipelago, to 10 inches in length.

409. *Platycephalus serratus,* Cuv. & Val. D. 1/8/12, A. 11, L. r. 84. Length of head 1/4, height of body 1/7 of the total length. Infraorbital ring serrated. Interorbital space 1/4 of the length of the orbit, which is 1¼ diameters from end of snout. Upper preopercular spine strong, the lower short: two more on the border of the interopercle. Reddish brown, with vertical bands: fins with black spots: dorsal with a black blotch: ventrals superiorly blackish. Ceylon and Arrakan.

410. *Platycephalus Quoyi.* D. 1/8/11, A. 11, L. l. 110. Length of head 2/7 to 1/4, height of body 1/9 of the total length. Eyes 1½ diameters from end of snout. Anterior nostril with a red barbel. Upper preopercular spine rather above twice the length of the lower. Rose coloured, marbled and dotted with brown or black. First dorsal with an oblique black band, rays and spines spotted: upper half of pectoral dark and spotted, lower half yellow: end of caudal blackish. Madras to the Malay Archipelago: it attains at least 9 inches in length.

411. *Platycephalus punctatus,* Cuv. & Val. D. 1/8/11, A. 11, L. l. 80-90. Length of head 4/13 in the total length. Upper preopercular spine much longer than the lower. Greyish brown, with five vertical brown cross bands and numerous scattered black dots: first dorsal yellow superiorly, and posteriorly black: pectoral and second dorsal rays with brown spots: caudal with large, irregular, black spots. Ceylon to the Malay Archipelago.

412. *Platycephalus Tasmanius,* Richardson. D. 1/7/14, A. 14, L. l. 115. Length of head 2/7 of the total. Lower preopercular spine the longer. Brown, the posterior portion of the caudal black. Madras and Australia.

---

* *Ulu parti,* Tam.

413. *Platycephalus insidiator*, Forsk. *Irrwa*, Tel.: *Ool-parthy*, Tam.: *Nga-paying-ki*, Mugh.: *A-ra-wud-dah* or *Chou-ur-dah*, And. D. 1/6-7/13, A. 13, L. r. 94, L. tr. 12/24, Cæc. pyl. 14 (8). Length of head nearly 1/4, height of body 1/7 of the total length. Lower preopercular spine the longer. Brownish above, becoming dirty white beneath: fins spotted. Caudal yellow, with a deep black band having a white border obliquely crossing its upper lobe, a second along its lower lobe. Red Sea, East coast of Africa, seas of India to the Malay Archipelago and beyond.

b. *Lateral line wholly or partially spined.*

414. *Platycephalus scaber*, Linn. *Irrwa*, Tel. D. 1/7-8/12, A. 12, Length of the head nearly 1/4 of the total. Superciliary ridge serrated. Upper preopercular spine very long, would reach the orbit if laid forwards. Anterior portion of lateral line spiny. Greyish brown above, white below: its first dorsal with a black band: the second, caudal and pectoral with brown spots, the last half of the caudal black: anal whitish: end of the ventrals stained with black. Seas of India to the Malay Archipelago and beyond.

415. *Platycephalus neglectus*, Trosch. D. 1/7/12, A. 12, L. r. 105. Length of head nearly 1/4 of the total length. Superciliary ridge serrated. Upper preopercular spine would reach 1/2 way to the orbit if laid forwards, the lower shorter. Lateral line with about 55 spines. Brown, the first dorsal blackish, the other vertical fins with black spots. Madras to China. [A specimen of this genus in my collection agrees partially with both these last species: the upper preopercular spine if laid forwards reaching 2/3 of the way to the orbit. Body banded, with red spots, &c.]

416. *Platycephalus tuberculatus*, Cuv. & Val. *Ool-putti*, Tam. D. 1/7-8/11-12, A. 11-12, L. l. 55, L. tr. 6/13. Length of head about 1/3, height of body 1/6 of the total length. Eyes 1/4 of a diameter apart, 1¼ diameters from end of snout. Ridges serrated. Upper preopercular spine the longer, in a few specimens there are three: the length varies with age. Lateral line spiny in its first 15 scales. Brown, with darker vertical bands: pectoral with brown spots in its upper two-thirds, and a black margin with an external white edge along its lower border: outer half of ventrals grey. Ceylon and Coromandel coast.

Family—CATAPHRACTI, *Cuv.*

Triglidæ, pt. & *Pegasidæ*, pt. Günther.

Pseudobranchiæ present or absent. Suborbital ring of bones articulated with the preopercle. Head and body more or less angular, cuirassed with plates or keeled scales entirely covering the body. The opercular pieces may or may not be anchylosed to one another. Teeth present or absent in the jaws, in one species present on the vomer. One or two dorsal fins: pectorals may be simple, with or without free rays, or they may be divided by a notch into two portions and elongated or not so: ventrals thoracic, with five or less rays. Lateral line present, or absent. Air-vessel present or absent. Pyloric appendages, when present, in small, moderate numbers, or numerous.

Genus—DACTYLOPTERUS, *Lacép.*

Head with its surfaces more or less flattened, both laterally and superiorly bony. The angle of the preopercle and the shoulder bone produced into the form of long spines. Granular teeth in the jaws only. Two dorsal fins of about equal length: pectorals much elongated, with the lower portion detached from and shorter than the upper. Scales on body keeled and of a moderate size. Lateral line absent. Air-vessel in two lateral portions, each furnished with a large muscle. Pyloric appendages in moderate numbers or numerous.

417. *Dactylopterus Orientalis*, Cuv. & Val. D. 1/1/5/8, A. 6, Cæc. pyl. 18, Vert. 9/13. The first ray is elongated and on the occiput: the second, which is shorter, is half way between it and the rest of the dorsal fin. Brownish red, pectorals with bluish blotches and white spots: other vertical fins brown spotted. Seas of India to China, &c.

Genus—*PEGASUS*, Linn.

Branchiostegal one: pseudobranchiæ absent. Gills four: gill-opening narrow, in front of the pectoral fin. Body broad and depressed, covered with bony plates, which are anchylosed on the trunk and moveable on the tail. Gill cover formed of one bony plate, and a small interopercle concealed by it. No teeth. One short dorsal and anal fin opposite to one another: pectorals horizontal and long, composed of simple rays, some of which may be spinous: ventral with one or two rays, the outer being elongated. Air-vessel absent.

418. *Pegasus draconis*, Linn. D. 5, A. 5, P. 11, V. 2. Tail with eight rings, the first, fourth and fifth with a spine directed backwards. Body with brown markings: snout and last caudal rings black: pectoral with a white edge and white band. East coast of Africa, seas of India to the Malay Archipelago and beyond.

Family—GOBIIDÆ.

Pseudobranchiæ present, sometimes rudimentary. Gill-openings varying from extremely narrow to wide: the gill. membranes attached to the isthmus: four gills. Body generally elongated. Eyes lateral, occasionally prominent. The infraorbital ring of bones does not articulate with the preopercle. Teeth of varying characters, canines present or absent: inferior pharyngeal bones may be separated (*), or coalesced with a median suture, as in *Euctenogobius*. A single rayed dorsal fin, sometimes divided into two portions, the spines are flexible, whilst this part of the fin has less rays than the remainder: anal similar to the soft dorsal: ventrals sometimes united so as to form a disk, or arising close together. Scales and lateral line present or absent. Air-vessel generally absent. Pyloric appendages, if present, few.

Genus—*GOBIUS*, Artedi.

Branchiostegals five: pseudobranchiæ. Gill-openings of moderate width. Body low and elongated. Opercles unarmed. Teeth in several rows in the jaws: canines sometimes present. Anterior portion of the dorsal fin, with from five to six flexible spines: the posterior more developed and of the same character as the anal: ventrals united, forming a disk, which is not attached to the abdomen, each has one spine and five rays. Scales present or absent, and either cycloid or ctenoid. Lateral line absent. Air-vessel, when present, generally small. Pyloric appendages usually absent.

(*) If the *Pharyngognathi* are to be considered as a distinct order, the genus *Euctenogobius* must be placed in such and removed from the *Gobiidæ*. I have not considered the order *Pharyngognathi* as one which can be retained, but this is not the place to enter upon arguments on the subject, much of this portion of my report being a compilation.

a. Canine teeth present.

419. *Gobius Bynoensis*, Richardson. D. $6/\frac{1}{16}$, A. $\frac{1}{15}$, L. l. 65, L. tr. 16*. Height of body 1/5 to 2/11 of the total length. Eyes, diameter 1/5 of length of head, 1/2 a diameter apart. Caudal rounded. Scales ctenoid. Greenish, back with from eight to ten cross bars. Two longitudinal bands, the superior from the snout, at first black becomes yellow on the body: the second also black from the mouth to the pectoral fin. A dark spot at the upper part of the base of the caudal. Anal with a greyish edge. Andamans, Malay Archipelago to Australia.

420. *Gobius acutipinnis*, Cuv. & Val. *Mang-moo-goo-da-lah-dah*, Andam. D. $6/\frac{1}{10}$, A. $\frac{1}{17}$, L. l. 25, L. tr. 6. Height of body 1/5, of first dorsal 1/3 of the total length. Eyes, diameter 1/3 of length of head, 1/2 a diameter apart. Anterior dorsal rays filamentous: caudal pointed. . Greyish, with four or five dull blotches on the body, and a dark mark at the root of the caudal: a brown band from the eye to the corner of the mouth. Dorsals and caudal spotted. Coasts of India and the Andaman Islands.

421. *Gobius venenatus*, Cuv. & Val. D. $6/\frac{1}{10}$, A. $\frac{1}{9}$, L. l. 35, L. tr. 10. Height of body 1/5 of the total length. Eyes, diameter 1/5 of length of head, less than one diameter apart. Head as broad as high: cheeks with pores. Dorsal rays not prolonged: caudal somewhat pointed. Reddish olive, with a series of ill-defined darker spots along the sides. Fins dusky or black. Coasts of India.

422. *Gobius ocellatus*, Day. D. $6/\frac{1}{10}$, A. 10, L. l. 32, L. tr. 8. Height of body 1/7 of the total length. Eyes, diameter 1/9 of length of head, 1½ diameter apart. Head slightly broader than high. Second dorsal spine may be elongated: caudal wedge-shaped. Scales, ctenoid on body, cycloid on the head. Olive, a dark green spot above angle of opercle, and six indistinct blotches along the sides. Some spots on fins and a yellow ocellus at the top of the caudal in its last half. Anal whitish, basal half covered with fine black dots: ventrals yellow. Bombay, to 5½ inches in length.

423. *Gobius polynema*, Bleeker. D. 6/11, A. 10, L. l. 28-30, L. tr. 8. Height of body 1/8 of the total length. Eyes, diameter 1/4 of length of head, 2/3 of a diameter apart. Head as broad as high, several small barbels under the lower jaw. First dorsal lower than the body, caudal pointed. Scales ctenoid. Purplish black, fins blackish: an ocellus edged with white or yellow on the upper portion of the base of the caudal fin. Seas of India to China and Japan.

424. *Gobius Masoni*, Day. D. $6/\frac{1}{10-11}$, A. $\frac{1}{9}$, L. l. 27, L. tr. 10. Height of body 1/5 of the total length. Eyes, diameter 1/6 of length of head, 1½ diameters apart. Head as wide as long, fine wart-like glands on it. First portion of dorsal fin low: caudal wedge-shaped. Scales ctenoid. Air-vessel large. Olive, with numerous brilliant blue spots on the nape and behind the pectoral fin: some blackish ones along the side. Dorsal, anal, ventral and caudal black, pectoral yellow margined with black. Bombay, to 4 inches in length.

425. *Gobius Andamanensis*, Day. D. $6/\frac{1}{10}$, A. 10, L. l. 26-29, L. tr. 9. Height of body 2/9 of the total length. Eyes, diameter 1/4 of

---

(*) By L. tr. is signified in the Gobies the number of rows of scales between the origins of the second dorsal and anal fins.

length of head, 3/4 of a diameter apart. Head 2/3 as wide as long. Dorsal spines elongated : caudal rounded. Scales ctenoid. Olive, spotted with rusty : fins likewise spotted. Andamans.

426. *Gobius brevirostris*, Günther. D. $6/\frac{1}{10}$, A. 10, L. l. 46, L. tr. 13. Height of body 1/4 to 2/9 of the total length. Eyes, diameter 1/5 of length of head, one diameter apart. Width of head equals its length posterior to the centre of the orbit. Dorsal spines flexible, the longest equals the length of the postorbital portion of the head : caudal wedge-shaped. Scales ctenoid. Olivaceous, with an irregular band from the mouth to the caudal fin : a second from the eye to the axil, where there is a large light-blue ocellus. Kurrachee, also China, to nearly $3\frac{1}{2}$ inches in length.

427. *Gobius viridi-punctatus*, Cuv. & Val. *Nga-bu*, Mugh. D. $6/\frac{1}{10}$, A. $\frac{1}{9}$, L. l. 28-34, L. tr. 13. Height of body 1/8 of the total length. Eyes, diameter 1/4 of length of head. Spines of anterior dorsal somewhat prolonged : caudal wedge-shaped. Scales ctenoid. Brownish, blotched with darker, the whole of the dark portion of the body with small, metallic green spots. Seas of India, to 6 inches in length.

428. *Gobius Bleekeri*, Day. D. $6\frac{1}{9}$, A. $\frac{1}{9}$, L. l. 33, L. tr. 11. Height of body 1/6 of the total length. Eyes, diameter 1/3 of length of head, very slightly apart. Height of head less than its length. Dorsal spines villiform, caudal wedge-shaped. Scales ctenoid. Olivaceous, clouded with darker : a large bluish spot from the first to the fourth dorsal spine : fins spotted : a blue ocellus on the upper part of the base of the pectoral : ventrals blackish. Madras, to nearly 3 inches in length.

429. *Gobius caninus*, Cuv. & Val. D. $6/\frac{1}{8-9}$, A. $\frac{1}{9}$, L. l. 27-30, L. tr. 9. Height of body from 2/11 to 1/6 of the total length. Eyes, diameter 1/4 of length of head, very slightly apart. Small warts on the cheeks. Second and third dorsal spines produced : caudal rounded. Light brown, with darker blotches : some have a dark violet shoulder spot. Seas of India to China, also east Africa : attaining 6 inches in length.

430. *Gobius Madraspatensis*, Day. D. $6/\frac{1}{8}$, A. 9, L. l. 27, L. tr. 7. Height of body 1/4 of the total length. Eyes, diameter 1/4 of length of head, 1/4 of a diameter apart. Dorsal spines flexible, one or more slightly prolonged : caudal rounded. Scales ctenoid. Olive, with irregular brown blotches and dots. Five to eight narrow vertical black bands. Dorsals and caudal spotted : ventrals tipped with black. Madras.

431. *Gobius Neilli*, Day. D. $6/\frac{1}{9}$, A. 9, L. l. 28, L. tr. 7. Height of body 1/4 of the total length. Eyes, diameter 2/7 of length of head, not 1/4 of a diameter apart. Spines of first dorsal prolonged : caudal somewhat pointed. Scales ctenoid. Ochreous colour, with rusty spots : first dorsal with a black mark between the first and fifth spines, to about half the height of the fin : caudal barred. Madras.

432. *Gobius gobiodon*, Day. D. $6/\frac{1}{8}$, A. 10, L. l. 22, L. tr. 9. Height of body 1/3 of the total length. Eyes 1/2 a diameter from end of snout. Gill-opening narrow. Warty tubercles on head. No elongated dorsal spines. Scales ctenoid. Brownish, ventrals nearly black. Andamans and Nicobars, to about 2 inches in length.

b. *Canine teeth absent.*

433. *Gobius grammepomus*, Bleeker, D. $6/\frac{1}{10}$, A. $\frac{1}{10}$, L. l. 50-55. Height of body from 2/11 to 2/15 of the total length. Eyes, diameter 1/4 to 1/5 of the length of the head. Height of head equals its width. Spinous dorsal somewhat lower than the body. No scales on head. Green, spotted with brown, a streak from the eye to the maxilla. Dorsal and caudal spotted, a blotch at the base of the pectoral. Seas of India to the Malay Archipelago.

434. *Gobius Stoliczkæ*, Day. D. 6/11, A. 11, L. l. 48, L. tr. 14. Height of body 2/9 of the total length. Eyes, diameter 1/6 of length of head, one diameter apart. Spinous dorsal without any prolongations. Scales ctenoid, some on head and cheeks. Olive, marbled with darker: head spotted with black: a dark ocellus at the base of the pectoral: dorsal spotted, and a black mark at the posterior portion of the base of the first: caudal barred. Andamans.

435. *Gobius planifrons*, Day. D. $6/\frac{1}{10}$, A. 10, L. l. ca. 46, L. tr. 15. Height of body 2/11 of the total length. Eyes, diameter 1/6 of length of head, $1\frac{1}{4}$ diameters apart. Head as broad as long, excluding the snout. First dorsal half as high as the body below it: caudal rounded. Scales finely ctenoid. Olive, fins very dark grey, second dorsal spotted: a black blotch at the base of the pectoral. Bombay, to 4 inches in length.

436. *Gobius elegans*, Cuv. & Val. D. $6/\frac{1}{10}$, A. $\frac{1}{8}$, L. l. 36, L. tr. 9. Height of body 1/6 of the total length. First dorsal rather lower than the body: caudal rounded. Buff dotted with brown, and three or four indistinct lines on the upper half of the side. A dark spot behind the orbit, another at the upper part of the root of the pectoral, and a third at its lower portion. Dorsal and caudal spotted. Bombay to the Malay Archipelago.

437. *Gobius macrostoma*, Steind. D. 6/11, A. 10, L. l. 33. Height of body 2/15 of the total length. Eyes, diameter 1/7 of length of head, nearly one diameter apart. Head broader than high. Spinous dorsal not so high as the body, caudal rounded. Vertical fins with dark streaks. Bombay.

438. *Gobius ornatus*, Rüpp. D. $6/\frac{1}{10}$, A. $\frac{1}{8-9}$, L. l. 26, L. tr. 7. Height of body 1/6 to 1/7 of the total length. Eyes, diameter 2/7 of length of head, very close together. Head rather broader than high. Dorsal lower than the body. Greyish brown, with three or four horizontal rows of oblong black patches. Fins, except the ventral, dotted with black. Red Sea, Andamans, to the Malay Archipelago.

439. *Gobius albo-punctatus*, Cuv. & Val. D. $6/\frac{1}{8}$, A. $\frac{1}{8}$, L. l. 35-40, L. tr. 10. Height of body 2/11 of the total length. Eyes, about one diameter from end of snout, very close together. Head broader than high. First dorsal somewhat lower than the body. Brownish, irregularly marbled, sides of head and body studded with white spots. Dorsal and caudal fins spotted. Red Sea, Andamans, Mauritius, Feejee Islands, and Port Essington.

440. *Gobius biocellatus*, Cuv. & Val. D. $6/\frac{1}{8}$, A. $\frac{1}{8}$, L. l. 35-38, L. tr. 9. Height of body 1/6 of the total length. Eyes, diameter 1/5 of length of head, 1 to $1\frac{1}{4}$ diameters apart. Height and breadth of head

equal half its length. First dorsal spine generally, but sometimes the fifth, the longest. Scales ctenoid. Greyish brown, with irregular dark blotches along the sides. Dorsal fin with several irregular whitish lines along its lower half: a black blotch with a white edge between its fifth and sixth spines. Coasts of India, to 6 inches in length.

441. *Gobius criniger*, Cuv. & Val. D. 6/$\frac{1}{5}$, A. $\frac{1}{5}$, L. l. 34, L. tr. 9. Height of body 1/6 of the total length. Eyes, diameter 1/4 of length of head. Width of head equals 3/4 of its height. Dorsal fin as high as the body, second spine sometimes prolonged. Pale ochreous, spotted and blotched with black, caudal and anal fins with black edges. Coasts of India to the Malay Archipelago.

442. *Gobius spectabilis*, Günther. D. 6/10, A. 9, L. l. 34, L. tr. 9. Height of body 1/7 of the total length. Eyes, diameter 1/6 of length of head, rather above one diameter apart. Head broader than high. Dorsal fin slightly higher than the body, spines filamentous. Yellowish brown, with indistinct blotches along the sides: fins spotted. This species is distinguished from *G. giuris*, because the caudal fin (in the single specimen procured at Ceylon) is elongated, pointed, and 1/4 of the total length. Ceylon, probably from fresh-water: attaining at least 10 inches in length.

443. *Gobius sadanundio*, Ham. Buch. *Ontoo-mossal*, Mugh. (Akyab.) D. 6/$\frac{1}{4}$, A. $\frac{1}{8}$, L. l. 28, L. tr. 8. Height of body 1/4 of the total length. Eyes, diameter from 1/3 to 2/7 of the length of the head, 1$\frac{1}{4}$ diameters apart. Head wide, equalling the length without the snout. Second and third dorsal spines elongated: caudal pointed. Scales ctenoid. Olive, with very large deep black, white-edged blotches on the body: first dorsal black, with a white ring on its two last rays: second dorsal and caudal spotted. Mouths of the Ganges, and along the Chittagong and Burmese coasts, attaining at least 3 inches in length.

444. *Gobius gutum*, Ham. Buch. D. 6/$\frac{1}{1}$, A. 11. Head and eyes small. Caudal rounded. Scales ctenoid. Greenish, with many black dots clustered into irregular spots resembling clouds in form: dorsal and caudal fins spotted. Lower portion of the Hooghly, to 3 or 4 inches in length.

445. *Gobius nunus*, Ham. Buch.? D. 6/11, A. 11, L. l. 25, L. tr. 9. Height of body 1/8 of the total length. Eyes in the anterior half of the head, one diameter from end of snout, 1/2 a diameter apart. Caudal wedge-shaped. Chestnut colour: three broad white bands across the head and nape: three short white bands descend from the base of the first dorsal fin. Andamans and mouth of Hooghly?

Genus—EUCTENOGOBIUS, Gill.

Branchiostegals six. Gill-openings rather narrow, not extending to the lower surface of the head. Body elongated. Eyes not prominent. Teeth in one row in the upper jaw, in several rows in the lower: no canines: palate edentulous. Inferior pharyngeal bones of an elongated triangular form, having a median longitudinal suture. Dorsal fin divided into two portions, the first with six flexible spines: ventrals united, forming a disk, but not adherent to the abdomen.

446. *Euctenogobius cristatus*, Day. D. 6/14, A. 14, L. l. 48, L. tr. 11. Height of body 1/6 of the total length. Eyes close together, diame-

ter 1/4 of length of head. Width of head equal to its length behind the middle of the eye: a low crest on the nape: several rows of warts on the cheeks. Caudal elongated and pointed. Scales cycloid, none on the head. Olivaceous: a light ocellus having a brown edge at the base of the pectoral fin: body blotched and spotted: some black bars on upper half of dorsal, and a badly defined violet ocellus edged with yellow on its last ray: caudal spotted in its upper half. Bombay and Madras: in the month of March, at the former place, they were breeding.

Genus—APOCRYPTES, Cuv. & Val.

Pseudobranchiæ rudimentary. Gill-openings of moderate width. Body elongated. Teeth conical in a single fixed row in either jaw, with canines in the lower (a pair being generally present above the symphysis posterior to the fixed row), and frequently in the upper as well. The first portion of the dorsal fin containing five or six spines, and either distinct from or continuous with the soft portion, which is similar to the anal. Ventrals united, forming a disk, and not adherent to the abdomen. Scales, when present, small, becoming larger posteriorly.

447. *Apocryptes macrolepis,* Bleeker. D. 6/29, A. 28. Height of body 1/11 of the total length. Eyes, diameter 1/5 of length of head. A pair of large canines near the symphysis of the lower jaw. Dorsal fins continuous at their bases. Scales small. Green, clouded with brown: dorsals spotted with black: caudal nearly black with green rays. Andamans and Malay Archipelago.

448. *Apocryptes serperaster,* Richardson. D. 6/27, A. 27. Height of body 1/9 to 1/11 of the total length. Eyes, one diameter from end of snout. A pair of small posterior canines near the symphysis of the lower jaw, another on each side. Dorsal fins not continuous. Greenish olive: caudal darker. Bengal and China.

449. *Apocryptes Borneensis,* Bleeker. D. 5/27-30, A. 26-27. Height of body 1/12 of the total length. Eyes close together, diameter 1/5 of length of head. A pair of posterior canines near the symphysis of the lower jaw. Scales exceedingly minute. Slate colour, tinged with violet along the abdomen: some black dots in the upper fourth of the dorsal, and upper half of caudal. Akyab and Malay Archipelago, to 4 inches in length.

450. *Apocryptes rictuosus,* Cuv. & Val. D. 6/24-27, A. 26-29, L. l. ca. 75. Height of body 1/13 of the total length. Eyes, diameter 1/7 of length of head. About 28 teeth in the lower jaw, and a pair of posterior canines near the symphysis. Dorsal fins continuous. Greyish, lighter towards the abdomen: some ill-defined oblique bands pass downwards and forwards from the base of the dorsal fin: caudal and last third of the dorsal dark grey, sometimes spotted with brown. Seas and estuaries of India: it attains 7 inches in length.

451. *Apocryptes Cantoris,* Day. D. 6/27, A. 26. Height of body 1/6 of the total length. Eyes in the second fifth of the head, 2/3 of a diameter apart. A pair of posterior canines near the symphysis of the lower jaw. Dorsal fins not continuous. Scales minute. Olive, cheeks and under surface of head with black spots: first dorsal dark with three black horizontal bands: the upper portion of the caudal dark and spotted. Andamans.

452. *Apocryptes Madurensis*, Bleeker. D. 6/24, A. 22-23, L. l. 53-55. Height of body 1/7 of the total length. Eyes in the second fifth of the length of the head, 1/2 a diameter apart. Teeth, 25 to 30 bilobate ones on either side of the lower jaw, and a pair of fine posterior canines near the symphysis. Scales cycloid. Olive brown, the upper third of the first dorsal between the second and fourth spines with a black mark: pectorals and caudal nearly black, with a white lower edge: anal black: ventrals white. Coasts of India and to the Malay Archipelago: it attains about 4 inches in length.

453. *Apocryptes glyphidodon*, Bleeker. D. 6/22, A. 22, L. l. 50. Height of body 1/9 of the total length. Eyes, diameter 1/6 of length of head. Teeth, about 25 bilobate ones on either side of the lower jaw, and a pair of fine posterior canines near the symphysis. Greyish, with five light brown spots along the sides, forming bands over the back: numerous fine dots over the head and body: pectorals deep olive with a white edge: the other fins dark. Bombay, also Malay Archipelago, to 4 inches in length.

454. *Apocryptes Andamanensis*, Day. D. 6/13, A. 13, L. l. ca. 60. Height of body 1/5 of the total length. Eyes, diameter 1/5 of length of head. A crest on the nape of the neck. A pair of posterior canines near the symphysis of the lower jaw. Dorsal fins separated by a notch. Deep green, with a dark mark at the base of the caudal: fins dark. Andamans, in brackish water, to 4 inches in length.

455. *Apocryptes dentatus*, Cuv. & Val. D. 5/32, A. 31. Height of body 1/14 of the total length. Eyes, diameter 1/6 of length of head. A pair of posterior canines in the lower jaw near the symphysis. Scales very small. Brownish, caudal dotted with brown. Coromandel coast of India.

456. *Apocryptes lanceolatus*, Bl. Schn. *Nulla ramah*, Tel.: *Pitallu*, Ooriah. D. 5/31-32, A. 29-30. Height of body 1/7 to 1/9 of the total length. Eyes small, 1/4 of a diameter apart, 1¼ from end of snout. A pair of small posterior canines in the lower jaw near the symphysis. Scales very small. Brown, blotched with darker: fins with dark spots and blotches. Shores of India and the Malay Archipelago.

457. *Apocryptes bato*, Ham. Buch. *Rutta*, Ooriah. D. $5/\frac{1}{20-21}$, A. $\frac{1}{22}$. Height of body 1/7 to 1/8 of the total length. Eyes, diameter 1/6 of length of head, 1½ diameters apart. A pair of moderately-sized posterior canines near the symphysis of the lower jaw. Scales minute. Greenish white, with about twelve ill-defined narrow bands, descending towards the abdomen: scales with brown points: fins white, but also with minute dots: a dark band at the base of the pectoral. Orissa and Lower Bengal, within tidal reach, attaining 5½ inches length.

Genus—$GOB^{I}ODON$, *Bleeker*.

Gill-openings of moderate width. Body oblong and compressed: head large. Teeth conical and fixed: a pair of canines generally present near the symphysis of the lower jaw. Two dorsal fins, the first with six spines and united at the base to the second; ventrals united. Scales absent.

458. *Gobiodon quinque-strigatus*, Cuv. & Val. D. $6/\frac{1}{10}$, A. $\frac{1}{8-9}$. Height of body 1/3 of the total length. Eyes small, 1 diameter apart.

and 1½ from end of snout. Canines in the lower jaw. Head with five vertical orange stripes: two irregular bands of the same colour pass along the body, breaking up into blotches, and a row of spots exists along the posterior third of the body. Andamans and Nicobars, to 2¼ inches in length: is also found in the Malay Archipelago.

459. *Gobiodon Ceramensis*, Bleeker. D. 6/$\frac{1}{10}$, A. ½. Height of body 2/7 of the total length. Eyes rather less than 1 diameter from end of snout. Small canines in the lower jaw. Dark brown: fins nearly black, except the caudal, which is almost white. Galle, also the Malay Archipelago.

Genus—PERIOPHTHALMUS, *Bl. Schn.*

Branchiostegals five: pseudobranchiæ rudimentary. Gill-openings rather narrow. Body sub-cylindrical. Eyes placed close together, very prominent, and the outer eyelid well developed. Teeth in both jaws, erect and conical. Two dorsal fins, the first with a varying number of flexible spines: base of pectoral muscular: ventrals united in their lower two-thirds: caudal with its inferior edge obliquely truncated. Air-vessel absent. Scales small or of moderate size, ctenoid, covering the body and the base of the pectoral fins.

460. *Periophthalmus Schlosseri*, Pall. D. 7/$\frac{1}{12}$, A. 14. (See F. W. F. report No. 21). Coasts of Bengal, Burma, and Andamans.

461. *Periophthalmus Koelreuteri*, Pall. *Chood-mud-dah*, And. D. 10-15/12-13, A. 11-14. L. l. 75-100. Height of body 2/9 to 2/11 of the total length. The first dorsal fin is very variously formed, sometimes produced, at other times not so. Second dorsal and anal generally banded, but the other colours are as diverse as the forms. Seas of India, ascending tidal rivers.

Genus—BOLEOPHTHALMUS, *Cuv. & Val.*

Branchiostegals five: pseudobranchiæ, a slit behind the fourth gill. Gill-openings narrow. Body sub-cylindrical: head oblong. Eyes very prominent, situated close together, the outer eyelid well developed. Cleft of mouth nearly horizontal, the upper jaw sometimes slightly the longer. Teeth in a single row, the anterior ones in the upper jaw enlarged and stronger than the others: those in the lower jaw of about equal size and in a single horizontal row, having a pair of posterior canines near the symphysis. Two dorsal fins, the anterior with five flexible spines: the second many rayed and about equal to the anal: pectoral with its basal portion muscular and generally free: ventrals united: caudal with its inferior edge obliquely truncated. Air-vessel absent. Scales when present rudimentary or small, generally largest posteriorly.

462. *Boleophthalmus Dussumieri*, Cuv. & Val. D. 5/$\frac{1}{26-27}$, A. $\frac{1}{25}$, L. r. ca. 125, L. tr. 12. Height of body 1/8 of the total length. Eyes, diameter 1/7 of length of head. Teeth three on either side of middle of upper jaw elongated and directed downwards: a pair of posterior canines in the lower jaw near the symphysis. Scales distinct on the body, but somewhat indistinct on the head. Grey, first dorsal purplish, covered with round black spots: the second with two or three rows of oblong white spots: caudal black. Bombay and coast of Sind to 6 inches in length.

463. *Boleophthalmus dentatus*, Cuv. & Val. D. $5/\frac{1}{20}$, A. $\frac{1}{25-26}$. Height of body 2/15 of the total length. Eyes, diameter 1/7 of length of head. Teeth three on either side of the middle of the upper jaw elongate, directed downwards, and slightly forwards. Scales only distinct in a narrow band on the abdomen, along either side of the anal fin, and a few towards the head, elsewhere they look like rough points. Olive grey, with dull vertical bands on the body, six or eight of which are continued to the lower half of the second dorsal fin. First dorsal purplish, covered with black spots, having whitish edges, whilst the upper margin of the fin is yellowish: second dorsal with about five rows of oblong white spots, and some black ones having white edges along the first-half of its base. Upper margin of the caudal with a white band and yellow spots between its black rays. Bombay and Kurrachee, to $7\frac{1}{2}$ inches in length.

464. *Boleophthalmus viridis*, Ham. Buch. D. $5/\frac{1}{26}$, A. $\frac{1}{26}$. Height of body 2/11 of its total length. Eyes, diameter 1/6 of length of head. Scales minute. Greenish, sides vertically banded. Caudal spotted. Bengal and Andamans to the Malay Archipelago and China.

465. *Boleophthalmus Boddaerti*, Pall. *Dahrm-brow*, Mugh. D. $5/\frac{1}{26}$, A. $\frac{1}{26}$, L. r. 70, L. tr. 21. Height of body 1/5 of the total length. Eyes, diameter 1/6 to 1/7 of length of head. Teeth, the six central ones in the upper jaw elongated: a pair of moderately sized posterior canines near the symphysis of the lower jaw. Inferior pharyngeal bones spoon-shaped, approximating along the inner side with a row of fine teeth merely at the opposed margins. Greenish blue, with seven or eight vertical black bands: body covered with opaque blue spots: first dorsal likewise blue-spotted, and three rows on the second, with four large series along its base. Pectoral orange with a black edge: anal and caudal blackish: ventrals purplish. Coasts and estuaries of India and Burma to the Malay Archipelago. It climbs up rocks and pieces of wood, when it resides in shallow estuaries. If kept damp it lives some time out of water, and is brought in considerable numbers to the Bombay markets in baskets, covered with a wet cloth.

466. *Boleophthalmus sculptus*, Günther. D. 5/25, A. 22, L. tr. 8. Scales said to differ from the last in being smaller. Colours much the same, except in wanting the blue spots. India.

467. *Boleophthalmus pectinirostris*, Linn. D. $5/\frac{1}{22-23}$, A. $\frac{1}{22-23}$, L. tr. 17. Height of body 2/9 of the total length. Eyes, diameter 1/4 of length of head. The body with small dark tubercles, and verdigris spots. Vertically placed lilac spots on the first dorsal fin, and six or seven transverse ones on the second: some also on the caudal: the other fins brownish. Coast of Burma to the Malay Archipelago.

Genus—ELEOTRIS, *Gronov.*

Philypnus, Cuv. and Val.: *Bostrichthys*, Dum.: *Culius, Butis, Valenciennea, Belobranchus & Eleotrioides*, Bleeker: *Lembus*, Günther.

Branchiostegals from four to six, occasionally terminating anteriorly in a spine: pseudobranchiæ present. Gill-openings of moderate width. Body sub-cylindrical: head oblong. Eyes lateral, not prominent, and of moderate size. Teeth small, present or absent on the vomer. Two dorsal fins, the anterior with few (5-8) spines, and these sometimes filamentous:

base of pectoral slightly muscular: ventrals placed close together but not united. Scales present. Air-vessel large. Anal papilla distinct. Pyloric appendages generally absent.

468. *Eleotris Jerdoni*, Day. D. 8/12, A. 12. Reddish fawn-colour: a row of small red spots on the back, followed by a second larger one, and this by a band of red edged with lilac on the sides. Several red spots on the head and cheeks, two on the first dorsal and five on the second: fins whitish. Madras: to 6 inches in length.

469. *Eleotris macrolepidota*, Bl. D. 7/8, A. $\frac{1}{10}$, L. l. 30. Height of body 1/4 of the total length. Eyes, diameter 1/4 of length of head. Interorbital space swollen. Teeth villiform. East Indies.

470. *Eleotris Sinensis*, Lacép. D. 6 $\frac{1}{10}\cdot\frac{1}{12}$, A. $\frac{1}{9}$, L. l. 140. Height of body 2/13 of the total length. Eyes small. Teeth, a semi-oval patch on the vomer. Scales on head and neck rudimentary. Dark brown, marbled: a black, white-edged ocellus at the upper part of the root of the caudal fin. Andamans to China.

471. *Eleotris sexguttata*, Cuv. & Val. D. 6/$\frac{1}{12}$, A. $\frac{1}{12}$, L. l. 75-80. Height of body 1/7 to 1/8 of the total length. Teeth in a single row, their sizes unequal. Second to fourth dorsal spines filamentous. Greenish shot with rosy: blue spots, with dark edges on the side of the head: occasionally a violet spot before the dorsal fin, which latter has a black superior margin: second dorsal with six longitudinal violet stripes, anal with two: caudal with pearl-coloured ocelli edged with violet. Ceylon to the Malay Archipelago.

472. *Eleotris macrodon*, Bleeker. D. 6/10-11, A. 9, L. l. 90-100. Height of body about 1/5 of the total length. Diameter of eye 1/9 of length of head, 3 diameters apart. A small barbel on either side of the upper jaw. A reddish brown ocellus, edged with white, on the upper part of the base of the caudal fin. Hooghly.

473. *Eleotris feliceps*, Blyth. D. 6/$\frac{1}{10}$, A. 11, L. l. 27, L. tr. 12. Height of body 1/5 of the total length. Dorsal spines filiform. Brownish white, irregularly spotted and blotched with a darker colour: bands pass downwards from the orbit: fins more or less spotted. Andamans.

474. *Eleotris ophiocephalus*, Cuv. & Val. *A-rig-dah* and *Mu-took-. dah*, Andamanese. D. 6/$\frac{1}{8-9}$, A. $\frac{1}{7}$, L. l. 32-36, L. tr. 12, Cæc. pyl. 2. Height of body 1/5 to 1/6 of the total length. Eyes, diameter 1/5 of length of head, 2 diameters apart. Palate edentulous. Fins without filamentous prolongation. Scales ctenoid, those on the top of head large. Olive brown, some irregular blotches along the sides, whilst three black bands radiate from the eye: a light ocellus edged with dark at the upper half of the base of the pectoral fin: vertical fins with light margins. Andamans, the coast of Africa, and Malay Archipelago: to at least 9½ inches in length.

475. *Eleotris cavifrons*, Blyth. D. 6/$\frac{1}{8}$, A. 8, L. l. 65, L. tr. 17. Height of body 1/5 of the total length. Eyes, diameters 1/6 of length of head, 2 diameters apart. A depression before the orbits. Light brown, with dark lines radiating from the eyes: fins barred in spots. Andamans: to 4 inches in length.

476. *Eleotris fusca*, Bl. Schn. D. 6/$\frac{1}{8}$, A. $\frac{1}{8}$, L. l. 60-65, L. tr. 16. Height of body 1/6 of the total length. Eyes, diameter 1/5 of length of head, one diameter from end of snout, and also apart. Angle of

preopercle with a blunt spine, projecting forwards. Leaden black, lighter on the abdomen, which sometimes has a yellow tinge : horizontal bar on the dorsal fins, sometimes vertical ones on the caudal. Coast of India to the Malay Archipelago, also the African coast, &c.

477. *Eleotris Soaresi*, Playfair. D. 6/$\frac{1}{8}$, A. $\frac{1}{8}$, L. l. 62-65, L. tr. 17. Proportions much the same as in *E. fusca*. Height of body 2/11 of the total length. Head broad and depressed, as is also the snout : a blunt spine exists at the angle of the preopercle. Dark brown or brownish black, fins with dark spots. Andamans and Mozambique : to 6 inches in length.

478. *Eleotris scintillans*, Blyth. D. 6/$\frac{1}{8}$, A. 8, L. l. 47, L. tr. 15. Height of body 2/9 of the total length. Eyes, diameter 1/5 of length of head, 1$\frac{1}{4}$ diameters from end of snout and apart. Brownish, dorsal, caudal, and anal spotted, having white edges. Akyab and the Andamans.

479. *Eleotris Cantoris*, Günther. D. 6/$\frac{1}{1}$, A. $\frac{1}{7}$, L. l 36-37, L. tr. 12. Height of body from 2/11 to 1/6 of the total length. Eyes, diameter 1/5 to 1/6 of length of head, 1/2 a diameter apart. Head obtuse, depressed. Scales on neck small. Deep blackish-brown marbled : second dorsal and caudal with brown spots. Andamans and Burma, to the Malay Archipelago.

480. *Eleotris caperata*, Cantor. *Ou-suf-foo*, Mugh. D. 6/$\frac{1}{8}$, A. $\frac{1}{8}$, L. l. 30, L. tr. 9. Height of body 2/9 of the total length. Eyes, diameter 2/9 of length of head, situated close together. Supraorbital margin serrated, likewise a serrated ridge on either side of the posterior limb of the intermaxillary. Scales ctenoid. Leaden brown, fins blackish, a deep scarlet spot edged with black on the base of the pectoral rays. Coasts of India and Burma to China, also the Andamans.

481. *Eleotris Amboinensis*, Bleeker. *Gagi-bala-kera*, Ooriah. D. 6/$\frac{1}{8}$, A. $\frac{1}{8}$, L. l. 28, L. tr. 9. Height of body 1/6 to 1/7 of the total length. Eyes, diameter 1/5 to 2/11 of the length of head, 1$\frac{1}{4}$ diameters from end of snout and apart. Supraorbital margin serrated and other similar ridges around nostrils and approximating to the intermaxillaries. Scales ctenoid. Generally of a leaden or brown colour, but occasionally blotched. A scarlet spot edged with black at the base of the pectoral rays. First dorsal nearly black, the second, also the anal and caudal, yellowish with irregular dark bands. Coasts of Orissa and Bengal to the Malay Archipelago : attaining 3 to 4 inches in length.

482. *Eleotris butis*, Ham. Buch. D. 6$\frac{1}{8}$, A. $\frac{1}{8}$, L. l. 28, L. tr. 9. Height of body 1/5 of the total length. Eyes, diameter 1/6 to 1/7 of length of head, 1/2 a diameter apart. Scales cycloid. Brownish, blotched with darker. A deep black spot on the lower half of the base of the pectoral rays : second dorsal spotted. Coasts of India to China, &c.

Genus—A*MBLYOPUS*, *Cuv. & Val.*

Gobioides, Lacép.

Branchiostegals four or five : gills four : pseudobranchiæ absent. Body elongated : head oblong : no cavity above the opercles. Lower jaw prominent, causing the cleft of the mouth to be directed upwards. Eyes lateral, minute or indistinct. Teeth in a band, with a single anterior row of large, curved, conical and distantly placed ones. The first portion of the dorsal fin consisting of five undivided rays, is separated by an interval from the soft portion, in the centre of which is a single sixth undivided ray. Second portion of dorsal and anal with many rays and more or less confluent with the caudal: ventrals united. Scales rudimentary or absent. Air-vessel, when present, small or large.

483. *Amblyopus gracilis*, Cuv. & Val. D. $\frac{6}{17-49}$, A. 47-49. Length of head from. 2/21 to 1/11 times in the total length. Lower jaw with small barbels. Dorsal and anal confluent with a short caudal: pectorals very short. Scales absent. Pondicherry, and perhaps the Hooghly.

484. *Amblyopus cirrhatus*, Blyth. D $\frac{6}{46-47}$, A. 46. Length of head $\frac{1}{10}$ of the total length. Eye exceedingly minute, sometimes almost invisible until the specimen has been some days in spirit. Short barbels on the chin. Vertical fins enveloped in skin and not confluent with the caudal: pectorals short, not quite half so long as the ventrals.

485. *Amblyopus cæculus*. Bl. Schn. *Gogee-ramah*, Tel. D. $\frac{6}{43-44}$, A. 45. Length of head 1/7 to 1/8 of the total length. Small barbels on the chin. Eyes minute. Vertical fins enveloped in skin and scarcely confluent with the caudal: pectorals short. Scales absent, except a few near the tail. Brownish, becoming dirty white tinged with red on the abdomen. Fins grey, central caudal rays black. Air-vessel large, oval. Seas and estuaries of India to the Malay Archipelago and China: attaining 14 inches or more in length.

486. *Amblyopus Buchanani*, Day. D. $\frac{6}{43}$, A. 36. Length of head 1/7 of the total length. Eyes distinct. A short pair of barbels behind the symphysis of the lower jaw, and a second smaller pair further back. Vertical fins continuous, their posterior portions enveloped in skin: caudal 1/8 of the total length: pectoral $\frac{1}{25}$. Air-vessel large and oval. Scales, a few crypts in the hinder part of the body contain some. Olive, outer halves of pectoral and ventrals black. Hooghly: attaining at least 11 inches in length.

487. *Amblyopus Hermannianus*, Shaw. D. $\frac{6}{37-40}$, A. 36-38. Eyes minute. No barbels. Vertical fins confluent and not enveloped in skin: caudal pointed: pectoral of moderate length. Scales present, rudimentary, and embedded in minute crypts in the skin. Pinkish, fins yellowish. Estuaries of India to the Malay Archipelago and China: attaining at least 8 inches in length.

488. *Amblyopus tænia*, Günther. D. $\frac{6}{34}$, A. 33. Length of head 1/11 of the total length. Eyes minute. No barbels. Vertical fins confluent and not enveloped in thick skin: caudal elongated and pointed: pectoral two-thirds as long as the head. Scales, a few crypts in the hinder part of the body contain some. Greenish olive: caudal blackish. This species requires comparing with *A. rubicundus*, Ham. Buch.

489. *Amblyopus roseus*, Cuv. & Val. $\frac{5}{43}$, A $\frac{1}{41}$. Length of head 1/8, of caudal 3/13, of pectoral 2/17 of the total length. Rose coloured with the vertical fins blackish. Bombay, where it is said to attain 18 inches in length.

Genus—TRYPAUCHEN, *Cuv. & Val.*

Branchiostegals four. A deep blind cavity above the opercle, and which is not in communication with that of the branchiæ. Body elongated and compressed: head likewise compressed. Eyes lateral, minute, not elevated. Teeth in a band: no canines. Dorsal fin single, the anterior portion consisting of six spines, the soft with many rays, as has also the anal, whilst both are confluent with the caudal: ventrals united, forming a disk. Scales small.

490. *Trypauchen vagina*, Bl. Schn. Na-vettee, Tam. D. $\frac{0}{40-40}$, A. 39-46, L. r. 60-80, L. tr. 15. The five lowest pectoral rays very short and unbranched. Reddish. Coasts of India through the Malay Archipelago to China: attaining at least 6 inches in length.

Family—CALLIONYMIDÆ.

Branchiostegals five to six: pseudobranchiæ. Gill-openings of moderate width or very narrow. Body mostly elongated. The infraorbital ring of bones does not articulate with the preopercle. Teeth in the jaws, none on the palate. Two dorsal fins, the anterior with from four to seven flexible spines: second dorsal and anal similar: ventrals wide asunder. Scales and lateral line present or absent. Air-vessel absent.

Genus.—CALLIONYMUS, *Linn.*

Branchiostegals five or six: pseudobranchiæ, a slit behind the fourth gill. Gill-openings very narrow, sometimes merely a round hole at the upper edge of the opercle. Head and anterior portion of the body depressed. Eyes of moderate size, usually directed somewhat upwards. Mouth narrow, upper jaw protractile. A strong, variously armed spine at the angle of the preopercle. Teeth in jaws minute: palate edentulous. Two dorsal fins, the anterior consisting of three or four flexible spines: ventrals with five rays, and widely separated one from the other. Lateral line single or double. Air-vessel absent.

491. *Callionymus sagitta*, Pall. D. 4/9-10, A. 9. Preopercular spine stout, having four large teeth directed inwards and slightly upwards, whilst a fifth at its base is directed forwards. Gill-openings small, on the upper surface of the head. Scales absent. Lateral line single. Yellowish brown, with many black ocelli, edged with yellow, and some dark spots below. First dorsal with some black upon it. Bombay, through the Seas of India and the Mauritius.

492. *Callionymus Goramensis*, Bleeker. D. 4/8-9, A. 7. Preopercular spine with five strong teeth internally and one at its base externally. Gill-openings small, on the upper surface of the head. First dorsal spine, and both its last ray and those of the anal elongated. Lateral line double. Head and body dotted with blue: first dorsal black edged: anal with its lower half black: caudal unspotted. Andamans and the Malay Archipelago.

493. *Callionymus altivelis*, Schleg. D. 4/8, A. 7. Preopercular spine bifurcated. Gill-openings small on the upper side of the neck, but close to the superior margin of the opercle. First dorsal with white bands and spots: the second dorsal brown, with white spots: anal dark, with two rows of dark blue spots: caudal with two dark bands: pectoral blackish, spotted: its edge white. The back banded. Madras to Japan.

494. *Callionymus opercularis*, Cuv. & Val. D. 4/9, A. 9. Preopercular spine with six teeth superiorly: none at its base. Gill-openings on the side of the neck not covered by the extremity of the opercle. Madras and Pondicherry.

495. *Callionymus lineolatus,* Cuv. & Val. D. 4/8, A. 7. Preopercular spine strong, with two teeth at the external third at its inner side. Gill-openings covered by the end of the opercle, which is not produced. Scales absent. Lateral line single. Five or six greyish bands cross the back, and are continued down the sides: one more crosses the head. Mouth and lower surface of head scarlet. Three or four irregular transverse white spots with dark edges, also some round ones over the upper half of the dorsal fin: three oblique brown bands on the second dorsal: two brown bands on the ventral and caudal. Anal red, with a dark edge, and having two or three blue, black-edged spots behind each ray. Madras and Bourbon.

496. *Callionymus Orientalis,* Bloch. Schn. Preopercular spine short, with three teeth. The first ray of the first dorsal equals 1/2 the total length, and the second little shorter: second dorsal and anal nearly double the height of the body: pectoral about 1/4, and caudal more than 1/5, of the total length. Orange with black spots, and a few smaller white ones sparsely scattered amongst them. Dorsal and anal with round brown spots between their rays, also white points on the first of these fins: anal tinged with blackish, especially towards its margin: three brown or black bands across the pectoral, and five on the caudal: ventrals grey. Tranquebar, to 6 inches in length.

Family—BLENNIIDÆ, *Müll.*

Pseudobranchiæ present. Gill-openings of varying extent. Body elongated, more or less cylindrical. The infraorbital ring of bones does not articulate with the preopercle. Teeth may be fixed in the jaws, or merely implanted in the gums: a posterior canine may be present, whilst some genera have molars. One, two, or three dorsal fins, occupying nearly the entire length of the back, the spinous portion when distinct being less, nearly equally or more developed than the soft: in some the whole fin is entirely composed of spines, whilst in others none are perceptible. Ventrals when present jugular (except in *Pseudoblennius*): they are sometimes rudimentary. Anal with a moderate or large number of rays. Caudal, when present, may be confluent, with or distinct from the vertical fins. Scales, when present, generally small. Air-vessel as a rule absent. Pyloric appendages absent.

This family can be divided thus:—

A.—The spinous and rayed portion of the dorsal fin of somewhat equal extent.

B.—The dorsal fin mostly composed of spines.

C.—The dorsal fin entirely composed of spines.

D.—The dorsal fin entirely composed of rays.

A.—*The spinous and rayed portion of the dorsal fin of somewhat equal extent.*

Genus.—BLENNIUS, *Artedi.*

Pholis, Cuv. & Val.

Branchiostegals six. Gill-openings wide. Body somewhat elongated, with a short snout. Cleft of mouth narrow. Generally a tentacle above the orbit. Teeth in a single row, fixed in the jaws: a posterior curved tooth usually present in one or both jaws. Dorsal fin single, the spinous portion being less or equally developed with the rayed: ventrals jugular, consisting of one spine and two rays: caudal distinct. Scales absent. Air-vessel and pyloric appendages absent.

497. *Blennius leopardus*, Day. D. $\frac{12}{18}$, A. (2 +) 13. Height of body about 1/3 of the total length. Eyes, diameter 2/7 of length of head, one diameter apart. A transverse crest of tentacles crossing the occiput, and a fringed orbicular one, half as long as the eye : one at the nostrils : two more bifid ones on either side of the symphysis of the lower jaw. No canines. A notch between the two portions of the dorsal fin : lower eight pectoral rays only connected by membrane in their lower halves. Brown, becoming white on the chest, and blotched all over with dark markings, leaving narrow interspaces of the lighter ground colour : fins spotted. Galle, to 4 inches in length.

498. *Blennius Steindachneri*, Day. D. $\frac{11}{13}$, A. (2 +) 16. Height of body 2/9 of the total length. A small fringed orbital tentacle about half the diameter of the eye in length : another at the anterior nostril : a line of about eight fringed tentacles from between the eye to the base of the dorsal fin. No canines. Dorsal fin not notched. Light olive, with six indistinct brownish bands, as wide as the ground colour, and extending to the base of the dorsal fin : the whole of the body and fins dotted with black. A large black white-edged ocellus between the first and second dorsal rays. Some vertical red bands on the pectoral and caudal fins. Anal black, having a narrow white edging. Kurrachee, to 4 inches in length.

Genus—SALARIAS, Cuv.

Branchiostegals six : pseudobranchiæ. Gill-openings wide. Body somewhat elongated. Mouth transverse, rather wide : generally a tentacle above the orbit, especially in the males. Sometimes a crest upon the head. A row of moveable small teeth in the gums, and usually a posterior curved canine on either side of the lower jaw : palate almost invariably edentulous. Dorsal fin single : a notch exists in some species between the spinous and rayed portions : ventrals jugular, with one spine and two or three rays. Scales absent. Air-vessel and pyloric appendages absent.

The fishes of this genus may be thus sub-divided :—
 A. Dorsal fin not distinctly notched.
 B. „ distinctly notched.

In some species the dorsal fin is more distinctly notched in the immature than in the mature. A crest on the head generally shows the specimen to be a male, and in these the dorsal fin is often comparatively higher than in the females. The presence or absence of an orbital tentacle is occasionally only a sexual distinction.

A. *Dorsal fin not distinctly notched.*

499. *Salarias tridactylus*, Bl. Schn. D. $\frac{13\text{-}14}{22\text{-}20}$, A, 26-27. Height of body 1/9 to 1/10 of the total length. Male with a crest on the head. A small, simple, orbital tentacle. Canine teeth, when present, small. Dorsal fin not distinctly notched, but slightly emarginate between the two portions, it does not extend on to the caudal. Bluish, with irregular white spots : dorsal with pale oblique streaks : other vertical fins nearly black. Coasts of Sind and Andamans.

500. *Salarias fasciatus*, Bl. D. $\frac{12}{18\text{-}19}$, A. 19-21. Length of head 1/6 to 2/13, height of body 1/4 to 1/5 in the total length. Anterior

profile nearly vertical. No crest. A bifid supraorbital tentacle, and one or two on the nape of the neck. Brown, with yellow and blue dots and spots, immature ones banded. Throat and chest with yellow bands. Red Sea, East Coast of Africa, through the seas of India to China, &c.

501. *Salarias frenatus*, Cuv. & Val. D. $\frac{12}{4}$, A. 18, V. 1/3. Length of head 1/5, height of body 1/7 of the total length. Anterior profile very oblique. No crest. A small fringed tentacle above the posterior angle of the orbit. Grey, four white blue-edged lines proceed from the throat to the eye: and one or two of the same colour pass from one eye to the other. Dorsal with oblique blue and black lines: these stripes are transverse and undulating on the caudal; anal bluish with the tips of the rays blackish. Malabar.

B. *Dorsal fin distinctly notched.*

502. *Salarias quadricornis*, Cuv. & Val. D. $\frac{13}{12}$, A. (2 +) 23-24. Length of head and height of body from 2/11 to 1/6 of the total length. Snout very obtuse. A crest present. A short simple supraorbital tentacle, and a fringed nasal one. No canines. Body greenish, with brown bands forming oblique streaks on the dorsal fin: anal with three longitudinal bands.

503. *Salarias æquipinnis*, Günther. D. $\frac{16}{13}$, A. 25. Height of body about 1/10, length of head 1/7 of the total length. No crest. A short tentacle over the posterior angle of the orbit. No canine teeth. Olive, with dark bands: dorsal and caudal rays black: anal greyish, having a black external band with a white margin. Kurrachee, Amboyna.

504. *Salarias lineatus*, Cuv. & Val. D. $\frac{12}{21\text{-}24}$, A. 23-25. Height of body 1/6 of the total length. A low crest on the crown of the head in the males: none in the females. A short bifid orbital tentacle. No canines. Brownish, with from five to seven longitudinal stripes on the sides, and also obliquely on the dorsal fin: anal yellowish, edged with brown. Kurrachee to the Malay Archipelago.

505. *Salarias cyanostigma (?)*, Bleeker. D. $\frac{12}{27}$, A. $\frac{22}{24}$. Length of head 2/13, height of body 1/8 of the total length. Snout obtuse. A crest. A simple tentacle over the orbit nearly as long as the eye: a bifid nasal one. A canine tooth in the lower jaw. Greenish, with cross bands, first dorsal spotted, it and the anal edged with black. Andamans.

506. *Salarias bellus*, Günther. D. $\frac{12}{20\text{-}22}$, A. 22-24. Length of head 1/4, height of body about 1/5 of the total length. Snout obtuse. A crest on the head. A simple supraorbital tentacle shorter than the eye, another nasal one. No canine tooth. Brownish, with bluish white dots most apparent in the posterior half of the body: both dorsals and caudal with black lines having white between: they are oblique on the dorsal and longitudinal on the caudal: anal black edged. Kurrachee, Andamans to China.

507. *Salarias Sumatranus*, Bleeker. D. $\frac{12}{26}$, A. 22. Length of head 1/6, height of body 1/5 of the total length. No crest on the head. A short, simple, supraorbital tentacle, and another bifid nasal one. No canine tooth. Greyish green, blackish longitudinal bands on the first dorsal: oblique yellow stripes on the second, with a row of black dots along its summit: two or three rows of dots on the anal. Andamans and Malay Archipelago.

508. *Salarias Dussumieri*, Cuv. & Val. D. $\frac{12}{20\text{-}21}$, A. 22. Height of body from 1/5 to 1/6 of the total length. No crest on the head. A long fringed orbital tentacle in the female, not in the male. No canine tooth. Body banded and spotted: dorsal fin darkish with oblique lines. East coast of Africa, Seas of India.

509. *Salarias periophthalmus*, Cuv. & Val. D.$\frac{12}{26}$, A. $\frac{21}{24}$. Length of head 4/25, height of body 1/7 of the total length. No crest. Snout very obtuse. A simple supraorbital tentacle, a fringed nasal one. A canine tooth in the lower jaw. Rose-coloured, with indistinct violet cross bands, and two rows of blue spots along the side: red spots on the head, and a red line behind the eye. The lower half of the dorsal bluish with oblique streaks, the upper half yellow with red spots: outer half of anal dark: caudal yellow. Andamans and Malay Archipelago.

510. *Salarias unicolor*, Ruppell. D. $\frac{13}{17}$, A. 18-19. The height of the body is 2/11 of the total length. The male has a crest on the head which is absent in the female. A long fringed orbital tentacle and a short nasal one. No canine teeth. Dorsal fin not continuous with the caudal. Colours in the *male* olive brown with indistinct blackish blotches along the sides and extended on to the dorsal fin: two or three rows of blue spots on the body and head. The *female* olive brown, becoming white beneath, and marbled all over with brown lines forming large insulated spaces. Head and upper two-thirds of the body dotted with light blue: fins yellowish: dorsal and anal, horizontally and sinuously banded in spots: caudal with five or six vertical dark bands. A brown band, divided by a light blue line at the base of the pectoral, which is yellowish barred with brown. Red Sea, and Kurrachee.

511. *Salarias vermiculatus*, Cuv. & Val. D. $\frac{12}{13\text{-}15}$, A. (2 +) 18. Length of head and height of body from 2/9 to 1/5 of the total. Snout obtuse. No crest on the head. A moderately long fringed supraorbital tentacle, and a small nasal one. Large canine in the lower jaw. Superiorly brownish, becoming bluish-white inferiorly, with a series of nine brown bars descending to the lateral line. The body, head, and fins reticulated with brown lines, enclosing circular or irregularly formed spaces. East Coast of Africa, Andamans.

512. *Salarias marmoratus*, Bennett. D. $\frac{12}{17}$, A. 18. Height of body 2/11 of the total length. No crest on the head. A long fringed supraorbital tentacle, having some filaments at its base: a small frontal and another nasal one. A strong canine tooth. Brownish, marbled with darker. Ceylon.

Genus—ANDAMIA, Blyth.

Differs from Salarias in that it possesses a broad adhesive sucker behind the symphysis of the lower jaw.

513. *Andamia expansa*, Blyth. D. $\frac{16}{18}$, A. 26. Height of body 1/9 of the total length. A short fringed orbital tentacle, and a small simple nasal one. No canines. Olive, banded with a darker shade: head spotted. Fins with dark edges, except the anal, which has a white margin. Andamans and Nicobars: to 3 or 4 inches in length.

Genus—*PETROSCIRTES*, Rüpp.

Blennechis, Cuv. & Val.: *Aspidontus*, Quoy & Gaim.

Branchiostegals six. Gill-openings reduced to a small orifice above the root of the pectoral fin. Body somewhat elongated. Snout short or of moderate length : cleft of mouth narrow : head sometimes with tentacles. Teeth, a single row of fixed ones in the jaws, generally with a strongly curved posterior canine. Dorsal fin single (a semi-detached portion has been recorded in one species): ventrals jugular, with two or three rays. No scales. Air-vessel present. Pyloric appendages absent.

514. *Petroscirtes punctatus*, Cuv. & Val. D. $\frac{12}{22}$, A. 23. Height of the body equals 1/6 of the total length. Upper canine half the size of the lower. The dorsal fin reaches to the root of the caudal. Grey, with three rows of small dark green spots. Bombay.

515. *Petroscirtes cyprinoides*, Cuv. & Val. D. 30. A. 19-20. Height of body 2/11 of the total length. Dorsal fin does not reach the root of the caudal. A white band along the side, above which are seven vertical bars : caudal yellowish. Bombay (?)

516. *Petroscirtes breviceps*, Cuv. & Val. Height of body 4 times and 2/3 in the total length. Dorsal fin reaches to the root of the caudal. A black lateral band : dorsal with black dots : caudal yellowish. Bay of Bengal.

Family—SPHYRÆNIDÆ, *Agassiz.*

Percoidei, pt. Cuv.

Branchiostegals seven: pseudobranchiæ. Body elongate, sub-cylindrical. Eyes lateral, of moderate size. Cleft of mouth deep. Teeth in the jaws large and cutting: present on palate, none on vomer. Two short dorsal fins remote from each other: anal similar to the second dorsal: ventral abdominal, situated opposite the first dorsal spine, and consisting of one spine and five rays. Scales small, cycloid. Lateral line continuous. Air-vessel present, bifurcated anteriorly. Pyloric appendages in moderate numbers or numerous.

Genus—*SPHYRÆNA*, Artedi.

Definition as in the family.

517. *Sphyræna jello*, Cuv. & Val. *Jellow*, Tel. D. 5/$\frac{1}{9}$, A. $\frac{1}{9}$, L. l. 120. L. tr. 35. Length of head 2/7 to 1/4, height of body 1/9 of the total length. Anteriorly a short fleshy appendage to the mandible. The first dorsal and ventrals commence on a vertical line opposite the end of the pectoral fin. Superiorly grey, becoming white on the abdomen. On the upper part of the side a festooned band intersecting the lateral line. Ventrals whitish : the other fins yellowish with fine black points, most numerous towards their margins. Red Sea, east coast of Africa, seas of India to the Malay Archipelago and beyond : attaining at least 4 feet in length.

518. *Sphyræna Forsteri*, Cuv. & Val. D. 5/$\frac{1}{9}$, A. $\frac{1}{9}$, L. l. 110. L. tr. 30. Length of head nearly 1/4, height of body 1/9 of the total length. Anteriorly a conical tubercle on the mandible, and about 19 strong trenchant teeth. The first dorsal commences on a line opposite the end of the pectoral fin : the ventrals arise under the posterior half of the pectorals and anterior to the first dorsal. Greenish above, silvery beneath. Dorsal and caudal violet, the other fins yellowish. Bay of Bengal, Malay Archipelago and beyond.

519. *Sphyræna Dussumieri*, Cuv. & Val. D. 5/1, A. 1/9, L. l. 95, L. tr. 10/16. Length of head 2/7, height of body 2/15 of the total length. A short fleshy appendage to the lower jaw, and about 22 trenchant teeth. The first dorsal and ventrals commence on a vertical line opposite the end of the pectoral fin. Bluish above, silvery beneath: dorsal, caudal, and anal black with white tips. Seas of India: attaining at least 4½ feet in length.

520. *Sphyræna obtusata*, Cuv. & Val. D 5/1, A. 1/9, L. l. 90. L. tr. 12/15, Cæc. pyl. 24. Length of head 3/10, height of body 1/6 to 1/7 of the total length. Lower jaw with a very small fleshy appendage anteriorly. Greyish-green superiorly, clouded beneath the lateral line: abdomen white: pectorals greyish, the other fins yellowish. Seas of India to the Malay Archipelago and beyond: attaining at least 2½ feet in length.

521. *Sphyræna Commersonii*, Cuv & Val. D. 5/1, A. 1/9, L. l. 80-90. Length of head from 2/7 to 1/4, height of body from 1/8 to 1/9 of the total length. Anteriorly a conical tubercle on the mandible, and from 15 to 18 strong posterior teeth. The first dorsal commences on a vertical line opposite the end of the pectorals, whilst the ventrals arise a little anterior to this line. Bluish green superiorly, silvery beneath: dorsal, caudal, and anal violet. Seas of India to the Malay Archipelago.

Family—ATHERINIDÆ, *Günther.*

Branchiostegals five or six: pseudobranchiæ. Four gills: gill-opening wide. Body more or less elongated and somewhat sub-cylindrical. Eyes lateral. Gape of mouth of moderate width: cleft not very deep. Teeth minute. Two dorsal fins, not conjoined, the spines of the first feeble, and less in number than the rays of the second, which is similar to the anal : ventrals abdominal, with one spine and five rays. Scales of medium size, cycloid. Lateral line indistinct. Pyloric appendages, if present, few. Air-vessel present. Vertebræ numerous in the abdominal and caudal portions.

Genus—*ATHERINA*, *Artedi.*

Body somewhat sub-cylindrical, with slightly compressed sides. Snout more or less obtuse, with the cleft of the mouth oblique, extending backwards to at least as far as to below the anterior edge of the orbit. Teeth very minute, but usually present on jaws and palate. Ventrals at some distance posterior to the pectorals. Scales of moderate size. Air-vessel present. Pyloric appendages, when present, few. Ova comparatively very large. A silvery lateral band.

522. *Atherina pinguis*, Lacép. D. 6/$\frac{1}{10}$, A. $\frac{1}{14+15}$, L. l. 42-45, L. tr. 7. Length of head 3/14, height of body 3/17 of the total length. Diameter of eye 2/5 in length of head, and much longer than the snout. Upper jaw overlapping the lower. Teeth distinct in jaws, vomer and palatines. The silvery lateral band includes the whole of the third and the upper quarter of the fourth rows of scales: a blackish mark on the posterior end of the pectoral. East coast of Africa, seas of India to the Malay Archipelago and beyond.

523. *Atherina Forskålii*, Rüpp. *Ko-re-dah*, Andam. D. 5-6/$\frac{1}{9-10}$, A. $\frac{1}{13-14}$, L. l. 40, L. tr. 7. Length of head 3/13, height of body 1/6 of the total length. Diameter of eye 2/5 in length of head and much longer than the snout. Jaws equal in front. Teeth distinct in jaws,

vomer, and palatines. Margins of scales smooth. The silvery lateral band includes the whole of the third and the upper half of the fourth rows of scales. Red Sea, east coast of Africa, seas of India to the Malay Archipelago.

524. *Atherina duodecimalis*, Bleeker. D. 5/$\frac{1}{9}$, A $\frac{1}{17}$, L. l. 35. Length of head 4/17, height of body 2/11 of the total length. Diameter of eyes nearly 1/2 of length of head, and twice the length of the snout. No black dots along the sides of the body. Ceylon and Malay Archipelago.

Family—MUGILIDÆ.

Branchiostegals from four to six: pseudobranchiæ. Gill-openings wide: gills four. Form of body oblong, compressed, whilst the head and anterior portion may be depressed. Eyes lateral, with or without adipose lids. Mouth narrow or of moderate width. Opercles unarmed. Teeth very fine, sometimes absent. Two dorsal fins, the first consisting of four stiff spines: anal slightly longer than the second dorsal: ventrals abdominal and suspended from an elongated shoulder bone, consisting of one spine and five rays. Scales cycloid, rarely ctenoid. Lateral line absent. Pyloric appendages generally few. Vertebræ 24.

Genus—MUGIL, *Artedi*.

Branchiostegals from four to six: pseudobranchiæ. Eyes with or without an adipose lid. Mouth more or less transverse, with a shallow cleft, and the anterior edge of the mandible sharp. Teeth, when present, minute. Pyloric appendages generally few (2—10). Upper portion of the stomach very muscular.

A. *Adipose eyelid well developed, sometimes covering at least the posterior third of the iris: upper lip usually not very thick.*

525. *Mugil carinatus?* (Ehreng), Cuv. & Val. D. 4/$\frac{1}{8}$, A. $\frac{3}{8}$, L. l. 40-42, L. tr. 12. Cæc. pyl. vi. Length of head 1/5, height of body nearly 1/5 of the total length. Eye with a broad adipose lid anteriorly and posteriorly, snout shorter than the eye: preorbital scaly, scarcely emarginate or denticulated: end of maxilla visible. Each scale with a raised line along its centre: 25 scales between the snout and the spinous dorsal: an elongated scale in the axil, and also along the side of the dorsal fin: vertical fins scaled. Caudal slightly emarginate; pectoral as long as the head without the snout. Cheeks golden: dorsal and caudal with minute dark spots, making the upper edge and posterior margin of the second dorsal blackish. Common along the coasts of India. Appears to be very similar to *M. Speigleri*, Bleeker.

526. *Mugil cunnesius*, Cuv. & Val. *Kunnesee*, Tel. D. 4/$\frac{1}{8}$, A. $\frac{3}{8}$, L. l. 42-43, L. tr. 13. Length of head 1/5, height of body nearly 1/5 of the total length. Eye with adipose membrane. Preorbital neither notched nor denticulated: end of maxilla visible. Twenty-two scales between the snout and the spinous dorsal: vertical fins not scaled. Dorsal and caudal fins with indistinct dark edges: anal with a dark mark along its centre: pectoral dark grey, with a white posterior and inferior edge, and a darkish superior one sometimes forming a dark spot in the axil. Seas of India and Malay Archipelago.

527. *Mugil parsia*, Ham. Buch.* *Tarrui*, Beng. D. 4/$\frac{1}{8}$, A. $\frac{3}{8}$, L. l. 35, L. tr. 12. Cæc. pyl. 5. Length of head 1/5, height of body

---

* Bleeker, a most accurate observer, gives L. l. 40 -45. Dr. Günther stated his description, accepted above, was drawn up from Hamilton Buchanan's 'types.' Arguing now

2/9 of the total length. Posterior adipose lid well developed. End of maxilla visible. Twenty-one scales between the snout and the dorsal fin. Second dorsal and anal scaly. Silvery, without marks. Seas of India.

528. *Mugil longimanus,*† Günther. D. 4/$\frac{1}{8}$, A. $\frac{3}{9}$, L. l. 35. Length of head from 1/5 to 4/21, height of body from 3/14 to 1/5 of the total length. Eye with both anterior and posterior adipose lids: upper lip thick. The maxilla is hidden when the mouth is closed. Silvery. Seas of India and Malay Archipelago.

529. *Mugil engeli,* Bleeker. D. 4/$\frac{1}{8}$, A. $\frac{3}{9}$, L. l. 33—34, L. tr. 10-11. Length of head 2/9, height of body 1/5 of the total length. Eye with a broad anterior and posterior adipose lid. The maxilla is quite hidden by the preorbital, which is scaly, slightly emarginate, and indistinctly denticulated. Eighteen rows of scales between the snout and the dorsal fin. Coloration uniform. Seas of India to the Malay Archipelago.

530. *Mugil planiceps,* Cuv. & Val. *Bangon,* Beng. D. 4/$\frac{1}{8}$, A. $\frac{3}{9}$, L. l. 33-35, L. tr. 11, Cæc. pyl. 5. Length of head 3/14, height of body 1/6 of the total length. Eye with a posterior adipose lid. Lips thin. Extremity of maxilla not covered by the preorbital. Twenty scales between the snout and dorsal fin: soft vertical fins scaly. A darkish line along each row of scales. This species I take to be *M. cephalus ?* or *M. bangon,* m. s. H. B. Coasts of India and China.

531. *Mugil Kelaartii,* Günther. D. 4/$\frac{1}{8}$, A. $\frac{3}{9}$, L. l. 32—33, L. tr. 12. Length of head 1/5, height of body 2/9 of the total length. Eye with a broad anterior and posterior adipose lid, one diameter from end of snout. Upper lip thin. Maxillary hidden by the preorbital, which is slightly notched anteriorly, and with a rounded denticulated extremity. Nineteen rows of scales between the snout and the dorsal fin. Silvery. Ceylon and Philippine Islands: to 4½ inches in length.

532. *Mugil Sundanensis,* Bleeker. D. 4/$\frac{1}{8}$, A. $\frac{3}{9}$, L. l. 30—32. Length of head 1/5, height of body 2/9 of the total length. Adipose eyelid well developed. Maxillary not concealed by preorbital, which latter is notched anteriorly. Anterior half of anal before the origin of the second dorsal: caudal emarginate. Silvery: caudal with a black edge. Bay of Bengal to the Malay Archipelago.

533. *Mugil lævis,* Ham. Buch. m. s. *M. Nepalensis?* Günther. D. 4/$\frac{1}{8}$, A. $\frac{3}{9}$, L. l. 29-30, L. tr. 11, Cæc. pyl. 4. Length of head 2/11, height of body 1/5 of the total length. Eye with an adipose anterior and posterior lid, the latter most developed. Extremity of maxilla not hidden by the preorbital, which is angularly bent, indistinctly denticulated at its extremity and scaled. Scales angular: no elongated one at the axilla, but one along the side of the first dorsal fin: soft dorsal and anal scaled. Sixteen rows between the snout and the dorsal fin. Darker above than below. Seas of India, ascending rivers: attains at least 6½ inches in length. This is figured in H. B's drawings, No. 69.

---

from observations in the preface to Catalogue of Fishes, Vol. III, the last three lines of page iv. of Preface, and Proc. Zool. Soc., 1872, p. 877, I cannot help thinking that there may still be room for examination and enquiry. H. B. only give 8 soft rays to the anal fin, and does not show that either it or the anal are scaly.

† The specimen in the British Museum is said to have come from the "East Indies, Presented by G. R. Waterhouse, Esq." If H. B's, type, it must be of his "*M. cephalus ?* or *M. bangon,* Ms.

534. *Mugil cephalotus*, Cuv. & Val. D. 4/$\frac{1}{8}$, A. $\frac{3}{(7)\,8}$, L. l. 38—40, L. tr. 14-15. Length of head 1/5, height of body nearly 1/5 of the total length. Eye with a broad adipose lid. Maxillary entirely hidden by the preorbital. Twenty-one scales between the snout and the dorsal fin. Silvery, with rather shining lines along the rows of scales. Red Sea, seas of India to China and Japan.

535. *Mugil subviridis*, Cuv. & Val. D. 4/$\frac{1}{8}$, A. $\frac{3}{8}$, L. l. 30, L. tr. 11. Length of head 3/16, height of body 2/9 of the total length. End of the maxillary not hidden by the preorbital, which latter is angularly bent and denticulated at its extremity. Eighteen scales between the snout and the dorsal fin. Greenish, with golden reflections. Seas of India.

536. *Mugil Cantoris*, Bleeker. D. 4/$\frac{1}{7-8}$, A. $\frac{3}{9-10}$, L. l. 33. Length of head from 2/9 to 2/11, height of body from 3/13 to 3/14 of the total length. Eye with anterior and posterior adipose lids. Maxillary entirely concealed by the preorbital, which is distinctly notched. Second dorsal much higher than long : anal scaled. Calcutta.

537. *Mugil poicilus*, Day. D. 4/$\frac{1}{8}$, A. $\frac{3}{8}$, L. l. 32, L. tr. 10, Cæc. pyl. 5. Length of head and height of body about 1/5 of the total length. Eye with anterior and posterior adipose lids. Maxillary not concealed by the preorbital, which latter is bent, having a rounded serrated margin, and scaled. Twenty rows of scales between the snout and the dorsal fin. Vertical soft fins largely covered by scales. Almost each scale on the body has an irregularly formed, usually round, central hole, which is deep black. From Bombay (where they appear to arrive about November and continue throughout the cold season) down the Malabar coast : attaining about 8 inches in length.

538. *Mugil cunnumboo*, Day. D. 4/$\frac{1}{7}$, A.   , L. l. 34, L. tr. 11. Length of head 1/5, height of body 1/5 of the total length. Eye with an anterior and posterior adipose lid. End of maxilla not hidden by the preorbital, the extremity of which latter is rounded and denticulated. Twenty-one rows of scales between the snout and the dorsal fin. Lower half of second dorsal and basal two-thirds of anal scaled. A dark mark along the centre of each scale : a dark axillary spot at base of pectoral : caudal darkest at its extremity.

B. *Adipose eyelids not developed : upper lip usually very thick.*

539. *Mugil cæruleo-maculatus*, (Lacép), Bleeker. D. 4/$\frac{1}{8}$, A. $\frac{3}{9}$, L. l. 38, L. tr. 12. Length of head from 3/17 to 1/5, height of body 2/9 of the total length. Maxillary concealed by the preorbital, which latter is not emarginate and hardly denticulated. Second dorsal and anal scaled and commence opposite one another: caudal forked. Silvery, with a black spot at the axil on the base of the pectoral. Bay of Bengal, Malay Archipelago.

540. *Mugil Ceylonensis*, Günther. D. 4/$\frac{1}{8}$, A. $\frac{3}{8}$, L. l. 32, L. tr. 12. Length of head and height of body each 3/14 of the total length. Maxillary concealed by the preorbital, which has a slight notch anteriorly, a rounded end and denticulated edge. Second dorsal commences above the first anal ray, both fins are scaled. Ceylon : to 5 inches in length.

541. *Mugil Troschelli*, Bleeker. D. 4/$\frac{1}{8}$, A. $\frac{3}{8}$, L. l. 31-32, L. tr. 10-11. Length of head and height of body each 1/5 of the total length. Maxillary not concealed by the preorbital, which latter is scaled, slightly emarginate and indistinctly denticulated. Eighteen or 19 rows of scales between the snout and the dorsal fin. Anterior third of second dorsal before the origin of the anal. Second dorsal and anal scaled. Silvery. Seas of India to the Malay Archipelago.

542. *Mugil Waigiensis*, Quoy. & Gaim. D. 4/$\frac{1}{7\cdot8}$, A. $\frac{3}{8}$, L. l. 26-27, L. tr. 9, Cæc. pyl. 10 and bifurcated. Height of body and length of head each 2/9 of the total length. Lips thin. The maxilla is not quite concealed by the preorbital, the extremity of which is rounded, with a denticulated margin. Sixteen rows of scales between the snout and dorsal fin: soft dorsal and anal scaled. Fins deep black, and a dark streak along the centre of each row of scales. Red Sea, seas of India to the Malay Archipelago and beyond.

543. *Mugil amarulus*, Cuv. & Val. D. 4/$\frac{1(2)}{8}$, A. $\frac{3}{8}$. Body compressed, and the head much shorter than the height of the body. A short axillary scale. Uniform, said to resemble a carp. Arrian-coupon river at Pondicherry during November and December. If it is identical with *M. oligolepis*, Bleeker, it has L. l. 26, L. tr. 10-11. Height of body 1/4, length of head 1/5 of the total length. Maxillary concealed by the preorbital, which latter is emarginate and denticulated.

544. *Mugil macrocheilus*, Bleeker. D. 4/$\frac{1}{7}$, A. $\frac{3}{8}$, L. l. 41-42. Length of head 5/27, height of body 4/21 of the total length. Lips fringed. Maxilla nearly concealed by the preorbital, which is not notched. A black spot superiorly at the base of the pectoral fin. Andamans to the Malay Archipelago.

Family—AULOSTOMATIDÆ, Cantor.

Fistuláridæ, pt., Müll.: *Aulostomatoidei*, pt. Bleeker.

Branchiostegals five to seven: pseudobranchiæ. Gills four, attached to the humeral arch. Form of body elongated: the anterior bones of the skull produced, forming a long tube, and having a small mouth at its anterior extremity. Teeth small. Spinous dorsal, when present, formed of isolated spines: soft dorsal and anal of moderate length: ventrals abdominal with six rays, no spine, and separated from the pubic bones which are attached to the humeral arch. Scales small or none, but parts of the skeleton or else dormal productions may be in the form of external plates. Air-vessel large. Pyloric appendages few. Vertebræ numerous.

Genus—*Fistularia, Linn.*

Solenostomus, sp. Klein and Gronov.: *Cannorhynchus*, Cantor.

Branchiostegals seven. Mouth slightly cleft. Dorsal and anal fins composed entirely of undivided rays: caudal forked, with its two central rays very elongated and filiform. No scales, but some bony casing behind the head above and below.

545. *Fistularia immaculata*, Comm. Goorum., Tel. D. 13-15, P. 13, A. 14-15, V. 6. Length of head to end of caudal (excluding the filaments) 2/5 of the total length. Eye, 1½ diameter in the postorbital portion of the head. A serrated ridge from the anterior superior angle of the orbit to the nostrils. Brown, becoming dirty white beneath:

sometimes with light spots. Seas of India to the Malay Archipelago and beyond.

Family—CENTRISCIDÆ, pt. *Bleeker.*

Fistularia, pt. Muller: *Amphisiloidei*, pt. Bleeker.

Branchiostegals three to four: pseudobranchiæ. Gills four. Form of body oblong or elevated and compressed : the anterior bones of the skull produced, forming a long tube and having a small mouth at its anterior extremity. Teeth absent. Two dorsal fins, the first short and having of one its spines strong : the soft dorsal and anal of moderate extent : ventrals abdominal, spineless and rudimentary. Scales, if present, small: the body either covered with a cuirass or ossifications which are not confluent with one another. Air-vessel large. Pyloric appendages absent. Vertebræ few.

Genus—AMPHISILE (*Klein*), *Cuv.*

Branchiostegals three or four: pseudobranchiæ. Gill-openings of moderate width. Body elongated and strongly compressed. A dorsal cuirass formed by portions of the skeleton. Teeth absent. Two dorsal fins situated far back: ventrals rudimentary. Air-vessel large. Pyloric appendages absent.

546. *Amphisile scutata*, Linn. *Marri kola*, Tam. B. III. D. 3/10-12, P. 10, V. 3, A. 12, C., 10 Vert. 6/11. The dorsal cuirass ends in a long spine not confluent with any in the dorsal fin. Whitish. Seas of India to China.

Family—TRACHYPTERIDÆ, *Günther.*

Tænoidei, pt. Cuv: *Gymnetridæ*, Gray.

Branchiostegals six: pseudobranchiæ present. Body elongated, and strongly compressed. Gill-openings wide : gills four. Eyes lateral. Cleft of mouth slight. Dentition feeble. A single dorsal fin occupying the whole length of the back, with a detached anterior portion, the whole composed of rays that are neither branched nor articulated : anal absent : caudal not in the longitudinal axis of the fish, or else rudimentary : ventrals thoracic. Pyloric appendages numerous. Vertebræ many. Bones soft.

Genus—REGALECUS, Brünn.

Gymnetrus, Bl. Schn.

Each ventral fin reduced to a long filament, dilated at the extremity : caudal rudimentary or absent.

547. *Regalecus Russellii.** Shaw. D. 4-5/320. Height of body 1/20 of the length. Teeth absent. Caudal rays distinct : ventrals in the form of two filaments as long as the rays of the crest. Silvery, with yellowish fins : second dorsal with a dark edge. Coromandel coast : to 2 feet 8 inches in length.

---

* Jerdon observes, "*Xiphichthys Russellii*, Sw. Russ. 39. I one day procured two specimens of this very curious species of Gymnetrus, which Swainson has named from Russell's figure, which however is very defective. Its tail ends in a long filament, and the dorsal and anal fins are much higher than is there represented." M. J. L. & S., 1851, p. 139. Jerdon's specimens must have been an entirely different fish, if no misprint has occurred.

Family—POMACENTRIDÆ,* *Richardson.*

Sciænoidei, pt., Cuv.: *Labroidei ctenoidei*, Müller: *Ctenolabridæ*, Owen. Branchiostegals from five to seven: pseudobranchiæ present. Gills three and a half. Eyes lateral. Body more or less short and compressed. Bones of head variously armed or smooth. Teeth in jaws feeble, palate edentulous. A single dorsal fin with the spines equal in number to or somewhat less than the rays, very rarely more: the soft anal similar to the soft dorsal, and with two or three spines: ventrals thoracic with one spine and five rays. Scales ctenoid. Air-vessel present. Pyloric appendages few.

Genus—AMPHIPRION (Bl. Schn.), *Cuv.*

Coracinus, sp. Gronov.

Branchiostegals five: pseudobranchiæ. All the opercles and preorbital denticulated, those on the opercle and subopercle being almost spinate. Teeth in the jaws in one row, conical and small. Scales of moderate or rather small size. Lateral line ceases below the end of the dorsal fin. Air-vessel present. Pyloric appendages few.

548. *Amphiprion ephippium*, Bloch. D. $\frac{10\text{-}11}{17\text{-}15}$, A. $\frac{2}{14}$, L. l. 55, L. tr. 7/8, Cæc. pyl. 2. The immature, *A tricolor*, Günther, are of a brownish colour, becoming of an orange tinge on the abdomen and free portion of the tail : a pearl-coloured band goes from the nape across the opercles : the ventral externally blackish. The adult becomes of a dirty yellow, the white band being lost, and the black part of the ventral changes to brown. Seas of India to the Malay Archipelago and beyond.

549. *Amphiprion percula*, Lacép. *Eá-ole-jo-do-dah*, Andam, or 'Turtle's stomach,' because they are generally found inside *Actiniæ*, which are looked upon as those organs. D. $\frac{11}{15}$, A. $\frac{2}{11\text{-}12}$, L. l. 55, L. tr. 7/23. Length of head 1/4, height of body 1/3 of the total length. Ground colour bright yellow, with three broad milk-white cross bands having a black edging, the anterior being convex, the convexity being forwards over the hind part of the head. The centre one from the middle of the dorsal fin to the vent, and the posterior one over the free portion of the tail. Seas of India to the Malay Archipelago and beyond.

550. *Amphiprion intermedius*, Schleg. D. $\frac{(10)\,11}{(14)\,15}$, A. $\frac{2}{(13)\,12}$, L. l. 50-55, L. tr. 6/21. In the two specimens I have examined from the Andamans, the length of the head 2/7, height of body 2/5 of the total length. Brownish, with a curved milk-white band from the nape over the opercle and sub opercle: a second from the last few dorsal spines to in front of the base of the anal, it is rather produced anteriorly : a third over the free portion of the tail. Fins brownish edged with black, and tipped with white. Andamans and Malay Archipelago.

551. *Amphiprion Clarkii*, Cuv. & Val. D. $\frac{10}{16\text{-}17}$, A. $\frac{2}{14}$, L. l. 54-55, L. tr. 6/19. Length of head 1/4, height of body 2/5 of the total length. Light brown, with three milk-white cross bands, the first, passing from

---

* The following families of Acanthopterygians are left together as they and the genera *Gerres* and *Euctenogobius* at least, form the Order, *Acanthopterygii pharyngognathi*, Mull: defined as differing from the remainder of the Acanthopterygians, owing to the inferior pharyngeal bones being coalesced, and with a median longitudinal suture.

in front of the dorsal fin, goes over the opercles just touching the posterior edge of the orbit: the second, commencing at the base of the last five dorsal spines, passes to the front of the base of the anal fin: the third crosses the free portion of the tail. Andaman Islands to the Malay Archipelago and beyond.

552. *Amphiprion bifasciatum*, Bloch. D. $\frac{11}{13\text{-}15}$, A $\frac{2}{12\text{-}13}$, L. l. 50-55, L. tr. $\frac{6\text{-}7}{17\text{-}19}$. Brownish black, with two milk-white cross bands: the anterior from the nape passes over the opercles just touching the posterior edge of the orbit: the second from the last three spines and first few dorsal rays is continued downward to the middle of the body, and backward to the summit of all the dorsal rays: caudal black with a white upper and lower edge.

553. *Amphiprion akallopisus*, Bleeker. The specimen in the Calcutta Museum is bleached, but appears to belong to this species. D. $\frac{9}{17}$, $\frac{9\text{-}10}{(20\text{-}15)}$, A. $\frac{2}{13}$, L. l. 60. Length of head 2/7, height of body nearly 1/2 of the total length. Apparently orange, lightest on the head, chest and base of caudal: a light band runs from the orbit along the base of the dorsal fin to the caudal. Andamans (to the Malay Archipelago).

Genus—PREMNAS, Cuv.

Branchiostegals five or six: pseudobranchiæ. All the opercles serrated: a long strong spine at the posterior edge of the preorbital. Teeth in a single row, conical, and small. Two anal spines. Scales of moderate size. The lateral line ceases below the end of the dorsal fin. An air-vessel. Pyloric appendages three.

554. *Premnas biaculeatus*, Bloch. D. $\frac{9\text{-}10}{16\text{-}18}$, A. $\frac{2}{13\text{-}15}$, L. l. 50-60, Cæc. pyl. 3, Vert. 12/14. Length of head 1/4, height of body 2/5 of the total length. Chestnut brown, fins edged with black. A broad white band margined with black, passes over the occiput and on to the opercles and sub-opercle: a second from the three last dorsal spines to in front of the base of the anal: a third over the free portion of the caudal. Seas of India to the Malay Archipelago, and beyond.

Genus—DASCYLLUS, Cuv.

Tetradrachmum, Cantor.

Branchiostegals five: pseudobranchiæ. Preopercle, and occasionally the preorbital are serrated. Teeth villiform in a narrow band the outer being somewhat the larger. Anal fin with two spines. Scales large or of moderate size. Lateral line ceases below the soft dorsal, but is continued along the central row of scales in the form of a circular hole in each. Air-vessel large. Pyloric appendages two or three.

555. *Dascyllus aruanus*, Linn. D. $\frac{12}{13}$, A. $\frac{2}{12}$, L. l. 25-27, L. tr. 3/11, Cæc. pyl. 3, Vert. 12/14. Length of head 1/4, height of body nearly 1/2 of the total length. Pearl white, with three vertical black bands, the anterior descending from the first three dorsal spines through the eye over the snout to the under surface of the lower jaw: the second from the sixth to the ninth dorsal spines to the ventral fins, which are black: the third from the base of the soft dorsal to the anal: caudal

dark, light posteriorly. Eastern Coast of Africa through the seas of India to Polynesia, &c.: it is much rarer in Western than in Eastern India, and numerous at the Andamans, Nicobars, and Burma.

Genus—*POMACENTRUS (Lacép), Cuv.*

Pristotis, Rüpp.

Branchiostegals five: pseudobranchiæ. Preopercle and usually infraorbital ring serrated. Teeth small, compressed: the crowns smooth or emarginated. Two anal spines. Scales rather large. Lateral line ceases below the soft dorsal fin. Air-vessel present. Pyloric appendages few.

556. *Pomacentrus Jerdoni,* (*) Day. D. $\frac{13}{13}$, A. $\frac{2}{14}$, L. l. 34, L. tr. 5/11. Length of head 1/5, height of body 2/7 of the total length. Preopercle finely serrated on its vertical margin, more coarsely at its angle. Infraorbital ring entire. Teeth compressed into a single row of about thirty. Olive, becoming light below: seven rows of light blue spots running across the gill cover: one row along the suborbitals, and one over the snout. A row of light lines along the centre of the scales on the sides. A black spot at the base of the pectoral. Fins dark coloured. Base of caudal barred in lines. Madras, to 5 inches in length.

557. *Pomacentrus trilineatus* (Ehren), Cuv. & Val. D. $\frac{13}{16}$, A. $\frac{2}{16}$, L. l. 28, L. tr. 3/9. Height of body 4/11 in the total length. Preorbital denticulated with two strong teeth anteriorly. Olivaceous: caudal yellow: each scale with one or two blue spots: three to five blue lines on the forehead, the outer of which are continued on to the nape, and sometimes along the base of the dorsal fin. A dark round spot on the opercle: a black spot margined with blue across the free portion of the tail. The immature have a similar spot on the anterior third of the soft dorsal. Red Sea, East Coast of Africa, Nicobars, and beyond.

558. *Pomacentrus punctatus,* Quoy. & Gaim. D. $\frac{12\text{-}13}{15\text{-}14}$, A. $\frac{2}{13\text{-}14}$, L. l. 27, L. tr. 3/9, Cæc. pyl. 3. Length of head 2/9, height of body 3/8 of the total length. Preorbital nearly as deep as long, denticulated. Brownish, head with irregular blue dots, and one on each scale. A black spot anteriorly edged with white across the free portion of the tail: opercle with a dark spot superiorly. Red Sea, Mauritius, and Andamans.

559. *Pomacentrus Bankanensis,* Bleeker. D. $\frac{13}{14}$, A. $\frac{2}{14\text{-}15}$, L. l. 26-28, L. tr. 3/9. Height of body 3/8 of the total length. Preorbital denticulated. Brownish: two blue lines along the forehead anteriorly converging on the snout and posteriorly extended on to the back: two more through the eye to the maxilla. A dark blue mark on the opercle: each scale with a blue spot, and a black white-edged ocellus at the base of the ninth to eleventh dorsal spines. Andamans, Nicobars, and Malay Archipelago to China.

560. *Pomacentrus vanicolensis,* Cuv. & Val. D. $\frac{13}{14}$, A. $\frac{2}{14}$, L. l. 26, L. tr. 2½/9. Length of head 2/7, height of body 2/5 of the total length. Preorbital and suborbital ring of bones serrated, the former

---

(*) Dr. Gunther has a genus *Lepidozygus* separated from *Pomacentrus* on account of the increased number of scales (L. l. 36), quite a peculiar physiognomy, and by a different dentition. The above seems 'the missing link.'

longer than deep. Preopercle denticulated: two opercular spines. Third and fourth dorsal spines the longest. Brownish: a black spot having a light anterior edge over the base of the free portion of the tail: a dark spot on the opercle. Dorsal, anal and caudal fins dark (spec. $3\frac{3}{10}$ inches long). The immature is said to have a dark spot in the middle of the soft dorsal. Andamans (and Vanicolo).

561. *Pomacentrus bifasciatus*, Bleeker. D. $\frac{13}{14}$, A. $\frac{2}{14}$, L. l. 25. Height of body 1/3 of the total length. Preorbital entire. Yellow, with a curved blue line on the preorbital: a black band from the nape over the opercles, and a second below the last dorsal spines. Fins yellow. Andamans and Malay Archipelago.

562. *Pomacentrus trimaculatus*, Cuv. & Val. D. $\frac{13}{13}$, A. $\frac{2}{14}$, L. l. 28. Height of body 1/3 of the total length. Preorbital serrated. Violet, with two transverse blue lines between the orbits: two or three large blackish spots along the back: bluish spots on the scales along the bases of the dorsal and anal fins: dorsal with two blue horizontal bands: anal with one. Andamans, Malay Archipelago to China.

563. *Pomacentrus albofasciatus*, Schleg. Height of body 1/3 in the total length. Suborbital ring of bones serrated, preorbital entire. Dark olive, with bluish spots on the cheeks: a curved blue line below the eye, and a broad vertical yellowish band underneath the last dorsal spines: sometimes a black spot on the base of the last few dorsal rays, and at the base of the pectoral. Andamans and Malay Archipelago.

Genus—GLYPHIDODON *(Lacép), Cuv.*
Euschistodus, Hypspops, sp., Gill. *Parma*, Günther.

Branchiostegals five or six: pseudobranchiæ. Body short, compressed. Cleft of mouth small. Opercles entire. Teeth compressed and in a single row, sometimes the alternate ones being similar. A single dorsal fin with the spines rather more or less, or equal in number to the rays: anal with two spines. Scales of moderate size. The tubular portion of the lateral line ceases below the end of the dorsal fin, but is continued in the form of minute circular orifices. Air-vessel present. Pyloric appendages few.

564. *Glyphidodon sordidus*, Forsk. *Chák-mud-dah*, Andam.: *Calamoia pota*, Tel. D. $\frac{13}{14\text{-}16}$, A. $\frac{2}{14\text{-}15}$, L. l. 29-30, L. tr. 4/12, Vert. 12/14. Length of head 1/5, height of body nearly 2/5 of the total length. Teeth very narrow. Vertical fins somewhat rounded: caudal deeply forked. Greenish-brown with five or six darkish vertical bands, broader than the ground colour: a dark spot across the summit of the free portion of the tail.' Immature with the anterior half of the spinous dorsal black. Red Sea, seas of India to the Malay Archipelago and beyond.

565. *Glyphidodon affinis*, Günther. D. $\frac{13}{13}$, A. $\frac{2}{15}$, L. l. 29, L. tr. 4/11. Height of body nearly 2/5 of the total length. Teeth very narrow. Vertical fins angular: caudal forked. Body with six dark vertical bands narrower than the ground colour, the last being over the free portion of the tail. Andamans and China.

566. *Glyphidodon cœlestinus* (Soland). Cuv. & Val. *Rahti potah*, Tel. D. $\frac{13}{13}$, A. $\frac{2}{12\text{-}14}$, L. l. 29-30, L. tr. 4/11, Vert. 11/15. Length of

head 1/4, height of body 2/5 of the total length. Soft dorsal produced into a point: caudal forked. Brownish, with five vertical dark cross bands which are not wider than the ground colour: caudal sometimes with a black upper and lower border. Red Sea through the seas of India and beyond.

567. *Glyphidodon notatus*, Day. D. $\frac{13}{13}$, A. $\frac{2}{12}$, L. l. 30, L. tr. 5/12. Brownish, each scale with a light centre. White vertical bands pass from the back, the anterior from the first dorsal spine to the base of the pectoral: the second from the middle spine to the base of the ventral: the third from the last three spines to the anal: and the fourth over the free portion of the tail: a black spot in the axil: caudal whitish. Andamans: to $3\frac{1}{2}$ inches in length.

568. *Glyphidodon unimaculatus*, Cuv. & Val. D. $\frac{13}{13}$, A. $\frac{2}{12\text{-}13}$, L. l. 26, L. tr. $2\frac{1}{2}$/8. Length of head nearly 1/4, height of body 2/5 of the total length. Teeth very slender, the alternate ones being posterior and narrower to the front row. Greenish, each scale having a bluish transverse mark: a jet black line along the upper part of the free portion of the tail. Bay of Bengal, Andamans, Borneo.

569. *Glyphidodon antjerius*, (Kuhl. & v. Hass.), Cuv. & Val. D. $\frac{13}{12\text{-}13}$, A. $\frac{2}{12\text{-}13}$, L. l. 26-28, L. tr. $2\frac{1}{2}$/9. Length of head 2/7, height of body 1/2 of the total length. Teeth slender, with a posterior narrower set that are alternate with the front row. Cœrulean-blue above the lateral line, and a black ocellus at the end of the dorsal fin: in one specimen two ocelli at this place. This fish is very variously marked, and most of the markings are absent in the adults. East coast of Africa, through the seas of India to the Malay Archipelago and beyond.

570. *Glyphidodon Batjanensis*, Bleeker. D. $\frac{12\text{-}13}{15\text{-}13}$, A. $\frac{2}{15}$, L. l. 25. Length of head 1/4, height of body 2/5 of the total length. Head and anterior portion of the body violet: cheeks dotted with blue, and two blue lines from the eye to the snout: sides of the body, its posterior extremity and fins, yellow, except the spinous dorsal which is violet: a brown spot in the axil. Andamans and Malay Archipelago.

571. *Glyphidodon Bengalensis*, Bloch. D. $\frac{13}{12}$, A. $\frac{2}{13}$, L. l. 30, L. tr. 4/11. The height of the body 4/7 of the total length, excluding the caudal fin. Soft dorsal produced into a point. Body with seven dark vertical bands, the last across the free portion of the tail. Seas of India to the Malay Archipelago.

572. *Glyphidodon Cochinensis*, Day. D. $\frac{13}{11}$, A. $\frac{2}{10\text{-}11}$, L. l. 25-28, L. tr. 3/8. Length of head 2/11, height of body 1/3 of the total length. Two opercular spines. Soft dorsal and anal very elongated and pointed. Purplish black, rather lighter on the abdomen, edges of scales sometimes with a lighter tinge: pectoral not so dark as the other fins. Cochin and the Andamans.

573. *Glyphidodon Sindensis*, Day. D. $\frac{13}{11}$, A. $\frac{2}{13}$, L. l. 26, L. tr. $3\frac{1}{2}$/10. Length of head nearly 1/4, height of body 3/8 of the total length. Eyes, diameter nearly 1/3 of length of head, 3/4 of a diameter from end of snout, and one apart. Depth of preorbital above the angle of the

mouth not 1/3 the diameter of the orbit. Upper profile of head convex. Anterior teeth notched. Dorsal spines increase to the 4th, from thence to the last nearly equal: caudal forked, the upper lobe the longer: pectoral rounded and as long as the head without the snout. Deep violet, which extends on to the dorsal and anal fins: pectoral hyaline with a dark mark at its base: ventral nearly black: the posterior extremities of the dorsal and caudal canary yellow: all the scales on the free portion of the tail with a light blue central spot. Some blue lines about the head in the young, which has also the free portion of the tail yellow and the chest light coloured. Kurrachee: very common at to $4\frac{2}{10}$ inches in length.

574. *Glyphidodon anabatoides*, Bleeker. D. $\frac{11\text{-}12}{10}$, A. $\frac{2}{10}$, L. l. 26, L. tr. $2\frac{1}{2}/9$. Length of head 1/4, height of body 2/5 of the total length, Width of preorbital 1/3 of that of the orbit: suborbitals very narrow. Olive, each scale with a blue dot: a blue spot at the commencement of the lateral line: dorsal dotted with blue, and with a dark edge: anal dotted with yellow and also with a dark margin: caudal with a brownish longitudinal band on either lobe. A black axillary spot on the pectoral. ventrals green. Andamans and Malay Archipelago.

Family—LABRIDÆ, Cuv.
Labroidei cycloidei, Mull.: *Cyclolabridæ*, Owen.

Branchiostegals five or six: pseudobranchiæ. Gills three and a half. Body oblong or elongated. Teeth in the jaws, palate edentulous: lower pharyngeal bones anchylosed into one without any median suture. A single dorsal fin with usually as many or more spines than rays: the anal rays similar to those of the dorsal: ventrals thoracic, with one spine and five rays. Scales cycloid. Lateral line complete or interrupted. Air-vessel present. No cœcal sac to the stomach nor pyloric appendages.

Genus—CHÆROPS, Rüppell.
Cossyphus, sp. Cuv. & Val.: *Hypsigenys*, Günther.

Branchiostegals five or six: pseudobranchiæ. Body oblong, compressed. Snout obtuse. Preopercle serrated or entire. The four anterior teeth conical and free, whilst the lateral ones are more or less confluent into an osseous ridge, a posterior canine tooth may be present; inferior pharyngeal teeth not confluent or pavement-like. A single dorsal fin with more spines (13) than rays (7): the anal with three spines and more rays than the soft dorsal (9-10). Scales large: cheeks high, covered with small scales, which usually are not imbricated: opercle scaled. Lateral line continuous.

575. *Chærops ancharago*, Bloch. D. $\frac{13}{7}$, A. $\frac{3}{8}$, L. l. 30. Preopercle serrated: a posterior canine tooth in adults. Yellow cheeks with large scarlet spots. A black vertical band extends from the interspace between the fourth dorsal spine and sixth ray to the middle of the depth of the body, divided anteriorly from another by a whitish ground colour: whilst nearer to the head exists a third dark band. Dorsal and caudal fins edged with orange. Andamans and Malay Archipelago.

Genus—COSSYPHUS, Cuv. & Val.
Harpe, and *Lepidaplois*, Gill.

Branchiostegals six: pseudobranchiæ. Body oblong, compressed. Snout more or less pointed. Preopercle serrated or entire. The four anterior teeth conical and free: the lateral teeth in a single row, and a posterior canine is as a rule present: inferior pharyngeal teeth, not confluent or pavement-like. A single dorsal fin with more spines (11-13) *than rays* (9-11): *the anal with three spines and more rays* (10-14) *than the soft dorsal. Scales of moderate size, those on the cheeks and opercles imbricated: the bases of the vertical fins scaled. Lateral line continuous.*

576. *Cossyphus axillaris*, Bennett: *Cul-moonjilli*, Tam. D. $\frac{12}{10\,(11)}$, A. $\frac{3}{12}$, L. l. 34, L. tr. 5/13. Snout pointed, 1/3 the length of the head. Purplish red anteriorly, becoming yellowish posteriorly, and on the fins: a black axillary spot, three more along the bases of the dorsal fin, and sometimes a black spot at the base of the caudal. Seas of India.

577. *Cossyphus Neilli*, Day. *Moonjilli*, Tam. D. $\frac{12}{10}$, A. $\frac{3}{12}$, L. l. 34, L. tr. 5/13. Snout pointed, about 1/3 of length of head. Scarlet without any dark marking: anal yellow: its central rays scarlet: dorsal red with its last few rays yellow: its sheath also with a yellow tinge: pectorals flesh-coloured: lips reddish. Madras.

Genus—*LABRICHTHYS*, Bleeker.

Body oblong, compressed. Snout more or less pointed. Preopercle entire. Teeth in a single row, the anterior ones being conical and free, a posterior canine present or absent: inferior pharyngeal teeth not confluent or pavement-like. A single dorsal fin with less spines (9) *than rays* (11): *anal with three spines, and less rays* (10) *than the dorsal. Scales large: opercles scaly, and cheeks more or less so. Lateral line continuous.*

578. *Labrichthys bicolor*, Day. D. $\frac{9}{11}$, A. $\frac{3}{10}$, L. l. 26, L. tr. 5/12. Length of head nearly 1/3, height of body 1/3 of the total length. No posterior canine. The body, behind a line from the commencement of the dorsal to the base of the anal, dark violet, some of the lower scales being blue spotted: anterior to this nearly white beneath, but darker along the top of the head, whilst most of the scales have more or less dark spots. Andamans: 4 inches in length.

Genus—*LABRIOIDES*, Bleeker.

Diproctacanthus, Bleeker.
Body oblong, compressed. Gill-membranes attached to the isthmus. Snout pointed, lips prominent, one of them having a notch anteriorly. Preopercle entire. A band of small teeth in the jaws, with a pair of curved canines in either jaw, the upper pair being received when the mouth is closed, between the lower ones: a posterior canine tooth: inferior pharyngeal teeth not confluent or pavement-like. A single dorsal fin with usually less spines (9) *than rays* (9-11): *anal with two or three spines, and rays nearly similar to those of the dorsal* (9-10). *Scales on body of moderate size, they are extended over the opercles, cheeks and base of the vertical fins. Lateral line continuous.*

579. *Labrioides dimidiatus*, Cuv. & Val. D. $\frac{9}{12\text{-}11}$, A. $\frac{3}{10}$, L. l. 46-50, L. tr. 4/15. Length of head and height of body each 2/7 of the total length. Nearly white, with a black band through the eye to the

upper third of the base of the caudal fin when it bends slightly to reach the centre of that fin : a black band along the anal, which passes to the caudal and joins the upper band at the end of that fin : a dark band along the dorsal fin. Red Sea, Andamans to the Malay Archipelago.

Genus—CHEILINUS (Lacép), Cuv.
Oxycheilinus et Crassilabrus, Gill.
Body oblong, compressed. Preopercle entire. Lower jaw not produced backwards. Teeth in one row, two canines in either jaw, none being directed outwards : no posterior canine : inferior pharyngeal teeth not confluent or pavement-like. A single dorsal fin, the number of spines (9-10) being about equal to the rays (9-11). Anal with three spines, the third being the longest, its rays rather less (8-9) than those of the dorsal. Scales large, two rows on the cheeks. Lateral line interrupted.

580. *Cheilinus chlorosus*, Bloch. D. $\frac{10}{9}$, A. $\frac{3}{8}$, L. l. 22. Olive brown, round yellow spots on the head, and body similarly dotted: spinous dorsal olive, with red edges, and spines with brown dots: soft dorsal reddish: anal, ventral, and caudal with small yellow dots. Bay of Bengal to the Malay Archipelago and beyond.

581. *Cheilinus trilobatus*, Lacép. D. $\frac{9}{10}$, A. $\frac{3}{8}$, L. l. 21. L. tr. 3/5, Vert. 10/13. Length of head and height of body each nearly 1/3 of the total length. Caudal fin rounded in the immature, trilobate in the adult. Green, red dots and stripes on the head, the latter being before and below the eye : each scale on the body with one or two vertical red streaks. Vertical fins green : dorsal and anal with red margins : soft dorsal sometimes red, and a black spot on the base of the last few rays. Andamans to the Malay Archipelago and beyond.

Genus—EPIBULUS, Cuv.
Body oblong, compressed. Preopercle entire. Mouth very protractile, the ascending processes of the intermaxillaries, also the mandibles and tympanics, being elongated. Teeth in one row, and two canines in either jaw, but no posterior canine : inferior pharyngeal teeth not confluent or pavement-like. A single dorsal fin, with less spines (9) than rays (10) : three anal spines, the rays less numerous (8) than those of the dorsal. Scales large, two rows on the cheeks. Lateral line interrupted.

282. *Epibulus striatus*, Day. D. $\frac{9}{10}$, A. $\frac{3}{8}$, L. l. 19, L. tr. 2/7. The posterior extremity of the lower jaw extends to below the hind edge of the orbit, and the ascending processes of the intermaxillaries to opposite the posterior third of the orbit. A white line between the orbits, and two more on the head, the first of which runs from the eye to the snout : the second descending from the orbit, meets one from the opposite side. Body greenish brown, with five narrow milk-white vertical bands, the first from the opercles to before the ventral fin, the next from the second dorsal spine to the end of the ventral fin, the third from the centre of the dorsal spine to the base of the anal, the fourth from the end of the dorsal fin to the end of the anal, the fifth round the free portion of the tail. Soft dorsal, and termination of the anal white. Andamans, a single specimen 1½ inches in length.

cclxvii

Genus—ANAMPSES, Cuv.

Body oblong, compressed. Preopercle entire. Teeth in jaws in one row, the two front ones in each being prominent, directed forwards, and compressed, with cutting edges: no posterior canine tooth: inferior pharyngeal teeth not confluent or pavement-like. A single dorsal fin, the spines (9) being less than the rays (12): anal with three spines, and the same number of rays as the dorsal. Scales rather large, none on the head. Lateral line continuous.

583. *Anampses cæruleo-punctatus*, Rüpp. D. $\frac{9}{12}$, A. $\frac{3}{12}$, L. l. 27, L. tr. 4/10, Vert. 11/15. Length of head 2/7, height of body 1/3 of the total length. Many bluish vertical lines radiating from the orbit. Reddish, each scale with a central blue spot: caudal with a white margin. Red Sea, Bay of Bengal. Two specimens to 7 inches in length exist in the Calcutta Museum, being those referred to in the J. A. S. for 1860, p. 448, as presented by Mr. Edwards.

Genus—HEMIGYMNUS, Günther.

Halichœres, sp. Rüppell : *Tautoga*, sp. Cuv. & Val.

Body oblong, compressed. Preopercle entire. Teeth in a single row, two canines anteriorly in either jaw, the lower ones being received when the mouth is closed between the upper pair, generally a posterior canine: inferior pharyngeal teeth not confluent or pavement-like. A single dorsal fin, with less spines (9) than rays (11) : anal with three spines and the same number of rays (11) as the soft dorsal. Scales of moderate size, none on the opercles, but a stripe of very small ones on the cheek. Lateral line continuous.

584. *Hemigymnus melapterus*, Bloch. D. $\frac{9}{11}$, A. $\frac{3}{11}$, L. l. 29, L. tr. 5/14. A posterior canine tooth concealed by the skin. Brownish above, becoming yellowish beneath: a dark mark behind the orbit: scales dotted with blue: dorsal and anal light at the external third, followed by a bluish band, external to which it is darker: caudal dark. Andamans to the Malay Archipelago and beyond.

Genus—STETHOJULIS, Günther.

Julis, sp. Cuv. & Val.

Body oblong and compressed. Preopercle entire: anterior teeth conical: a posterior canine: inferior pharyngeal teeth not confluent or pavement-like. A single dorsal fin with less spines (9) than rays (11): anal with three spines and the same number of rays as the soft dorsal. Scales of moderate size: none on the head. Lateral line continuous.

585. *Stethojulis strigiventer*, Benn. D. $\frac{9}{11}$, A. $\frac{3}{11}$, L. l. 26, L. tr. 2/9. Length of head 2/7, height of body 1/4 of the total length. Greenish, a black band from the mouth passes below the eye to the opercle, several longitudinal yellow lines and some black dots along the sides in the lower half of the body: a black spot on the last dorsal ray, another at the base of the caudal fin, which last mark is apparently sometimes absent. From East Africa through the seas of India to the Malay Archipelago.

586. *Stethojulis Finlaysoni*, Cuv. & Val. Greenish. Commencing at the corner of the mouth and passing below the eye is a brown lateral band covered with deeper coloured spots. The dorsal, bordered with red, is dotted with this colour: the anal is similar. Three oblique orange bands bordered with blue on either lobe of the caudal. Under surface of the head silvery, with a tinge of orange below the brown band in its entire length. Ceylon, from a drawing by Major Finlayson.

Genus—PLATYGLOSSUS (*Klein*), Bleeker.

Halichœres, sp. Rüppell: *Julis*, sp. Cuv. & Val.: *Macropharyngodon, Güntheria:* and *Hemitautoga*, Bleeker: *Chœrojulis*, Gill.

Body oblong, compressed. Anterior teeth conical, but neither bent outwards nor backwards: a posterior canine: inferior pharyngeal teeth not confluent or pavement-like. A single dorsal fin with less spines (9) than rays (10-14): anal with three spines and about the same number of rays (11-14) as the soft dorsal. Scales rather large: those on the thorax smaller than on the remainder of the body: none on the head (except a few rudimentary ones behind the eye in two or more species). Lateral line continuous.

587. *Platyglossus notopsis* (Kuhl. & v. Hass.), Bleeker. D. $\frac{9}{13}$, A. $\frac{3}{11}$, L. l. 27-28, L. tr. 3/11. Length of head 1/4, height of body 2/7 of the total length. Purplish brown, with four or five red longitudinal bands. Two black ocelli on the dorsal fin, largest in the young: the smallest one between the first and second spines: the larger between the third and sixth rays: caudal with a yellow band at its base, and yellow edges. Andamans to the Malay Archipelago.

588. *Platyglossus marginatus*, Rüppell. D. $\frac{9}{13}$, A. $\frac{3}{11}$, L. l. 27, L. tr. 3/10. Length of head 2/9, height of body 2/7 of the total length. Blackish-green: head and anterior part of the body with undulating grass-green streaks edged with blue. Numerous red blue edged streaks and spots on the vertical fins, which have blue margins: a large green crescentic-shaped mark on the basal half of the caudal. Red Sea, through the seas of India to the Malay Archipelago.

589. *Platyglossus Dussumieri*, Cuv. & Val. *Sahnee moia*, Tel.: *Kullaray meen*, Tam. Length of head 1/4, height of body 2/7 of the total length. Green, with some irregular violet bands between the eyes, and one from it to the snout: two or three pass irregularly over the opercles: back with five or six bands of deep blue, which are chequered with dull purplish red. A black spot at the base of the pectoral, with a light anterior edge. Dorsal and anal greenish, with an oval yellow spot between each ray, the soft portions with a broad purplish external margin: a blackish ocellus, with a yellow edge between its fifth and seventh spines. Caudal tipped with blue, its angles being orange with a red base and transverse red bands. Seas of India to the Malay Archipelago and beyond.

590. *Platyglossus leparensis*, Bleeker. D. $\frac{9}{?}$, A. $\frac{3}{12}$, L. l. 26. Length of head 1/4, height of body 2/9 of the total length. A broad brown streak from the eye to the maxilla, and a brown spot behind the orbit: four curved yellowish lines on the upper portion of the opercle.

A silvery line from the eye to the caudal fin, and below it four or five more: many of the scales with brown spots. Two black ocelli on the dorsal fin, the anterior between the first two spines, and the posterior between the first two rays: caudal fin yellowish red. Andamans to the Malay Archipelago.

591. *Platyglossus hortulanus*, Lacép. D. $\frac{9}{11 \cdot 12}$, A. $\frac{3}{11}$, L. l. 28, L. tr. 2½/9, Vert. 10/15. Length of head 1/4, height of body nearly 1/4 of the total length. Two rows of minute scales behind the orbit. Yellowish brown: broad bluish longitudinal bands on the head, and the anterior portion of the back with bluish spots. One (may be two or three more) yellow spot on the back below the fourth dorsal spine, and sometimes a black spot behind it. Oblique brown streaks on the dorsal fins, sometimes enclosing spaces: a black axillary spot: usually another at the upper part of the base of the caudal, on which brown spots may be present or absent: anal with longitudinal bands. Eastern coast of Africa, through the seas of India to the Malay Archipelago and beyond. A specimen from the Andamans in the Calcutta Museum is 8½ inches in length.

592. *Platyglossus nebulosus*, Cuv. & Val. D. $\frac{9}{11}$, A. $\frac{3}{11}$, L. l. 28. Length of head rather more than the height of the body, which is 3/13 of the total length. Olivaceous: several violet bands on the head, the one on the cheek being curved, but not completing a circle: the opercular lobe violet: a dark band passing from it to the tail, in its course it gives off several superior and inferior short processes, between which are silvery spots, and there are more along the base of the dorsal fin: some oblique silvery streaks covered by the pectoral fin. A minute spot sometimes present between the first two dorsal spines: and a larger one between the first two rays: two or three red ocelli margined with blue between the dorsal spines and oblique violet bands ascending forwards: anal with a band along its centre and ocelli at its base: caudal occasionally with black dots. Red Sea, seas of India, to the Malay Archipelago.

593. *Platyglossus bimaculatus*, Rüpp. D. $\frac{9}{11}$, A. $\frac{3}{11}$, L. l. 27, L. tr. 2/10. Length of head and height of body each 1/4 of the total length. Green, light violet streaks radiate from the eye: a blue longitudinal band passes across the opercle, and is continued direct to the upper part of the base of the caudal fin. A black spot on the ninth and tenth scales of the lateral line. Dorsal and anal fins with their basal halves reddish. Violet, with blue dots: their external halves with red, black, blue, and green longitudinal bands: caudal with blue spots. Or else the dorsal may have two longitudinal and the caudal three transverse rows of ocelli. Red Sea, Ceylon.

594. *Platyglossus kawarin*, Bleeker. D. $\frac{9}{11}$, A. $\frac{3}{11}$, L. l. 27, L. tr. 2½/9. Length of head and height of body each 2/7 of the total length. Caudal rounded. Head with a blue band from the eye to the snout: a second from the angle of the mouth along the posterior margin of the orbit to the upper part of the head: a third from the interopercle: a blue blotch on the centre of the opercle, and a light-blue wide band above: some blue spots on the summit of the head. Colours below the

lateral line are blue, each scale having a rosy central spot, forming seven longitudinal bands. Dorsal and anal fins blue, with three rows of round reddish spots : caudal yellowish, with a dark edge. Andamans, 3½ inches in length.

595. *Platyglossus scapularis*, Bennett. D. $\frac{9}{\mathrm{T}\mathrm{T}}$, A. $\frac{3}{\mathrm{T}\mathrm{T}}$, L. l. 25, L. tr. 2½/10. Length of head and height of body each 1/4 of the total length. A broad red blue-edged band passes from the snout to the eye : a second irregular one goes obliquely upwards from the eye to the wide lateral band, which, of a brownish violet colour, passes from the shoulder to the back of the tail : a red streak extends from the axil to the abdomen : dorsal and anal margined with a green blue-edged band. Caudal with reddish-violet transverse bands. From the east coast of Africa, through the seas of India, to the Malay Archipelago.

596. *Platyglossus Ceylonensis*, Bennett. D. $\frac{9}{\mathrm{T}\mathrm{T}}$, A. $\frac{3}{\mathrm{T}\mathrm{T}}$. Yellowish. "Head grey rivulated with yellow : a yellow interrupted band along the lateral line : another lateral band of the same colour, edged with blue, runs along the side and emits numerous short processes towards the belly : an oblique yellow streak across the base of the pectoral to the belly : vertical fins yellow : a band along the base of the dorsal edged with blue inferiorly : an oblique blue streak behind the base of each dorsal ray, a second behind its middle, and a spot of the same colour behind its top : anal fin with two blue bands : caudal with three irregular transverse rivulated blue bands." Ceylon.

597. *Platyglossus purpureo-lineatus*, Cuv. and Val. Green, with violet streaks on the back, on the middle of the body and on the borders of the dorsal and anal fins : pectorals yellow : head yellowish, spotted with violet : the caudal with small red streaks : the base of the dorsal spotted with very vivid red. Ceylon.

Genus—*NOVACULA*, Cuv. & Val.

Xyrichthys, sp. Cuv. & Val.

Body oblong, compressed: the head compressed, with its upper edge short or obtuse. Anterior teeth conical : no posterior canine : inferior pharyngeal teeth not confluent or pavement-like. Dorsal fin single with less spines (9) than rays (12) : the two first dorsal spines sometimes more or less separated from the others. Scales rather large : head may be entirely scaleless, or the cheeks with two or more rows of small ones. Lateral line interrupted.

This Genus has been sub-divided into the following *Sub-genera*.

A. Cheeks without any or only a few rudimentary scales : no anterior division of the dorsal fin. *Xirichthys et Malacocentrus*, Gill.

 a. Upper edge of the head sharp. *Novacula*, Bleeker.
 b.   ,,   ,,   ,,   ,, rather obtuse. *Novaculichthys*, pt. Bleeker.

B. Cheeks with two or three rows of scales : no anterior division of the dorsal fin. *Novaculichthys*, pt. Bleeker.

cclxxi

C. Cheeks scaled: no anterior division of the dorsal fin.

D. Cheeks scaleless: the two first dorsal spines more or less separated from the remainder of the fin. *Iniistius*, Gill.

a. Upper edge of the head sharp. *Xirichthys*, Bleeker.
b. „   „   „   „   rather obtuse. *Novaculichthys*, Bleeker.

E. Cheeks scaled: the two first dorsal spines more or less separated from the remainder of the fin. *Hemipteronotus*, Bleeker.

598. *Novacula cyanifons*, Cuv. & Val. D. $\frac{9}{12}$, A. $\frac{3}{12}$, L. l. 26. Two first dorsal spines a little longer than the others: ventrals produced, reaching the origin of the anal: scales said to be similar to *N. cultrata*, which has none on the cheeks. Rose coloured: a blue band along the upper ridge of the head to the dorsal fin: dorsal fin with oblique and caudal with transverse blue streaks: pectorals blue: ventrals yellow. Pondicherry.

599. *Novacula rufa*, Day. D. $\frac{9}{12}$, A. $\frac{3}{12}$, L. l. 26, L. tr. 5/13. Length of head 2/7, height of body nearly 1/3 of the total length. Body strongly compressed, and the upper edge of the head sharp: some small scales behind and below the orbit. Two first dorsal spines produced, and a deep notch in the interspinous membranes separating them from the remainder of the fin. Rose colour, becoming yellowish on the abdomen: the fins yellowish, except the caudal. which has dark greyish reticulated bands. Madras.

Genus—*JULIS*, sp. Cuv & Val.

Body oblong and compressed: snout not produced. Anterior teeth conical: no posterior canine tooth; inferior pharyngeal teeth not confluent or pavement-like. A single dorsal fin with less spines (8) than rays (11-14): anal with two or three spines and about the same number of rays as the soft dorsal (11-14). Scales large: none on the head. Lateral line continuous.

600. *Julis dorsalis*, Quoy. and Gaim. D. $\frac{8}{13}$, A. $\frac{3}{11}$, L. l. 29. L. tr. 2½/9, Vert. 11/14. Length of head and height of body each rather above 1/4 of the total length. Caudal lobes somewhat produced. Broad red bands radiating from the eyes: six vertical cross bars along the back, sometimes extended on to the dorsal fin: a black spot in the axil: a red band along the side of the tail: usually a black band on the dorsal fin, and the anal with a black spot anteriorly. East Coast of Africa, seas of India, to the Malay Archipelago and beyond.

601. *Julis lunaris*, Linn. D. $\frac{8}{13}$, A. $\frac{2}{11}$, L. l. 27-28, L. tr. 2½/11, Vert. 11/14. Length of head 2/9, height of body nearly 1/4 of the total length. Head violet, with several oblique reddish bands: body green, each scale with a vertical red streak, forming bands: an oblong reddish-violet spot on the pectoral. Dorsal red, with a blue and yellow margin: anal violet, with a yellow edge: caudal yellow: its base and lobes green.

Red Sea, East coast of Africa, seas of India, to the Malay Archipelago and beyond.

602. *Julis trilobata*, Lacép. D. $\frac{9}{13}$, A. $\frac{3}{11}$, L. l. 29, L. tr. 3½/11. Length of head and height of body each a little more than 1/4 of the total length. Caudal lobes slightly produced. Green or blue; a red band passes across the opercle to the caudal fin : a second of a brownish-violet colour goes along the back to the upper margin of the caudal fin: and a third along the abdomen to the lower margin of the caudal, the rays of which latter are green and the membrane red and violet. Dorsal fin green, with a broad band along its middle, and having a black spot anteriorly: anal green, with a dark basal band: the posterior half of the pectoral blackish. From the Red Sea and east coast of Africa, through the seas of India to the Malay Archipelago and beyond.

603. *Julis amblycephalus*, Bleeker. D. $\frac{9}{13}$, A. $\frac{3}{11}$, L. l. 27, L. tr. 2/9. Length of head 4/19, and height of body 4/21 of the total length excluding the caudal fin. Caudal lobes produced. The upper two-thirds of the head and body olive-brown, each scale having a vertical streak : the lower third of the body white: a narrow blue dark-edged line passes from the lower angle of the orbit to the axil : a second across the angle of the preopercle : a broad triangular spot exists in the axil of the pectoral, which fin is yellow with a blackish extremity. Dorsal fin black anteriorly, becoming a little lighter posteriorly, the soft portion having a narrow white edge: caudal lobes dark olive. Ceylon to the Malay Archipelago.

604. *Julis purpurea*, Forsk. D. $\frac{9}{13}$, A. $\frac{3}{11}$. Length of head and height of body each 1/4 of the total length. Caudal lobes slightly produced. Bluish or greenish, five red bands, one of which passes through the eye over the base of the pectoral, and joins its fellow from the opposite side: a red band from the opercle to the caudal: a second above and a third below this one. Fins greenish : dorsal and anal with a red longitudinal band. Red Sea and Bombay.

Genus—GOMPHOSUS, Lacép.

Body oblong : compressed. Snout produced, tubiform. Gill membranes attached to the isthmus. Anterior teeth conical : no posterior canine tooth : inferior pharyngeal teeth not confluent or pavement-like. A single dorsal fin with less spines (8) than rays (13) : anal with two or three spines and less rays (11) than the soft dorsal. Scales rather large, none on the head. Lateral line continuous.

605. *Gomphosus cœruleus*, Lacép. D. $\frac{8}{13}$, A. $\frac{2}{11}$, L. l. 29, L. tr. 3/10. Caudal fin emarginate. Violet, vertical fins yellow, the dorsal and anal with blue edges : the upper and lower margins of the caudal blue. Seas of India to the Malay Archipelago.

606. *Gomphosus melanotus*, Bleeker. D. $\frac{8}{13}$, A. $\frac{2-3}{11}$, L. l. 26, L. tr. 3/10. Caudal fin cut square or rounded. Upper part of head and back deep brown, becoming lighter on the sides : each scale darkest at its base : cheeks pinkish. Pectorals yellow: vertical fins dark-coloured, becoming deep brown externally, having a very narrow light edge : caudal the same, with a black margin and a rather wider white border:

ventrals whitish, the outer ray brown. Andamans and Malay Archipelago.

Genus—CORIS, Lacép.

Hologymnosus et Labrus, sp. Lacép : Halichœres, sp. Rüppell : Julis, sp. Cuv. & Val. : Pseudocoris, Hemicoris, et Ophthalmolepis, Bleeker.

Body oblong, compressed. Anterior teeth conical : a posterior canine may be present : inferior pharyngeal teeth not confluent or pavement-like. A single dorsal fin with less spines (9) than rays (11-12) : anal with three spines, and about the same number of rays (11-13) as the soft dorsal. Scales rather small : none on the head (except in C. lineolata). Lateral line continuous.

607. *Coris formosa*, Bennett. D. $\frac{9}{12}$, A. $\frac{3}{12}$. Caudal rounded. "Bluish-grey, with circular black spots : head yellow, with two oblique blue bands ascending towards the origin of the dorsal, one commencing from the snout and passing through the eye, the other parallel to the first, running below the eye. Dorsal and anal fins brown, the former with a red margin and with two green lines running within the red : black dots between the rays. Anal with a narrow green edge, and a narrow green intermarginal line : a series of green dots within the margin. The inner half of the caudal red, the outer yellowish white." Ceylon.

608. *Coris cingulum*, Lacép. D. $\frac{9}{12}$, A. $\frac{3}{12}$, L. l. 60, L. tr. 6/28. Length of head and height of body each 1/4 of the total length. No posterior canine. Caudal rounded. Head and anterior part of the body olive brown, occasionally with two brown blotches, also small blue spots over the head, and a black spot on the opercle : a black axillary spot, a pale cross band opposite the end of the pectoral, behind which the colours are brown, sometimes with red blotches on the back of the tail. Vertical fins dark with black spots and white edges, or with a wide yellow margin. Red Sea and seas of India.

Genus—CYMOLUTES, Günther.

Xyrichthys, sp. Cuv. & Val.

Body oblong, compressed. Snout rather elevated. Anterior teeth free : no posterior canine. A single dorsal fin with less spines (9) than rays (12-14) : anal with three spines, and an equal number of rays to the soft dorsal. Scales small. Lateral line interrupted.

609. *Cymolutes prætextatus*, Quoy & Gaim. D. $\frac{9}{12}$, A. $\frac{3}{12}$, L. l. 73. Greenish, with a dark-blue-edged band across the shoulder. Ceylon, Mauritius, and Malay Archipelago.

Genus—PSEUDODAX, Bleeker.

Odax, sp. Cuv. & Val.

Body oblong, compressed. Each jaw with two pairs of broad incisors, having cutting lateral edges : teeth in the inferior pharyngeals confluent and pavement-like. A single dorsal fin having nearly the same number of spines (11) as rays (12) : anal with three spines and more rays (14) than the second dorsal. Scales of moderate size, extended over the cheeks and opercles. Lateral line continuous.

610. *Pseudodax Moluccanus*, Cuv. & Val. D. $\frac{11}{12}$, A. $\frac{3}{14}$, L. l. 32, L. tr. 4/12. Back reddish, becoming white on the abdomen: dorsal and anal with black reticulated lines: anal with from two to four undulated dark longitudinal bands: caudal brown, white at its base. Nicobars to the Malay Archipelago.

Genus—CALLYODON (Gronov.), *Cuv. & Val.*

Body oblong, rather compressed. Teeth in jaws soldered together into one deep-cutting lamina: the anterior ones are imbricate and more or less distinct: the inferior pharyngeals, where the teeth (which are pavement-like) are present, broader than long. A single dorsal fin with less spines (9) than rays (10): anal with two spines and less rays (8) than the soft dorsal. Scales large: a row on the cheeks. Lateral line strongly bent or interrupted below the posterior end of the dorsal fin.

611. *Callyodon viridescens*, Rüpp. D. $\frac{9}{10}$, A. $\frac{2}{8}$, L. l. 24. Green, with dark brown spots along the side: two red streaks on the snout: dorsal and anal with oblique brown bands, and a black spot between the two first dorsal spines. Red Sea, Andamans.

Genus—PSEUDOSCARUS Bleeker.

Body oblong, somewhat compressed. The upper jaw projecting beyond the lower: the upper lip double in its whole extent. The anterior teeth soldered together, arranged in quincuncial order: the teeth in the inferior pharyngeal bones pavement-like, and the space they cover longer than broad. A single dorsal fin with less spines (9) than rays (10): anal with two spines and less rays (8-9) than in the soft dorsal.

612. *Pseudoscarus chrysopoma*, Bleeker. D. $\frac{9}{10}$, A. $\frac{2}{8}$, L. l. 24. Three rows of scales on the cheeks, the inferior of which cover the lower limb of the preopercle. A pointed tooth at the angle of the jaws. Green. Upper lip with one, lower with two cross bands, passing into a subtriangular spot between the eye and the angle of the mouth: three short lines radiate from the eye. Scales with a reddish margin. Dorsal and anal fins rosy, with a narrow band along their bases, and green margins: caudal green. Seas of India and Malay Archipelago.

613. *Pseudoscarus rivulatus*, Cuv. & Val. *Ah-dah*, Andam. D. $\frac{9}{10}$, P. 14, A. $\frac{2}{8}$, L. l. 22, L. tr. 2/7½. Two rows of scales on the cheeks, and two scales on the lower limb of the preopercle. Two small pointed teeth at the angle of the jaws. Green, each scale with a reddish edge. Snout with several undulating green lines, its ground color reddish. A narrow green band along the base and edge of the dorsal fin, with an intermediate row of spots: anal green, becoming lighter towards its margin, which is edged with dark green: caudal with green spots. Andamans and Malay Archipelago.

614. *Pseudoscarus pyrrhostethus*, Richardson. D. $\frac{9}{10}$, A. $\frac{2}{8}$. Two rows of scales on the cheeks, and two scales on the lower limb of the preopercle. Head reddish: jaws whitish: a blue cross band on the lips: a curved blue streak from the angle of the mouth to below the eye: scales on the body with a blue edge: vertical fins red, with blue bases

and margins: caudal sometimes with blue spots, and occasionally they are also present on the dorsal. Ceylon, Malay Archipelago, and China.

615. *Pseudoscarus Troschelli*, Bleeker. D. $\frac{9}{10}$, A. $\frac{3}{9}$, L. l. 23-24., Two rows of scales only on the cheeks. Eyes small. Edge of jaws denticulated. Pointed teeth at the angle of the upper jaw. The colours of this fish differs widely, being pink, green, or brown, and variously marked. East Coast of Africa from Beluchistan, through the Seas of India to the Malay Archipelago.

616. *Pseudoscarus Russellii*, Cuv. & Val. *Sahnee moya*, Tel. D. $\frac{9}{10}$, A. $\frac{3}{9}$, L. l. 24. Eyes diameter 1/5 of length of head. Sea green. Each scale, except those on the chest, having a semilunar reddish base equal to about half its width : the scales between the bases of the ventral and anal fins reddish : head rosy : a few short blue lines radiate from the lower edge of the orbit, anteriorly one passes to the upper edge of the lip, which it skirts, another goes to its lower margin, which it encircles. Dorsal fin bluish-green edged with light blue : an irregular reddish band extends along its centre, another at its base, and a third just below its blue margin : caudal reddish, with four or five vertical bluish-green bands: anal of a light blue : pectoral and ventral reddish, the outer ray being blue : eyes hazel. Coromandel coast. A female, 18 inches long, captured July 12th 1868.

Order—ANACANTHINI.

178. All the rays of the vertical and ventral fins articulated, the latter, when present, being jugular and tho-
Fishes without any spinous racic. Air-vessel, if existing, not having a rays. pneumatic duct.

Sub-Order—ANACANTHINI—GADOIDEI.
Structure of head symmetrical on the two sides.

Family—GADIDÆ, *Cuv.*
Gadoidei, pt. Cuv.

Pseudobranchiæ, when present, glandular and rudimentary. Body more or less elongated. Gill-openings wide: gill-membranes, as a rule, not being attached to the isthmus. From one to three dorsal fins, occupying nearly the entire length of the back, the rays of the last being well developed: one or two anal fins: caudal usually free, but sometimes united to the dorsal and anal The dorsal with a separate anterior portion. Ventrals jugular, consisting of several rays, or should they be reduced to a filament, the dorsal fin is divided into two. Scales cycloid, of moderate or small size. Air-vessel and pyloric appendages usually present.

Genus—BREGMACEROS, *Thompson.*
Calloptilum, Richardson.

Branchiostegals seven : pseudobranchiæ absent. Body fusiform, posteriorly compressed. Gill-openings very wide, the gill-membranes being united beneath the throat, but not attached to the isthmus. Eyes lateral. Mouth anterior and oblique. Teeth in jaws minute and moveable, also on vomer ; none on palate. Two dorsal fins, the anterior consisting of an elongated ray arising from the occiput : the second and the anal having each a central dwarfed portion almost forming a distinct fin. Ventrals jugular, consisting of five or six rays, the outer of which are elongated. Scales cycloid, of moderate size. Lateral line continuous. Air-vessel present. Pyloric appendages few.

617. *Bregmaceros McClellandi*, Thomp. B. VII, D. 1/16+X+15, P. 25, V. 5-6, A. 22 + X + 15, C. 15, L. l. 64, L. tr. 6/8. Silvery: back shaded with green : occiput, upper half of pectoral, and first half of dorsal spotted with black. Coasts of India to China and the Philippines : attaining three inches in length.

618. *Bregmaceros atripinnis*, Day. B. VII, D. 1/20 + XV + 22, P. 21, V. 6, A. 22 + X + 26, C. 17, L. l. 70, L. tr. 18, Cæc. pyl. 2. Rich brown, becoming lighter on the abdomen : fins black, except the ventral, which is of a dirty white. Bombay, coasts of India, Burma, and the Andamans : up to 5 or more inches in length.

Family—OPHIDIIDÆ, *Müller.*

Pseudobranchiæ present or absent. Body more or less elongated. Gill-openings wide: the gill membranes not being attached to the isthmus. Eyes of moderate size, rudimentary or absent. Barbels present or absent. Sometimes canine teeth in the jaws, the vomerine and palatine ones absent or present. Vertical fins usually confluent, without any distinct anterior dorsal or anal. The dorsal occupies the greatest portion of the length of the back. Pectorals may be absent: ventrals when present rudimentary and jugular, except in *Brotulophis*, where they are situated opposite to the pectorals. Scales present or absent: lateral line when present may be single, double, or interrupted. The vent may be at the throat. Air-vessel usually present. Pyloric appendages when present of small numbers.

Genus—BROTULA, *Cuv.*

Branchiostegals eight. Body elongated, compressed. Eyes of moderate size. Barbels present. Villiform teeth on jaws, vomer, and palatines. Vertical fins confluent: ventrals reduced to a single filament, which is sometimes bifurcated: the fin is attached to the humeral arch. Scales present, minute. Air-vessel large, either rounded posteriorly or with two horns. A single pyloric appendage.

619. *Brotula maculata*, Day. B. VIII, D. 115, V. 1, A. 107. Six pairs of barbels : 2 on the snout : 2 pairs on maxilla : 3 pairs on mandible. Eyes diameter 1/3 of length of head and 2/3 of a diameter from end of snout. Air-vessel large and rounded posteriorly. Light dirty greenish colour, with a few reddish brown spots about the body and a dark round mark behind the eye. Muzzle and rostral barbels black. Fins greyish, black externally.

Genus—XIPHOGADUS, *Günther.*

Xiphasia, Swainson.

Branchiostegals five. Body elongated and compressed. Eyes of moderate size. Barbels absent. A single row of teeth in the jaws and a pair of strong canines. Vertical fins confluent, the dorsals commencing above the eyes. Ventrals reduced to a single filament, and the fin attached to the humeral arch. Scales absent.

620. *Xiphogadus setifer*, Swainson. B. V, D. 223, A. 112, C. 10. Two of the caudal rays filamentous. The single specimen found by Russell at Vizagapatam was 14 inches long.

Genus—BLEEKERIA, Günther.

Branchiostegals six: pseudobranchiæ composed of lamellæ. Body low elongated and compressed. Gill-openings very wide: the gill-membranes not united. Gills four: a cleft behind the fourth. Eyes of moderate size. Barbels absent: lower jaw very prominent. Teeth absent. Dorsal single and long: anal of moderate size. Lateral line single. Air-vessel absent. Vent remote from the head.

621. **Bleekeria kallolepis,** Günther. B VI., D. 40, A. 15, L. l. 100, L. tr. 3/14. Back olive coloured : sides silvery. A single specimen, 55 lines long, was sent from Madras by the late Dr. Jerdon to the British Museum.

Sub-Order—ANACANTHINI—PLEURONECTOIDEI.

Structure of head unsymmetrical on the two sides.

Family—PLEURONECTIDÆ, Flemm.

Heterosomata, Bonaparte.

Pseudobranchiæ well developed. Gills four. Body strongly compressed, flattened, with one of its sides coloured, the other being destitute of colouration, or having merely some spots. Both eyes (except in the very young) placed on the superior or coloured surface, sometimes rudimentary. The two sides of the head not equally developed, one remaining almost rudimentary. The jaws and dentition may be nearly equally developed on both sides, or more so on the blind than the coloured. A single long dorsal and anal fin. Pectorals, if present, may be rudimentary. Scales present or absent. Lateral line on the coloured side single, double, or triple, curved or straight. Air-vessel absent.

Genus—PSETTODES, Bennett.

Branchiostegals seven. Cleft of mouth deep, the maxilla being about half the length of the head. Gill-membranes scarcely united at the throat: gill-rakers replaced by groups of minute spines. Eyes on the right or left side. Jaws and dentition nearly equally developed on both sides. Two rows of curved, slender, sometimes barbed, distant teeth in either jaw, the anterior of the inner row in the mandible being received into a groove anterior to the vomer: teeth present on vomer and palate. Dorsal fin commencing on the nape. Most of the dorsal and anal rays branched. Scales small, ctenoid. Lateral line gradually descends to the straight portion.

622. **Psettodes erumei,** Bloch. Schn. D. 47-56, P. 16, A. 35-41, C. 17, L. l. 70-75. This fish may be coloured on either the left or the right side. Brownish or blackish, with the vertical fins edged with white : a lightish band usually present across the free portion of the tail. Cross bands have likewise been observed. Seas of India, attaining at least 16 inches in length.

Genus—PSEUDORHOMBUS, Bleeker.

Branchiostegals six. Cleft of mouth deep, the maxilla being nearly half as long as the head. Gill membranes united beneath the throat but not attached to the isthmus. Gill-rakers lanceolate. Eyes on the left side: interorbital space not concave. Jaws and dentition nearly equally developed on both sides : teeth in both jaws of unequal sizes and in a single row : vomer and palate edentulous. Dorsal fin commences on the snout: its rays and those of the anal are simple. Scales of moderate size, or rather small. Lateral line having a strong curve anteriorly.

623. *Pseudorhombus Russellii*, Gray and Hard. B. VII, D. 70-77, P. 12, V. 6, A. 56-60, C. 17, L. l. 75. Scales ctenoid. Lateral line smooth : its anterior curve equals half its length. Reddish brown, sometimes spotted or blotched with a darker tint. Seas of India and Burma, attaining a foot in length.

624. *Pseudorhombus nauphala*, Ham. Buch. D. 73, P. 13, V. 6, A. 58, C. 17, L. l. 76. Greenish-brown superiorly, with some scattered black spots. Estuaries of India and Burma, to eight inches in length.

625. *Pseudorhombus arsius*, Ham. Buch. D. 75-81, P. 12, V. 6, A. 55-60, C. 15, L. l. 65. Greenish-brown, with darker blotches: two black ocelli edged with white exist on the straight portion of the lateral line. Found from Sind throughout the Coasts of India and Burma: it attains at least 7 inches in length.

626. *Pseudorhombus triocellatus*, Bl. Schn. D. 61-69, P. 12, V. 6, A. 49, C. 19, L. l. 65. Scales ctenoid, smallest in the anterior portion of the body. Height of curve on lateral line equals half its length. Of a rich brown colour, with three large heart-shaped purplish ocelli having light centres and yellowish edges, placed in the form of a triangle, the two anterior being on a vertical line slightly anterior to the middle of the length of the fish, and the third midway between them and the base of the caudal. Seas of India: to 6 or 8 inches in length.

Genus—PLATOPHRYS, *Swains.*

Bothus, Bp.: *Rhomboidichthys*, Bleeker.

Branchiostegals six. Cleft of mouth moderate or of small extent, maxilla being about one-third or less in the length of the head. Eyes on the left side separated by a concave interorbital space of a varying extent. Jaws and dentition nearly equally developed on both sides. Teeth minute, of an equal size and in a single or double row : vomerine and palatine ones as a rule absent. The dorsal fin commences on the snout: its rays and those of the anal are simple. Scales ctenoid. Lateral line having a strong curve anteriorly.

This genus has been divided into *sub-genera* as follows :—(1) scales very small, *Rhomboidichthys* : (2) of moderate size (40 rows in lateral line) and deciduous, *Platophrys* : (3) of moderate size and not deciduous, *Engyprosopon*.

627. *Platophrys leopardinus*, Günther. D. 80-86, A. 67-68, L. l. 76-80. Brownish, with numerous ocellated spots. Andaman Islands.

628. *Platophrys pantherinus*, Ruppell. D. 85-93, P. 11, V. 6, A. 65-69, C. 17, L. l. 84-90. Brown, with rich chestnut-coloured blotches, and numerous bluish-white spots. Andaman Islands.

Genus—*SOLEA* (*Lacép*), Cuv.

Cleft of mouth narrow, twisted round to the left side. Eyes on the right side, the upper being partially or entirely in advance of the lower. Nostrils variously formed. Dentition most developed on the blind side, where the teeth in the jaws are in villiform rows, none on vomer or palate. The dorsal fin commences on the snout, and is not confluent with the caudal: pectorals present or absent. Scales small, ctenoid. Lateral line straight.

This genus has been sub-divided:—

A.—Pectorals developed, *Microbuglossus*.
   a.—Nostrils on blind side not dilated, *Solea*.
   b.—One of nostrils on blind side dilated and broadly fringed, *Pegusa*.
B.—Pectorals on both sides rudimentary or small, *Buglossus*.
Pectorals absent, *Aseraggodes*.

629. *Solea heterorhina*, Bleeker. D. 88-94, P. 9-8, A. 80-82, L. l. 110-112. Nasal tube not dilated, longer than the eye. Both pectorals developed. Olive, with from 20 to 25 vertical dark bands. Andamans.

630. *Solea Indica*, Günther. D. 55, P. 4, A. 44, L. l. ca. 85. Left pectoral absent, the right not much longer than the eye. Colours brownish: dorsal and anal fin dark. One specimen, 22 lines long, was sent from Madras by Dr. Jerdon to the British Museum.

Genus—*ACHIRUS*, sp. (*Lacép*) Cuv.

Pardachirus, Günther.

Gill openings narrow, the membranes being broadly united below the throat: gill-rakers rudimentary. Eyes on the right side, the upper in advance of the lower. Mouth narrower on the left side. Teeth minute and only on the blind side. The dorsal fin commences on the snout and ends close to the root of, but is not confluent with the caudal; each dorsal and anal ray scaly and with a pore at its base. Two ventrals. Pectorals absent. Scales small, cycloid, or very indistinctly and partially ctenoid. Lateral line straight, on the blind side there are two, the superior commencing at the snout passes along the upper profile of the neck.

631. *Achirus pavoninus*, Lacép. D. 66-68, P. 4, A. 50-56, C. 15, L. r. 94-96. A few scales on head and anterior portion of the body, ctenoid. Greyish-brown, body and fins with various sized white spots, each having a black border, and some a central black dot. Andamans.

Genus—*SYNAPTURA*, Cantor.

Achiroides, Bleeker: *Æsopia*, sp. *Euryglossa* et *Eurypleura*, Kaup.

Branchiostegals six. Eyes on the right side, the upper in advance of the lower. Cleft of the mouth narrow, twisted round to the left side: minute teeth on the left side only: palate edentulous. One of the nostrils on the blind side dilated in some species, not so in others, whilst amongst the latter both pectorals may be present, the right being somewhat the longer, amongst which some have the nasal tube small and simple (*Synaptura*), or bifid (*Euryglossa*). Secondly, the left pectoral may be longer than the right (*Anisochirus*). Thirdly, the left pectoral may be rudimentary (*Æsopia*). Fourthly, both pectorals may be absent (*Achiroides*). Vertical fins confluent. Scales ctenoid, small. Lateral line straight

Synaptura pan, Ham. Buch. (see No. 50, F. W. F. Report.)

632. *Synaptura foliacea*, Richards. D. 62-65, P. 7-9, V. 5, A. 46-51, C. 16-20, L. l. 80-99. No dilated nostril on the blind side: the tube small and simple. Pectorals on both sides. Greyish brown, with several vertical irregular blackish bands. Seas of India, extending to China; attains at least 9 inches in length.

633. *Synaptura jerreus*, Cuv. D. 65-68, P. 9-10, V. 6, A. 52-56, C. 16, L. l. 96, L. tr. 25/31. Scales on neck not larger than those on the body. Greyish brown, with ten reddish-brown vertical bands, commencing on the dorsal and continued on to the anal fin, where they become nearly black: caudal black, with some irregular white markings towards its edges. Ceylon and Coromandel: attaining at least 6 inches in length.

634. *Synaptura albomaculata*, Kaup. D. 74, P. 7-8, A. 59, C. 16, L. l. 105 (?) L. tr. 38/45. A barbel between two prominent nostrils. Body with five rows of yellowish-white dots. Attains probably above 12 inches in length. Coromandel.

635. *Synaptura Commersoniana*, Lacép. D. 71-81, P. 9, A. 61-66, C. 12, L. l. 155. Leaden grey, vertical fins black, with a broad white edge. Some have minute white spots over the body. Seas of India and Malayan Peninsula.

636. *Synaptura Orientalis*, Bl. Schn. D. 66-67, P, 7-6, A. 48-52, C. 14, L. l. 78. Nasal tube bifid. Two pectoral fins. Deep grey, with blackish blotches, and in specimens 5 or 6 inches in length short narrow black bands go to the lateral line. Seas of India: attaining 12 inches in length.

637. *Synaptura multifasciata*, Kaup. D. 93, A. 67, P. 7-10, L. l. 110. Nasal tube longer than the eye. Left pectoral fin rudimentary. Head and body with 27 narrow cross bands.

638. *Synaptura zebra*, Bloch. D. 77-89, A. 66-79, C. 13-15, L. l. 126-128. Nasal tube short. Left pectoral fin rudimentary. Olive, with twelve pairs of brown bands, three of which are on the head. Caudal with some whitish marks. Seas of India.

639. *Synaptura quagga*, Kaup. D. 66-68, P. 9, A. 58-60, C. 18, L. l. 90-100. Nasal tube not elongated. Eyes, each generally having a small filament. Left pectoral fin rudimentary. Yellowish, with 11 brown cross bands wider than the interspaces, three being on the head. Caudal brownish, with a pair of black spots edged with yellow on its posterior half. Seas of India and China to 6 or 8 inches in length.

Genus—ÆSOPIA, sp. Kaup.

Eyes on the right side, the upper slightly in advance of the lower. Cleft of mouth narrow twisted round to the left side. A short tubular nostril on the coloured side. First dorsal ray thick and prolonged, the succeeding few low. Both pectorals rudimentary. Vertical fins confluent.

Æsopia cornuta, Cuv. D. 72-75, P. 10, V. 3-4, A. 61-62, C. 17, L. l. 100. The first dorsal ray is considerably thickened and prolonged, the succeeding few are lower than the remaining ones in the fin. Twelve to thirteen vertical chestnut bands on a light ground colour, the anterior of which is on the snout. Fins black, the caudal with white markings, dorsal having a white edge. Coromandel coast: attaining at least 6 inches in length.

Genus—PLAGUSIA, pt. Cuv.

Gill-openings very narrow. Eyes on the left side. Anterior portion of snout prolonged and curving downwards and backwards in the form of a hook covers the mandible. Mouth rather narrow and unsymmetrical: lips on the coloured side fringed. A single nostril on the left side before the angle of the lower eye, but none in the interorbital space. Vertical fins confluent. Pectorals absent. Scales ctenoid, small. Lateral line on the coloured side, double or triple.

641. *Plagusia marmorata*, Bleeker. D. 99-106, V. 4, A. 75-85, C. 10, L. l. 100. Rostral hook long, reaching to some distance behind the lower eye. Two lateral lines on the left side, separated by 17 rows of scales. Brown, marbled with darker. India to China.

642. *Plagusia bilineata*, Bloch. *Nah-lah-ku*, Tel. D. 96-101. A. 70-74, C. 8, L. l. 84-90. Rostral hook reaches to below the hind edge of the eye. Two lateral lines on the left side, separated by from 12 to 14 rows of scales. Brownish, each scale lightest in its centre, fins dull orange. Seas of India to the Malayan Archipelago.

Genus—CYNOGLOSSUS, pt. Ham. Buch.
Cantoria, Arelia, Trulla, et Icania, Kaup.

Gill-openings very narrow. Eyes on the left side. Anterior portion of snout prolonged, and curves downwards and backwards in the form of a hook. Mouth rather narrow and unsymmetrical. Lips not fringed. There may be two nostrils on the coloured side, one of which is in the interorbital space (Arelia): or two nostrils before the lower eye, the inferior of which is broader than the superior (Cantoria): or only one nostril which is in front of the lower eye (Trulla): or no conspicuous nostril (Icania). Minute teeth on the right side only. Vertical fins confluent: pectorals absent. Scales ctenoid (in C. dubius, cycloid). Lateral line on the coloured side, double or triple.

643. *Cynoglossus potous*, Cuv. B. vi., D. 107, V. 6, A. 87, C. 12, L. l. 87. Rostral hook ends below the vertical from the anterior edge of the upper orbit. Three lateral lines on the left side. Darkish brown superiorly: vertical fins with a light edge. Coromandel coast.

644. *Cynoglossus lingua*, Ham. Buch. D. 142, V. 4, A. 110, C. 8, L. l. 105. Two nostrils (*Arelia*). Cycloid scales on right side. Two lateral lines on the left side, separated by 13 rows of scales. Brownish, some with cloudy markings. Seas of India: attaining at least 18 inches in length.

645. *Cynoglossus Borneensis*, Bleeker.? D. 108-112, V. 4, A. 88, C. 12, L. r. 93. Two nostrils (*Arelia*). Scales ctenoid on both sides. Two lateral lines on the left side, separated by 16 rows of scales. Brownish, without markings. Seas of India, to 20 inches in length.

646. *Cynoglossus macrolepidotus*, Bleeker. D. 111-116, V. 4, A. 86-89, L. l. 54. Two nostrils (*Arelia*). Scales cycloid on the right side. Two lateral lines on the left side, separated by 6 rows of scales. Brownish, with an ill-defined bluish band along the bases of the dorsal and anal fins: a bluish blotch on opercles. Seas of India, extending to the Malayan Archipelago.

647. *Cynoglossus dubius*, Day. D. 110, V. 4, A. 88, C. 12, L. r. 104. Rostral hook does not extend backwards so far as to below the orbit. Two nostrils (*Arelia*). Scales cycloid on both sides. Two lateral lines on the left side, separated by 21 rows of scales. Brown, without marks. Seas of Sind, the largest specimen, 20 inches in length, taken at Gwadur.

648. *Cynoglossus quadrilineatus*, Bleeker. D. 102-112, V. 4, A. 83-86, C. 10, L. l. 95. Rostral hook does not extend backwards so far as to below the orbit. Two nostrils (*Arelia*). Scales cycloid on the right side. Two lateral lines on the left side, separated by 14 rows of scales. Brown, fins yellowish: a darkish band on opercle. Seas of India to the Malayan Archipelago.

649. *Cynoglossus Bengalensis*, Bleeker. D. 103, V. 4, A. 80, C. 10, L. l. 90. Rostral hook extends to behind the mandibular symphysis. Two nostrils (*Arelia*). Two lateral lines on the left side, separated by 13 rows of scales. Brown, with some dark vertical blotches: fins dark, with light edges. Seas of India.

650. *Cynoglossus brevis*, Günther. D. 95-98, V. 4, A. 76, L. l. 100. Rostral hook extends to behind the mandibular symphysis. Two nostrils (*Arelia*). Two lateral lines on the left side, separated by from 16 to 18 rows of scales. Brownish, with blackish vertical bands which are continued on to the fins. Seas of India.

651. *Cynoglossus Hamiltonii*, Günther. D. 100-102, V. 4, A. 76-78, C. 10, L. l. 80. Rostral hook short. No perceptible nostril. Two lateral lines on the left side, separated by 13 rows of scales. Reddish brown, with irregular brownish black spots, sometimes banded. Ganges to Malayan Archipelago.

Order—*PHYSOSTOMI, Müller.*

179. All the fin rays articulated, with the exception of the first in the dorsal and pectoral, which are frequently more or less ossified. Ventral fins, when present, abdominal and spineless. Air-vessel, if existing, having a pneumatic duct, except in the family *Scombresociae.*

Fishes in which as a rule a communication exists between the air-vessel and the Pharynx.

*Family—*S I L U R I D Æ.

Margin of the upper jaw formed by the intermaxillaries: the maxilla rudimentary, often constituting the base of a barbel: no sub-opercle. Either the rayed or adipose dorsal fins, may be present or absent. Skin scaleless, and either smooth or covered with osseous plates. Air-vessel, when present, either free in the abdominal cavity (*Silurinæ*), or more or less enclosed in bone (*Amblycepinæ*): it communicates with the organs of hearing by means of the auditory bones.

*Sub-family—*S I L U R I N Æ.
Air-vessel not enclosed in bone.

Genus—MACRONES, Dumeril.

Bagrus, pt. Cuv. & Val.: *Hypselobagrus, Hemibagrus, Pseudobagrus* and *Aspidobagrus*, Bleeker: *Batasio*, pt. Blyth.

Branchiostegals from about six to twelve. Eyes with free circular margins. A separate interneural shield on the nape (Macrones), or no such shield (Hemibagrus). Mouth terminal, transverse: upper jaw generally the longer. Barbels eight, one nasal, one maxillary, and two mandibular pairs. Villiform teeth in the jaws, and in a more or less uninterrupted curved band on the palate. First dorsal fin with one spine and from five to seven rays: adipose dorsal of varying length: pectoral spine serrated: anal short or of moderate length: ventral with six rays: caudal forked. Air-vessel of moderate or large size, attached to the under surface of the bodies of the anterior vertebræ.

A. No separate interneural shield on the nape.
Macrones gulio, Ham. Buch. (see no. 61, F. W. F. Report.)

Genus—ARIUS, Cuv & Val.

Sciades, sp. et *Ariodes,* Müll & Trosch: *Hexanematichthys, Guitinga, Hemiarius, Cephalocassis, Netuma* et *Pseudarius,* Bleeker.

Branchiostegals from five to six. Head osseous superiorly, or covered with very thin skin. Eyes with free orbital margins. Mouth anterior: upper jaw generally the longer. Anterior and posterior nostrils placed close together, the latter being provided with a valve. Barbels six, one maxillary, and two mandibular pairs. Teeth in the jaws villiform: always palatine, and sometimes vomerine ones: these may be villiform or granular. First dorsal with one spine and seven rays: the adipose of moderate length or short: pectoral spine strong and serrated: ventral with six rays, situated behind the vertical from the posterior margin of the first dorsal fin: caudal forked or emarginate. Air-vessel not enclosed in bone.

A. Vomerine teeth present, forming a continuous or but slightly interrupted band, more or less confluent with those on the palatines.

(a). *Palatine teeth villiform.*

652. *Arius nenga,* Ham. Buch. B. vi., D. ⅟0, A. 20. Occipital process nearly as broad at its base as it is long: basal bone narrow and crescentic in form. Maxillary barbels reach the hinder third of the pectoral fin. Teeth in two broad triangular patches, approximating

anteriorly. Dorsal spine strong, as long as the head, granulated in front, serrated behind : pectoral spine rather shorter, externally granulated, internally serrated: upper two-thirds of adipose dorsal deep black. Hooghly : attaining at least 13 inches in length.

653. *Arius sagor*, Ham. Buch. D. ⅟0, A. 17-18. Occipital process rather wider than long : the basal bone large and somewhat butterfly-shaped. Maxillary barbels reach the end, the external mandibular the middle of the pectoral fin. Teeth in two pairs of confluent villiform patches, each of which has a rounded posterior free edge. Dorsal spine strong, four-fifths as long as the head, granulated or serrated in its anterior upper fourth, and moderately serrated posteriorly : pectoral spine stronger than the dorsal, serrated on its external fourth and along its whole inner margin. Greyish-brown, whitish below. Seas of India, ascending large rivers, is found in the Malay Archipelago : it attains at least 3 feet in length.

654. *Arius sona*, Ham. Buch. D. ⅟0, A. 16-17. Occipital process keeled, granulated, wider at its base than it is long, and convex at its posterior extremity, where it meets the basal bone, which is narrow in the centre, and somewhat of a S-shape. Eyes in the anterior half of the head. Maxillary barbels reach the end of the head, or even further. Teeth in two triangular patches with their apices converging, their bases emarginate, the vomerine ones in two small patches almost confluent with one another and also with the palatine ones. Dorsal spine strong, as long as the head without the snout, and posteriorly serrated : pectoral stronger, but slightly shorter, serrated internally. Brownish above, bluish on the sides glossed with gold, dirty white below. Seas of India, entering the mouths of large rivers : it attains 3 feet in length.

655. *Arius thalassinus*, Rüpp. D. ⅟0, A. 15-17. Occipital process slightly keeled, rather longer than broad at its base : basal bone narrow and crescentic in form. Maxillary barbels reach to about the root of the pectoral fin, the external mandibulars are slightly shorter. Teeth in three spots on either side of the palate, coalescing anteriorly so as to appear as if there were one large triangular patch. Dorsal spine as long as the head without the snout, granulated in front, serrated behind : the pectoral stronger but a little shorter, granulated externally, feebly serrated internally. A black spot on the upper third of the adipose dorsal. Found from the Red Sea through those of Africa and India to the Malay Archipelago,

656. *Arius doroides*, Cuv. & Val. D. ⅟0, A. 17. Occipital process twice as broad at its base as it is long : basal bone large and reniform. Maxillary barbels reach the posterior border of the opercles. Teeth, the palatine patches ovate and not triangular. The anterior edge of the dorsal and the external one of the pectoral spines are granulated : ventrals shorter than pectorals. Seas and estuaries of the Coromandel and Bengal coasts of India : attaining 2 feet in length.

b. Palatine teeth granular or obtusely conical.

657. *Arius Dussumieri*, Cuv. & Val. D. ⅟0, A. 14. Occipital process keeled, longer than broad. Maxillary barbels reach the root of the pectoral. Teeth on palatines in four patches, a small one also on

either side of the vomer. Dorsal spine strong, as long as the head without the snout, and finely serrated on both edges: pectoral spine equally strong. Malabar and Ceylon.

B. The teeth on the palate usually in two widely separated patches; if the vomerine ones exist they are in two distinct and separate spots, which may be confluent with those on the palatines.

a. Palatine teeth villiform.

658. *Arius Burmanicus*, Day. *Nga-young*, Burm. D. $\frac{1}{0}$, A. 20-22. Mouth spatulate. Occipital process as long as wide at its base. Maxillary barbels reach the base or middle of the pectoral fin; the external mandibulars are nearly as long. Teeth in palatines in two small, oval, and widely separated patches. Dorsal spine strong, half as long as the head: pectoral of equal length, stronger and serrated on both sides. Purplish above, white beneath. Dorsal fin externally stained with black. Rivers of Burma, and taken within tidal influence.

659. *Arius æquibarbis*, Cuv. & Val. D. $\frac{1}{0}$, A. 22. Maxillary and outer mandibular barbels of equal length. Caudal lobes equal one-fourth of total length. Rangoon and Bengal.

660. *Arius subrostratus*, Cuv. & Val. D. $\frac{1}{0}$, A. 20. Occipital process rather longer than broad at its base: basal bone narrow. Maxillary barbels scarcely reach the eye, the external mandibular are shorter. Teeth in two widely separated patches. Dorsal spine half as long as the head, nodulated anteriorly, serrated posteriorly: pectoral spine somewhat shorter, serrated internally in its anterior half. Silvery, fins with fine black spots. Western coast of India.

661. *Arius cælatus*, Cuv. & Val. D. $\frac{1}{0}$, A. 18-20. Occipital process as long as wide at its base, it is slightly keeled, and a little broader at its base than at its concave anterior extremity, where it meets a rather narrow V-shaped basal bone. Maxillary barbels reach the middle of the pectoral fin, the external mandibular are one-fifth shorter. Teeth in two widely separated triangular patches, those on the vomer confluent with those on the palate. Dorsal spine nearly as long as the head, serrated superiorly on both sides: pectoral spine as strong, rather shorter, and serrated internally. Adipose dorsal with a large black spot. Seas of India to Malay Archipelago: attaining a large size.

662. *Arius Sumatranus*, Bennett. D. $\frac{1}{0}$, A. 19. Occipital process somewhat keeled, slightly longer than broad at its base: basal bone rather narrow and V-shaped. Maxillary barbels reach the posterior end of the head, the external mandibular the base of the pectoral fin. Teeth in two widely separated triangular patches. Dorsal spine as long as the head without the snout, serrated posteriorly: pectoral rather shorter and internally serrated. Leaden colour, lighter beneath.

663. *Arius rostratus*, Cuv. & Val., D. $\frac{1}{0}$, A. 17. Occipital process somewhat keeled, wider at its base than it is long: basal bone narrow. Eyes in the commencement of the posterior half of the head.

None of the barbels reach so far as to the posterior edge of the orbit. Teeth in two triangular widely separated patches not so long as the diameter of the eye. Dorsal spine half as long as the head, serrated posteriorly: pectoral a little longer and serrated internally. Silvery, a series of fifteen minutely spotted vertical lines exist along the body: vertical fins stained with grey. Malabar coast: to at least 11 inches in length.

b. Palatine teeth granular or obtusely conical.

664. *Arius maculatus*, Thunb. D. $\frac{1}{0}$, A. 20-22. Occipital process keeled, rather longer than broad. Maxillary barbels much shorter than the head. Teeth in palate in two separate oval patches. Dorsal spine strong, not so long as the head, serrated on both edges: the pectoral slightly shorter. Adipose dorsal with a large black spot. East Indies.

665. *Arius gagora*, Ham. Buch. *Nga-young* and *Nga-yeh*, Burm. D. $\frac{1}{0}$, A. 18-19. Occipital process keeled, slightly longer than broad: basal bone narrow. Maxillary barbels somewhat longer than the head. Teeth in palate in widely separated, somewhat rhomboidal patches, and slightly convergent posteriorly. Dorsal spine nearly as long as the head without the snout, slightly serrated anteriorly, more strongly so posteriorly: pectoral spine stronger, a little longer, rugose and slightly serrated externally, strongly so internally. Greyish, silvery above, becoming white beneath. Fins stained with grey, a blackish mark on the adipose dorsal. Seas of India and Burma, ascending rivers, often far above tidal influence: it attains upwards of 8 inches in length.

666. *Arius macronotacanthus*, Bleeker. D. $\frac{1}{0}$, A. 18. Occipital process shorter than broad at its base: basal bone small. Maxillary barbels extend to the middle, the external mandibular ones to the base of the pectoral fin. Teeth, or two separate elliptical patches, often as long as the eye, and slightly divergent posteriorly. Dorsal spine very strong, as long as the head, slightly serrated on both edges: pectoral spine strong, shorter than the dorsal. Adipose dorsal with a large black spot, the other fins yellow.

667. *Arius jatius*, Ham. Buch. D. $\frac{1}{0}$, A. 19-20, very similar to *A. gagora*, but with a wider mouth. Occipital process keeled, broader at its base than long. Maxillary barbels reach the base of the pectorals. Teeth granular, frequently almost imperceptible on the palate, if distinct, in oval patches, somewhat convergent posteriorly. Dorsal spine with a soft termination, it equals the length of the postorbital portion of the head, granulated anteriorly, serrated posteriorly: pectoral spine two-thirds as long as the head. Bluish superiorly, becoming white inferiorly: maxillary barbels black: a black spot on the soft dorsal fin. Estuaries of the Ganges.

Genus—BATRACHOCEPHALUS, Bleeker.

Branchiostegals five. Gill-membranes united at the throat, without a notch. Cleft of mouth wide: lower jaw the longer. Eyelids with a free circular margin. Barbels two, rudimentary, and inserted at the chin: nostrils approximating, the posterior provided with a valve. Teeth obtusely conical in either jaw, in a broad longitudinal band on the palate: none on the vomer. Dorsal fin with one spine and seven rays, inserted anterior to the ventrals: adipose fin short. Anal of moderate length, not confluent with the caudal, which is forked. Ventral with six rays. Air-vessel in the abdominal cavity not enclosed by bone.

668. *Batrachocephalus mino*, Ham. Buch. B. V, D. 1/0, A. 20. Occipital process longer than broad at its base. Silvery. Seas and estuaries of India and Burma, entering rivers.

Genus—KETENGUS, Bleeker.

Branchiostegals five. Gill-membranes united, and also grown to the isthmus, but having a free posterior edge, which is notched. Head osseous superiorly. Eyelids with a free circular margin. Barbels six, small, no nasal ones. Nostrils approximating, the posterior provided with a valve. A single row of small compressed teeth, which are sub-truncated or almost tricuspid in the jaws: palate edentulous. Dorsal fin with one spine and seven rays, inserted anterior to the ventrals: adipose fin short: anal of moderate length and not continuous with the caudal, which latter is forked. Ventral with six rays.

669. *Ketengus typus*, Bleeker. D. 1/0, A. 20. Gape of mouth wide: upper jaw the longer: dorsal spine as long as the head. Silvery. Andamans and Malay Archipelago.

Genus—OSTEOGENIOSUS, Bleeker.

Branchiostegals five. Gill-membranes united at the throat, emarginate and overlapping the isthmus: upper surface of the head covered with very thin skin: mouth anterior: upper jaw the longer. Barbels, a single pair of semi-osseous maxillary ones. Nostrils approximating, the posterior provided with a valve: teeth in the jaws villiform: obtusely conical on the palate, where they form two widely separated patches. Dorsal with one spine and seven rays inserted anterior to the ventrals: adipose fin short. Anal of moderate length not united with the caudal, which is forked. Ventral with six rays. Air-vessel in the abdominal cavity not enclosed in bone.

670. *Osteogeniosus militaris*, Linn. D. 1/0, A. 20-23. Head granular superiorly, its depth being equal to two-thirds of its length. Diameter of eye 1/4 of width of interorbital space, which last is more than 1/2 of length of head. Silvery. Seas of India and Burma, entering rivers.

671. *Ostegeniosus Valenciennesii*, Bleeker. D. 1/0, A. 19-22. Head nearly smooth superiorly, its depth being equal to rather more than half its length. Diameter of eye 1/3 or 2/7 of interorbital space, which last is less than 1/2 of length of head. Silvery. Seas of Burma, entering rivers.

Genus—PLOTOSUS, Lacép.

Branchiostegals twelve. Gill-membranes separated and not attached to the isthmus. Head depressed. Gape of mouth transverse. Eyes small, lids with a free orbital margin: nostrils remote from one another, the anterior being on the front edge of the snout. Barbels eight, the nasal pair situated before the posterior nostril: *one maxillary and two mandibular pairs. Teeth conical in the upper, mixed in the lower jaw, and molar-form on the vomer. Two rayed dorsal fins, the first with one spine and four or five rays, the second many rayed and confluent with caudal and anal, the latter of which is also elongated. Ventral fins many rayed* (12). *Air-vessel of moderate size, with a thick tendinous covering, but not enclosed in bone.*

672. *Plotosus canius,* Ham. Buch.: *Irung-kellettee,* Tam. B. XII, 1 D. $\frac{1}{5}$, 2 D. + C. + A, 224 to 271. Nasal barbels reach the nape, the maxillary ones the end of the opercles. Teeth, intermaxillary band twice as long as broad, a crescentic vomerine band. Brownish, fins with black edges. Estuaries of India, Burma, and Malay Archipelago: it attains 3 feet and upwards in length.

673. *Plotosus Arab,* Fosk. B. XII, 1 D. $\frac{1}{4\text{-}5}$, 2 D. + C. + A. 169-190. Nasal barbels reach the eye: the maxillary ones are half the length of head. Chesnut colour, with two white longitudinal bands: the vertical fins with black edges. From the East Coast of Africa to Japan and Polynesia.

Family—SCOPELIDÆ, *Müller.*

Branchiostegals as a rule numerous. Pseudobranchiæ well developed. Gill-openings very wide. Opercular pieces sometimes incomplete. Margin of the upper jaw formed by the intermaxillaries. Barbels absent. Adipose dorsal fin present. Scales present or absent. Ova enclosed in sacs in the ovaries, and excluded by oviducts. Intestinal canal short. Pyloric appendages, when present, few in number. Air-vessel absent.

Genus—SAURUS, *sp. Cuv.*

Branchiostegals from eight to sixteen. Gill-openings very wide: the gill-membranes not attached to the isthmus. Body elongated, subcylindrical: head oblong, muzzle short. Eyes of moderate size, lateral. Gape of mouth wide, cleft very deep: *edge of the upper jaw entirely formed by long and thin intermaxillaries*: *the maxillary likewise elongated, thin, and adherent to the intermaxillaries. Teeth numerous, pointed, some of which are elongated and slender, and can be laid downwards and inwards*: *they exist on the jaws, tongue, and palatine bones, a single row being on the palate and usually none or only a few on the vomer. First dorsal, with a moderate number of rays, situated nearly in the middle of the length of the fish*: *adipose fin small. Ventral with eight rays, the internal being the longest*: *it is inserted anterior to the origin of the dorsal, and not far behind the pectorals, which are short: anal of moderate length or short. Caudal forked. Body covered with moderately sized scales. Lateral line complete. Pyloric appendages few.*

674. *Saurus Indicus,* Day. B. XV, D. 13/0, P. 13, A. 9, L. l. 55, L. tr. 3½/7. Upper jaw slightly the longer. Width of snout equals the length of the concave interorbital space. Brownish in the upper two-thirds, dirty white beneath. Numerous bluish irregular spots or blotches along the back and sides, in places almost forming horizontal bands.

cclxxxix

Dorsal and caudal white, with greyish spots, forming irregular horizontal bands. Madras, to 7 inches in length.

675. *Saurus myops*, Cuv. B. XVI, D. 12/0, A. 15-16, L. l. 56-58, L. tr. 3½/7. Lower jaw slightly the longer. Upper half of body of a golden colour, with four longitudinal blue lines having black edges, and nearly as wide as the ground colour. A black spot at shoulder. Abdominal surface silvery. Dorsal fin with three rows of yellow spots : a yellow band along the middle of the pectoral : outer half of anal yellow. Seas of India, &c.

Genus—SAURIDA, Cuv.

Differs from *Saurus* as follows.—*Teeth cardiform, the inner ones being the longest and slender, all can be laid downwards and inwards, they exist on the jaws, tongue and palatine bones : those on palate are in a double or treble band on either side, the inner ones being the shortest. Ventral fin with nine rays, the inner not much longer than the outer ones. Pectoral short or of moderate length.*

676. *Saurida tombil*, Bloch. *Arranna.* Mal.: *Oolooway*, Tam. D. 11-12/0, A. 10, L. l. 53-64, L. tr. 4½/7. Brownish-grey along the back, white beneath, the whole with yellow reflections. Seas of India, mostly during the cold months, is not much esteemed as food : it attains a foot in length.

Genus—HARPODON, Lesueur.

Branchiostegals from twenty-three to twenty-five : pseudobranchiæ. Gill-openings wide. Body elongated and compressed. Eyes small. Snout short : bones of head soft and partly modified into wide muciferous channels. Cleft of mouth deep : margin of the upper jaw formed by the intermaxillaries, which are thin and tapering, maxillaries absent. Teeth cardiform, recurved and of unequal size, the largest being on the mandible and barbed. Teeth exist on the intermaxillaries, mandibles, palatines, the tongue and hyoid, those on the palatines are large and can be laid downwards. Dorsal fin in the middle of the length of the body, with a moderate number of rays : adipose fin small. Pectoral and ventrals long, the latter with nine rays, and inserted below the anterior dorsal ones, at some distance behind the pectorals. Anal of moderate length. Scales thin and deciduous, especially in the anterior portion of body. Air-vessel absent. Pyloric appendages sixteen.

677. *Harpodon nehereus*, Ham. Buch. *Cucah sawahri*, or *Coco mottah*, Tel. Bummaloh or Bombay duck. D. 12-13/0, A. 14-15. The scales commence opposite the origin of the dorsal fin. Of a brownish colour. Seas of India and Burma, ascending rivers : it attains a foot in length, and is esteemed as food either fresh or dried, in which latter form it is extensively employed as a relish with curries, and known as the "Bombay duck."

Family—SCOMBRESOCIDÆ.

Pharyngognathi malacopterygii, Müller.

Pseudobranchiæ concealed, glandular. Margin of the upper jaw formed, mesially by the intermaxillaries, laterally by the maxillaries. Barbels present or absent. Lower pharyngeals united into a single bone. Dorsal fin rayed, with or without finlets posterior to it, situated opposite the anal, and in the caudal portion of the vertebral column: no adipose dorsal. Scales present, frequently a keeled row along either side of the free portion of the tail. Air-vessel generally present, sometimes cellular, and destitute of a pneumatic duct. Stomach and intestines in one straight undivided tube. Pyloric appendages absent.

Genus—*BELONE, Cuv.*

Branchiostegals rather numerous. Gill-openings wide. Body elongated, sub-cylindrical or compressed. Eyes lateral. The jaws prolonged into a beak, the upper of which is formed by the intermaxillaries. Fine teeth, or rugosities in both jaws, with a single row of long, widely-set conical ones: palate toothed or toothless. The anterior dorsal rays may or may not be elevated, forming a lobe to the fin, whilst the middle and posterior ones may be short or elongated: no finlets: caudal usually forked. Scales small. Lateral line on free portion of tail, with or without a keel.

This Genus has been thus sub-divided:—

A. Anterior dorsal rays elevated, forming a lobe, *(Belone.)*
B. ,, ,, ,, not forming a lobe, *(Potamorrhaphis,)* Günther.

A. *Anterior dorsal rays elevated, forming a lobe, Belone.*

678. *Belone schismatorhynchus,* Bleeker. D. 24-25, A. 26-27. Length of head about 1/4 of the total. A wide shallow groove on the head, scaled in its anterior half: superciliary edge striated. Vomerine teeth absent. Free portion of tail moderately depressed, without a distinct keel. Posterior dorsal rays nearly reach the base of caudal fin, which is forked. Pectoral black in its outer two-thirds: most of the other fins with black stains. Seas of India: it attains at least 18 inches in length. I have taken specimens full of roe in February.

679. *Belone annulata,* Cuv. & Val. *Pahmum kolah,*" Tam. D. 23-24, A. 21-22. Length of head about 1/4 of the total. Diameter of eye 2/5 of length of postorbital portion of the head. A wide shallow groove on the head: superciliary edge striated. Vomerine teeth absent: tongue rough from tubercles. Free portion of tail moderately depressed, with an ill-defined lateral keel. Posterior dorsal rays nearly or quite reach the base of the caudal fin, which is forked: ventral arises midway between anterior edge of the orbit and the root of the caudal. Centre of caudal stained blackish. Seas of India: to at least 2 feet in length.

680. *Belone choram,* Forsk. D. 22-23, A. 20-21. Length of head nearly 1/4 of the total. Diameter of eye 1/2 of length of postorbital portion of the head. A wide shallow groove on the head: superciliary edge striated. Vomerine teeth absent: tongue rough with tubercles. Free portion of tail depressed so as to be as wide as high, and having a narrow keel along its side. Posterior dorsal rays nearly or quite reach the

base of the caudal, which is forked. Ventral arises nearer the head than the root of the caudal.

681. *Belone melanostigma*, Cuv. & Val. D. 23-24, A. 25. Length of head nearly 1/4 of the total. Superciliary edge striated. Teeth small. Posterior dorsal rays nearly reach the base of the caudal. Ventral arises midway between the posterior edge of the orbit and the end of the base of the anal. Caudal forked. Sides with from one to seven large black blotches, which seem occasionally to be absent in the females. Bombay and Red Sea : attaining 2 feet or more in length. Females were taken full of roe in November and also in February.

682. *Belone leiurus*, Bleeker. D. 19, A. 22. Length of head 2/7 of the total length. A wide shallow groove on the head, broadest anteriorly. Superciliary and parietal regions striated. Vomerine teeth absent: tongue smooth. Free portion of tail compressed, higher than wide, without any distinct lateral keel. Posterior dorsal rays do not extend nearly so far as the root of the caudal fin, which last is not forked. A silvery lateral band : caudal stained. Coasts of India to the Malay Archipelago : attaining at least 23 inches in length. Female full of roe captured in February.

683. *Belone strongylurus* (V. Hass.), Bleeker. *Cun-gur*, Sind: *Ooshee collarchee* and *Coco meen* "long nosed fish," Tam.: *Thook-o-doo-noo-dah*, And. D. 13-14, A. 15-16. Length of head nearly 1/3 of the total length(*). A shallow median groove on the head: superciliary region scarcely striated. Vomerine teeth absent. Free portion of tail compressed, higher than wide without any distinct lateral keel. Posterior dorsal rays do not extend nearly so far as the root of the caudal fin, which last is rounded. Ventral fin arises midway between the orbit and the base of the caudal. Pectoral longer than the distance from the orbit to the margin of the opercle. A round deep black spot near the centre of the root of the caudal fin. Coasts of India to the Malay Archipelago : attaining 2 feet or more in length.

Genus—HEMIRAMPHUS, *Cuv.*

Hyporhamphus, Euleptorhamphus, Zenarchopterus, and *Oxyporhamphus*, Gill: *Dermatogenys* (K. and v. Hass.), Peters : *Hemiramphodon*, Bleeker.

Branchiostegals rather numerous. Gill-openings wide. Body sub-cylindrical and elongated. Eyes lateral. Upper jaw, which is formed by the intermaxillaries, is more or less triangular in form and short : whilst the lower jaw, in the mature, is elongated far beyond the upper. Teeth villiform in both jaws. The number of dorsal and anal rays may be about equal, or either may be in excess of the other: the dorsal fin may commence anterior to, above, or behind the origin of the anal: no finlets posterior to the dorsal fin. Pectoral may or may not be prolonged : caudal mostly forked or emarginate, sometimes rounded. Scales of moderate or large size: air-vessel large, occasionally cellular. Dorsal and anal rays may be modified: some are viviparous. No pyloric appendages.

(*) Dr. Günther considers there are two species for the second of which he reserves the specific term *strongylurus*, known by its eye being a little larger and its head slightly longer. For the above he retains the specific name of *caudimaculata*.

This genus has been thus divided :—
A. Pectorals short.
B.   „   long.
A. Pectorals short.

684. *Hemiramphus Russelli,* Cuv. and Val. D. 17, A. 13. Length of head without the beak 1/6, of the beak rather more than 1/6 of the total length. Eyes diameter 1/4 of length of head. Ventrals in the posterior 1/3 of the distance between apex of intermaxillaries and base of caudal fin, which latter is forked. A silvery band. Coromandel Coast to the Malay Archipelago.

685. *Hemiramphus Reynaldi,* Cuv. and Val. D. 16, A. 15. Length of beak 2/11 of the total length. Upper jaw pointed: caudal forked: dorsal higher than long. Silvery band not very distinct: dorsal and caudal edged with black. Intermediate between *H. Dussumieri* and *H. Gaimardi.* Seas of India, to 9 inches in length.

686. *Hemiramphus Georgii,* Cuv. and Val. D. 16, A. 15, L. 1. 58-60. Length of entire head 2/5, of beak 1/4 of the total length. Eyes, diameter one and two-thirds in the postorbital portion of the head, and 3/4 of the interorbital space. Dorsal and anal fins scaleless, the latter commencing under the sixth dorsal ray. Ventral in the posterior 1/3 of the distance between the anterior margin of the orbit and the base of the lower caudal lobe, which latter fin is lobed, its central rays being much longer than the eye. A silvery band: upper two-thirds of the anterior portion of the dorsal black: caudal grey in its posterior half: each scale spotted on its edges. Seas of India to the Malay Archipelago, attaining at least 8 inches in length.

687. *Hemiramphus leucopterus,* Cuv. and Val. D. 16, A. 14. Fins white: all the body is transparent white and silvery: the lateral band very brilliant: beak black. Bombay, to 5 inches.

688. *Hemiramphus xanthopterus,* Cuv. and Val. D. 15, A. 16. The beak is shorter and the eye larger than in *H. limbatus.* White back, with the edges of the scales bordered with greenish and having nacreous reflections: the single fins are yellow, and the paired ones transparent: the skin of the beak is black, and its extremity scarlet. Alleppey, to 6 inches in length.

689. *Hemiramphus Dussumieri,* Cuv. and Val. D. 15, A. 14, L. 1. 52. Length of entire head from 1/3 to 5/16, of beak from 2/11 to 1/6 in the total length. Eyes 1 diameter apart, 1¼ diameters in the postorbital portion of the head. Triangular portion of upper jaw wider than long. Dorsal commences anterior to the anal: its upper edge concave: ventral arises slightly nearer the base of the caudal than the axil of the pectoral: caudal deeply forked, lower lobe the longer, its central rays being as long as the postorbital portion of the head: no scales on vertical fins. Back dark: lateral band well developed: a dark mark on the anterior part of the dorsal fin. Seas of India to the Malay Archipelago.

690. *Hemiramphus plumatus,* Blyth. D. 15, A. 13, L. 1. 66. Upper jaw 1/3 longer than wide. Eyes 1¼ diameters apart, and rather nearer the posterior extremity of the head than the anterior end of the upper jaw. Dorsal commences anterior to the anal: its upper edge

concave : caudal forked : the lower lobe the longer : the ventral commences in the last third of the distance between the end of the snout and the base of the caudal fin. A lateral band. Ceylon, attaining at least 8 inches in length.

691. *Hemiramphus Gaimardi*, Cuv. and Val. D. 14-16, A. 15-16, L. l. 50. Length of entire head about 1/3, of beak 2/11 of the total length. Eyes one diameter apart, and two-thirds the length of the postocular portion of the head. Upper jaw broader than long. Base of dorsal a little longer than that of the anal : both with some very fine scales anteriorly : ventral arises slightly nearer the eye than to the base of the caudal fin, which is moderately forked, its central rays being much longer than the eye. A silvery band as wide as a scale : a scarlet tip to the mandible. Seas of India to the Malay Archipelago.

692. *Hemiramphus far*, Rüppell. *Verrikolah*, Tam. D. 13-14, A. 11-13, L. l. 52-54. Length of head, including beak, 1/3 of the total. Eyes 1½ diameters in the postorbital portion of the head. Ventral arises midway between the end of the pectoral and the base of the caudal. Dorsal and anal scaled in their anterior portions : caudal forked, its central rays being shorter than the eye. Air-vessel cellular. A silvery streak : four black blotches along the side. Seas of India, attaining at least 13 inches in length. This appears to be identical with *H. Commersonii*, Cuv. and Val.

693. *Hemiramphus angulatus*, Ham. Buch. *Gungaturi*, Ooriah. D. 13-14, A. 13-15, L. l. 50-52. Length of entire head nearly 1/3, of beak 1/6 of the total length. Eyes 1 diameter apart and 1¼ diameters in the postorbital portion of the head. Upper jaw keeled, 1/3 wider at the base than it is long. Dorsal commences slightly in advance of the anal, the anterior portion of each being the higher : ventrals commence midway between the posterior margin of the eye and the base of the caudal fin, which last is lobed, the lower being the longer. Pectoral pointed, two-thirds as long as the head : no scales on the vertical fins. A brilliant silvery longitudinal band nearly covering the depth of one scale. Estuaries of the Ganges and rivers in Orissa. This is the species I described as *Esox ectuntio*, H. Buchanan. The specimens in the Calcutta Museum were labelled *H. brachynotopterus*, Bleeker, which designation is incorrect.

694. *Hemiramphus marginatus*, Forsk. *Kuddera*, Tel. D. 12-14, A. 11-12, L. l. 50-52. Upper jaw as broad as long. Length of head, including beak, from 3/8 to 1/3 of the total length. Diameter of eyes 4/7 of the length of the postorbital portion of the head. The ventral arises somewhat nearer the caudal than the axil of the pectoral. Dorsal and anal fins scaleless. Caudal deeply forked, its central rays being shorter than the orbit. An indistinct silvery band. Red Sea, Seas of India to the Malay Archipelago.

695. *Hemiramphus limbatus*, Cuv. and Val. D. 13, A. 13-14, L. l. 51. Length of head nearly 1/3, of beak 1/6 of the total length. Eyes nearly 1 diameter apart and 3/4 of the length of the postorbital portion of the head. Dorsal arises anterior to the anal, than which its base is much longer : root of the ventral rather nearer to the base of the caudal than the posterior margin of the orbit : caudal lobed but not deeply, the lower being the longer, its central rays being as long as the

postorbital portion of the head. A well-marked silvery longitudinal band nearly as wide as the depth of one scale : beak tipped with coral red. Seas of India.

696. *Hemiramphus cirrhatus*, Day. D. 14, A. 12, L. l. 42, L. tr. 7/3. Length of entire head 2/5, of beak nearly 1/4 of the total length. Eyes $1\frac{1}{2}$ diameters in the postorbital portion of the head. A barbel at the posterior nostril half the length of one diameter of the orbit. Pectoral pointed, rather longer than the head without the snout: ventral arises in the last third of the distance between the front end of the snout and the base of the caudal, which last is rounded or very slightly emarginate: anal with a very short base commencing slightly behind the origin of the dorsal, but the length of its base is only 1/3 of that of the latter fin. No scales on the vertical fins. A narrow longitudinal silvery streak until nearly below the base of the dorsal, when it becomes one-third as wide as a scale: upper half of dorsal black. Bombay. This appears somewhat similar to *H. amblyurus*, Bleeker, but the *H. cirrhatus* has its intermaxillaries only 1/4 wider at the base than long, instead of being twice as wide.

Males with the anal rays modified: viviparous.

697. *Hemiramphus ectunctio,** Ham Buch.: *H. ectunctio*, Blyth.: *H. neglectus*, Day. D. 14, A. 10-12. Length of entire head about 1/3, of beak 1/5 of the total length. Eyes $1\frac{1}{4}$ diameters apart and $2\frac{1}{4}$ in the length of the postorbital portion of the head. Upper jaw twice as long as wide at its base. The ventrals commence in the last third of the distance between the orbit and the base of the caudal fin, which latter is rounded. Silvery: lateral stripe distinct: the tip of the mandible milk white. Ganges and large rivers of Burma, mostly within tidal influence.

698. *Hemiramphus Buffonis*, Cuv. & Val. *Koo-door-rock-o-dah*, And. D. 10-13, A. 10-12. Length of entire head 2/5, of the beak 2/9 of the total length. Eyes 2/3 of a diameter apart and from 3/5 to 2/3 in the length of the postorbital portion of the head: a nasal barbel. Upper jaw a little broader at its base than it is long. Ventral arises in the last third of the distance between the snout and the base of the caudal, which latter is somewhat rounded: anal commences under about the third dorsal ray, its base is only half the length of that of the former fin. A silvery streak: the upper half of the dorsal fin either very dark or black. Andamans and Malay Archiplego: said to ascend brackish waters.

699. *Hemiramphus dispar*, Cuv. & Val. D. 11-12, A. 11-12, L. l. 40. Length of entire head 2/5, of beak 1/4 of the total length. Eyes rather above 1 diameter apart, from 2/3 to 4/7 of the length of the postorbital portion of the head. Upper jaw as broad as long. A small nasal barbel. Dorsal fin in the posterior 1/4 of the distance between the eye and the base of the caudal, which is cut square or rounded: anal commences somewhat behind the dorsal, and some of its rays modified: the ventrals commence in the beginning of the last third of the distance between the anterior extremity of the upper jaw and the base of the tail.

---

* In my fresh-water fishery report I gave this species as *H. amblyurus*, Bleeker, with which it appears to agree in every respect except that *it has no nasal barbel*. As the existence of one was not referred to in the description of the specimens in the British museum, it may be that they are not invariably present, never I think in Indian examples.

No scales on dorsal or anal fins. A silvery lateral band. Seas of India,

700. *Hemiramphus brachynopterus*, Bleeker. D. 9, A. 16, L. l. 43. Length of head nearly 1/3, of beak 4/19 of the total length. Eyes nearly 1 diameter apart and 2 diameters in the postorbital portion of the head. Anal commences somewhat anterior to the dorsal: the caudal convex. Hooghly river.

B. *Pectorals elongated.*

701. *Hemiramphus longirostris*, Cuv. & Val. D. 22, A. 20. Length of the beak 1/4 of the total length. Length of the head without the beak 1/6 of that of the trunk. Eyes 1 diameter apart and in the length of the postorbital portion of the head. Pectorals very long, equalling 3/11 of the length of the trunk: ventrals very small: caudal deeply forked. A silvery band. Coromandel coast.

Genus—EXOCŒTU^S, *Artedi.*

Cypsilurus, Swains.: *Halocypselus*, Weinland.: *Parexocætus* Bleeker.

Body moderately oblong. Gill-openings very wide. Jaws short: the intermaxillaries and maxillaries separate. Barbels present or absent. Mandible in some species with a cutaneous appendage. Teeth, when present, minute and rudimentary. Pectorals elongated used as an organ of flying: the dorsal fin may be much or only moderately elevated: ventrals long, short, or of moderate length. Air-vessel large. Pyloric appendages absent.

702. *Exocætus mento*, Cuv. & Val. D. 10-11, A. 10-13, L. l. 38. Length of head and height of body each 2/9 of the total length. Eyes diameter 2/5 the length of the head and 1/2 a diameter from end of snout. No barbels. Dorsal fin elevated* as high as the body, and its membrane deeply cleft: 16 to 17 rows of scales between the occiput and origin of this fin: ventrals in some specimens reaching to the anus, in others to the anal fin, the fin arises midway between the anterior edge of the orbit and the base of the caudal fin: pectoral reaching to opposite the middle or end of the dorsal. Bluish, becoming silvery along the abdomen: dorsal fin black: upper half of pectoral black, having a white edge: ventral white: anal white with a dark mark along its base: upper lobe of caudal white with a black bar along its base: lower lobe greyish. East coast of Africa, seas of India to the Malay Archipelago. In the cold season it is occasionally numerous at Madras, but I never obtained it above $4\frac{3}{10}$ inches in length.

703. *Exocætus furcatus*, Mitchell. D. 13, A. 9-10. A pair of long barbels at the symphysis of the lower jaw. Dorsal fin moderately high, but its longest rays do not extend to the caudal. The ventrals reach the root of the caudal and the pectorals nearly as far: the former fin arises slightly nearer the base of the caudal than to the anterior extremity of the snout. The posterior part of the ventrals and the lower part of

---
* As many of the species of *flying fish* are only known from young specimens, the size of the eye, and the length of the fins must be accepted with a reservation. It has also been remarked of barbels "that one may well suspect that in some species at least they are peculiar to the young state only."

the anal black. *Young*, with three broad vertical bands* across the anterior half of the abdomen: opercles and pectorals marbled with black: lower caudal lobe with three dark cross bands. India and the Atlantic Ocean.

704. *Exocœtus micropterus*, Cuv. & Val. D. 15, A. 14-16. Length of head 2/9, pectoral 2/7, height of body 1/6 of the total length excluding caudal fin. A pair of short barbels at the symphysis of the lower jaw, which is prominent: mandible with a cutaneous appendage. The ventral, which does not extend to the anal, arises midway between the axil and the base of the caudal. Dorsal fin not so high as the body. Bluish above, silvery on the sides and below: fins dark. Seas of India to the Malay Archipelago and beyond.

705. *Exocœtus evolans*, Linn. D. 12-14, A. 13-15, L. l. 42. Length of head 2/9, height of body 1/6, of dorsal fin 1/14 of the total length. Eyes diameter 2/7 of length of head, 3/4 of a diameter from end of snout and 1½ apart. No barbels. Dorsal commences slightly in advance of the anal: pectoral reaches the base of the caudal: the ventral arises midway between the anterior edge of the orbit and the posterior end of the base of the dorsal fin, it only reaches half way to the anal: caudal deeply forked, lower lobe the longer. East coast of Africa, Seas of India to the Malay Archipelago, and beyond.

706. *Exocœtus obtusirostris*, Günther. D. 13, A. 13, L. l. 40. Said to be "closely allied to *E. evolans*, but has the snout shorter and the head more elevated." India and tropical and sub-tropical seas, to at least 9 inches in length.

707. *Exocœtus pœcilopterus*, Cuv. & Val. *Parravay-kolah*, Tam. D. 12-15, A. 9, L. l. 45. Length of head 1/5, of caudal 2/7 of the total length. Eyes diameter 1/3 of length of head, 3/5 of a diameter from end of snout and 1 apart. No barbels. Dorsal commences some way in front of the anal, the height of its anterior rays 1/10 of the total length, 27 scales between occiput and origin of the dorsal fin: ventral arises somewhat nearer the posterior edge of the orbit than to the base of the caudal, and reaches to the end of the base of the root of the anal: the pectoral reaches as far as the end of the dorsal. Pectoral with many rounded and oval spots, sometimes transversely arranged in bands, in others irregularly placed. Out of 3 specimens up to 13 inches in length taken together, two have black spots above the dorsal fin: ventral white or with spots. Seas of India to the Malay Archipelago. *Exocœtus spilopterus*, Bleeker, appears very similar to the adult of this species.

708. *Exocœtus Bahiensis*, Ranzani. D. 12-13, A. 9-10, L. l. 50. Height of body 2/13, length of head 2/11 of the total length. Eyes diameter 1/3 of length of head, nearly 1 diameter from end of snout. No barbels. Dorsal commences someway in front of the anal, and midway between the posterior edge of the orbit and the end of the lower caudal lobe, the height of its anterior rays being 2/3 as long as the head: ventral arises midway between the end of the snout and that of the upper lobe of the caudal fin: it reaches to the centre of the base of the anal: the pectorals reach to the end of the base of the dorsal fin.

---

* Vertical bands in *Hemiramphi, Exocœti*, and many of the *Scombresocidæ*, &c., are usually a sign that the specimen is immature.

Dorsal with a dark mark on its summit: the pectoral appears sometimes to be darker than the remainder of the fins. Seas of India to the Malay Archipelago and beyond.

709. *Exocætus brachysoma*, Bleeker. D. 11-13, A. 8-9, L. l. 44. Height of body 2/11, length of head 1/5 of the total length. Eyes, diameter nearly 1/3 of length of head, not quite 3/4 of a diameter from end of snout. No barbels. Dorsal commences someway in front of anal and midway between the anterior edge of the orbit and the end of the lower caudal lobe: there are 24 scales between the occiput and its commencement, the height of its anterior rays being 2/3 as long as the head: ventral arises midway between the middle of the eye and the base of the caudal fin: it reaches to the centre of the base of the anal: the pectorals reach to the middle of the base of the dorsal fin. Pectoral dark, lighter posteriorly. From the East Coast of Africa through the Seas of India to the Malay Archipelago and beyond.

Family—PSEUDOCLUPEIDÆ, *Bleeker.*

Albulina, Elopidæ, and *Chanina,* Günther.

Branchiostegal in varying numbers, may be numerous (35): pseudobranchiæ present. Body more or less compressed: abdominal edge not spinate. Head scaleless, or merely a few fine ones behind the maxillary region. Eyes lateral. Opercular pieces four: mouth bordered mesially by the intermaxillaries, laterally by the maxillaries. No barbels. A single dorsal fin with weak rays placed opposite or nearly so to the ventrals: anal having about the same number of rays as the dorsal. Scales sometimes deciduous. Lateral line present. Stomach with a blind sac. Air-vessel present, more or less simple. Pyloric appendages in moderate numbers or numerous.

Genus—CHANOS, *Lacép.*

Lutodeira, (Kuhl). Rüpp.

Branchiostegals four: pseudobranchiæ present. Gill-membranes entirely united below and not attached to the isthmus. Body moderately elongated and compressed: abdomen rounded. An accessory branchial organ in a recess behind the true gill-cavity. Mouth small, anterior and transverse. Teeth absent. Ventral fin opposite the dorsal, which last has more rays than the anal: caudal deeply cleft. Scales rather small. Lateral line distinct. Air-vessel with a constriction. Pyloric appendages numerous.

710. *Chanos salmoneus,* Bl. Schn. *Tulu candal,* Tam.: *Palah bontah,* Tel.: *Hu-meen,* Canar.: *Pu-meen,* Tulu. D. 13-17, A. 9-11, L. l. 75-80, L. tr. 12/15, Vert. 19/26. Length of head and height of body each about 1/5 of the total length. Attains three feet at least in length: is called the *milk fish* by Europeans in south Canara, where Hyder Ali introduced it from the sea into tanks of fresh water and there it still thrives. Red sea, Seas of India to the Malay Archipelago and beyond.

Genus—ELOPS, *Linn.*

Branchiostegals numerous: pseudobranchiæ. Gill-membranes entirely separated. Body rather elongated and compressed: abdomen rounded. An osseous gular plate is attached to the symphysis of the mandible, and covers the intermediate part. Mouth wide, anterior, upper jaw the shorter: the maxilla forming the lateral edge of the mouth. Villiform teeth in the jaws, vomer, palatine, and pterygoid bones, also on the tongue and base of the skull. Ventrals opposite to the dorsal, which last has a few more rays than the anal. Scales small. Lateral line distinct. Pyloric appendages numerous.

711. *Elops saurus*, Linn. *Jinnagow*, Tel.: *Ullahti*, Tam. D. 22-24, A. 15-17, L. l. 95-100, L. tr. 12/14. Length of head and height of body each 1/5 of the total length. Silvery: fins yellowish, with a greenish tinge. Red Sea, East Coast of Africa, through the Seas of India to the Malay Archipelago and beyond.

Family—CLUPEIDÆ, *Cuv.*

Gill-openings usually very wide: pseudobranchiæ, when present, well developed. Abdomen mostly compressed, generally into a sharp edge, and usually serrated. Opercular pieces four. Eyes lateral, with or without adipose lids. Margin of the upper jaw formed mesially by the intermaxillaries, laterally by the maxillaries, which are composed of three pieces not ossified together. Mouth may have a deep cleft, with small intermaxillaries and the maxilla elongated, and either the upper or lower jaw projecting, or the mouth may be transverse. Barbels absent. A single dorsal with a few or moderate number of weak rays: anal sometimes many-rayed. Scales on the body: none on the head. Lateral line absent. Stomach with a blind sac. Air-vessel more or less simple. Pyloric appendages, when present, numerous.

Genus—CHIROCENTRUS, *Cuv.*

Branchiostegals eight: pseudobranchiæ absent. Body much elongated and compressed: abdomen with a sharp but not serrated margin. Gill-membranes united for a short distance: gill-opening wide. Eyes with adipose lids. Cleft of mouth oblique and deep: the lower jaw the longer. A row of canines in the mandible, and a horizontal pair in the intermaxillaries: minute teeth on the palatines, pterygoids, and tongue. A single short dorsal fin placed far backwards opposite to a long anal: an elongated osseous appendage in the axil: ventrals very small. Scales thin, small, and deciduous. Air-vessel cellular. Pyloric appendages absent. Intestines short, the mucous membrane forming a spiral fold.

712. *Chirocentrus dorab*, Forsk. *Mooloo-alley* and *kiru-wahlah*, Tam.: *Wahlah*, Tel.: *Kunda*, Ooriah. D. 16-17, A. 33-34, Vert. 29/46. Length of head about 1/6, height of body 1/7 of the total length. Bluish-green along the back: silvery sides and abdomen. Red Sea, through the Seas of India to the Malay Archipelago and beyond, attaining 12 feet in length.

Genus—DUSSUMIERA, *Cuv. and Val.*

Branchiostegals numerous: pseudobranchiæ well developed. Gill-membranes entirely separate. Body somewhat elongated, compressed: abdomen rounded. Snout pointed: upper jaw not projecting: cleft of mouth of moderate width. Eyes without adipose lids. Small fixed teeth in the jaws, and villiform ones on the palatines, pterygoids, and on the tongue. Dorsal fin opposite to the ventral: anal of moderate length. Scales of medium size, very deciduous. Pyloric appendages numerous.

713. *Dussumiera Hasseltii*, Bleeker. B. XV, D. 18-20, A. 15-16, L. l. 52-60. Length of head 2/9, height of body 1/6 to 2/13 in its total length. Eyes diameter 1/4 of length of head, 1 apart, and rather more from end of snout. Back deep bluish-green, divided by a narrow golden band from the silvery sides and abdomen. Seas of India to the Malay Archipelago and beyond, attaining at least 8 inches in length.

714. *Dussumiera acuta*, Cuv. and Val. *Poonduouringa*, Tam.: *O-pul-dah*, Andam. B. XIV—XV., D. 18-21, A. 15-17, L. l. 40-42. Length of head 1/5, height of body 1/5 to 4/21 of its total length. Eyes, diameter 1/3 to 2/7 of length of head, 2/3 of a diameter apart, and slightly more from the end of snout. Back deep bluish-green, divided by a narrow bronze-coloured band from the silvery sides and abdomen. Seas of India to the Malay Archipelago.

Genus—SPRATELLOIDES, *Bleeker*.

Branchiostegals six: pseudobranchiæ well developed. Gill-membranes entirely separated: no osseous gular plate. Body elongated, moderately compressed or sub-cylindrical: abdomen rounded. Snout compressed: the mouth anterior with a lateral cleft: the upper jaw not overlapping the lower. Eyes without adipose lids. Teeth small and deciduous, but may be present on the jaws, vomer, pterygoids, and tongue. Dorsal fin placed opposite the ventrals: anal of moderate length or short. Scales of medium size, rather deciduous. Pyloric appendages in moderate numbers.

715. *Spratelloides Malabaricus*, Day. D. 14, A. 18, V. 8, L. l. 38, L. tr. 9. Pale silvery without any marks. It is a small species found down the Western Coast of India: it ascends rivers and enters estuaries.

Genus—CLUPEA,* sp. *Artedi*.

Body oblong or sub-elongate, with the serrature of the abdomen extending anteriorly into the thoracic region. Upper jaw not projecting beyond the lower. Mouth anterior or antero—superior. Teeth, when present, rudimentary and deciduous. Dorsal fin situated opposite the ventrals: anal with a moderate, or large number of rays: caudal forked. Scales large, of moderate or more rarely of small size.

716. *Clupea albella*, Cuv. and Val. B. VI, D. 19, A. 23. Length of head 1/5, height of body 2/7 of the total length. Small teeth on the jaws and pterygoids only. Ventrals below the middle of the dorsal fin. Brilliant silvery, with a bluish tint, and darkest on the back: in certain reflections there are some longitudinal bands. Pondicherry, from whence two examples 3½ inches long were brought.

717. *Clupea fimbriata*, Cuv. and Val. *Kich-uk-louar*, Sind: *Cuttay charlay*, Malabar: *Poondu ringa*, Tam.: *Charree-addee*, Hind. D. 18-20, A. 16-20, L. l. 46, L. tr. 11. Length of head 1/5, height of body 1/4 or more of the total length. Eyes, diameter 1/4 of length of head, 1 diameter from end of snout. Deciduous teeth in the jaws: an oval patch on the palatines, and a minute central band on the tongue. Ventrals below the middle of the dorsal fin. Scales in regular rows, adherent, their edges striated, jagged, or indented. Greenish with silvery sides. Seas of India and probably Malay Archipelago. It abounds in certain years on the Malabar Coast.

718. *Clupea Moluccensis*, Bleeker. D. 17-19, A. 17-18, L. l. 45 (40), L. tr. 12. Length of head 1/5 height of body 2/9 in the total

---

* Dr. Bleeker, in his admirable and truly magnificent *Atlas Ichthyologique*, considers Genus *Alosa*, Cuv., as distinct, not only due to the absence of teeth on the jaws and inside the mouth, but principally by the conformation of their jaws, the upper being notched to receive the more or less compressed tip of the mandibles. I propose deferring it is and several other similar points for the present. Nos. 726 to 728 would be *Alosa's*, if this genus is accepted, as I think it ought to be.

length. Lower jaw projecting. Eyes, diameter 2/7 of length of head, 1 diameter from end of snout. A narrow row of teeth on the palatines, pterygoids, and on the tongue. Ventrals inserted below the middle of the dorsal fin. Scales regularly arranged, with their edges roughened or crenulated, Bluish along the back: sides and abdomen silvery. Ceylon and Malay Archipelago, where it abounds. This appears to *very* closely resemble *C. kowal*.

719. *Clupea longiceps*, Cuv. and Val. D. 16, A. 16. Length of head 2/7, of caudal 1/7. Teeth on the palatines, pterygoids, and tongue. Pondicherry, to six inches in length, said to have been received in a bad state.

720. *Clupea kowal*, Rüppell. *Kowal*, Tel.: *No-na-li*, Tam. D. 17-18, A. 17-19, L. l. 44-46, L. tr. 12. Length of head 1/5, height of body 1/4 of the total length. Lower jaw projecting. Eyes, diameter 3/10 of length of head, 1 diameter from end of snout. A few rudimentary teeth along the centre of the tongue and in a band on the palate. Ventrals inserted below the middle of the dorsal fin. Scales regularly arranged, each crossed by several vertical lines. Bluish along the back: sides and abdomen silvery. Red Sea, East Coast of Africa, Seas of India, and perhaps the Malay Archipelago.

721. *Clupea Neohowü*, Cuv. and Val. *Lee-gur*, Beluch.: *Louar*, Sind: *Mutthi*, Canerese and North Malabar: *Charlay*, South Malabar. D. 17, A. 15-16, L. l. 45-48, L. tr. 13. The proportions differ widely, and are affected by the times of breeding. Length of head 2/7 to 1/4 of the total length. Eyes diameter 1/5 of length of head, 1 diameter from end of snout. Minute teeth on the tongue, palatines, and pterygoids, but are easily lost. Lower jaw projecting. Ventrals inserted below the last half of the dorsal fin. Scales regularly arranged, and their edges indistinctly serrated. Bluish along the back: sides and abdomen silvery, with a golden line dividing the two colours. I have obtained them on the western coast, from whence ' Cannanore' Dussumier brought them. They come in shoals to the shores of Beluchistan, Sinde, all the way down Malabar and Ceylon, also to the Andaman Islands; but I have not yet seen a specimen from the Coromandel Coast or Burma ; they may, however, sometimes go there. This appears to resemble very closely *C. lemuru*, Bleeker.

722. *Clupea leiogaster*, Cuv. and Val. D. 17-18, A. 18-20, L. l. 40, L. tr. 11-12. Length of head 1/5 to 4/21, height of body 2/11 to 1/6 in the total length. Lower jaw not projecting. Eyes, diameter 1/4 of length of head, more than 1 diameter from end of snout. Fine teeth on the palatines, pterygoids, and along the middle of the tongue. Ventrals inserted below the second fourth of the dorsal fin. Scales regularly arranged. Greenish along the back: whitish shot with purple along the abdomen, the two colours being divided by a longitudinal silvery band, with about 14 or 15 black spots in its course. Ceylon and Malay Archipelago.

723. *Clupea lineolata*, Cuv. and Val. *Sardinella lineolata*, C. & V. D. 18, A. 17. Height of body 3/14 of the total length. Teeth on the palatines, pterygoids, and on the tongue. Bluish on the back, becoming silvery below, with two or three plumbaceous coloured lines dividing these two colours. Ceylon to 4 inches in length.

724. *Clupea lile,* Cuv. and Val. B. V, D. 14-15, A, 17, L. l. 38-40, L. tr. 9-10. Height of body 2/7, length of head 1/5 of the total length. Profile of dorsal surface nearly straight: of abdomen very convex. Teeth on pterygoids, and a central band along the tongue. Ventrals inserted below the anterior dorsal rays. Scales adherent. A silvery longitudinal band along the side. Bombay and down the Western Coast of India. This species appears to very closely resemble or be identical with *C. argyrotænia,* Bleeker.

725. *Clupea melanura,* Cuv. and Val. D. 14-16, A. 17-18, L. l. 35-38, L. tr. 11. Length of head 1/5, height of body 2/7 of the total length. Eyes diameter 1/3 of length of head, nearly 1 diameter from end of snout. No teeth inside the mouth. Ventrals inserted opposite the commencement of the dorsal fin. Scales smooth, in regular rows. Back blue, shot with purple : a fine yellow line dividing it from the silvery sides and abdomen : outer third of caudal lobes black. Seas of India and the Malay Archipelago.

726. *Clupeai palasah,* Cuv. and Val. (see F. W. Fishery Report No. 373). D. 17-19, A. 18-19, L. l. 45-49, L. tr. 17. Length of head 1/5, of caudal 1/6, height of body 1/4 of total length. Width of opercles 2/3 of depth. Caudal peduncle, as deep as long. 14 scutes behind ventral fin, 16 anterior to it. No spots along the body, except occasionally one behind the opercles. All the large rivers of India and Burma. H. Buch's. description must have been intended for this species, but Dr. Günther considers his figure to represent the following, and which he believes identical with *C. kanagurta,* Bleeker.

727. *Clupea ilisha,* (H. B.) Günther. B. VI, D. 18-19, A. 19-21. L. l. 40-44, L. tr. 15. Height of body a little more than 1/3, length of head 2/7 of the total length without the caudal fin. Width of opercle about 1/2 its depth. Caudal peduncle as deep as long. 12 or 13 scutes behind the ventral fin. The young are said to have a dark shoulder spot. From East Coast of Africa to the Malay Archipelago. Ham. Buch's. figure does not, according to Dr. Günther, represent the common Hilsa fish of Bengal, but the one here described, and he marks with a doubt Russell's *Keelee* as perhaps this fish. Bleeker considers the *Keelee* identical with *C. kanagurta,* so it may be that H. B. figured the adult *Kanagurta* as the young of the *palasah.*

728. *Clupea toli,* Cuv. and Val. D. 16-18, A. 17-19, L. l. 41, L. tr. 14. Length of head 1/5, of caudal 2/9, height of body 4/13 of the total length. Eyes, diameter 1/4-1/5 of length of head, 1 diameter from end of snout. Opercle 1/2 higher than wide. Dorsal commences nearer snout than base of caudal. Caudal peduncle, rather deeper than long. 18 scutes before ventral fin, 12 behind it. A dark shoulder spot in the young. Seas of India to the Malay Archipelago. This species does not appear ever to ascend rivers to breed like the *C. palasah.*

Genus—*PELLONA, Cuv. & Val.*

Platyaster, Swains.: *Ilisha,* (Gray) Richardson and Bleeker.

Branchiostegals six. Body rather elongated and strongly compressed, with the thoracic and abdominal edges serrated. Mouth of moderate size, upper jaw generally emarginate and shorter than the lower.

Fine sharp teeth in the jaws, palatines, and pterygoid bones, also on the tongue, but none on the vomer. Dorsal fin small, medial. Ventrals small, inserted anterior to the dorsal: anal elongated. Scales large, or of moderate size, rarely small.

A. *Dorsal fin wholly or almost entirely anterior to base of anal.*

729. *Pellona Indicus*, Swains.: *Ditchoa*, Tel. : *Poo-na-no-dah*. And. D. 16-17, A. 37-39, L. l. 45, L. tr. 13-14. Length of head 2/9, height of body nearly 1/3 of the total length. Eyes diameter nearly 1/3 of length of head, 2/3 of a diameter from end of snout. Origin of dorsal fin midway in the distance between the snout and the posterior end of the base of the anal, it is situated entirely in front of the anal, the length of the base of this latter being equal to 1/3 of the total length excluding the caudal fin. There are 8 or 9 spines along the abdominal profile posterior to the base of the ventral fin, and 18 anterior to it. Silvery. From the East Coast of Africa through the Seas of India to the Malay Archipelago.

730. *Pellona megalopterus*, Swains. *Jangarloo*, Tel.: *Paunia puiee*, Ooriah. D. 18-19, A. 42-44, L. l. 48-50 (54), L. tr. 14-15. Length of head 2/9, of caudal 2/11, height of body 1/4 of the total length. Eyes, diameter scarcely 1/3 of length of head, and 3/4 of a diameter from end of snout. Origin of dorsal fin midway in the distance between the snout and the posterior end of the base of the anal, it is not situated entirely in front of the anal, the length of the base of this latter being equal to 2/7 of the total length, or 1/3 excluding the caudal fin. There are 8 to 9 spines along the abdominal profile posterior to the base of the ventral, and 20 to 21 anterior to it. Golden, glossed with purple. Seas of India to the Malay Archipelago. It is very common in India, and appears to be identical with P. *Dussumieri*, C. & V.; and P. *motius*, C. & V. may be the young of this species.

731. *Pellona motius*, Ham. Buch. *Ursi or Alise*, Ooriah. D. 16-18, A. 39-46, L. l. 43-45, L. tr. 13. Length of head from 1/5 to 2/11, of caudal nearly 1/5, height of body 1/4 of the total length. Eyes, diameter 1/3 of length of head, 3/4 of a diameter from end of snout, and 1/2 a diameter apart. Origin of dorsal fin slightly nearer to the snout than to the base of the caudal, and is situated entirely in front of the anal, the length of the base of this latter fin being equal to 1/3 of that of the total length excluding the caudal fin. There are 9 spines along the abdominal profile posterior to the base of the ventral, and 14 or 15 anterior to it. Silvery, with a burnished lateral band, and some fine dots on the fins. Coasts of Bengal and Orissa : ascending rivers high above tidal influence : it does not appear to attain much above 4 inches in length. *Pellona micropus*, Cuv. & Val., with D. 17, A. 42, L. l. 47, and 27 spines along the abdominal profile may be this species.

732. *Pellona ditchela,?* Cuv. & Val. D. 14-15, A. 40, L. l. 36, L. tr. 9. Length of head 1/5,* of caudal 1/5, height of body 2/7 of

---

* The proportions given to Russell's fish by Cuv. and Val. are height of body and length of head each 2/9 of the total length. The specimen, if it is the same species described above, was scarcely 3½ inches in length. Russell observes his species attains to 6 inches.

the total length. Eyes diameter about 1/3 of length of head, 1 diameter from end of snout and also apart. Origin of dorsal fin nearer the snout than the base of the caudal, and entirely in front of the anal, the length of the base of this latter fin being equal to 1/7 of the total length. There are 11 spines along the abdominal profile posterior to the base of the ventral, and 22 anterior to it. Silvery, with a burnished lateral band. Coromandel Coast.

733. *Pellona melastoma*, Bl. Schn. D. 17, A. 38. Length of head 2/9, height of body 3/11 of the total length. Eyes, diameter above 1/3 of length of head. Origin of dorsal fin on a vertical line with that of the ventral: anal low and extended to below the whole of the length of the free portion of the tail. Twenty-eight spines along the lower profile. Coromandel coast.

734. *Pellona Leschenaultii*, Cuv. and Val. D. 21, A. 42, L. l. 70. Height of body a little less than 1/4 of the total length. Pectorals large, posteriorly rounded, and reaching beyond the insertion of the ventral, which is 8 or 9 scales before the dorsal. Pondicherry, described from a dried specimen 20 inches long.

735. *Pellona filigera*, Cuv. and Val. D. 21, A. 52. Height of body 1/4 of the total length. The rays of the fins, especially of the caudal, are prolonged into filaments. Coasts of India, to 4 inches in length.

Genus—OPISTHOPTERUS, *Gill*.

Pristigaster, sp. Cuv., &c.

Branchiostegals six. Body oblong, compressed. The lower jaw projecting: the maxilla elongated posteriorly. Small sharp teeth in the jaws, palatines, pterygoids, and tongue: none on the vomer. Dorsal fin situated behind the commencement of the anal, which has many rays: ventrals absent. Scales of moderate or small size, very deciduous: serrature along the abdominal profile well developed.

736. *Opisthopterus tartoor*, Cuv. and Val. D. 13 (17), A. 58-61, L. l. 50. Length of head nearly 1/5 (5/24), length of caudal 1/6, of pectoral 1/5, height of body 1/4 of the total length. Eyes diameter 1/3 of length of head, 2/3 of a diameter from end of snout, 1/3 of a diameter apart. Opercle nearly 3/4 higher than wide, without any emargination. Origin of dorsal fin rather nearer the base of the caudal than the axil: pectorals reach to above the first anal ray, whilst the latter fin commences midway between the snout and the posterior extremity of its own base. Spines along the abdominal profile 28 to 32. Silvery. Seas of India. Dr. Bleeker questions whether his Malay Archipelago species is identical with the Malabar one, and I must concur in his opinion, for in the latter the pectorals are longer than the head reaching to the commencement of anal fin, the body is higher, the caudal shorter, and the opercle differently formed.

Genus—RACONDA, *Gray*.

Apterygia, Gray.

Branchiostegals six. Body oblong, compressed. The lower jaw projecting: the maxilla may be elongated posteriorly or truncated. Small teeth on the jaws, palatines, pterygoids, and tongue: none on the vomer. Dorsal and ventral fins absent: anal elongated. Scales rather small and deciduous. Serrature along the abdominal profile developed but weak.

737. *Raconda Russelliana*, Gray. A. 83-92, L. l. 60-64, L. tr. 12. Length of head 3/19, height of body slightly exceeds 1/5 of the total length. Eyes, diameter 3/10 of length of head, and more than the distance from the end of the snout. In adults the maxillary is said to extend to the gill-opening. From 31 to 38 weak spines along the abdominal edge. Silvery, with a dark spot behind the upper part of the opercle. Bay of Bengal to the Malay Archipelago.

Genus—ENGRAULIS, Cuv.

Branchiostegals short and rather numerous. Gill-openings wide: the membrane connecting the two sides being short, thus leaving the isthmus uncovered. Body oblong or elongated, compressed, and serrated along the abdominal edge. Cleft of mouth lateral: snout conical: the upper jaw the longer: maxillaries of varying length, but always long, having a membraneous attachment to the cheeks. Teeth small, sometimes of unequal size, usually present on the jaws, vomer, palatine and pterygoid bones. The dorsal fin may be wholly or partially in advance of or entirely posterior to the origin of the anal: the upper pectoral rays may or may not be prolonged: anal with many or a moderate number of rays. Scales large or of moderate size.

A.—*Dorsal fin entirely in front of anal.*

738. *Engraulis Bælama*, Forsk. B. XI, D. 14-16, A. 29-32, L. l. 40-42, L. tr. 9-10. Length of head 2/9, height of body 1/4 of the total length. Snout much projecting over the lower jaw. Maxilla extending to the gill-opening: it is somewhat enlarged opposite the mandibular joint. Fine teeth in both jaws. Origin of dorsal fin rather nearer snout than base of caudal: anal commencing just posterior to its termination: 21 or 22 gill-rakers on the lower branch of the outer branchial arch. Weak spines on the abdominal edge posterior to the ventral fin. Black venules behind the upper angle of the opercle on the shoulder. Red sea, east coast of Africa through the seas of India to the Malay Archipelago.

739. *Engraulis Malabaricus*, Bloch. *Monangoo*, Mal.: *Poor-relan*, Tam.: *O-pul-dah*. And. D. 14, A. 40-42, L. l. 39-41, L. tr. 11-12. Length of head 1/5, of caudal 2/11, height of body 4/15 of the total length. Eyes, diameter 1/4 of length of head, 3/4 of a diameter from end of snout, and 1½ apart. The maxilla reaches to just beyond the gill-opening. Origin of dorsal fin midway between snout and base of caudal: anal commences behind the last dorsal ray: pectoral as long as the head. Gill-rakers from 21 to 25 on the horizontal limb of the outer branchial arch. 10 spines on the abdominal edge posterior to the ventral fin, and 17 anterior to it. Black venules behind the upper angle of the opercle on the shoulder: dorsal and end of caudal blackish: pectoral sometimes entirely black. Abundant on the Malabar coast, found also in the Bay of Bengal.

740. *Engraulis Hamiltonii*, Gray. D. 13, A. 36-40, L. l. 47. L. tr. 12-13. Length of head from 1/5 to 4/21, of caudal 1/6, height of

body 1/4 of the total length. Eyes, diameter nearly 1/4 of length of head, 3/4 of a diameter from end of snout. The maxilla reaches to just beyond the gill-opening. Origin of dorsal fin rather nearer snout than the base of the caudal : the anal commences behind the last dorsal ray : pectoral not so long as the head. Gill-rakers 13 on the horizontal limb of the outer branchial arch. The abdominal edge strongly spined. Black venules behind the upper angle of the opercle on the shoulder. Seas of India and Malay Archipelago.

741. *Engraulis mystacoides*, Bleeker. D. 13-15, A. 35-40, L. l. 42-43 (45), L. tr. 12. Length of head and of caudal each 1/5, height of body 2/9 of the total length. Eyes, diameter 2/7 of length of head, 1/2 a diameter from end of snout, which projects. The maxilla extends backwards to about as far as the posterior end of the root of the pectoral fin. Origin of dorsal fin midway between the anterior edge of the orbit and the base of the caudal fin : the anal commences behind the last dorsal ray : the pectorals reach to the last third of the small ventrals. Gill-rakers 13 on the horizontal limb of the outer branchial arch. 11 strong spines on the abdominal edge behind the ventral fin, and 17 anterior to it. Black venules over the scapular region. Seas of India to China.

742. *Engraulis mystax*, Bl. Schn. D. 13-15, A. 34-36, L. l. 40-42, L. tr. 9. Length of head 2/9, of caudal 2/11, height of body 2/9 of the total length. Eyes, diameter 2/7 of length of head, 2/3 of a diameter from the end of the snout, which projects. The maxilla extends backwards almost to the ventral fins. Dorsal commences nearer snout than the base of the caudal, the anal beginning a short way behind its last ray. 7 to 8 strong spines on the abdominal edge behind the ventral fin, and 13 to 14 anterior to it. Coppery colour along the back : a dark shoulder mark formed of black points : caudal with a dark extremity. Seas of India to the Malay Archipelago.

743. *Engraulis rhinorhynchus*, Bleeker. *Tampuri*, Ooriah. D. 13, A. 32-37, L. l. 37, L. tr. 9. Length of head nearly 1/4, of caudal 1/5, height of body 2/7 of the total length. Eyes, diameter 2/7 of length of head, 3/4 of a diameter from end of snout, and 1 apart. Snout pointed, overhanging the mouth : the maxilla produced backwards as far as the gill-opening. Origin of dorsal fin midway between the end of the snout and the base of the caudal : the anal commencing at a short distance posterior to the last dorsal ray. 8 to 9 spines on the abdominal edge behind the ventral fin, and 15 or 16 anterior to it. Greenish along the back, becoming silvery-white on the sides and below : a blackish band over the nape, extending down to the shoulders. Ceylon and Coromandel coast to the Malay Archipelago.

744. *Engraulis setirostris*, Brouss. *Yeka-poorawah*, Tel. D. 14-16, A. 34-38, L. l. 36 (44), L. tr. 10-11. Length of head 1/6, of caudal 1/6, height of body 4/17 to 1/5 of the total length. Eyes, diameter 2/7 of length of head, 1/2 a diameter from end of snout, and 1 apart. Snout hardly projecting. The maxillary very prolonged, extending to the posterior end of the ventral fin or even beyond. Origin of dorsal midway between end of snout and base of caudal fin : the anal commences just posterior to the last dorsal ray. Abdominal edge spinate. Gill-rakers 10 on the horizontal limb of the outer branchial arch. Black

venules in the region of the scapula. Seas of India to the Malay Archipelago.

B.—*Dorsal fin wholly or in part above the anal.*

745. *Engraulis Indicus*, (v. Hasselt): Bleeker. *Nattoo*, Tel.: *Nettellee*, Tam.: *Zoo-roo-cart-dah*, And. D. 14-16, A. 19-21, L. l. 40, L. tr. 8-9, Cæc pyl. 16. Length of head 1/5, of caudal 1/6, height of body 1/6 in the total length. Eyes, diameter 2/7 of length of head, 2/3 of a diameter from end of snout, and nearly 1 apart. Snout much projecting beyond the mouth. The maxilla extending to just behind the mandibular joint. Teeth fine. Origin of dorsal fin rather nearer base of caudal fin than the anterior extremity of the snout: the anal commences below its centre. Spines on abdominal edge slender. A lateral silvery band. Seas of India to the Malay Archipelago.

746. *Engraulis Commersonianus*, Lacép. D. 15-16, A. 20-21, L. l. 36-40, L. tr. 8-9. Length of head 1/5, of caudal 2/11, height of body from 2/9 to 1/6 in its total length. Head 1/3 longer than high. Eyes, diameter from 1/3 to 2/7 in the length of the head, 1 diameter apart, and 1/2 from the end of snout, which much projects beyond the mouth. The maxilla, which is posteriorly pointed, reaches the angle of the preopercle. Origin of dorsal fin midway between the posterior edge of the orbit and the base of the caudal: the anal commences below the last third of the dorsal. Abdominal edge with weak spines. A silvery longitudinal band along the sides. From the East Coast of Africa through the Seas of India to the Malay Archipelago, and beyond.

Genus—COILIA, Gray.

Mystus, Lacép : *Trichosoma*, Swains.: *Chætomus*, McLelland : *Collia*, Schleg. : *Leptonurus*, Blkr. (1)

Branchiostegals nine to eleven. Body elongated, compressed and tapering to a pointed tail: abdomen trenchant and serrated along the abdominal edge. Snout pointed and projecting. Mouth cleft to behind the eye: the maxilla produced posteriorly. Teeth on the jaws, vomer, palatine, and pterygoid bones, also on the tongue. A single rather short dorsal fin placed in the anterior portion of the back: anal elongated and confluent with the caudal: from four to seven of the upper pectoral rays produced into moderately thick filaments. Scales of moderate or small size.

747. *Coilia ramcarati*, Ham. Buch. D. 13, P. 6 + VI, A. 94-110, L. l. 70. Length of head 1/6, height of body 1/5 in the distance from the end of snout to the commencement of the dorsal fin, 3/11 of the total length, excluding the caudal fin. Eyes, diameter 1/5 of length of head. The maxilla reaches the mandibular joint. The upper six pectoral rays elongated: abdomen not serrated anterior to the ventral fin. 28 gill-rakers on the horizontal limb of the outer branchial arch. Golden colour, with the edges of the anal and caudal black. Estuaries of the Ganges.

748. *Coilia Dussumieri*, Cuv. and Val. *Oorialli*, Ooriah. D. 14-15, P. 12 + VI, A. 80 to 110, L. l. 70-80, L. tr. 10-11. Length of

---

(1) *Toika*, Tel.

head 1/6 to 2/13, of caudal 2/17, height of body 1/5 to 2/11 in the distance from end of snout to the commencement of the dorsal fin, 8/10 of the total length. Eyes, diameter 1/4 of length of head, 1/2 a diameter from end of snout, and 1½ apart. Maxilla extends to the branchial opening. Abdomen strongly serrated, 8 spines behind the ventral fin and 3 anterior to it. 25 gill-rakers on the horizontal limb of the outer branchial arch. Golden colour, with two or three rows of round burnished golden spots along the lower half of the side. Seas and estuaries of India to the Malay Archipelago.

749. *Coilia Reynaldi*, (*) Cuv. & Val. D. 14, A. 110. The anal longer than in *C. ramcarati*. The distance from the end of the snout to the anal fin is less than a third of the total length. Eyes smaller than in *C. ramcarati*. Six pectoral filaments. Irrawaddi, to 4 inches in length.

750. *Coilia quadragesimalis*, Cuv. & Val. B. X. D. 15, P. 6+VI, A. 42. Height of body 4/15 of the total length. Mouth obtuse. Eye small, and the maxillary does not pass the angle of the jaw. Silvery, shot with gold, having nacreous reflections: fins yellowish. Ganges.

Genus—CHATOESSUS, *Cuvier*.

Branchiostegals from four to six. Body oval, short, deep, and moderately compressed, with a sharp, serrated abdominal edge. Snout overhanging a rather narrow, transverse mouth. The superior combs of the first branchial arch unite with those of the opposite side, forming two angles, one pointing forwards, the other backwards, the fourth branchial arch having an accessory respiratory organ. Teeth absent. A single dorsal fin, having the last ray prolonged in some species: ventrals anterior to or below the dorsal fin: anal commencing posterior to the dorsal and with many rays: caudal forked.

A. *The last dorsal ray elongated.*

751. *Chatoëssus nasus*, Bloch.: *Noonah*, Mal.; *Muddu candai*, Tam. : *Kome*, Tel. and Ooriah. D. 15-16, A. 20-25, L. l. 46-50, L. tr. 18-19. Length of head 1/5 to 2/11, height of body 1/3 of the total length. Eyes, diameter 1/3 of the length of head, 3/5 of a diameter from end of snout, and 1 apart. Snout much projecting over the mouth. Origin of dorsal fin nearer the end of the snout than the root of the caudal fin. The scales in the upper rows are darkest at their bases, thus forming dark lines: a bluish spot on the upper portion of the opercle posteriorly. Seas of India to the Malay Archipelago.

B. *None of the dorsal rays elongated.*

752. *Chatoëssus chacunda*, Ham. Buch.: *Muddeeru*, Tel.: *Korepaig-dah*, And. D. 17-19, A. 20, L. l. 40, L. tr. 13-14. Length of head from 2/9 to 1/5, of caudal 1/5, height of body from 2/5 to 1/3 of the total length. Eyes, diameter 1/3 of length of head, 2/3 of a diameter from end of snout, and 1 apart. Snout much projecting over

---

(*) Bleeker's *Atlas* having only been received up to p. 140 of the volume containing the Clupeoids, his definition of this species is not available. But as he states there are 6 species in the Malay Archipelago and 6 are figured, whereas the text of the last, which is wanting, is termed *C. quadrifilis*, Günther, and the sixth figure is named *Reynaldi*, Cuv. and Val., with 4 pectoral filaments, he may consider the 2 identical as did Cantor.

the mouth. Origin of dorsal fin much nearer snout than base of caudal. Scales regularly arranged. 28 spines along the abdominal edge, 11 of which are behind the ventral fin. A dark mark on the shoulder. Seas of India to the Malay Archipelago, attaining at least 8 inches in length.

753. *Chatoëssus champole*, Ham. Buch. D. 15, A. 21, L. l. 46, L. tr. 19. Length of head a little above 1/4, height of body 2/5 of the total length, excluding the caudal fin. Snout not much projecting beyond the mouth. Origin of the dorsal fin much nearer snout than base of caudal. A black shoulder spot, sometimes succeeded by several other smaller ones. Estuaries of Bengal and also in fresh water. Ham. Buch. gives a *C. gagius* like it, but larger, from N. Behar.

754. *Chatoëssus manminna*, Ham. Buch. D. 14, A. 22-25, L. l. 58-60, L. tr. 22. Length of head 1/5, of caudal 2/11, height of body 2/7 of the total length. Snout projecting. Origin of dorsal fin between the snout and base of the caudal fin. 32 spines along the abdominal edge, 12 being behind the ventral fin. A black spot generally present on the shoulder. The fresh waters and estuaries in Bengal and Burma, attaining 8 inches in length.

Family—SYMBRANCHIDÆ.

Gill-openings confluent into a single slit, which is situated on the abdominal surface. Gills well developed or rudimentary. Body elongated. The humeral arch may or may not be attached to the skull. Margin of the upper jaw formed by the intermaxillaries, the maxillaries being internal and parallel to them. Barbels absent. Palatine teeth, when present, in a single row or a narrow band. Vertical fins in the form of mere folds of skin, and no paired ones. Scales if present minute. Vent far behind the head. An accessory breathing sac present or absent. Air-vessel absent. Ribs present. Stomach destitute of a blind sac. No pyloric appendages. Ovaries with oviducts.

Genus—MONOPTERUS, Lacép.

Fluta, Bl. Schn.: *Ophicardia*, McLelland : *Apterygia*, Basil.

Branchiostegals five or six. Gill-membranes almost entirely attached to the isthmus, having a single transverse opening. Three branchial arches, with the laminæ rudimentary, having moderately wide intermediate slits. Palatine teeth in a narrow band. Scales absent. No accessory breathing sac.

755. *Monopterus Javanensis*, Lacép. Of a cylindrical form, with its greatest diameter at the occiput, from whence it tapers to the end of the tail. Brownish above, becoming of a dirty yellow below, with numerous deep brown or black spots. Estuaries of India to the Malay Archipelago and beyond.

Family—MURÆNIDÆ, *Müller*.

Body elongated, cylindrical, or band-shaped: the humeral arch not attached to the skull. The branchial openings in the pharynx may be wide or narrow slits, and either separate or united. Margin of the upper jaw is constituted anteriorly by the intermaxillaries, which are more or less coalescent with the vomer and ethenoid, whilst laterally the sides of the upper jaw are formed by the maxillaries. Vertical fins, when present, are confluent or separated by a projecting tail: pectorals and ventrals present or absent. Scales, when present, rudimentary. Vent may be situated close to the root of the pectoral fins, or a long distance posterior to the head. The heart may be situated just, or a long distance behind the gills. Stomach with a blind sac. No pyloric appendages. Ovaries destitute of oviducts.

Genus—MURÆNESOX, McLelland.

Cynoponticus, Costa: *Brachyconger,* Bleeker.

Gill-openings wide, approximating to the abdomen. Snout rather elongated, the upper jaw the longer. Two pairs of nostrils, the posterior of which are opposite to the upper part or centre of the orbit. Teeth in the jaws rather fine, with some canines anteriorly: vomer with several rows of teeth, the middle of which are conical or compressed. Dorsal fin commencing above the gill-opening; it, the anal, caudal and pectoral well developed. Anus a long distance from the gill-opening.

756. *Murænesox talabon,* Cuv.: *Tala-bon & Culim-poun,* Tel.: *Kotah,* Tam. Eyes, diameter from 2/7 to 1/4 in the length of the snout. The vomerine teeth are conical and widely set, none of them with basal lobes. Golden green superiorly becoming yellowish-white below. Vertical fins with blackish margins. Seas of India to the Malay Archipelago. This eel attains a very large size: they are common in the Bombay Bazars to 10 feet or more in length.

757. *Murænesox talabonoides,* Bleeker. Snout long and narrow. The vomerine teeth are straight, the posterior ones being compressed and having basal lobes: the external mandibular teeth directed outwards. Estuaries of the Ganges to the Malay Archipelago.

758. *Murænesox cinereus,* Forsk. Eyes, diameter 1/2 the length of the snout. The vomerine teeth are compressed, and have an anterior and posterior basal lobe: the external mandibular teeth are not directed outwards. Seas of India to the Malay Archipelago and beyond.

Genus—MURÆNICHTHYS, Bleeker.

Body elongated and cylindrical. Gill-openings narrow. Eyes small. Nostrils on the edge of the upper jaw. Dorsal fin low or rudimentary, commencing a long distance posterior to the gill-opening. Pectorals absent.

759. *Murænichthys vermiformis,* Peters. Angle of the mouth slightly posterior to the eye. Teeth in the jaws and on the vomer in a single row. The origin of the dorsal fin behind the vent. Ceylon.

760. *Murænichthys Schultzii,* Bleeker. Angle of mouth considerably posterior to the eye. Teeth mostly in two rows. The origin of the dorsal fin opposite the vent. Andaman Islands and Malay Archipelago.

Genus—OPHICHTHYS, (Ahl. sp.), Günther.

Gill-openings may be close together. Snout greatly or moderately produced. Cleft of mouth wide or of medium width: lips may or may not be fringed. Teeth in jaws and on vomer, either pointed, granular, or small and conical: in the maxilla they may be in from one to four rows, or in bands, whilst in the mandibles they may be in one or two rows: canines present or absent. Dorsal fin, when present, commences either in advance of or nearly above the gill-opening, or behind the root of the pectoral: the pectorals, when present, may be rudimentary, or only developed in the adult, or of good size: anal present or absent: extremity of tail free.

A. *Teeth pointed: pectorals developed in the adult.*

761. **Ophichthys hyala**, Ham. Buch. Eye small. Cleft of mouth extending slightly behind the orbit. Teeth of nearly equal size, pointed, those in the intermaxillaries the strongest: the posterior mandibular teeth are in one row, the remainder in two. Dorsal commences at a short distance behind the end of the pectoral. Colours uniform. Estuaries of the Ganges.

762. **Ophichthys ornatissimus**, Kaup. Cleft of mouth very deep: snout moderately produced. Teeth pointed, the maxillary and anterior vomerine in two, the remainder in a single row: no large canines. Dorsal commences behind the end of the pectoral. 16 to 17 large round spots along the lateral line, which are separated by another band of spots of different sizes. Irregular spots on the head: a transverse and two longitudinal rows of white spots on the occiput: curved whitish lines between the eyes: dorsal with black marginal spots and stripes. Malabar.

B. *Teeth obtuse or granular: pectorals developed or rudimentary.*

763. **Ophichthys boro**, Ham. Buch. Eyes small. Cleft of mouth extending to behind the eyes. Teeth granular, in several rows. The dorsal commences behind the base of the pectoral. Colouration uniform. Seas of India and Malay Archipelago.

764. **Ophichthys colubrinus**, Bodd. Eyes very small. Cleft of mouth of moderate width, extending to just behind the eyes. Teeth conical, in two rows. The dorsal fin commences in front of the gill-opening just behind the nape; it and the anal are rather low: pectoral rudimentary. Numerous (25 to 35) brown rings surround the body. Andamans, Malay Archipelago, and beyond.

C. *Teeth conical and of about equally small size. Gill-openings contiguous longitudinal slits on the ventral surface: pectorals sometimes rudimentary, but, as a rule, absent.*

765. **Ophichthys Orientalis**, McLelland. Eyes small. Cleft of mouth of moderate width. Teeth pointed and in one row. The dorsal fin commences at a short distance behind the gill-opening, it and the anal being low: pectoral absent. A row of round whitish spots across the occiput, having a short, forward—directed line. Ceylon and Bay of Bengal.

Genus—MORINGUA, Gray.

Raitaboura, Gray: *Ptyobranchus*, McLell.: *Aphthalmichthys*, Kaup: *Pseudomoringua*, Bleeker.

Body sub-cylindrical, with the trunk considerably longer than the tail. Gill-openings rather narrow and inferior: heart far posterior to the branchiæ. Cleft of mouth narrow. The posterior nostril situated in front of the eye. Teeth in a single row. Vertical fins limited to the tail: pectorals, if present, small. Scales absent.

766. **Moringua raitaboura**, Ham. Buch. The dorsal and anal fins occupying the greater part of the tail, and both interrupted in the middle, the last commences at some distance behind the head: pectorals present. Purplish above, with black dots. Estuaries of the Ganges to the Malay Archipelago, attaining at least 22 inches in length.

767. **Moringua macrocephala**, Bleeker. Length of head 1/9 of the total length. A few rays at the end of the tail, otherwise the

vertical and pectoral fins are reduced to mere cutaneous folds. India and Malay Archipelago.

Genus—*MURÆNA*, sp. Artedi.

Gymnothorax, Bl.: *Murænophis*, Lacép.: *Echidna*, Forst.: *Thærodontis, Strophidon,* and *Locodontis,* McLell.: *Sidera, Eurymyctera, Thyrsoidea, Limamuræna, Polyuranodon, Pæcilophis, Gymnomuræna, Priodonophis,* and *Tæniophis,* Kaup.: *Pseudomuræna,* Johnson.

Body moderately or exceedingly elongate. Gill-openings narrow. Two nostrils on either side of the upper surface of the snout, the posterior a round foramen, which may or may not be furnished with a tube: the anterior tubular. Teeth well developed and acute or molariform: the maxillary teeth may be in one or two rows. Dorsal fin elevated or not so.

A. Teeth acute. Posterior nostril not tubular.

768. *Muræna punctata,* Bl. Schn.: *Calamaia paum,* Tel. Teeth in a single row: canines moderately developed. Tail rather longer than the body. Blackish brown, with dark white-edged ocelli. Seas of India.

769. *Muræna Rüppellii,* McLelland. Teeth in a single row in the adult: canines moderately developed. Tail longer than the body. Yellowish, with from 18 to 20 black rings encircling the head and body: they are narrower than the ground colour: the first 3 are on the head, the others sometimes become less distinct as age increases. Andamans and Malay Archipelago.

770. *Muræna reticularis,* Bl. Teeth in a single row, some being slightly serrated: canines very small. Tail longer than the body. Buff, with about 16 cross bands on the body, which are somewhat broader than the interspaces and most defined on the ventral surface: head and back brown spotted: sometimes a band on the former. Andamans, Malay Archipelago, and China.

771. *Muræna tessellata,* Richardson. Teeth in a single row in the adult: canines moderately developed. Tail nearly as long as the body. Dark polygonal or rounded spots on the head, body, and fins, either separated by narrow light lines or interspaces, the spots being the widest. From the East Coast of Africa through the seas of India to the Malay Archipelago, and China.

772. *Muræna tigrina,* Rüppell. Teeth in a single row, except at the anterior portion of the vomer. Tail nearly twice as long as the body. Yellowish, with three longitudinal rows of round brown spots, the middle being the largest. Red Sea and Andaman Island.

773. *Muræna undulata,* Lacép. Teeth in a single row, with sometimes two additional teeth forming an inner maxillary row: canines strong, normally 4 pairs in mandibles and 2 in maxillaries. Tail longer than the body. Brownish, with spots and blotches on the head, and pale yellowish undulating or reticulating lines over the body and fins: no black spot at gill-opening, nor a white edge to the fins. East Coast of Africa, Andamans to the Malay Archipelago, and beyond.

774. *Muræna tile,* Ham. Buch. Teeth in two rows, except the lateral mandibular ones, which are single; canines small. Tail slightly shorter or as long as the body. Umber colour, with numerous small yellowish specks on the body and fins. Seas of India and Malay Archipelago.

775. *Murœna picta*, Ahl. Teeth in one row, except on the vomer and in front of the mandibles, where anteriorly there are two: no distinct canines. Tail about as long as the body. Greyish, with many small dark spots separated by a light net-work or marbling. East Coast of Africa, Seas of India to the Malay Archipelago, and beyond.

776. *Murœna flavomarginata*, Rüpp. Teeth in one row, except in the anterior part of the vomer, where there are two canines of moderate size. Tail as long as the body. Head and end of tail black, the remainder brown, marbled with black, or with large dark spots sometimes forming three rows: gill-opening in the middle of a black spot, and fins occasionally with a white edge. Red Sea, East Coast of Africa, Seas of India to the Malay Archipelago, and beyond.

777. *Murœna sathete*, Ham. Buch. Teeth in one row, except in the maxilla, vomer, and front mandibular, which are in two rows: canines badly developed. Tail rather longer than the body. Superiorly brownish back, becoming dirty yellow beneath. Estuaries of Ganges and Penang.

778. *Murœna macrurus*, Bleeker. Teeth in the maxilla and mandibles in two rows: canines badly developed. Tail twice as long as the body. Deep brown externally being blackish. Seas of India to the Malay Archipelago.

B. *Teeth generally obtuse or molarform.*

779. *Murœna zebra*, Shaw. Teeth in molarform bands. Tail half as long as the trunk. Deep brown, encircled with five yellow whole or half-bands, which are more irregular in shape on the head. Andamans, Malay Archipelago, and beyond.

780. *Murœna nebulosa*, Ahl. Teeth molarform. Yellowish, with black vermiculated lines along the back, and two rows of black spots along the sides: white dots may be included within the black spots. East Coast of Africa, seas of India to the Malay Archipelago, and beyond.

781. *Murœna nigra*, Day. Teeth molarform: those in the maxillary in a double row and pointed. Tail nearly half the total length. Uniform black. Andamans.

Genus—*GYMNOMURÆNA*, Lacép.

Murænoblenna, Lacép.: *Ichthyophis*, Lesson: *Uropterygius*, Rüpp.: *Channamurœna*, Richardson.

Gill-openings of moderate width or narrow. Two pairs of nostrils on the upper surface of the snout, the posterior being a round foramen, or with a short tube. Teeth small, pointed, and numerous. Scales absent.

782. *Gymnomurœna tigrina*, Lesson. Maxillary and front mandibular teeth in two rows: no canines. Brownish, with various sized black spots and blotches. East Coast of Africa, Seas of India to the Malay Archipelago, and beyond.

Order—*LOPHOBRANCHII*, Cuv.

180. Fishes having a dermal, segmental skeleton, with the opercular pieces reduced to a single plate. Gill-openings small: gills consisting of small rounded tufts attached to the branchial arches. Muscular system very slightly developed. Snout produced: mouth terminal, but small. Teeth absent. Air-vessel stated to be destitute of a pneumatic duct.

Order of fishes with small tuft-like gills, attached to the branchial arches: dermal skeletons and apparently no pneumatic duct.

Family—SYNGNATHIDÆ, Kaup.

Gill-openings small, round, and situated at the posterior-superior angle of the gill-cover. A single dorsal fin. Ventrals, and occasionally one or more of the other fins, absent.

Genus—*SYNGNATHUS*, Artedi.

Corythoichthys, Trachyrhamphus et Halicampus, Kaup.
Body with more or less distinct ridges : the dorsal edge of the trunk not being continuous with that on its caudal portion : humeral bones firmly united into the breast ring. The opercle may be entirely crossed by a distinct ridge, or it may be only at its base, or the ridge absent. Dorsal fin either opposite or nearly so to the vent: its base may be raised or not so : pectorals well developed : caudal present. An egg-pouch on the tail of the males, the eggs being covered by cutaneous folds.

A—A bony ridge entirely crossing the opercle.

783. *Syngnathus spicifer*, Rüppell. *Ea-de* or *Lah-a-tha-dah*, And. D. 23-27. Rings 15-16 + 39-42. A sharp median ridge along the centre of the snout, crown of head, and nape. Dorsal commences behind the vent. Abdomen with about 13 black bars, as wide as the interspaces : sometimes uniform brown. Red Sea, East Coast of Africa through the Seas of India to the Malay Archipelago.

784. *Syngnathus cyanospilus*, Bleeker. D. 20-23. Rings 13-15 + 33-35. A low median ridge along the snout, crown, and nape : a second along the side of the head. Dorsal commencing on the anal ring. Body with irregular brown cross bars, each with a white posterior edge : black dots on the dorsal. East Coast of Africa, Seas of India to the Malay Archipelago.

B—Bony ridge on opercle absent or only on its base.

785. *Syngnathus serratus*, Schleg. *Cul pamboo*, Tam. D. 26-27, Rings 21-25 + 45-47. Snout less than half the length of the head, and with a serrated crest. Base of dorsal fin elevated. Tail black with a white lower edge. Seas of India to China.

786. *Syngnathus Ceylonensis*, Günther. D. 26. Rings 24+46. Snout more than half the length of the head, and with a slight superior ridge. Base of dorsal fin elevated. Zanzibar and Ceylon : may be a variety of the last.

Genus—*NEROPHIS*, Kaup.

Netasomata, Eichwald : *Scyphius*, Risso.
Body rounded and ridges when present very indistinct : the tail tapering to a point without possessing any or only a rudimentary caudal fin. Dorsal fin of moderate length placed opposite to the vent : the pectorals absent. Ova attached to the loose integument of the abdomen in the males, and not covered by lateral cutaneous folds.

787. *Nerophis Dumerilii*, Stein. D. 37. 27 body rings, the dorsal fin placed on the last 7 and first 3 of the tail, this last portion being twice as long as the trunk. Forehead concave : snout half as long as head. Bombay.

Genus—*GASTROTOKEUS*, Kaup.

Body depressed, with smooth shields and a prehensile tail, which latter is not so long as the body. Dorsal fin of moderate length placed nearly opposite the vent: pectorals present. The lateral line passing along the margin of the abdomen. Ova imbedded in soft substance on the abdomen of the males, but destitute of any lateral cutaneous folds.

788. *Gastrotokeus biaculeatus*, Bl. D. 40-45, P. 17-23. Rings 18 + 45-55. Superciliary edge more or less spinate posteriorly: old individuals sometimes with filaments. East Coast of Africa, Seas of India to the Malay Archipelago, and beyond.

Genus—*ACENTRONURA*, Kaup.

Body rather compressed, with shields without tubercles, and a prehensile, finless tail: occiput compressed into a crest without any coronet. Dorsal fin with rather few rays: pectorals present. Ova carried by the males in a sac, which is situated below the tail and opening near the vent.

789. *Acentronura gracillima*, Schleg. D. 16, Rings 13+41. Snout short 2/5 of length of head. Dorsal fin stands on 4 rings, 2 of which belong to the tail. Andamans and Japan.

Genus—*HIPPOCAMPUS*, Leach.

Trunk compressed, the shields being furnished with tubercles or spines: the tail prehensile, finless, and longer than the trunk. Occiput compressed and forming a coronet at its posterior superior corner, which is usually surmounted by spines or knobs. Dorsal and pectoral fins present. Ova carried by the males in a sac, which is situated below the tail and opening near the vent.

790. *Hippocampus trimaculatus*, Leach. D. 20, Rings 11 + 36. Tubercles not much developed: coronet low, with 4 or 5 spines: supraorbital spine and one on either side of the throat claw shaped. Ochre: two rows of blackish spots on the dorsal fin. Burma and Malay Archipelago.

791. *Hippocampus guttulatus*, ([1]) Cuv. *Coodara meen*, Tam. D. 17. Rings 11+33. Tubercles blunt: coronet low: supraorbital spine obliquely truncated. Colouration various. Red Sea, East Coast of Africa, Seas of India to the Malay Archipelago, and beyond.

792. *Hippocampus hystrix*, Kaup. D. 17-18. Eleven body rings. Tubercles developed into acute spines. Light coloured, with numerous brown dots: snout with broad darkish rings: each spine with a black tip. Zanzibar, Aden, Andamans, and Japan.

Order—*PLECTOGNATHI*, Cuv.

181. Fishes with the bones of the head completely ossified, whilst those in the remainder of the body are incompletely so: vertebræ few. Gill-openings narrow, situated in front of the pectoral fins: gills pectinate.

Fishes having united jaws.

([1]) I marked H. *comes*, Cantor, as obtained at the Andamans, the specimen not being at present available for re-examination, it may have been H *comes*, Kaup., or H *guttulatus*, Cuv., consequently Cantor's fish is not included in this list.

Head generally large. Mouth narrow: the bones of the upper jaw mostly united, sometimes produced into the form of a beak. Teeth may be distinct in the jaws or absent. There may be a single soft-rayed dorsal fin belonging to the caudal portion of the vertebral column, and situated opposite the anal: in some a rudimentary spinous dorsal is also present: ventrals, when existing, in the form of spines. Skin either smooth, or with rough scales, or ossified in the form of plates or spines. Air-vessel destitute of a pneumatic duct.

Family--S C L E R O D E R M I, *Cuv.*

Body compressed or angular: snout somewhat produced. Distinct teeth in the jaws of various characters. A barbel in one genus. The elements of a spinous dorsal and ventral fins generally present, but mostly variously modified. Skin rough or spinate, or the scales in the form of a firm carapace.

Genus—TRIACANTHUS, Cuv.

Body compressed, ending in a somewhat elongated caudal portion. Eyes lateral. Teeth in two rows in both jaws, the outer 10 in number, being incisor-like, the inner, from 2 to 4, being more molarform. First dorsal fin consisting of a long and strong spine, followed by from 3 to 5 smaller ones : ventrals formed by a pair of strong spines articulated by a joint or ossified to the pelvic bones. Scales minute and rough. Air-vessel strong.

793. *Triacanthus biaculeatus*, Bl. Satura, Ooriah.: *Moolahral*, Tam.: *Ko-tah-thoo-lay-po-dah*, And.: *Turgoorch*, Beluch. D. 5/22-25, A. 16-19. Second dorsal spine not much longer than the third. Silvery, with or without a black spot on the first dorsal fin. Seas of India to the Malay Archipelago and beyond, attaining at least 9½ inches in length. A popular term for it is 'file fish.'

794. *Triacanthus strigilifer*, Cantor. D. 5/22, A. 16. Second dorsal spine two or three times as long as the third: scales larger than in the other species. No black on spinous dorsal.

Genus—BALISTES, sp. Artedi.

Baliastapus, Tiles: *Xenodon, Erythrodon, et Pyrodon*, Rüpp.: *Melichthys, Leiurus*, Swain.: *Balistapus*, Kaup. Sub-genera, *Parabalistes, Pseudobalistes, Canthidermis*, Bleeker.

Branchiostegals six. Body compressed. Barbels absent. Sometimes a groove before the eyes. Upper jaw with a double series of incisor-like teeth, 8 in the outer, 6 in the inner row: mandible with 8 similar teeth in one row, these teeth may be white, uneven, and more or less notched: or white, even, and incisor-like, or of a burnt sienna colour, and the supero-lateral pair projecting. The first dorsal fin consisting of a strong spine succeeded by two weak ones: ventrals as an osseous appendage. There may or may not be oval, flattened osseous productions on the scales behind the gill-opening. Scales forming a carapace: in some species there are rows of spines or tubercles on the side of the free portion of the tail, which is either compressed or depressed.

A. *Free portion of the tail depressed*—(*Leiurus.*)

795. *Balistes stellatus*, Willugh. D. 3/27, A. 25, L. l. 44, L. tr. 24.([1]) Two obtuse ridges along either side of the free portion of the

---

([1]).—The lateral transverse is counted on a line from the origin of the dorsal fin to the vent.

tail. A patch of enlarged scales posterior to the gill-opening. Posterior margin of caudal fin undulated, and the lobes with filamentous prolongations in the adult. In the immature grey with some large bluish white blotches along the head and back, and smaller ones over the side of the body: in the adult a whitish band passes along the side and there are dark longitudinal stripes along the dorsal and anal fins. Red Sea, Seas of India to China.

B. Free portion of the tail compressed: the teeth white, uneven and more less notched.

796. *Balistes maculatus*, D. 3/26, A. 24, L. l. 55-56, L. tr. 31-39. A groove in front of the eye below the nostrils: cheeks entirely scaly: no patch of enlarged scales behind the gill-opening. Scales granulated, having a prickle at their base, most apparent in the young: neither spines nor tubercles on the side of the free portion of the tail. Caudal truncated. Uniform brown or black, with white oval or round spots. Seas of India to China, and beyond.

797. *Balistes vetula*, Linn. D. 3/30-32, A. 29, L. l. 61-63, L. tr. 37. A groove in front of the eye below the nostrils: cheeks entirely scaly, a patch of enlarged scales behind the gill-opening. No spines or tubercles on the side of the tail. In adults the anterior dorsal rays and caudal lobes may be filamentous. In the immature some oblique black lines go along the rows of scales. In the adult there are two curved blue dark-edged bands along the side of the head, and a black one with a light margin under the eye, from which other similar ones radiate: caudal with its upper and lower edge blue, and a bluish band near its posterior extremity: transverse bluish bands on the dorsal and anal fins. Coromandel Coast of India, and beyond

798. *Balistes niger*, Mungo Park. D. 3/26-28, A. 23, L. l. 47, L. tr. 24-26. A groove in front of the eye below the nostrils: cheeks entirely scaly: a patch of enlarged scales behind the gill-opening. From 6 to 8 rows of rather small recurved spines on the side of the tail. Caudal truncate. Brownish-black, caudal with white edges: a light ring round the lower jaw. Red Sea, Seas of India to the Malay Archipelago, and beyond.

799. *Balistes mitis*, Bennett. *Rahtee-yellakah*, Tel.: *Cul korawa*, Tam. D. 3/30-31, A. 27-28, L. l. 55-65, L. tr. 34. A groove in front of the eye below the nostrils: cheeks entirely scaly: a patch of enlarged scales behind the gill-opening. Most of the scales on the side of the tail with a small round smooth tubercle. Caudal slightly emarginate or truncated. Blackish or yellowish brown: lower surface of mandible yellow, and a yellow stripe generally goes towards the base of the pectoral fin. Fins darkest at their edges. East Coast of Africa, Seas of India to the Malay Archipelago, and beyond.

800. *Balistes conspicillum*, Bl. Schn. *Lama-yellakah*, Tel. D. 3/25-26 A. 21-22, L. l. 46, L. tr. 29. A groove in front of the eye below the nostrils: cheeks entirely scaly: a patch of a few enlarged scales behind the gill-opening. Two and a half rows of tubercles on the scales on the side of the tail. Caudal truncate. Colours vary, generally greyish brown or brownish black with white marks: and a light band across the posterior part of the snout from eye to eye, and a white one

across its anterior extremity : caudal white with black edges and a brown band across its base. Seas of India to the Malay Archipelago, and beyond.

801. *Balistes viridescens*, Bl. Schn. D. 3/24-25, A. 23, L. l. 29, L. tr. 18. A groove in front of the eye below the nostrils : cheeks entirely scaly : a patch of a very few enlarged scales behind the gill-opening. Four and a half rows of recurved spines on the side of the tail. Caudal rounded. A light ring round the muzzle joining one from below, and dividing the black lip from a black band on the forehead : body brownish olive, each scale darkest in the centre : a wide blackish band from the eye to the base of the pectoral fin : vertical fins yellowish with dark margins : large blue blotches on the first dorsal fin. Red Sea, Andamans to the Malay Archipelago.

802. *Balistes flavimarginatus*, Rüpp. D. 3/26-27, A. 23-24, L. l. 30-35, L. tr. 20. Anterior part of snout partly covered with tubercular rudimentary scales, neither in the adult are the cheek entirely scaled : a few enlarged scales behind the gill-opening. Four to six rows of rather small recurved spines on the side of the tail. Caudal varies from rounded in the immature to deeply lunated in the adult with elongated lobes. Immature with dark spots in the centre of some of the scales : vertical fins with a black and white margin. Adults become of a nearly uniform colour. Red Sea, Andamans to the Malay Archipelago.

803. *Balistes aculeatus*, Linn. D. 3/25, A. 22, L. l. 40, L. tr. 23. No groove in front of eye below the nostrils : a patch of enlarged scales behind the gill-opening. Two and half rows of recurved spines on the side of the tail. Caudal rounded in the immature, undulated in the adult. Greyish with two pair of oblique white bands passing from the middle of the body to the anal fin : a black interocular band : three blue lines pass from the eye to the base of the pectoral, and an orange band from the angle of the mouth to the same place : base of caudal spines black. West Coast of Africa, Seas of India to the Malay Archipelago, and beyond.

804. *Balistes undulatus*, Mungo Park. D. 3/26-27, A. 24, L. l. 41-44, L. tr. 24-30. No groove in front of the eye : cheeks entirely scaly : a patch of a few enlarged scales behind the gill-opening : six strong spines on either side of the tail in two rows. Caudal fin truncated. Brownish with numerous undulating yellow lines from the eye and back to the anal and caudal fins : three or four likewise pass from the angle of the mouth and lips to between the ventral and anal fins : a dark band along the base of the second dorsal and anal. Red Sea, East Coast of Africa, Seas of India to the Malay Archpelago, and beyond.

805. *Balistes rectangulus*, Bl. Schn. D. 3/23-24, A. 20-21, L. l. 42-46, L. tr. 28-35. No groove in front of the eye : a patch of enlarged scales behind the gill-opening. Three and a half rows of recurved spines on either side of the tail. Caudal fin slightly convex. Olive with a black band in front of the eyes and a second between them, which is continued to the vent, where it increases in width extending along the anterior two thirds of the base of the anal fin : another black band passes from behind the gill-opening towards the end of the soft dorsal : fins light coloured : pectoral with a black base. East Coast of Africa, Seas of India to the Malay Archipelago, and beyond.

C. *Teeth brownish red: the superior lateral pair projecting* (*Erythrodon.*)

Balistes erythrodon, Günther. D. 3/35, A. 30, L. l. 33. A patch of enlarged scales behind the gill-opening, slightly raised lines along the scales on the sides of the tail. Caudal lobes elongated. Black with the hind edge of the caudal fin white. East Coast of Africa, Seas of India to the Malay Archipelago.

Genus—MONACANTHUS, Cuv.

Alutarius, Cuv.: *Stephanolepis*, Gill: *Chætodermis*, *Paramonacanthus*, *Amanses* Gray: *Pseudomonacanthus*, *Liomonacanthus*, *Oxymonacanthus*, *Branchaluteres*, *Acanthaluteres*, *Ceratacanthus*, *Paraluteres*, *Pseudaluteres*, and *Aluteres*, Bleeker.

Body compressed: in some species the side of the tail peculiarly armed in adult males, but less apparently so or not at all in the females. Barbels absent. Incisor-form teeth in both jaws, in two rows in the upper with six in the outer row, and a single row of six in the mandibles. The first dorsal fin composed of a strong spine, occasionally a second rudimentary one: ventral fin, when present, reduced to a single osseous process, sometimes rudimentary and either moveable or fixed. Scales minute and rough. Vetebræ 7/11-14.

A. *Dorsal spine with 2 rows of barbs.*

807. *Monacanthus setifer*, Bennett. *Kora-wan*, Tam. D. 28-34, A. 29-33. Dorsal spine above hind edge of orbit: ventral spine small, moveable: caudal rounded. Sides of tail with fine short bristles in the adult. Brown, with badly-marked spots and streaks. East Coast of Africa, Seas of India to China, and beyond.

B. *Dorsal spine rough but barbless.*

808. *Monacanthus monoceros*, Osbeck: *Korawan*, Tam. D. 48, A. 50. Dorsal spine above the middle of the orbit: no ventral spine: caudal short, truncated. Brownish. East coast of Africa, seas of India to the Malay Archipelago, and beyond.

809. *Monacanthus scriptus*, Osbeck: *Chipi-kora-wan*, Tam.: *Mullah purroah*, Mal. D. 44-48, A. 47-52. Dorsal spine very weak, above the middle of the orbit: no ventral spine: caudal long, rounded. Yellowish, covered with small brown spots, and blue blotches. East coast of Africa, seas of India to the Malay Archipelago, and beyond: attaining at least 10 inches in length.

Genus—OSTRACION, Artedi.

Lætophrys, *Tetrosomus*, Swains.: *Acanthostracion*, Bleeker: *Cibotion*, Kaup.: *Aracana*, Günther.

Branchiostegals six. Body shortened and angular, with the integuments modified into a solid carapace composed of angular osseous plates in juxtaposition with one another, but leaving the snout, bases of the fins and the hind portion of the tail covered by soft skin. They may be destitute of spines or have them variously situated. The carapace from three to five ridged, and open or closed behind the anal fin. Mouth small, maxillaries and intermaxillaries coalescent. Teeth slender and in one row. A single spineless dorsal fin placed opposite the anal. Ventrals absent.

A. Carapace closed behind the anal fin. (*Ostracion*.)

810. *Ostracion gibbossus*, Linn. *Cul plaachee*, Tam. Carapace three-ridged, the dorsal one terminating in an elevated, compressed, triangular spine: four small backwardly directed spines along each ventral ridge: a small supraorbital spine. Caudal rounded. A blue spot in the centre of each osseous plate. Red Sea, East coast of Africa, seas of India to the Malay Archipelago.

811. *Ostracion cubicus*, Linn. *Cul plaachee*, Tam. Carapace four-ridged, spineless: an obtuse hump above the snout in the adults. Greenish or yellowish olive, having one—rarely more—large dark-edged blue spots, generally in the centre of each plate along the side: a round brown mark at the base of the dorsal fin, and generally a few scattered spots or blotches over the caudal. Red Sea, East coast of Africa, seas of India to the Malay Archipelago, and beyond.

812. *Ostracion punctatus*, Bl. Schn. Carapace four-ridged, pineless. Covered with numerous white dots, some being confluent into lines. East coast of Africa, seas of India to the Malay Archipelago, and beyond.

813. *Ostracion cornutus*, Linn. *Martoo plaachee*, Tam. Carapace four-ridged: a long anteriorly-directed spine above each orbit: a posteriorly directed one at the termination of each ventral ridge: a prominence in the middle of the dorsal ridge. Carapace and tail with rather large round dark blue spots. Red Sea, east coast of Africa, seas of India, to the Malay Archipelago and beyond.

Family—GYMNODONTES, Cuv.

Body more or less short. Some possess the means of dilating an elastic portion of the œsophagus, or an abdominal sac, with air: or this power may be entirely wanting. Bones of the upper and lower jaw in the form of a beak, having a cutting edge, and being covered with a layer of ivory-like substance in which a median suture may be present or absent. A spineless dorsal, anal and caudal exist: pectorals present: ventrals absent. Dermal covering modified into small or large spines or laminæ. Pelvic bones and air-vessel present or absent.

Genus—TRIODON, (*Reinw*). Cuv.

Branchiostegals six. The upper jaw with, the lower without, a median suture. Two separate nasal orifices on either side. Tail elongated, terminating in a bilobed fin: dorsal and anal with few rays. Abdomen possessing a large sac, the upper portion of which can be dilated with air, whilst is kept distended by elongated pelvic bones. The dermal covering consisting of spinate, osseous laminæ which are not imbricate. Air-vessel present.

814. *Triodon bursarius* (Rein.), Cuv. D. 10, A. 10, C. 18. A large, irregularly shaped, black yellow-edged spot on the upper portion of the sac. Seas of India to the Malay Archipelago.

Genus—TETRODON, (1) Linn.

Back broad, or compressed into a ridge. Either jaw with a median suture. Should a conspicuous nasal organ exist:—there may be two on

---

(1) *Globe* or *puff fishes*. *Plaachee*, Tam. : *Kappa*, Tel.

either side in a papilla *(Tetraodon,* Bleeker*)* : or a single tubular one (*Crayracion,* Bleeker) : or an imperforate one having a fringed edge and spiny body (*Chelonodon,* Müller) : or a simple round cavity and smooth body (*Monotretus,* Bib.) : or two imperforate tentacles on either side, (*Arothron,* Müll.) : or the nasal organs may be inconspicuous, and the back compressed into a keel (*Anosmius,* Peters). Dorsal and anal fins with few rays. Body wholly or partially covered with fine dermal spines, or such may be absent. There may be a more or less distinct fold along the lower part of the tail, and very apparent nasal organs *(Gastrophysus,* Müll.*)* : or the fold be absent, but the body be spinate and the nasal organs very distinct *(Cheilichthys,* Müll.*)* : or the fold be absent and the skin smooth (*Liosaccus,* Günther). A portion of the œsophagus dilatable and able to be distended with air. Air-vessel present.

A. Back broad: two nasal openings in a papilla: a fold of skin along the lower side of the tail (Gastrophysus.)

815. *Tetrodon lunaris,* Bl. Schn.: *Kappa koorawah,* Tel. : *Cha-modah,* Andam. The back is entirely, or only anteriorly spined, abdomen similarly protected : snout, sides and tail spineless. Greenish olive above, sides and abdomen of a white satin, having a yellowish line from the eye to the centre of the caudal fin : end of caudal dark. Red Sea, seas of India to the Malay Archipelago, and beyond : attaining at least a foot in length.

816. *Tetrodon sceleratus,* (Forst) Gm. Linn. *Weldiah plachay,* Tam. Head and back finely shagreened : abdomen with three rooted spines : sides spineless. Olive-green, superiorly with some brown spots : whitish laterally and below, with a silvery longitudinal band : a white spot in front of the eye : a brown band round the mandible : gill-openings deep black. East Coast of Africa, Seas of India to the Malay Archipelago, and beyond, attaining at least 2½ feet in length.

817. *Tetrodon oblongus.* Bl. Back and abdomen with two rooted spines : sometimes bands of spines before and behind the pectoral fins. Superiorly brown with white spots, occasionally with badly marked brown transverse bands on the sides : there is often a large black humeral spot. Seas of India to the Malay Archipelago and beyond, attaining at least 15 inches in length.

B. Back broad: a single nasal opening, sometimes with two lips at its extremity on either side: body spinate. (*Crayracion,* Bleeker).

818. *Tetrodon viridipunctatus,* Day. Anterior two thirds of the distance between the snout and the dorsal fin, also the abdomen, spinate. Superiorly light green with emerald spots : a similar interocular bar, and a second posteriorly across the back : abdomen white : four black spots under the throat. Cochin.

C. Back broad: a simple circular nasal cavity: body smooth, (*Monotretus,* Bib.)

819. *Tetrodon immaculatus,* Bl. Schn. Generally spinate, except the lips and last half of the tail. Greenish superiorly : brownish along the

middle of the side, and dirty white below : upper and lower edges of the caudal black. In some there is a white spot over the eye : occasionally there are several parallel greyish longitudinal bands on the sides. From the Red Sea, through the Seas of India to the Malay Archipelago, and beyond, attaining at least 12 inches in length.

820. *Tetrodon mappa*, Less. Generally spinate, except the lips and last half of the tail. Brownish superiorly, becoming white below. Upper half of caudal fin and the abdomen in the young covered by a net-work of lines, which are indistinct on the dark back. A blotch in the adults below the pectoral fin : brown lines radiate from the eye, and the vent is in the centre of a black spot. East Coast of Africa, Seas of India to the Malay Archipelago, attaining at least 12½ inches in length.

821. *Tetrodon stellatus*, Bleeker. Generally spinate, except the lips and the posterior portion of the tail. Colours differ : the upper portion of the body with black or brown dots, which are confluent in the young. In the immature there are usually oblique bands more or less broken up on the abdomen or sides, but which generally disappear in the adult. Vent in a black ring : some black spots round the base of the pectoral : brown spots on the fins, which are sometimes absent from the dorsal and anal. East Coast of Africa, Seas of India to the Malay Archipelago, and beyond, attaining at least 2 feet in length.

822. *Tetrodon reticularis*, Günther : *Paatha*, Mal. : *Ko-pud-dah*, And. Generally spinate, except on the lips. Upper surface of body deep grey, becoming white below, with from eight to ten longitudinal black stripes, which, under the eye, as well as round the mouth and pectoral fin, are concave : on the back are darker blotches anteriorly, (where blackish band.surround spots of white or grey), and chequered with black posteriorly : caudal reticulated with black on a white ground. Seas of India to the Malay Archipelago, attaining at least 17 inches in length.

823. *Tetrodon hispidus*, Lacép. Generally spinate, except the lips and hind portion of the tail. Brown superiorly, with a moderate number of bluish-white spots : one or two bluish bands round the orbit, gill-opening and pectoral fin: sometimes there are lines or spots of black on the abdomen. From the Red Sea, through the Seas of India to the Malay Archipelago, and beyond, attaining at least 20 inches in length.

824. *Tetrodon bondarus*, Cantor : *Bondaroo kappa*, Tel. Generally spinate, except the lips and the hind portion of the tail. Olive superiorly, with an interrupted black net-work surrounding brownish-white spots : a large black spot on either side of the head : another on the throat, from whence a dark line ascends to either pectoral fin : a yellow spot in front of the pectoral fin ; inside and behind it a large black one, in which is a yellow halfring. Coromandel Coast of India to the Malay Archipelago.

Genus—DIODON, (Linn.) Cuv.*

Body nearly globular. Jaws without median suture. Nasal tentacle simple, with a pair of lateral openings. Body covered with stiff and erectile dermal spines, each having a pair of lateral roots. A portion of the

---

* Sun-fishes.

œsophagus dilatable, and can be distended with air. No pelvic bones. Air-vessel present.

825. **Diodon hystrix**, Linn.: *Moollu plaachay*, Tam. Frontal spines of moderate size: those on the body strong, with a pair of grooves on their dilated base: those behind the pectoral the longest: at the posterior portion of the back broad and short. The back, sides, and fins with numerous brown spots. Seas of India to the Malay Archipelago and beyond.

Sub-class—CHONDROPTERYGII.

182. Skeleton cartilaginous: no cranial sutures. Rarely a single gill-opening, as the gills are attached by their outer edges to the skin, and there exists an intervening gill-opening between each: no gill cover. Three series of valves at the bulbus arteriosus. Optic nerves, although united, do not decussate. Body with vertical and paired fins, the posterior pair abdominal: caudal with an elongated upper lobe. Intestines with a spiral valve. Male sex with prehensile organs attached to the vertical fins. Ovaries containing large ova, which are fertilised, and in some likewise developed internally. Embryo with external deciduous gills. No air-vessel.

Sub-class of cartilaginous fishes.

Order—PLAGIOSTOMATA.

183. Body more or less cylindrical or depressed: the trunk may or may not pass into the tail. From five to seven gill-openings, which may be lateral or inferior. Jaws distinct from the skull.

An order of cartilaginous fish with transverse mouths.

Sub-order—SELACHOIDEI, OR SHARKS.

Body more or less cylindrical, gradually merging into the tail. Gill-openings lateral.

Family—CARCHARIIDÆ.

The snout may be produced longitudinally (*Carchariidæ*, 'true sharks') or laterally (*Zygænina*, 'hammer-headed sharks'). Spiracles absent or present. Eye with a nictitating membrane. A small pit may or may not exist above the root of the tail, and a second behind the angle of the mouth may be present or absent. Mouth crescentic, inferior. Teeth may be erect or oblique, with a single cusp, having sharp and smooth or serrated edges: or they may be small, the cusps being obsolete: or with one in the centre and one or two lateral ones: or even obtuse. The first dorsal fin, destitute of a spine, is placed opposite the interspace between the pectoral and ventral: anal fin present.

Genus—CARCHARIAS, Müll. & Henle.

No spiracles. A pit before the root of the caudal fin. Snout longitudinally produced. Mouth crescentic: the labial fold or groove rarely extends beyond the angle of the mouth. Teeth with a sharp triangular cusp, sometimes dilated. The first dorsal fin, destitute of a spine, is placed opposite the interspace between the pectoral and ventral: caudal with a distinct lower lobe.

A. *Teeth entire, the upper and lower oblique and not swollen at the base, (Scoliodon).*

826. *Carcharias laticaudus*, Müll. & Henle: *Dun-da-nee*, Sind. Length of preoral portion of snout equal or slightly above the distance between the eye and the gill-opening: a short groove at the angle of the mouth not extending on to the upper jaw. The length of the base of the anal fin is nearly or quite equal to its distance from the ventral: the pectoral does not extend so far as to below the origin of the dorsal: its posterior edge is nearly straight. Uniform colour, with the pectoral of a deep grey or black. Seas of India to the Malay Archipelago and beyond.

827. *Carcharias acutus*, Rüpp. *Purrooway sorah*, Tam. Length of preoral portion of snout equal to or slightly less than the distance between the eye and the gill-opening: a short groove at the angle of the mouth not extending on to the upper jaw. The length of the base of the anal fin is one-half or less than its distance from the ventral: the pectoral extends to at least below the origin of the dorsal; its posterior edge is slightly concave. Bronze colour above, white below: posterior edge of pectoral pure white: upper edge of caudal dark. Red Sea, seas of India to the Malay Archipelago, and beyond: said on the Malabar Coast to attain a large size, but the largest I obtained was 25 inches long, whilst all were females.

828. *Carcharias Walbeehmii*, Bleeker: *Ei-dah*, And. Length of preoral portion of snout more than the distance between the eye and the gill-opening: a short groove at the angle of the mouth extending a little distance on to both jaws. The length of the base of the anal fin is equal to about two-fifths of the space between it and the ventral: the pectoral extends a little beyond the origin of the dorsal: its posterior edge is slightly concave. Back greenish, becoming white below: anterior edge of dorsal fin dark, and a similar tinge along the upper edge of the caudal. Seas of India to the Malay Archipelago.

B. *Teeth entire, the central lower, smaller than the lateral ones, having swollen bases and slender points: those on the upper jaw flat and oblique, (Physodon).*

829. *Carcharias Mülleri*, Müll. and Henle. Snout elongated and pointed. Ventral commencing below the end of the base of first dorsal: anal in advance of the second dorsal, which is very small. Bengal.

C. *The upper teeth only with the bases serrated.*

830. *Carcharias Macloti*, Müll. and Hen. Snout long, pointed: its preoral portion nearly equalling the distance from the eye to the gill-opening, the nostrils being nearer the mouth than the end of the snout. The bases of the teeth in the upper jaw denticulated on both sides: those in the lower jaw erect. Seas of India to New Guinea.

831. *Carcharias hemiodon*, Müll. and Henle. Snout rounded: nostrils midway between its end and the mouth: distance between the nostrils

equals the length of the snout. Seas of India, to at least 28 inches in length.

D. *Some or all the teeth serrated on their bases and also on the edges of the cusps* (*Prionodon.*)

832. *Carcharias Malabaricus*, Day. Preoral portion of snout nearly as long as the mouth is wide: a very small groove at the angle of the mouth. Upper teeth 28 in number, oblique, triangular, with rather enlarged bases, and serrated in their whole extent: lower teeth slender, erect. The entire pectoral fin not so long as the head, but 1/4 longer than broad, scarcely emarginate: the posterior end of the dorsal is at an equal distance from the ventrals as its anterior extremity is from that of the pectoral. Grey above, white below: upper half of the anterior and two-thirds of the second dorsal deep black. Malabar Coast, to 16 inches in length.

833. *Carcharias Gangeticus*, Müll. and Henle. Snout very short and obtuse: nostrils close to its extremity. Teeth 27-30 on either jaw, and all serrated. First dorsal commences behind the base of the pectoral. Of a grey colour. Seas of India to Japan, &c., ascending rivers to above tidal influence, and attaining from 5 to 6 feet in length.

834. *Carcharias melanopterus*, Quoy. and Gaim. Snout short and obtuse: nostrils nearer its extremity than the mouth. Teeth serrated, 25-31, upper ones oblique, the outer edge notched, the inner straight: the lower narrower. The pectoral with its upper edge 3 times as long as its lower, reaches the end of the base of the dorsal fin, which latter is slightly nearer the root of the pectoral than that of the ventral. Ends of all the fins deep black. Seas of India to the Malay Archipelago.

835. *Carcharias Bleekeri*, Duméril. Snout of moderate length, the nostrils being between its extremity and the mouth. Teeth finely serrated,$\frac{2\frac{4}{5}}{2\frac{5}{5}}$, upper ones oblique, the outer edge notched, the inner straight: the lower narrow, with a broad base and nearly erect. Pectoral with its upper edge 4½ to 5 times as long as its lower. A deep black spot at the lower edge of the end of the pectoral fin: a second at the end of the inferior lobe of the caudal. East Coast of Africa, Seas of India, attaining at least 27 inches in length.

836. *Carcharias limbatus*, Müll. and Henle. Snout somewhat pointed, the length of its preoral portion being less than the width of the mouth: nostrils between its extremity and the mouth. Teeth serrated $\frac{25\text{-}29}{27\text{-}30}$. Pectoral with its upper edge nearly four times that of the lower. The lower edge of the end of the pectoral, second dorsal, anal, and lower caudal lobe black. Seas of India, and beyond.

837. *Carcharias Temminckii*, Müll. and Henle. Preoral portion of snout almost as long as the width of the mouth: nostrils nearer the mouth than the end of the snout. Teeth $\frac{36\text{-}38}{38\text{-}40}$, upper rather narrow, with a broad base erect and serrated: the lower entire. Pectoral with its upper edge nearly three times that of the lower. Second dorsal opposite anal, and nearly as large as the first. Of an uniform colour. India.

Genus,—G*ALEOCERDO,* *Müll. & Henle.*

Spiracles small. *Membrana nictitans* present: pupil of eye round. Mouth crescentic. Teeth oblique, serrated on either edge, and with a deep notch on the outer margin. The first dorsal spineless, placed opposite the interspace between the pectoral and ventral fins: caudal with a double notch. A pit on the tail, both above and below, at the base of the caudal fin.

838. *Galeocerdo Rayneri*, McDonald and Barron. *Wulluven sorah*, Tam. Preoral portion of snout of considerably less length than the distance between the inner angles of the nostrils: a long labial fold along the upper jaw. Caudal fin a little above 1/4 of the total length, and not quite equal to the interspace between the two dorsals. Obscure spots and vertical stripes on the body. Seas of India to Australia.

839. *Galeocerdo tigrinus*, Müll. and Henle. Preoral portion of snout of less length than the distance between the inner angles of the nostrils: a long labial fold along the upper jaw. Caudal fin 1/3 of the total length and considerably exceeding the interspace between the two dorsals: second dorsal slightly in advance of the anal. Numerous brown spots over the body. Seas of India to Japan and beyond.

Genus—ZYGÆNA,* *Cuv.*

Cestracion, Klein: *Cestrorhinus*, Blainv. : *Sphyrna*, (Rafin) Müll. and Henle: *Eusphyra* and *Reniceps*, Gill.

The anterior portion of the head is broad, flattened, and laterally elongated with the eyes situated at its lateral extremities and the nostrils at its fore border. A membrana nictitans present. Spiracles absent. Mouth crescentic. Teeth similar in both jaws, placed obliquely and notched. The first dorsal fin spineless, situated above the interspace between the pectoral and the ventral: caudal with one notch, and a pit at the commencement of the fin.

840. *Zygæna Blochii*, Cuv. The two lateral expansions of the head from twice to thrice as long as broad, with a deep groove along the anterior edge. Nostril much nearer mouth than the eye. Mouth angular. Greyish. Seas of India to China.

841. *Zygæna malleus*, Risso. : *Koma sorah*, Tel. The hind edge of the lateral expansion of the head nearly equals its width near the eye, and with a groove along almost its entire anterior edge. Nostrils close to the eye. Body and fins slaty-grey : undersurface white. Red Sea, Seas of India to Malay Archipelago, and beyond.

842. *Zygæna tudes*, Cuv. Anterior edge of head much curved, but not continuous with the lateral one : the length of its hind margin less than its width near the eye. Nostril close to the eye and a groove running along the anterior edge of the head. East Coast of Africa, Seas of India to the Malay Archipelago, and beyond.

Genus—MUSTELUS, *Cuv.*

Membrana nictitans present. Small spiracles behind the eyes. Mouth crescentic with long labial folds. Teeth small, numerous, similar in both jaws, pavement-like, obtuse or with indistinct cusps. The first dorsal fin spineless, situated above the interspace between the pectoral and ventral: the second nearly as large as the first : caudal without a distinct lower lobe, and no pit at the commencement of the fin.

---

* *Combun sorah*, Tam.

843. *Mustelus manazo*, Bleeker. Preoral portion of snout rather less than the distance between the angles of the mouth. Teeth without distinct cusp. Origin of dorsal fin opposite the inner angle of the pectoral. No placenta. Grey, sometimes with white spots. Japan and perhaps Ceylon.

Family—SCYLLIIDÆ.*

Spiracles distinct. Eye without any nictitating membrane. Nasal and buccal cavities confluent or distinct. Mouth inferior. Teeth small, several rows being generally in use. The first dorsal fin spineless, placed above or behind the ventrals: an anal present, which may be in front of, below, or behind the second dorsal.

Genus—SCYLLIUM, *Cuv.*

Scylliorhinus, *Halælurus*, *Poroderma*, and *Cephaloscyllium*, Gill. *Spiracles behind the eyes. Nasal and buccal cavities distinct. Teeth small, in several rows, usually with a central and one or two lateral cusps. Origin of anal fin in advance of that of the second dorsal: upper edge of caudal not serrated.*

844. *Scyllium marmoratum*, Bennett. Nasal valves confluent, having a single transverse uninterrupted flap, and a well developed labial fold. Brown spots, bands, or ocelli superiorly. Seas of India to the Malay Archipelago, and beyond.

845. *Scyllium Capense*, (Smith) Müll. and Henle. Nasal valves not confluent : a short labial fold superiorly. With whitish spots and darker and lighter vertical bands. Seas of India and the Cape.

Genus—STEGOSTOM, *Müll. & Henle.*

Spiracles about the size of the small eyes, behind which they are situated. Fourth and fifth gill-openings close together. Nasal and buccal cavities confluent. Snout obtuse: upper lip thick, ending in a barbel on either side. A well-developed labial fold. Teeth small, sometimes trilobed, and forming an almost quadrangular plate. Two dorsal fins, the first above the ventral, the second anterior to the anal, which is near the caudal, the latter being very elongate.

846. *Stegostoma tigrinum*, Gm. Linn.: *Pollee-makum*, Tel. : *Corungun sorah*, 'monkey-mouthed shark,' Tam. Of a buff colour with dark spots or brown vertical bands. East Coast of Africa, Seas of India to the Malay Archipelago, and beyond, attaining at least 6 feet in length.

Genus—CHILOSCYLLIUM, *Müll. & Henle.*

Hemiscyllium, Müll. and Henle.

Spiracles below the eyes. Fourth and fifth gill-openings close together. Nasal and buccal cavities confluent: nasal valve with a barbel. Lower lip continuous or interrupted in the centre. Teeth small, triangular, with or destitute of lateral cusps. Two dorsal fins, the first above or behind the ventral, the second considerably anterior to the anal, which is near the caudal.

847. *Chiloscyllium Indicum*, Gm. Linn.: *Corungun sorah*, Tam.: *Poos-hee*, Beluch.: *Bokee-sorah* and *Ra-sorah*, Tel. Lower labial fold

* *Dog-fishes.*

continuous. Origin of first dorsal fin behind the base of the ventral. Sometimes smooth or tubercular ridges along the back. The colours vary exceedingly : they may be simply of a reddish brown, or with dark cross-bands, which may or may not include white or black spots : sometimes the cross-bands break up into spots or bands. Seas of India to the Malay Archipelago and beyond.

Sub-order—BATOIDEI.

Spiracles present : gill-openings in five pairs, and on the ventral surface of the body. Body depressed, forming, due to largely-developed pectoral fins, a more or less flat disk, and having usually a thin and slender tail. Dorsal fin, when present, on the tail : anal absent.

Family—PRISTIDÆ.

Saw-fishes.

Snout much produced, flattened, and having a saw-like appearance, due to large teeth existing on its lateral edges.

Genus—*PRISTIS, Latham.*

Body elongate and depressed. Gill-openings inferior and of moderate width. Spiracle wide and posterior to the eye, which latter has no nictitating membrane. Nostrils inferior. Teeth minute and obtuse. Dorsal fins spineless, the first quite or nearly opposite the ventrals : front edge of pectoral free.

848. *Pristis pectinatus*, Latham. From 24 to 32 pairs of teeth, the anterior placed close to one another, the intervening space not exceeding double the base of a tooth : the three posterior teeth are twice as far asunder. Dorsal fin arises opposite the ventral : second dorsal scarcely smaller than the first : no lower caudal lobe. Red Sea, seas of India to the Malay Archipelago, and beyond.

849. *Pristis zysron*, Bleeker. From 26 to 32 pairs of teeth, the anterior placed close to one another, the intervening space not exceeding double the base of a tooth : the three posterior teeth are thrice as far asunder. Dorsal fin arises opposite the middle of the root of the ventral : second dorsal scarcely smaller than the first ; it extends backwards nearly to the root of the caudal, which has no lower lobe. Seas of India to the Malay Archipelago.

850. *Pristis cuspidatus*, Latham. *Yahla*, Tel. From 23 to 34 pairs of broad teeth, which do not commence so far forwards as in the other species : in the young their hinder edge is barbed. Dorsal fin arises behind the root of the ventral : caudal with a lower lobe. Seas of India to the Malay Archipelago : ascending rivers.

Family—RHINOBATIDÆ.

The disk not much dilated laterally : the rayed portion of the pectoral fin not continued on to the snout. Tail thickened, and moderately elongated with two well-developed dorsal fins : likewise a caudal and a longitudinal fold on either side. No electric organs.

Genus—*RHYNCHOBATUS*, Müll. & Henle.
Rhina, sp. et *hinobatus*, sp. Bl. Schn.: *Rhamphobatis*, Gill.

Body depressed and elongated. Gill-openings inferior, narrow, and internal to the base of the pectoral fin. Spiracles wide and behind the eyes, which latter have no nictitating membrane: snout rather elongated and acute: nostrils inferior, oblique wide slits. Teeth obtuse, ridged, the dentary plate having an undulated surface. Dorsal fins spineless, the first opposite the ventrals: front edge of pectoral free, not extending to the head: caudal with a well-marked lower lobe.

851. *Rhynchobatus ancylostomus*, Bl. Schn. Snout very broad with a semicircular outline. Longitudinal rows of tubercles one on either side of the head, continued to the sides of the body, and a median one along the back: a few round the front edge of the eye and below the spiracle. Dental plate deeply undulated. Reddish-grey. East Coast of Africa, seas of India to the Malay Archipelago, and beyond.

852. *Rhynchobatus Djiddensis*, Forsk.: *Walawah tenkee*, Tel. Snout pointed and elongated. The tubercles are arranged somewhat as in the last species, but they are generally absent from below the spiracle, and some usually exist along the supraorbital margin. Dental plate slightly undulated. A round black ocellus generally present on the shoulder, with about six small white ones around it: four or five rows of large white spots may or may not exist along the sides, and also be present on the pectoral fin. Red Sea, seas of India to the Malay Archipelago, and beyond.

Genus—*RHINOBATUS*, sp. Bl. Schn.
Syrrhina, Müll. and Henle.

Body depressed and elongated. Spiracles wide and behind the eyes. Snout elongated, the cranial cartilage being produced, and the interval between it and the pectoral fin being filled by a membrane. Nostrils oblique and wide: the anterior nasal valves not confluent. Teeth obtuse, ridged. Dorsal fins spineless: both far behind the ventral: no lower caudal lobe.

853. *Rhinobatus granulatus*, Bl. Schn: *Purrungun*, Tam.: *Cun-da-ree*, Sind. Anterior nasal valve with no lateral dilatation. The distance between the external angles of the nostrils equals about 3/5 in the preoral portion of the snout: the two rostral ridges narrow and closely approximating: mouth transverse. Tubercles on the back, and a row of compressed spines along its middle, which may become obsolete with age: some spines on the edge of the orbit and on the shoulder. Seas of India to the Malay Archipelago and beyond.

854. *Rhinobatus obtusus*, Müll. and Henle. Anterior nasal valves with no lateral dilatation. The distance between the external angles of the nostrils equals 2/3 in the preoral portion of the snout: the two rostral ridges widely divergent posteriorly and confluent in their anterior third. Mouth transverse. Back with rough scales, but no tubercles or spines at the orbit or shoulder. Seas of India.

Family—TORPEDINIDÆ.

Trunk broad and disk smooth. Anterior nasal valves confluent and forming a quadrangular flap. Tail with a rayed dorsal (except in Temera) and caudal, also a longitudinal fold along either side. An electric organ situated between the pectoral fin and the head.

Genus—*NARCINE*, Henle.

Disk distinct from the tail, which has a lateral fold on either side and is longer than the disk. Spiracles close behind the eyes: nasal valves confluent, forming a quadrangular flap. Teeth nearly flat, with a central point. Two dorsal fins, the anterior behind the ventrals and usually smaller than the posterior. Electric apparatus present.

855. *Narcine timlei*, Bl. Schn. Posterior edge of caudal confluent with the lower and rounded. Reddish brown, with chocolate-coloured spots: white below. Seas of India to the Malay Archipelago and beyond.

Genus—*ASTRAPE*, Müll. & Henle.

Tail with a fold on either side. Spiracles close behind the eyes, which last are minute. Anterior nasal valves confluent, forming a large flap. Teeth pointed, extending slightly beyond the outer edge of the jaws. A single dorsal fin on the tail: caudal well developed: an electric apparatus on the side of the head between it and the pectoral fin.

856. *Astrape dipterygia*, Bl. Schn. Spiracles not fringed: Vent slightly nearer the head than the posterior extremity of the caudal fin. Dull reddish-olive above, whitish below. A white spot near the posterior end of the head, another above the end of the ventral, and generally a third at the root of the caudal: ventral with a white edge. Seas of India to the Malay Archipelago and beyond.

Family—RAJIDÆ.

Disk broad, rhombic: tubercles or spines usually present: the pectorals extend to the snout: tail with a longitudinal fold on either side: no serrated caudal spine: electric organs absent.

Genus—*PLATYRHINA*, Müll. & Henle.

Disk rhombic with a fold on either side: tail distinct. Nasal valves distinct. Two dorsal fins on the tail: caudal well developed: ventrals separated one from the other. Body covered with rough asperities and spines.

857. *Platyrhina Schönleinii*, Müll. and Henle. Disk sub-circular: snout obtuse: tail as long as the disk. Dental plate undulating: three elevations in the lower and three in the upper jaw. Nostrils wide apart. A row of strong spines, having smaller lateral ones along the median line of the back and tail: some more along the edge of the orbit and on the shoulder. Brown, covered with light blotches. Coromandel Coast of India.

Family—TRYGONIDÆ.

Disk wide: the pectorals are continued to the extremity of the tail, which is long and slender without any lateral fold: no vertical fins, unless modified into the form of a serrated spine.

Genus—*UROGYMNUS*, Müll. & Henle.

Anacanthus, Ehren. *Rhachinotus*, Cantor.

Disk sub-circular: tail long and distinct, destitute of any spine, but with a narrow inferior fold: pectorals united anteriorly. Teeth flattened. Body covered with osseous tubercles, amongst which are sharp conical spines.

858. *Urogymnus asperrimus*, Bl. Schn. *Moollan tiriki*, Tam. On the pectoral fins are numerous small conical spines, irrespective of those

over the body amongst the tubercles. East Coast of Africa, Seas of India to the Malay Archipelago.

Genus—*TRYGON, Adanson.*
Himantura, Hemitrygon and *Hypolophus*, Müll. and Henle. *Paratrygon*, Duméril.
Disk oval or rhomboidal: tail elongated and tapering. Nasal valves coalescent, forming a quadrangular flap. Teeth flattened or with a central point or transverse ridge. Pectoral fins united anteriorly: tail destitute of a fin, or if with a cutaneous fold, such does not extend to its extremity: it is armed superiorly with one or two lanceolate spines serrated on both sides. Body smooth or with tubercles.

A.—Dental laminæ transverse; if undulating, very slightly so: no caudal cutaneous fold. (Himantura.)

859. *Trygon uarnak,* Forsk. *Sona kah tiriki,* Tam. Disk about as wide as long: tail about three times as long as the disk. Snout somewhat pointed. One or more large tubercles in the middle line of the back: in the adult the back is shagreened, but almost smooth in the immature: no large tubercles in the median line of the tail. Brown, sometimes with darker spots: tail of the young annulated with white and brown. Red Sea, Seas of India to the Malay Archipelago and beyond, often ascending rivers.

860. *Trygon Gerrardi,* Gray. Disk broader than long: tail about three times as long as the disk. Snout somewhat obtuse. One or more large tubercles in the middle line of the back, round which some smaller ones are grouped. Brown with round yellow spots: young with the tail annulated. Bay of Bengal to the Malay Archipelago and beyond.

861. *Trygon Bleekeri,* Blyth. Tail from about 1/3 to 1/4 times as long as the disk. Snout prolonged and pointed. A large round tubercle in the centre of the back, and commonly three smaller triangularly disposed before it, and three similarly placed behind it. Tubercles continued along the upper surface of the tail to the caudal spine, from whence, in adults, they are continued to its extremity. Brown above and below, with a narrow white median longitudinal patch on the abdomen. Bengal.

862. *Trygon walga,* Müll. & Henle. Disk nearly as broad as long: tail rather longer than the disk. Snout prolonged and pointed. No large tubercles on the back, but several short spines along the tail, anterior to the usual large one. Of an uniform colour. Bay of Bengal to the Malay Archipelago.

863. *Trygon polylepis,* Bleeker. Disk slightly longer than broad: tail about as long as the disk. Snout prolonged and pointed. Interorbital space tuberculated, with a narrow band of tubercles along the median line of the back, and widening in the scapular region, where it has a cruciform appearance. No tubercles on the tail. Seas of India.

864. *Trygon nuda,* Günther. Disk about as broad as long: tail half longer than the disk. Snout rather pointed. No tubercles. Seas of India to China.

865. *Trygon marginata,* Blyth. Disk slightly longer than broad: tail half longer than the disk. Tubercles not only over the upper surface,

but also on the broad dark margin of the lower parts: an irregular row of pointed tubercles on either side of the middle line of the back: tail tuberculated as far as its spine. Grey above, buffy-white below, with a dark border, except in front.

B.—Dental laminæ transverse; if undulating, usually slightly so: tail with an upper or lower cutaneous fold. (Trygon.)

866. *Trygon Bennettii*, Müll. & Henle. Disk about as broad as long : tail about 3 times as long as disk, and with a low cutaneous fold along its inferior surface. Snout somewhat pointed. A tubercle in the middle of the back in adults, with some flat ones around it, and which extend backwards to the caudal spine. Seas of India to China and beyond.

867. *Trygon imbricata*, Bl. Schn. Disk as broad as long : tail about as long as the body, with low upper and inferior cutaneous folds. Small tubercles on the nape and back, with a row of conical spines along the shoulder and back : along the tail as far as the spine are large tubercles intermixed with smaller ones. Seas of India.

868. *Trygon zugei*, Müll. & Henle. *Chumbara kah*, Tam. Disk about as broad as long: tail twice as long as disk, and with an upper and lower cutaneous fold. Snout much produced and pointed. Dental laminæ much undulating. About six or more pointed tubercles commencing opposite the ventral fin in the middle line of the back and continued to the caudal spine. Brownish superiorly. Seas of India to the Malay Archipelago and beyond.

C.—Lower dental lamina somewhat pointed, the upper being angularly bent for its reception. (Hypolophus.)

869. *Trygon sephen*, Forsk. *Wolga tenkee*, Tel. *Aart wallan tiriki*, Tam. Disk rhombic: angles rounded : tail about three times as long as the disk, with a broad inferior cutaneous fold. Some large globular tubercles in the scapular region, elsewhere the whole upper surface covered with flat ones. Red Sea, through the seas of India to the Malay Archipelago and beyond.

Genus—*PTEROPLATEA, Müll. & Henle.*

Ætoplatea, Müll & Henle.

Body at least twice as broad as long: tail thin, generally shorter than the body, with or without a rudimentary fin, but having a serrated spine: spiracles with or without a tentacle: nasal valves confluent, and forming a quadrangular flap. No papilla at bottom of the mouth. Teeth with from one to three cusps. Pectoral fins united in front. Skin smooth or tubercular.

A.—Tail destitute of any fin.

870. *Pteroplatea micrura*, Bl. Schn. *Tenkee-kunsul*, Tel. No tentacle to spiracle. Two small spines on caudal fin. Skin smooth. Reddish brown : tail annulated with white and brown : in the centre of each light ring superiorly, generally a brown spot. Seas of India to the Malay Archipelago and beyond.

Family—MYLIOBATIDÆ.

Pectoral fins large, developed along the sides of the body, occasioning it to appear very broad: these fins are not present on the sides of the head, but re-appear at the end of the snout as a pair of detached fins.

Genus—MYLIOBATIS, Cuv.

Head distinct from disk: snout with a soft prolongation, internally supported by fin rays: nasal valves coalescent, forming a quadrangular flap. Teeth hexagonal, flat, the central ones being broader than long: the external rows narrow. Tail very long and whip-like, having a dorsal fin near its base, and usually a serrated spine posterior to it. Body smooth or tuberculated superiorly.

871. *Myliobatis Nieuhofii*, Bl. Schn. Disk twice as broad as long. Dark above, white below. Seas of India to the Malay Archipelago and beyond.

Genus—ÆTOBATIS, Müll. & Henle.

Stoasodon, Cantor.

Head distinct from disk: snout with a soft prolongation, internally supported by fin rays. Nasal valves distinct, each forming a long flap. Teeth hexagonal, broad, flat, with the lower dental laminæ projecting beyond the upper. Tail very long and whip-like. Dorsal fin present, and a serrated spine posterior to it.

872. *Ætobatis narinari*, Bl. Schn. *Eel-tenkee*, Tel. *Currooway tiriki*, Tam. *Ra-ta-charm-dah*, Andam. Dorsal fin situated on a vertical between the ventrals. Superiorly of a dark leaden colour, but often with round bluish-white spots. Red Sea, Seas of India to the Malay Archipelago and beyond.

Genus—RHINOPTERA, Kuhl.

Mylorina and *Micromesus*, Gill.

Head distinct from the disk, but with a pair of rayed appendages on the lower edge of the snout. Nasal valves confluent, forming a broad flap. Teeth broad, flat, in five or more rows, the central ones being the broadest. Tail whip-like, with a dorsal fin and a serrated spine posterior to it.

873. *Rhinoptera adspersa*, Müll. and Henle. *Mutta tiriki*, Tam. Teeth in nine rows in the upper, and seven in the lower jaw. Upper surface rugose. Said to have come from the East Indies.

Genus—DICEROBATIS, Blainv.

Cephaloptera, Dumeril. *Mobula*, A. Dumeril.

Pectoral fin not extended on to the sides of the head, which latter is pruncated in front, whilst on either side is a forwardly-pointing horn-like Trojection, which is internally supported by fin rays. Nostrils not confluent. Teeth in jaws very small, flat, or tuberculated, and in many row Tail whip-like, with a dorsal fin situated above and between the ventrals, and with or destitute of a serrated spine.

874. *Dicerobatis eregoodoo*, Cantor. *Eregoodoo tenkee*, Tel. Teeth in from 80 to 90 rows in the upper jaw, about twice as wide as long, and having one or two points posteriorly. Body and tail smooth. No spine. Seas of India.

www.ingramcontent.com/pod-product-compliance
Lightning Source LLC
Chambersburg PA
CBHW051733300426
44115CB00007B/551